CELL MEMBRANE TRANSPORT

Experimental Approaches and Methodologies

CELL MEMBRANE TRANSPORT
Experimental Approaches and Methodologies

Edited by

David L. Yudilevich
University of London
London, United Kingdom

Rosa Devés
University of Chile
Santiago, Chile

Salvador Perán
University of Málaga
Málaga, Spain

and

Z. Ioav Cabantchik
The Hebrew University of Jerusalem
Jerusalem, Israel

SPRINGER SCIENCE+BUSINESS MEDIA, LLC

Library of Congress Cataloging-in-Publication Data

Cell membrane transport : experimental approaches and methodologies / edited by David L. Yudilevich ... [et al.].
p. cm.
"Proceedings based on an international course ... held September 16-24, 1988, in Lanjarón, Spain"--T.p. verso.
Includes bibliographical references and index.
ISBN 978-1-4757-9603-2 ISBN 978-1-4757-9601-8 (eBook)
DOI 10.1007/978-1-4757-9601-8
1. Cell membranes--Congresses. 2. Biological transport--Congresses. I. Yudilevich, D. L.
[DNLM: 1. Biological Transport--physiology--congresses. 2. Cell Membrane--physiology--congresses. QH 601 C3915]
QH601.C352 1991
574.87'5--dc20
DNLM/DLC
for Library of Congress 91-2948
CIP

Proceedings based on an international course on New Methods in the Study of Transport Across the Cell Membrane, held September 16–24, 1988, in Lanjarón, Spain

ISBN 978-1-4757-9603-2

Originally published by Plenum Press, New York in 1991
Softcover reprint of the hardcover 1st edition 1991

PREFACE

Experimental science is a complicated creature. At the head there is a Gordian knot of ideas and hypotheses; behind is the accumulated mass of decades of research. Only the laboratory methods, the legs which propel science forward, remain firmly in touch with the ground. Growth, however is uneven; dinosaurs develop by solid means to give a vast body of results, but few ideas. Others sprint briefly to success with brilliant, though ill-supported, ideas. The problems which this book addresses is to maintain an organic unity between new ideas and the current profusion of innovative experimental tools. Only then can we have the framework on which our research thoughts may flourish.

The contributors are outstanding scientists in their respective fields and they record here in a clear manner the methodology with which they perform their experiments. They also illustrate some of their most exciting findings. In all chapters the emphasis is on the critical analysis of the methodology which is often avoided in refereed Journals. These techniques are explained in this book in adequate detail. Each chapter is extensively referenced and contains the most recent material available from author's laboratory at the time of going to press.

Lanjarón is a lovely small Andalucian town in Spain, sixty kms south of Granada, in the Alpujarra mountains. It was there that the idea of writing this book originated. The occasion was an **International Advance Course on New Methods in the Study of Membrane Transport**. Scientists from various countries met for eight days with almost the same number of young researchers partially disguised as students, to discuss and critically analyze the techniques that allow progress to be made in membrane transport research. It was through this experience that it was realised that we could extend the impact of our ideas by writing a book. We hope that the result of this effort will be fruitful.

To make the material clearer to interested readers chapters are divided into sections with generous use of subtitles. There are 23 chapters grouped as follows: Part 1: general methodologies. Part 2: isotopic methods for flux determination in various membrane preparations. Part 3: electrophysiological methods. Part 4: spectroscopic methods. Part 5: molecular approaches to structure and function of transporters. Part 6: transport regulation.

The field of transport of solutes across cell membranes is of wide interest and at the cross roads of basic and applied Medical Sciences. The discipline covers a wide range of science from Physiology to Molecular Biology. This book provides an authoritative coverage of the subject and this will be useful to Physiologists, Biochemists, Biophysicists, Cell Biologists, Pharmacologists and Clinicians at all levels from postgraduate student to experienced researcher.

ACKNOWLEDGMENTS

The International Advanced Course on **New methods in the Study of Cell Membrane Transport** was sponsored by the following organizations: Dirección General de Investigación Científica y Tecnológica, Fondo de Investigaciones Sanitarias de la Seguridad Social, Fundación Ramón Areces, Consejería de Salud Junta de Andalucía, Consejería de Educación y Ciencia Junta de Andalucía, Delegación de Cultura de Granada, Ayuntamiento de Lanjarón, C.I.T. Lanjarón, The British Council, Wellcome Trust, U.K. The meeting was organized under the auspices of the Universidad de Málaga and King's College, University of London.

David L. Yudilevich

London, October 1990

CONTENTS

PART I: GENERAL METHODOLOGIES

PART II: ISOTOPIC METHODS FOR FLUX DETERMINATION IN VARIOUS MEMBRANE PREPARATIONS

PART IV : SPECTROSCOPIC METHODS

PART VI: TRANSPORT REGULATION

Part 1: GENERAL METHODOLOGIES

1

KINETICS OF TRANSPORT: CHARACTERIZING THE INTERACTION OF SUBSTRATES AND INHIBITORS WITH CARRIER SYSTEMS

Rosa Deves

Department of Physiology and Biophysics
Faculty of Medicine, University of Chile
Casilla 70055, Santiago, Chile

INTRODUCTION

Unlike soluble enzymes, the proteins involved in membrane transport are vectorial catalysts that operate between two compartments and as a result the kinetic study of transport requires specialized theoretical and practical approaches which do not have a counterpart in enzymology.

Several comprehensive treatments have dealt with the problem of testing and characterizing transport systems, using the kinetic approach (Wilbrandt and Rosenberg, 1961; Stein, 1986, 1989; Andersen, 1989) and rejection criteria have been proposed to distinguish between alternative transport models such as different kinds of channels or carriers. This problem will not be addressed here.

The aim of this article is to discuss experimental strategies that may be used to disclose the molecular interactions of substrates and inhibitors with a transport protein. Emphasis will be placed on the interpretation of the transport parameters frequently used in specificity or inhibition studies. The objective is not to provide a exhaustive treatment of the subject, but rather to illustrate the nature of the problem by means of examples. No kinetic derivations are included as they can be found in the references cited. Intuitive explanations are privileged throughout.

THE CARRIER MODEL

The analysis will be based on the four state carrier or "classical" carrier illustrated in Fig. 1. According to this model the transporter alternates between two conformational states that differ with respect to the orientation of its binding site, which can become exposed either to the inside or outside of the cell. Transport depends on four consecutive events: substrate binding, translocation of the carrier-substrate complex, substrate dissociation and return of the free carrier to the initial state.

The model has been chosen for two reasons: first, because it has proved capable of explaining a variety of experimental observations (Hoare, 1972; Krupka and Deves, 1981;

Cell Membrane Transport, Edited by D. L. Yudilevich *et al.*
Plenum Press, New York, 1991

Jarvis et al., 1983; Stein, 1986) and second, because it is a general model whose character changes drastically depending on which step is rate limiting. For example, if the translocation of the free carrier is assumed to be very rapid with respect to the other steps, the model turns into a three state channel. The kinetic treatment of the classical carrier, therefore, can easily be applied to other models by making appropriate assumptions about rate limiting steps.

In this discussion it will be assumed that the rate limiting step in the process is translocation and not association or dissociation of substrate from the carrier. This assumption is supported by the observation of accelerated exchange in several systems such as the leucine (Hoare, 1972), choline (Deves and Krupka, 1979b) and glucose (Levine et al., 1965) transporters of erythrocytes.

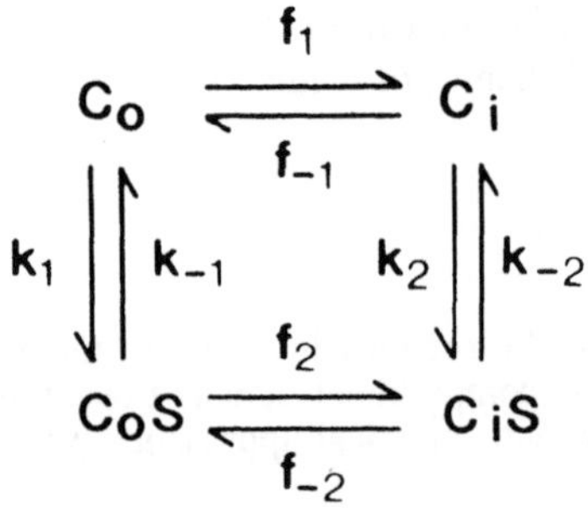

Fig. 1. Transport scheme for the classical carrier model. The conformation of the carrier alternates between inward-facing and outward-facing forms, C_i and C_o respectively. Substrate in the external solution adds to C_o to form C_oS; internal substrate binds to C_i. The substrate dissociation constants on the outer and inner faces of the membrane are $K_{So} = k_{-1}/k_1$ and $K_{Si} = k_{-2}/k_2$.

TRANSPORT EXPERIMENTS

The rates of substrate transport can be measured in several types of experiment, that differ from one another in the initial distribution of substrates across the cell membrane. The possibilities are so diverse that it has been necessary to invent a special terminology to describe the conditions of an experiment. By convention, the terms cis and trans refer to the relative locations of substrates or inhibitors across the membrane; cis designates the compartment occupied by the reference substrate, the one whose transport is followed, while trans designates the compartment on the opposite side of the membrane.

In this chapter we will deal with 3 different types of transport experiment: (1) zero-trans experiments, where the substrate is initially present in only one compartment and its flux across the membrane is determined, (2) equilibrium exchange experiments, where the substrate is present in both compartments at the same concentration, but only one is labelled, and the unidirectional flux of labelled substrate is measured and (3) infinite-trans experiments, where the unidirectional flux of substrate is measured in the presence of a saturating concentration of unlabelled substrate in the trans compartment.

CHARACTERIZING A SUBSTRATE

Two questions may be asked with regard to the interaction of a substrate with a carrier protein: what is its affinity for the transport site and how easily does it undergo translocation.

The translocation step

The transport capacity of a carrier is reflected by its maximum transport rate (the rate at saturating substrate concentrations), because under this condition the system is working at its full potential. The maximum rate of transport, however, does not always directly report the rate of translocation of the substrate-carrier complex (f_2, f_{-2}) as the process also depends on the rate at which the free carrier returns to the original conformation (f_1, f_{-1}). Expressions for the maximum rates for different experimental arrangements (zero trans exit and entry, equilibrium exchange and infinite-trans entry), in terms of the microscopic rate constants, are listed in Table 1.

The maximum transport rate of zero-trans entry is seen to depend on two constants: the rate constant for the inward translocation of the complex (f_2) and the rate constant for the outward movement of the free carrier (f_{-1}). The same is true for zero-trans exit with the directions reversed. As the two rate constants may have different magnitudes, it is possible that with certain substrates, transport may be limited by the rate of movement of the free carrier. The situation is different for the exchange mode. Here the maximum rate depends exclusively on the rate of translocation of the carrier-substrate complex. Differences in the rates of zero-trans fluxes and exchange have been reported in several transport systems. Exchange-only systems such as the chloride-bicarbonate exchanger represent an extreme example of this condition (Cabantchik, 1983; Passow, 1987).

Table 1. Maximum Rates of Transport (Vmax)

Experiment	Vmax[a] Symbol	Vmax[a] Expression	Limiting values of Vmax[b] $f_1 = f_2$	$f_1 >> f_2$	$f_1 << f_2$
Zero-trans entry	$\bar{V}_{So}$	$\frac{f_2 f_{-1} Ct}{f_2 + f_{-1}}$	$\frac{f_2\ Ct}{2}$	$f_2\ Ct$	$f_{-1}\ Ct$
Zero-trans exit	$\bar{V}_{Si}$	$\frac{f_1 f_{-2} Ct}{f_1 + f_{-2}}$	$\frac{f_{-2}\ Ct}{2}$	$f_{-2}\ Ct$	$f_1\ Ct$
Equilibrium exchange	$\bar{\bar{V}}_S$	$\frac{f_2 f_{-2} Ct}{f_2 + f_{-2}}$	$\frac{f_2\ Ct}{2}$	$\frac{f_2\ Ct}{2}$	$\frac{f_2\ Ct}{2}$
Infinite-trans entry	$\tilde{V}^s_S$	$\frac{f_2 f_{-2} Ct}{f_2 + f_{-2}}$	$\frac{f_2\ Ct}{2}$	$\frac{f_2\ Ct}{2}$	$\frac{f_2\ Ct}{2}$

[a] The expressions listed have been derived under the assumption of rapid equilibrium. General expressions for maximum rates of transport can be found elsewhere (Deves and Krupka,1979a).

[b] For the sake of simplicity in the presentation it is assumed that $f_1 = f_{-1}$ and $f_2 = f_{-2}$. Ct is the total carrier concentration.

The table also shows limiting values of the maximum velocity, as a function of the relative magnitudes of the rates of translocation of the free carrier and the carrier-substrate complex, for three cases: i) the rates of translocation of the complex and the free carrier are equal ($f_2 = f_1$), ii) the free carrier translocates much faster than the substrate-carrier complex ($f_1 >> f_2$) and iii) the complex translocates faster than the free carrier ($f_2 >> f_1$).

Clearly a zero-trans arrangement cannot be used to compare relative translocation rates in structure activity studies, since the f_2/f_1 ratio is likely to vary with different substrates. An alternative is to measure exchange rates instead of zero-trans rates. Exchange experiments, however, are comparatively difficult to set up, specially when a large number of substrates must be studied. A methodological approach that offers a solution to this problem, and which does not require the use of labelled substrates, is explained below.

The binding step

The apparent affinity of a substrate is obtained from the concentration dependence of the transport rate. A half-saturation constant can be defined that corresponds to the substrate concentration that can sustain a rate equal to one half of the maximum transport rate. The apparent affinity is dependent on the true dissociation constant of the substrate-carrier complex (K_{So} or K_{Si}), but it is also influenced by the microscopic rate constants. Half-saturation constants applying to different kinds of experiments are shown in Table 2. When the expressions are solved for the three limiting cases (as was done for Vmax in Table 1) it becomes evident that with the exception of equilibrium exchange, the half-saturation constant measured varies as a function of the relative rates of translocation of the free carrier and the substrate carrier complex (f_1 and f_2).

Table 2. Half-saturation constants (Km)

Experiment	Km[a] Symbol	Km[a] Expression	Limiting values of Km[b] $f_1 = f_2$	$f_1 >> f_2$	$f_1 << f_2$
Zero-trans entry	$\bar{K}_{So}$	$\frac{K_{So}\,(f_1+f_{-1})}{(f_{-1} + f_2)}$	K_{So}	$2K_{So}$	$\frac{2f_1\,K_{So}}{f_2}$
Zero-trans exit	$\bar{K}_{Si}$	$\frac{K_{Si}\,(f_1+f_{-1})}{(f_{-2} + f_1)}$	K_{Si}	$2K_{Si}$	$\frac{2f_{-1}\,K_{Si}}{f_{-2}}$
Equilibrium[c] exchange	$\bar{\bar{K}}_{S}$	$\frac{K_{So}(1+f_1/f_{-1})}{(1 + f_2/f_{-2})}$	K_{So}	K_{So}	K_{So}
Infinite-trans entry	$\tilde{K}^{s}_{So}$	$\frac{K_{So}\,(f_1+f_{-2})}{(f_{-2} + f_2)}$	K_{So}	$\frac{f_1\,K_{So}}{2f_2}$	$\frac{K_{So}}{2}$

[a] The expressions listed have been derived under the assumption of rapid equilibrium. General expressions for the half-saturation constants can be found elsewhere (Deves and Krupka,1979a).
[b] For the sake of simplicity in the presentation it is assumed that $f_1 = f_{-1}$ and $f_2 = f_{-2}$.
[c] Labelled substrate in the external medium.

Relative affinities therefore, cannot be estimated from the half-saturation constants directly in zero-trans entry experiments unless an experimental strategy is devised to eliminate the contribution of the translocation rate constants.

Determination of relative translocation rates and affinities for unlabelled substrate analogues

The relative translocation rates of the carrier-substrate complex and the free carrier (f_2/f_1) may be determined by comparing the unidirectional flux of labelled substrate in the presence or absence of unlabelled analogue in the trans compartment. The most convenient design is to measure the exit of labelled substrate (at a low concentration) into a medium that contains unlabelled substrate (Deves and Krupka, 1979b). The analogue alters the substrate exit rate if the carrier-analogue complex translocates at a different rate than the free carrier. If the translocation of the complex is faster, addition of analogue causes acceleration. The effect is concentration dependent and a maximum is reached at saturating concentrations. If the rate of movement of the complex is slower than that of the free carrier, the rate is inhibited and at saturating concentrations it reaches a minimum. The magnitude of the limiting values depends on the relative rates of translocation of the complex and the free carrier (f_2/f_1). The analogue exerts its effect by altering the distribution of the free carrier and either increasing or reducing the concentration of the inward-facing conformation. From the concentration dependence of the effect, a half-saturation constant for the analogue can be calculated.

The application of this method to a specificity study of the choline carrier of human erythrocytes (Deves and Krupka, 1979b) is shown in Fig. 2. The ratio of the choline exit rate in the presence (v) and absence ($\bar{v}$) of varying concentrations of analogue outside is plotted against the reciprocal value of the concentration of unlabelled analogue ($1/[T_o]$), multiplied by ($v/\bar{v}-1$). The equation that defines the straight line is also shown in the figure. This equation has been derived assuming that the concentration of the labelled substrate inside is very low in relation to its half-saturation constant.

The intercept on the ordinate represents the effect of the analogue at saturating concentration ($\tilde{v}^T/\bar{v}$), which in terms of microscopic constants equals:

$$\frac{\tilde{v}^T}{\bar{v}} = \frac{1 + f_{-1}/f_1}{1 + f_{-1}/f_2} \qquad (1)$$

and the slope of the line corresponds to the half saturation constant outside ($\bar{K}_{To}$). This constant is the same that would be obtained if the entry of the analogue was measured directly (Table 2). The ratio of the $\tilde{v}^T/\bar{v}$ values for two analogues is identical to the ratio of their maximum rates,

$$\frac{(\tilde{v}^T/\bar{v})_1}{(\tilde{v}^T/\bar{v})_2} = \frac{(\bar{V}_{To})_1}{(\bar{V}_{To})_2} \qquad (2)$$

Experimental parameters for a series of choline analogues that differ in the lenght of an alkyl substituent on the quaternary ammonium group are listed in Table 3.

If we assume that the carrier is equally distributed between the inward and outward conformations ($f_1 = f_{-1}$), the relative rates of translocation of the analogue-carrier complex and the free carrier (f_2/f_1) can be calculated from the $\tilde{v}^T/\bar{v}$ values using Equation 1. Once values for f_2/f_1 have been asigned the half-saturation constants ($\bar{K}_{To}$) may be corrected to eliminate the contribution of the translocation constants. (See Krupka and Deves, 1980, for a method to determine the equilibrium distribution of the free carrier (f_1/f_{-1})).

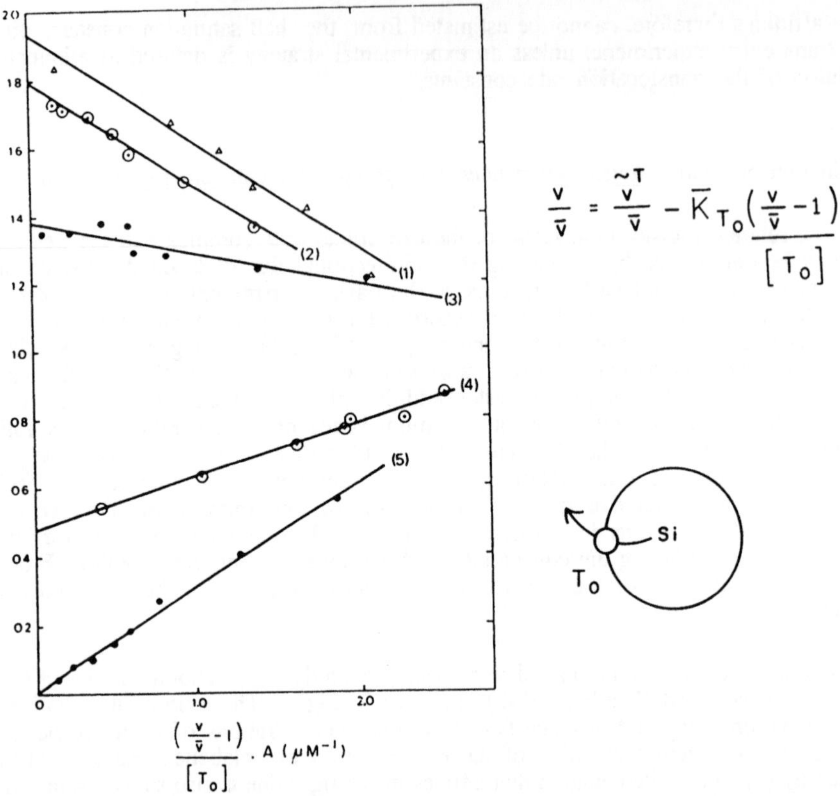

Fig. 2. Effect of choline analogues on the unidirectional efflux of choline in human erythrocytes. Relative exit rates in the presence (v) and absence of analogue ($\bar{v}$) are plotted against the reciprocal value of the analogue concentration in the external medium ($1/[T_o]$), multiplied by ($v/\bar{v}$ - 1). The factor A is introduced so that the various analogues can be shown on the same scale. (1) Choline chloride (trimethyl analogue), A=20; (2) Dimethyl ethyl analogue, A = 40; (3) Diethyl methyl analogue, A=400; (4) Dimethyl isopropyl analogue, A=2000; (5) Dimethyl n-decyl analogue, A=1. The experimental parameters determined are: $\tilde{v}^T/\bar{v}$ = (1) 1.93± 0.04; (2) 1.79 ± 0.01 ;(3) 1.38 ± 0.02 ;(4) 0.48 ± 0.02 (5) 0.01 ±0.01. $\bar{K}_{To}$(μM) = (1) 6.3± 0.62; (2) 12.4 ±0.61; (3) 35.2 ±7.7; (4) 303 ±19; (5) 0. 31 ±0.009. (Data taken from Deves and Krupka, 1979b).

Table 3. Transport parameters for substrate analogues of choline entering the human erythrocyte.

R-group		$\bar{K}_{To}$ (μM)	$\tilde{v}^T/\bar{v}$
	Methyl	6.3 ± 0.62	1.93 ± 0.04
CH_3	Ethyl	12.4 ± 0.61	1.79 ± 0.01
|	n-Propyl	33.2 ± 6.40	1.32 ± 0.03
$R\text{-}N^+\text{-}CH_2CH_2OH$	n-Butyl	30.1 ± 1.80	0.12 ± 0.02
|	n-Pentyl	7.3 ± 1.20	0.04 ± 0.08
CH_3	n-Decyl	0.3 ± 0.01	0.01 ± 0.01
	n-Dodecyl	0.095± 0.01	0.02 ± 0.02

The transport parameters were calculated from the effect of various analogues of choline, present in the external medium, on the efflux of labelled choline. $\bar{K}_{To}$ is the analogue half-saturation constant and $\tilde{v}^T/\bar{v}$ represents the maximum effect of the analogue and its related to its rate of transport (Eqn. 2). (Data taken from Deves and Krupka, 1979b).

The transport parameters were calculated from the effect of various analogues of choline, present in the external medium, on the efflux of labelled choline. $\bar{K}_{To}$ is the analogue half-saturation constant and $\tilde{v}^T/\tilde{v}$ represents the maximum effect of the analogue and its related to its rate of transport (Eqn. 2). (Data taken from Deves and Krupka, 1979b).

Trends in affinities and transport rates for the analogues listed in Table 3 are plotted in Fig. 3. A discrepancy can be appreciated between true and apparent values, the discrepancy is greater the higher the rate at which the analogue is transported. Consider, for example, the relationship between choline and its propyl analogue. Whereas the maximum rates of transport differ by only 8%, the relative translocation rate constant of the carrier complexed with choline is at least twice that of the propyl analogue. The result is more dramatic for the affinities. If apparent affinities are compared the carrier appears to be able to discriminate well among the three first members of the series, with the corrected constants, however, no difference in affinity is seen in going from methyl to propyl. These results very clearly illustrate the dangers involved in deducing structure activity relationships directly from measured experimental parameters.

The method described above, has a limitation in that it can only be applied to the study of transport systems that exhibit a significant degree of asymmetry, for if the analogue-carrier complex and the free carrier translocate at comparable rates, no effect on substrate movement will be observed. An extension of this approach that overcomes this difficulty, and thus enhances its applicability, has been described (Deves and Boyd, 1989).

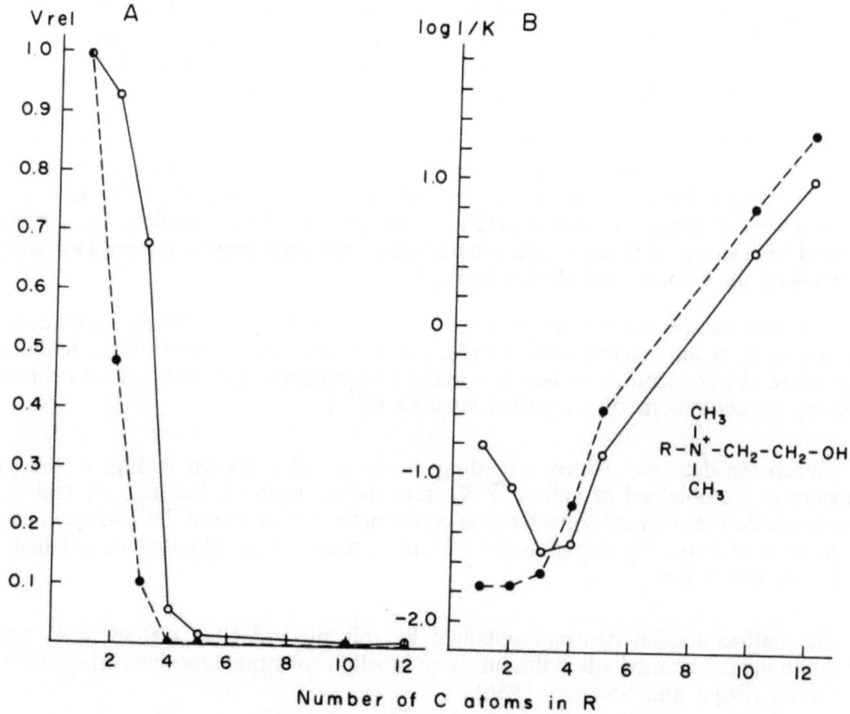

Fig. 3. Trends in affinities and transport rates for a series of choline analogues as a function of the length of an alkyl substituent on the quaternary ammonium ion.(A) Relative maximum transport rates ($\bar{V}_{To}$) calculated from Eqn.2 (o) and relative rates of translocation for the carrier complex with different analogues (f_2) calculated from Eqn. 1. (●). (B) Logarithm of apparent ($1/\bar{K}_{To}$) (o) and corrected affinity constants ($1/K_{To}$) (●). (Data taken from Table 3).

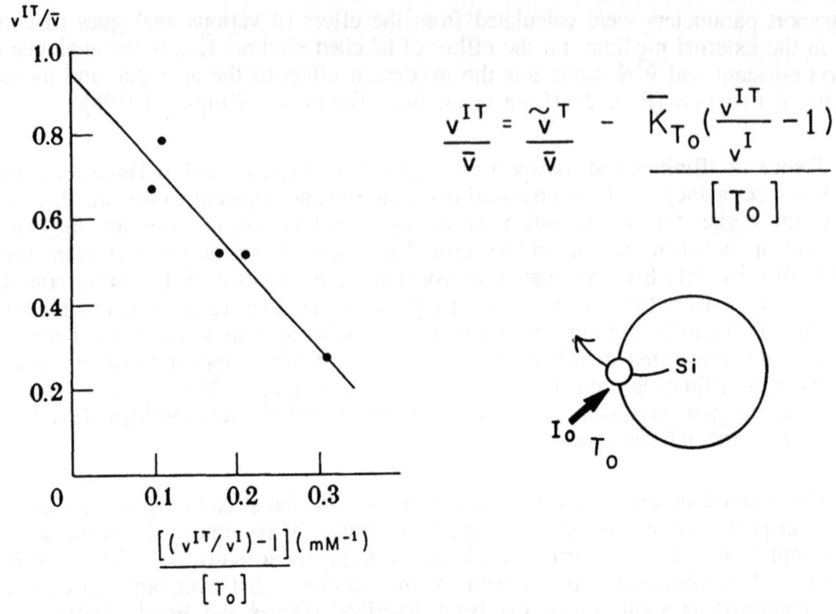

$$\frac{v^{IT}}{\bar{v}} = \frac{\tilde{v}^{T}}{\bar{v}} - \frac{\bar{K}_{T_o}\left(\frac{v^{IT}}{v^{I}} - 1\right)}{[T_o]}$$

Fig 4. Determination of transport parameters for sulphate in placental brush-border vesicles. $\bar{v}$ is the first order rate constant for exit under zero trans conditions (0.037$\pm$ 0.001 min^{-1}),v^{I} the rate of exit in the presence of the reversible inhibitor chromate (0.2 mM) in the external medium(0.01$\pm$ 0.0005 min^{-1}) and v^{IT} the rate in the presence of chromate + varying concentrations of sulphate. The following parameters are obtained: K_{To} =2.0 $\pm$ 0.38mM and $\tilde{v}^{T}/\bar{v}$ = 0.95 $\pm$ 0.075. (Data taken from Deves and Boyd, 1989).

The method involves determining the effect of an analogue added to the external solution on the unidirectional efflux of a labelled substrate which has been partly inhibited by adding a competitive inhibitor to the external medium. The ability of the analogue to relieve the inhibition can be directly related to its inward rate of translocation. This approach was applied to a study of divalent anion transport in placental membrane vesicles and the results obtained for sulphate are shown in Fig.4.

Sulphate-loaded vesicles were tested for exit under three different conditions: (i) with no additions in the external medium ($\bar{v}$), (ii) with chromate a competitive inhibitor of sulphate, added at a concentration enough to cause 74% inhibition (v^{I}) and (iii) with chromate plus varying concentrations of unlabelled sulphate (v^{IT}).

When the data are plotted according to the equation shown in Fig. 4, the same two parameters are obtained as before: $\tilde{v}^{T}/\bar{v}$, from the intercept on the ordinate and $\bar{K}_{To}$, from the slope. As the sulphate concentration on the outside is increased the inhibition caused by chromate is relieved. By extrapolation at high substrate concentrations of sulphate the ratio becomes nearly one.

The half-saturation constant obtained by this method (2.0 $\pm$ 0.38 mM) agrees well with the value obtained when the influx of labelled substrate was determined directly (1.8-2.0 mM) (Boyd and Shennan,1986).

CHARACTERIZING AN INHIBITOR

The observed inhibition behavior in transport experiments depends heavily on the details of the experimental design and does not always coincide with the underlying molecular mechanism. The correct interpretation of inhibition experiments therefore requires a thorough understanding of the kinetic principles involved. There are three possible ways in which the substrate, the inhibitor and the carrier may interact, these are illustrated in Fig 5.

The inhibitor may bind to the carrier whether free or complexed with the substrate (noncompetitive), it may bind to the free carrier only (competitive), or it may bind to the carrier substrate complex only (uncompetitive). We would expect therefore that in the first case, the substrate should have no effect on the potency of the inhibitor, in the second the substrate should protect by displacing the inhibitor and in the third the substrate should facilitate the inhibition by promoting the binding of the inhibitor.

The observed behaviour, however, may not reflect these interactions, because the ability of the substrate to displace the inhibitor in a given experimental arrangement depends not only on the inhibition mechanism itself, but on several other factors as well: 1) The relative location of the inhibitor and the substrate, 2) The relative affinity of the inhibitor for different carrier forms and 3) The transport mechanism (Krupka and Deves, 1983; Deves and Krupka, 1989).

Our aim here is to illustrate by means of examples different kinds of inhibition behaviour and in doing this to explain at an intuitive level why the differences between observed behaviour and mechanism arise.

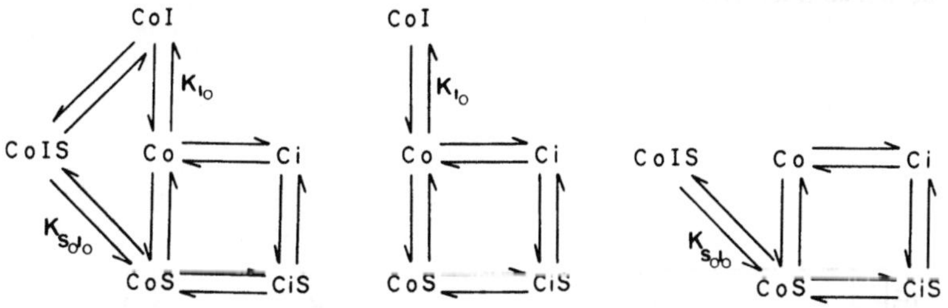

Fig. 5. Kinetic scheme for the carrier model involving a substrate (S) and a reversible inhibitor (I) that binds to the external carrier. Three forms of interaction are possible: noncompetitive, competitive and uncompetitive. K_{Io} and K_{SoIo} are equilibrium dissociation constants that characterize the interaction of the inhibitor with the free carrier and the carrier-substrate complex respectively.

The dependence of the observed pattern of inhibition on the relative location of substrates and inhibitors across the membrane.

Several experimental arrangements that differ in the initial distribution of substrates with respect to the two membrane faces were mentioned above. In the same way, inhibitors can be placed either outside, inside, or on both sides when their effects are tested. The relative location of substrates and inhibitors influences the observed pattern of inhibition. This problem will be illustrated by considering the effect of a competitive inhibitor present in the external medium on the entry and exit rates of labelled substrate. It will be shown that whereas the entry experiment discloses the true inhibition mechanism the exit experiment is misleading.

The experiment involves determining the effect of the inhibitor on the transport rate measured at varying substrate concentrations. Companion experiments are run in the presence or absence of the inhibitor (at a fixed concentration) and the data are analyzed on the basis of the Michaelis-Menten equation.

The effect of dimethyl(2-hydroxyethyl)ammonium iodide (or dimethyl decyl analogue) on the rate of choline entry in red cells is shown in Fig. 6. The results are plotted in the form of a Lineweaver-Burk plot (1/v against 1/[S]).

From the scheme in Fig. 5, it is clear that with this experimental arrangement the substrate will displace a competitive inhibitor and release the inhibition at high concentrations. The result is predicted by the corresponding kinetic equation also shown in Fig. 6. The velocity equation is written in terms of experimental parameters: $\bar{V}_{So}$ the maximum velocity for entry, $\bar{K}_{So}$, the half-saturation constant for the substrate outside and $\bar{K}_{Io}$, the apparent inhibition constant. At very high substrate concentrations the term on the right hand-side is small compared to one and the rate becomes independent of the inhibitor. As a result, the two lines intersect at a common point on the ordinate. The apparent inhibition constant is the inhibitor concentration that doubles the slope (or that reduces the rate to one half at very low substrate concentration).

The situation differs if substrate and inhibitor are present initially on opposite sides. The effect of the dimethyldecyl analogue on choline exit is shown in Fig. 7 and the behavior is seen to be noncompetitive.

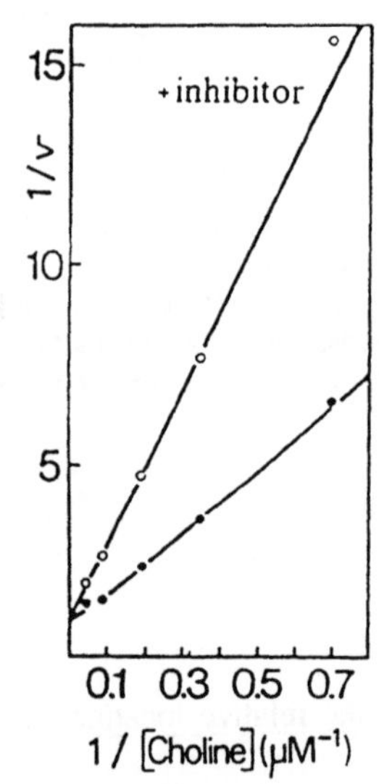

$$\frac{1}{v} = \frac{1}{\bar{V}_{So}}\left\{1 + \frac{\bar{K}_{So}}{[S_o]}\left(1 + \frac{[I_o]}{\bar{K}_{Io}}\right)\right\}$$

$$\bar{K}_{Io} = K_{Io}\left(1 + \frac{f_1}{f_{-1}}\right)$$

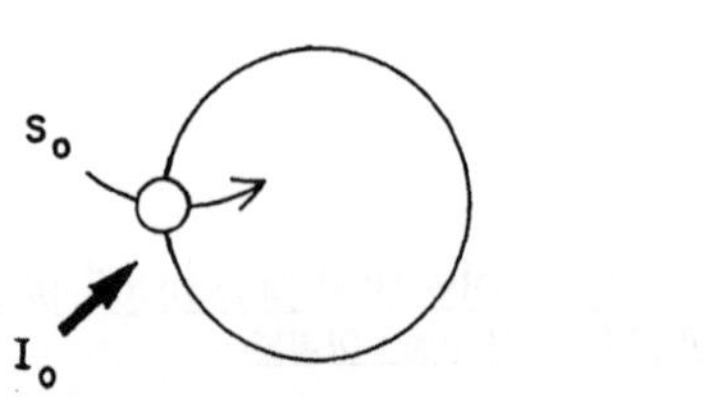

Fig. 6. Inhibition of choline transport by dimethyl-n-decyl (2-hydroxyethyl) ammonium bromide (0.29 µM) in zero-trans entry. Rates of choline entry in the presence (o) and absence (•) of inhibitor (I_o) are shown. Results are presented in the form of a Lineweaver-Burk plot. The rates are expressed in relative units. The maximum rate of entry is 0.14 µmolmin^{-1} per liter of packed cells. $\bar{K}_{Io}$ the competitive inhibition constant is 0.23 ± 0.02 µM. The corresponding kinetic equation is also shown. (Data taken from Deves and Krupka, 1983)

Although the underlying mechanism is the same as before (competitive), the new experimental constraints prevent competition, because substrate and inhibitor have access to different carrier forms (Ci and Co respectively). In consequence, the observed behaviour is noncompetitive. In accordance with this result the corresponding kinetic equation shows that both the substrate-dependent and substrate- independent terms are affected by the inhibitor. A new inhibition constant, ($\tilde{K}^S_{Io}$) equivalent to the concentration of inhibitor that doubles the intercept on the ordinate (or reduces the Vmax to one half), must be considered.

Some inhibitors, because they readily diffuse across the cell membrane, cannot be restricted to one compartment and are inevitably present on both sides. Lipid soluble substances, such as alcohols and anaesthetics are common examples. In cases such as this, it becomes obvious that the required experiment is one in which the substrate concentration

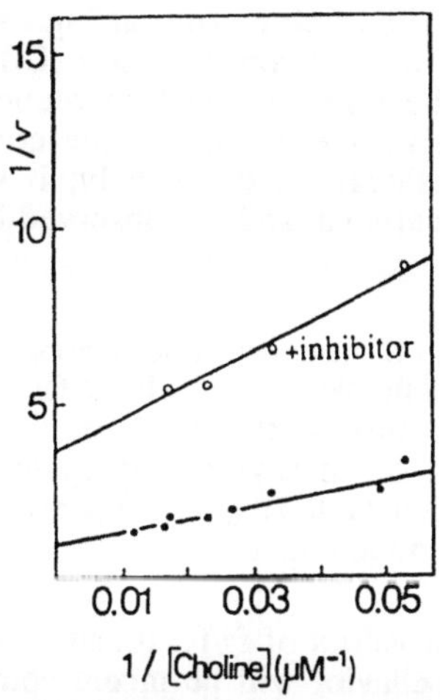

$$\frac{1}{v} = \frac{1}{\bar{V}_{Si}}\left\{1 + \frac{[I_o]}{\tilde{K}^S_{Io}} + \frac{\bar{K}_{Si}}{[Si]}\left(1 + \frac{[I_o]}{\bar{K}_{Io}}\right)\right\}$$

$$\tilde{K}^S_{Io} = K_{Io}\left(1 + \frac{f_1}{f_{-2}}\right)$$

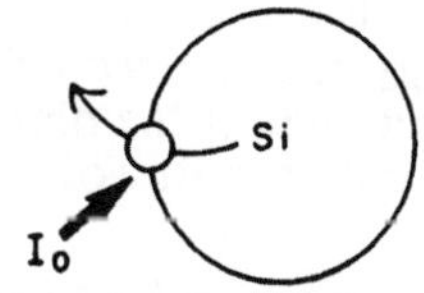

Fig. 7. Inhibition of choline transport by dimethyl-n-decyl (2-hydroxyethyl) ammonium bromide (0.29 μM) in zero-trans exit. Rates of choline exit in the presence (o) and absence (•) of inhibitor are shown. The rates are expressed in relative units. The maximum rate of entry is 0.23 μmol x min^{-1} per liter of packed cells. Two inhibition constants are determined, $\bar{K}_{Io}$ (from the slope): 0.19 ± 0.057 μM and $\tilde{K}^S_{Io}$ (from the intercept): 0.10 ± 0.024 μM. The corresponding kinetic equation is also shown. (Data taken from Deves and Krupka, 1983)

is simultaneously varied on both sides of the membrane, i.e equilibrium exchange. One is tempted therefore, to enunciate the general rule that the inhibition mechanism can be unambiguously determined provided the experiment is designed so as to have the inhibitor and the substrate (whose concentration is varied) in the same compartment.

This precaution, however, is not sufficient in the case of systems that exhibit asymmetry between the rate of movement of the free carrier and the carrier substrate complex.

If, at this point, we reexamine the results shown in Fig. 7 we will notice that the inhibition of choline exit by the dimethyl n-decyl analogue is not strictly noncompetitive and that an uncompetitive component appears, i.e. the inhibitor has a greater effect on the

Inhibition patterns in asymmetric systems

As discussed in the earlier sections, when the rates of reorientation of the free carrier and its substrate complex are unequal, addition of substrate to one side of the membrane alters the distribution of the carrier between its inner and outer forms. Thus, if the complex moves across the membrane more rapidly than the carrier returns, the presence of substrate on one side causes the carrier to accumulate on the opposite side (trans). On the other hand, if the complex moves more slowly than the free carrier, the carrier accumulates on the cis side. In an asymmetric system the pattern of inhibition will depend therefore, on the transport properties of the substrate, for in shifting the carrier partition the substrate will alter the accessibility of the carrier to an inhibitor that binds predominantly on one side of the membrane, and in this way strengthen or weaken the inhibition.

This problem comes across very clearly in the case of a noncompetitive inhibitor present in the same compartment as the substrate. We shall assume that entry is being measured in the presence of a noncompetitive inhibitor outside. If the substrate used is one whose complex reorientates faster than the free carrier, at high substrate concentrations the carrier will exist mainly in the internal conformation (C_i), a form that is inaccesible to an external inhibitor and the potency will be reduced. The inhbition pattern will not be strictly noncompetitive and a competitive component will appear. On the contrary if the substrate translocates more slowly, it will pull the carrier to the outside, the inhibition will be potentiated and the observed behavior will be partly uncompetitive.

The corresponding rate equation is shown in Fig. 8. As expected for a noncompetitive mechanism the inhibitor increases both the intercept and the slope, i.e., it affects the maximum velocity and the substrate half-saturation constant. The first effect depends on the apparent inhibition constant $\bar{K}_{SoIo}$ and the second on $\bar{K}_{Io}$. In an asymmetric system these two constants will differ even if the dissociation constants (K_{Io} and K_{SoIo}) are equal and the pattern of inhibition will not be strictly noncompetitive.

Theoretical plots are given for three different values of f_2/f_1. It can be seen that depending on the value of this ratio the observed behavior will be mixed competitive-noncompetitive, noncompetitive or mixed noncompetitive-uncompetitive. The only experiment that could disclose the mechanism in this case is an equilibrium exchange experiment, for in this type of experiment the distribution of the carrier is not significantly changed by the substrate.

The inactivation of the choline-carrier of erythrocytes by N-ethylmaleimide (NEM) is a good example of this kind of behavior. NEM inactivates the carrier by reacting with an SH group exposed on the inward-facing conformation only and located outside the substrate site (Martin, 1971; Edwards, 1973; Deves and Krupka, 1981; Krupka and Deves, 1988).

For the purpose of this discussion NEM can be considered as an inhibitor present inside only. The rate at which NEM reacts with the system, therefore, is directly proportional to the concentration of the inward-facing carrier and substrates and inhibitors will affect the rate of inactivation if they alter the carrier distribution. Rates of inactivation measured under different conditions are given in Table 4.

Similar values of the inactivation rate constant are obtained in the presence or absence of equilibrated choline and this argues in favour of a noncompetitive mechanism. A reversible inhibitor located outside protects against inactivation by trapping the carrier in the external conformation. On the contrary when choline is placed outside the rate is increased. External choline accelerates the inactivation by increasing the concentration of the internal carrier form.

$$\frac{1}{v} = \frac{1}{\bar{V}_{So}} \left\{ 1 + \frac{[I_o]}{\bar{K}_{SoIo}} + \frac{\bar{K}_{So}}{[S_o]} \left(1 + \frac{[I_o]}{\bar{K}_{Io}} \right) \right\}$$

Constant affecting the slope $\bar{K}_{IO} = K_{IO} \left(1 + \frac{f_1}{f_{-1}} \right)$

Constant affecting the intercept $\bar{K}_{SoIo} = K_{SoIo} \left(1 + \frac{f_2}{f_{-1}} \right)$

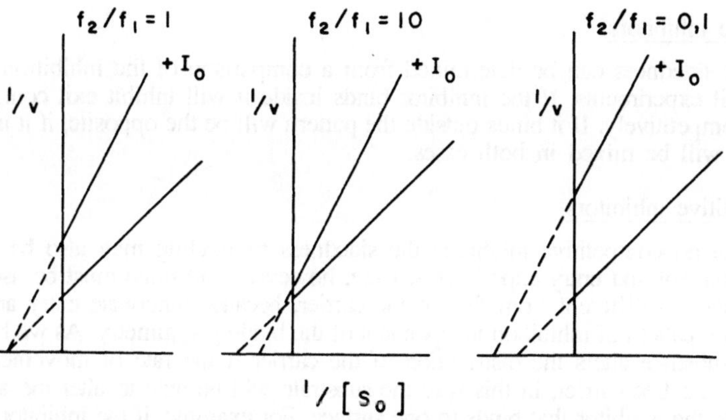

Fig. 8. Effect of an external noncompetitive inhibitor on substrate entry. Theoretical plots predicted by corresponding kinetic equation assuming different values of the f_2/f_1 ratio are shown.

Table 4. Inactivation of the choline carrier by N-ethylmaleimide (1 mM).

Condition	Inactivation rate constant k (min^{-1})
Control	0.20 ± 0.02
Choline at equilibrium (50 μM)	0.23 ± 0.02
Choline outside (50 μM)	0.33 ± 0.02
Dimethyl-n-decyl(2hydroxyethyl) ammonium iodide outside (18 μM)	0.019 ± 0.004

(Data taken from Deves and Krupka, 1981 a,b)

intercept than on the slope. This means that the substrate increases the potency of a competitive inhibitor present in the trans compartment! This behaviour is expected for a system with $f_2 >> f_1$ such as the choline carrier, for the substrate shifts the carrier distribution to the opposite side where the inhibitor is located (compare the expressions for the apparent inhibition constants $\overline{K}_{Io}$ and $\tilde{K}^{S}_{Io}$).

Sidedness of inhibition

We have shown above how the pattern of inhibition is affected by the relative location of substrate and inhibitor. Since what really matters is the effective concentration of inhibitor, that is the concentration in relation to the affinity of the inhibition site, it follows that the observed behaviour will also be dependent on the relative affinity of the inhibitor on either side. For example, a competitive permeant inhibitor which binds to both sides is not expected to produce competitive behaviour in either exit or entry experiments. However if the carrier recognizes the inhibitor only from the inside, competitive inhibition of exit will be observed. The inhibition pattern therefore can be used to predict the sidedness of inhibition of penetrating inhibitors which are unavoidably present on either side of the membrane.

Competitive inhibitors

The sidedness can be determined from a comparison of the inhibition behaviour in entry or exit experiments. If the inhibitor binds inside it will inhibit exit competitively and entry noncompetitively. If it binds outside the pattern will be the opposite, if it is symmetrical the pattern will be mixed in both cases.

Noncompetitive inhibitors

With noncompetitive inhibitors the sidedness of binding may also be inferred from the effects on exit and entry experiments. Here, however a substrate must be used whose rate of reorientation is different from that of the carrier, because otherwise entry and exit would give the same pattern of inhibition independent of the binding symmetry. As we have explained above the substrate alters the distribution of the carrier if the rate of movement is different than that of the free carrier, in this way the substrate will be able to alter the accessibility of the carrier to the inhibitor that binds to one surface. For example, if the inhibitor binds outside only, and the substrate carrier complex translocates faster than the free carrier, the inhibitor will be more potent at high substrate concentrations in exit than in entry, the opposite is true for an internally bound inhibitor. A symmetrical inhibitor will be unaffected by the carrier distribution and will inhibit noncompetitively in both entry and exit.

Recently another method for determining the sidedness of action of nonpolar inhibitors has been proposed which offers the advantage of being experimentally simpler than the one described above (Deves and Krupka, 1990a). Sidedness is deduced from the effect of an inhibitor on the flux ratio for the substrate, that is the ratio of infinite-trans and zero-trans rates. If the inhibitor reduces one rate more than the other, the flux ratio changes. Consider the case of a system that exhibits accelerated exchange: in an exit experiment, addition of unlabelled substrate to the external compartment increases the rate by shifting the carrier inward. If the potency of an inhibitor present on both sides is compared in zero-trans and infinite-trans experiments the result will depend on the binding assymetry. An externally bound inhibitor will be more potent in a zero-trans arrangement, because addition of unlabelled substrate will shift the carrier away from the inhibitor, the opposite will be true of an internally bound inhibitor. In this case the substrate increases the proportion of carrier available to the inhibitor. A symmetric inhibitor has the same effect on zero-trans or infinite-trans rates. This method can be applied to competitive, noncompetitive (Deves and Krupka, 1990b) or uncompetitive inhibitors.

INHIBITION BEHAVIOUR BY COMPETING SUBSTRATES

In this section we will briefly discuss the very peculiar effects produced by competing substrates, present at equilibrium on both compartments, on the zero trans flux of a labelled substrate. The effect of imidazole on the rate of choline entry in erythrocytes is shown in Fig. 9.

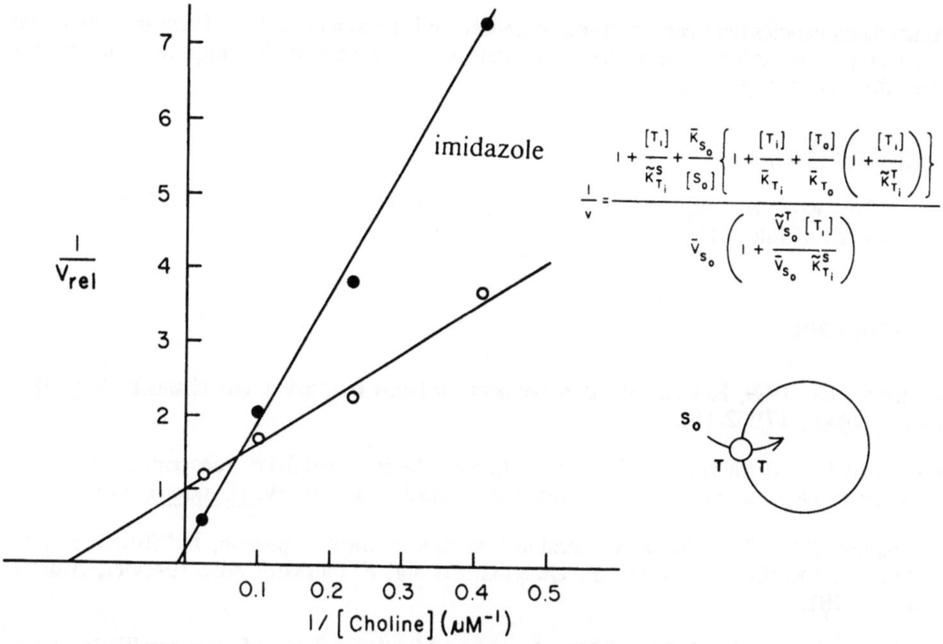

Fig.9. Effect of imidazole (4.6mM) on the rate of choline entry . The results are presented in the form of a Lineweaver-Burk plot. Maximum rate in the control is 0.156 ± 0.003 μmol min^{-1} per liter of cells. The equation that defines the behaviour is also shown. Nomenclature of the transport parameters as in Tables 1 and 2 (Deves and Krupka, 1987).

Imidazole is a low affinity substrate of the choline carrier that rapidly enters into the red cell by passive diffusion. Rates of choline entry were measured in the absence and presence of equilibrated imidazole (4.6mM). The Lineweaver-Burk plot of the data shows that imidazole inhibits the rate when the substrate concentration is low, it has no effect at intermediate choline concentrations (the lines intersect) and activates the rate when choline is saturating.

This surprising result can be explained as follows. A competing substrate will always behave as an inhibitor at the cis side (it displaces the labelled substrate from the transport site), but it can act as an activator from the trans side, if the rate of translocation of the carrier-substrate complex is faster than that of the free carrier (accelerated exchange). Which of the two effects predominates with an equilibrated competing substrate, depends on the relative concentrations of the two substrates. At high concentration of the labelled substrate only the trans effect of the competing substrate becomes evident and the rate is accelerated (the internal competing substrate, although present has been displaced by the labelled substrate), at intermediate concentrations of the labelled substrate the activation caused from the trans side is cancelled by an emerging inhibitory effect on the cis side; as the concentration is decreased further inhibition will be the predominant effect. The observation of activation and inhibition resulting from the action of a single ligand is predicted by the carrier model and can be explained without invoking separate inhibition and activation sites.

CONCLUSIONS

Because of the vectorial nature of the transport process the kinetic analysis of transport must consider particular relationships that do not have analogies in the experimental paradigm of enzyme kinetics. Even the simplest models, not involving multiple or allosteric sites, predict discrepancies between observed behaviour and actual mechanism. These complications, which may seem unfortunate at first sight, can be exploited to gather important information about the transport process.

ACKNOWLEDGEMENTS

This work was supported by Grants from the University of Chile (2674B) and from FONDECYT (Chile) (156/88).

REFERENCES

Andersen, O.S., 1989, Kinetics of ion movement mediated by carriers and channels. Methods in Enzymology, 171:62-112.

Boyd, C.A.R. and Shennan, D.B. 1986, Human placental sulphate transport: studies on chorionic trophoblast brush border membrane vesicles. J. Physiol. (London), 377:15-24.

Cabantchik, Z.I., 1983, Structure-function relations in Band 3 protein, in:"Structure and Function of Membrane Proteins",E. Quagliariello and F. Palmieri, eds., Elsevier, Amsterdam, 271-281.

Deves, R. and Krupka, R.M., 1979a, A general kinetic analysis of transport.Tests of the carrier model based on predicted relations among experimental parameters. Biochim. Biophys. Acta, 556:533-547.

Deves, R. and Krupka, R.M., 1979b. A simple experimental approach to the determination of carrier transport parameters for unlabelled substrate analogs. Biochem. Biophys. Acta, 556:524-532.

Deves, R. and Krupka, R.M., 1981a, Evidence for a two state mobile carrier mechanism in erythrocyte choline transport: Effects of substrate analogs in inactivation of the carrier by N-ethylmaleimide. J. Membrane Biol., 61:21-30.

Deves, R. and Krupka, R.M., 1981b, Reaction of internal forms of the choline carrier of erythrocytes with N-ethylmaleimide:evidence for carrier conformational change on complex formation. J. Membrane Biol., 63:99-103.

Deves, R. and Krupka, R.M., 1983, Apparent noncompetitive inhibition of choline transport in erythrocytes by inhibitors bound at the substrate site. J. Membrane Biol., 74:183-189.

Deves, R. and Krupka,R.M., 1987, Effects on transport of rapidly penetrating, competing substrate: activation and inhibition of the choline carrier of erythrocytes by imidazole. J. Membrane Biol., 99:13-23.

Deves, R. and Boyd, C.A.R., 1989, The determination of kinetic parameters for carrier-mediated transport of non-labelled substrate analogues: a general method applied to the study of divalent anion transport in placental membrane vesicles. Proc. R. Soc. Lond., B237:85-97.

Deves, R. and Krupka, R.M., 1989, Inhibition kinetics of carrier systems. Methods in Enzymology, 171:113-132.

Deves, R. and Krupka, R.M., 1990, A simple test for the sidedness of binding of transport inhibitors, Biochim. Biophys. Acta, in press.

Deves, R. and Krupka, R.M., 1990, Inhibition of choline transport in erythrocytes by n-alkanols. Biochim. Biophys. Acta, in press.

Edwards, P.A.W., 1973, Evidence for the carrier model of transport from the inhibition by N-ethylmaleimide of choline transport across the human red cell membrane. Biochim. Biophys. Acta, 311:123-140.

Jarvis, S.M., Hammond, J.R., Paterson, A.R.P. and Clanachan, A.S.,1983, Nucleoside transport in human erythrocytes. A simple carrier with directional symmetry in fresh cells, but with directional asymmetry in cells from outdated blood. Biochem. J., 210:457-461.

Krupka, R.M. and Deves, R.,1980, The choline transport system of erythrocytes. Distribution of the free carrier in the membrane. Biochim. Biophys. Acta, 600: 228-232.

Krupka, R.M., and Deves, R., 1981, An experimental test for cyclic versus linear transport models. The mechanism of glucose and choline transport in erythrocytes. J. Biol. Chem., 256:5410-5416.

Krupka, R.M. and Deves, R., 1983, Kinetics of inhibition of transport systems. Int. Rev. Cytol., 84:303-352.

Krupka, R.M. and Deves, R., 1988, The choline carrier of erythrocytes: location of the NEM-reactive thiol group in the inner gated channel. J. Membrane Biol., 101:43-47.

Hoare, D.G., 1972, The transport of L-leucine in human erythrocytes: a new kinetic analysis. J. Physiol. (London)., 221:311-331.

Levine, M., Oxender, D.L. and Stein, W.D., 1965, The substrate-facilitated transport of the glucose carrier across the human erythrocyte membrane. Biochem. Biophys. Acta., 109:151-163.

Martin, K, 1971, Some properties of an SH group essential for choline transport in human erythrocytes. J. Physiol (London)., 213:647-667.

Passow, 1987, Molecular aspects of band 3 protein-mediated anion transport across the cell membrane. Rev. Physiol. Biochem Pharmacol., 103:62-85.

Stein, W.D., 1986, "Transport and Diffusion across Cell Membranes", Academic Press, Orlando, Florida.

Stein, W.D., 1989, Kinetics of transport: analyzing,testing and characterizing models using kinetic approaches. Methods in Enzymology, 171:23-61.

Wilbrandt, W., and Rosenberg, T., 1961, The concept of carrier transport and its corollaries in pharmacology. Pharmacol. Rev., 13:109-183.

2

CARRIER KINETICS SHOW HOW THE SODIUM PUMP USES ATP TO RENDER PUMPING OF BOTH SODIUM AND POTASSIUM EFFECTIVE, AND SUGGEST A MODEL FOR THE ACTION OF THE F_0F_1 ATP-ASES

WILFRED D. STEIN

Dept of Biochemistry
Hebrew University
JERUSALEM, 91904, ISRAEL

INTRODUCTION

The counter- and co-transport of many metabolites into and out of the living cell are brought about by coupling of such flows with the flow of sodium or of hydrogen ions (Harold, 1986). By these means, the electrochemical gradient of the pumped metabolite is coupled to the existing gradient of the sodium ion or proton. But how is the electrochemical gradient of the sodium or proton itself established? To understand this process is the task of the present contribution. We shall see that these ions are pumped out of the cell or organelle by primary active transport systems which use metabolic energy directly. The source of metabolic energy in the different primary active transporters may be the splitting of ATP, the harnessing of the flow of electrons in an oxidation-reduction reaction, or the absorbtion of the energy of a photon in the light-driven pumps. In this paper, I shall consider mainly the sodium pump of animal cell plasma membranes, but the principles raised will also be applied to the F_1F_0 ATPases of mitochondria, chloroplasts and bacteria.

THE SODIUM PUMP OF THE PLASMA MEMBRANE

We all know that human red blood cells can be withdrawn from the body, stored under refrigeration for several weeks, and then transfused into a patient where they will function adequately in maintaining life processes in the recipient. It took medical researchers many years of patient exploration to work out the best methods of blood storage. Blood cells freshly drawn from the body have a sodium ion concentration which is about one-tenth of that of the extracellular medium, the blood plasma. In contrast, their potassium ion concentration is more than ten-fold that of the plasma. On cold storage, the cells lose potassium and gain sodium ions, until cellular and extracellular levels of these ions become almost equal, at some 15 mM for potassium and some 120 mM for sodium.

If the cells are then incubated in the presence of a metabolisable substrate, sodium is ejected from the cell and potassium taken up, until the ion concentrations normally present in the cell are restored. Thus cells for transfusion must be stored with an adequate supply of

Cell Membrane Transport, Edited by D. L. Yudilevich *et al.*
Plenum Press, New York, 1991

metabolisable substrate and then incubated before transfusion takes place. Now, the subsequent movement of both sodium and potassium ions is against their concentration gradients. Hence, both cannot be driven by the mutual coupling of their flows. Something other than an anti- or sym-port is required here. We find that the system needs the direct consumption of metabolically-produced ATP for the pumping of both ions The system is thus a primary active transporter, the so-called sodium pump.

A major step forward in the study of this ion pump came from an experiment performed by the Hungarian biochemist, George Gardos (Gardos, 1954). If red blood cells are swollen osmotically, pores open up in the plasma membrane. The cells lose their internal haemoglobin and, of course, their red colour. They are then termed "ghosts". During the process of swelling of the cells, as the haemoglobin leaves, normally non-permeable substances can be induced to enter the cells through the pores. Gardos loaded red cells, in this way, with ATP and arranged the ion concentrations of the cells so that sodium was the same level at the two faces of the membrane, as was potassium. Gardos showed, by direct measurement, that the cells hydrolysed the ATP as they pumped sodium out of, and pumped potassium into, the cells. ATP added outside the cells could not be used; only that present at the intracellular face of the cell membrane could. Ian Glynn showed that pumping out of sodium at an appreciable rate requires potassium ions to be present outside the cell (Glynn, 1956; for reviews of more recent work see Glynn and Karlish, 1975; 1990). Careful quantitative work showed that 3 sodium ions are pumped out of the cell, and 2 potassium ions are pumped in, for each molecule of ATP that is split. Finally it was shown that the split products of the ATP, namely ADP and inorganic phosphate ion (P_i) are liberated at the intracellular face of the membrane. Pumping requires that sodium and ATP be present inside the cell, potassium outside, and under these conditions hydrolysis of ATP brings about the coupled ion pumping.

Now this hydrolysis of ATP is the typical action of an ATPase enzyme. The finding that the pump splits ATP in the presence of sodium and potassium ions meant that cell membranes must contain an ATPase that is activated by sodium and potassium. Sure enough the Danish Physiologist, Jens Skou (Skou, 1957; more recent studies in Skou, 1983), soon found such an enzyme, originally isolating it from crab nerves (although it was later shown to be present in nearly all animal cell plasma membranes). This enzyme is the sodium + potassium - activated ATPase (the Na,K-ATPase). The enzyme has been prepared in pure form, shown to be composed of two polypeptide chains, alpha and beta, of relative molecular mass some 112 KD and 34 KD, respectively. The genes coding for these two polypeptides have been cloned and sequenced. The purified, lipid-free enzyme can be reconstituted back into phospholipid vesicles. In this state, it is fully active as an enzyme and can transport cations into and out of the vesicles.

An important advance was made by Hans-Jurgen Schatzmann (Schatzmann, 1953; more recent studies in Schatzmann, 1983) who showed that the enzyme could be totally, and specifically, inhibited by the drug ouabain, found in the Strophanthus plant, one of the cardiac glycosides still used today in the treatment of heart conditions. (Related to ouabain is another cardiac glycoside known as digitalin, being found in the foxglove plant *Digitalis*.) The identification of this powerful inhibitor meant that experiments could easily be conducted to study the role of the pump-enzyme, since its action could always be blocked by ouabain.

One such experiment was performed by Robert Post (Post, 1969). He showed that cell membranes, isolated from dog kidney, became phosphorylated when they were incubated with ATP (radioactively labelled in the terminal phosphate residue) in the presence of sodium ions. This transfer of labelled phosphate from ATP to the membrane proteins was inhibited by ouabain and hence a function of the Na,K-ATPase. The addition of potassium to the phosphorylated membranes released from them their covalently-bound phosphate, the process again inhibited by ouabain. this meant that the ATP splitting must be taking place in two steps, each step being activated by one of the two cations, phosphorylation by sodium, dephosphorylation by potassium - to give an overall hydrolysis of ATP, activated by the combination of Na and K. Post found that treatment of kidney membranes with the sulfhydryl reagent, N-ethylmaleimide (NEM) an inhibitor of this ATPase, enabled the two steps to be functionally dissected. Phosphorylated membranes in the absence of NEM were, of course, dephosphorylated by potassium. But if NEM was added, potassium failed to activate dephosphorylation, and only by adding ADP could the phosphate be released from the enzyme (and transferred back to reform ATP!). NEM blocked the transformation of the enzyme to the potassium-sensitive form. Post hypothesised that two conformations of the enzyme must exist. The first to form is ADP-sensitive, the second to form, potassium-sensitive, while NEM paralyses the transformation of the first to the second. These two forms of the phosphorylated pump-enzyme are now known as E_1P and E_2P (P symbolising the covalently bound phosphate residue). E_1P is ADP-sensitive, E_2P potassium-sensitive. Identification of these two forms enabled Post, together with Albers (Albers, 1967), to put forward a scheme for linking the biochemical events of ATP splitting and the transport events of sodium/potassium exchange. Their scheme is known as the Albers-Post model and an updated version of is depicted in Fig. 1.

In this scheme, E_1 and E_2 are two forms of the pump-enzyme that have their cation-binding sites facing the intracellular surface and the extracellular surface of the membrane, respectively. In the presence of intracellular sodium ions (three in number) and ATP, E_1 is phosphorylated, releasing ADP and yielding E_1PNa. This then undergoes a conformation change to yield E_2PNa, the form that has its cation-binding sites facing the extracellular surface. The three sodium ions dissociate from this form to yield E_2P and two potassium ions from the extracellular surface take their place, yielding E_2PK. This form now undergoes dephosphorylation to E_2K, finally interconverting to E_1K. The potassium ions, now bound to a site that is accessible from the intracellular surface, dissociate and release free E_1 one more, to restart the whole cycle. In a single cycle, three sodium ions are transferred in the opposite direction, while a molecule of ATP is split to ADP and P_i. The movements of sodium and potassium and the splitting of ATP are strictly coupled, although we do not know how the coupling is controlled. This remains a crucial problem for future investigators to solve.

The two forms E_1 and E_2, whether phosphorylated or not, differ in their conformation. Thus the two behave differently towards proteolytic digestion. Peter Jorgensen from Denmark (for a recent review see Jorgensen and Andersen, 1988) showed that the former is split in two places by chymotrypsin, the latter only in a third site, coinciding with neither of the first two. Steven Karlish (Karlish, 1980) showed that the two forms differed in their fluorescent properties, a finding which has enabled many studies to be undertaken on the rates of interconversion between the two conformations. We now know that the scheme of Fig. 1 is a simplification of the true picture since a major, intermediate form that is bound to potassium has its cation -binding and -releasing sites accessible at neither side of the membrane (Post et al., 1972; Glynn and Richards, 1982). The potassium in this form is said to be "occluded" within the enzyme. Sodium can also be "occluded" in separate form, also an intermediate between the two sodium-bound forms shown in Fig. 1.

Interestingly, the Na,K-ATPase recognises an analogue of phosphate, the anion vanadate, at the site on the enzyme that normally binds phosphate. This vanadate binds to the enzyme relatively well, but its dissociation from the enzyme, under the influence of potassium, is exceedingly slow. Hence, once bound, it remains bound to the enzyme blocking the phosphate-binding centre, and inhibiting the enzyme from taking part in any further cycle of phosphorylation and dephosphorylation. It seems that all the cation-activated ATPases that demonstrate an enzyme intermediate with phosphate bound to the enzyme can be inhibited by vanadate in such a nearly irreversible reaction. The members of this family of enzymes are referred to as the "vanadate-inhibitable" ATPases, or following Peter Pedersen and Ernesto Carafoli's suggestion : "P ATPases" - since they form a phosphorylated intermediate during their action.

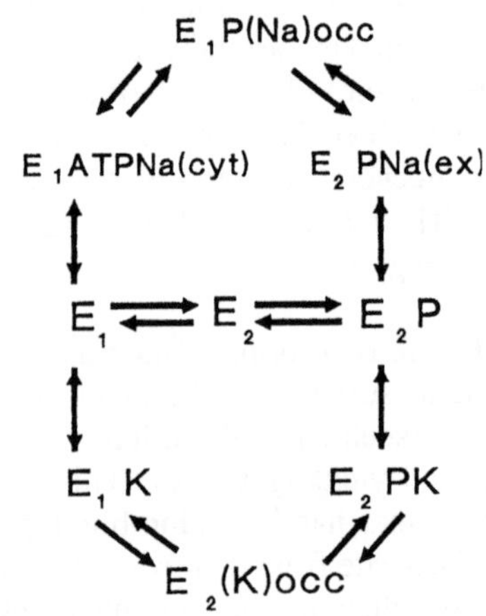

Fig. 1 The conventional scheme for Na- and K-pumping linked to ATP splitting. ("occ" signifies an occluded form of ion).

ADDITIONAL MODES OF ACTION OF THE SODIUM PUMP

Other major features of the pump-enzyme that can be inferred directly from Fig. 1 concern the enzyme's ability to catalyse the exchange of intracellular and extracellular potassium. (Most such experiments have been performed using human red blood cells, which can easily be loaded with radioactive ions and the efflux of the preloaded cation, or the influx from an external solution, followed as a function of time. For a fine review of these studies see Glynn and Karlish, 1975).

Both the exchange reactions involve the reversal of partial reactions of the pump-enzyme. Consider Fig. 1 carefully to convince yourself: if the pathways involving sodium are reversible, then when no potassium is present to bind to the enzyme, sodium ions will be exchanged across the membrane (as Patricio Garrahan and Ian Glynn showed - Garrahan and Glynn, 1967). This reaction needs the presence of ATP and ADP (the latter to accept P from E_1PNa) but results in no overall splitting of the ATP. Similarly, potassium ions (when little or no sodium is present) can be exchanged in a reaction that is speeded up by ATP and P_i but results in no hydrolysis.

Some other properties of the system do not follow from Fig. 1 and, indeed, show that the scheme is an over-simplification. There is a slow, but definite, exchange of potassium ions in the absence of any phosphate or possibility of phosphorylation, and there is also a net transport of potassium ions in such conditions. Finally, there is a slow hydrolysis of ATP in the complete absence of potassium ions, accompanied by the transfer of sodium from intracellular to extracellular surface. This means that there is a route for the (slow) breakdown of E_2P in the absence of potassium ions. All these reactions are true "slippages" of the pump-enzyme.

Just as has been discussed in earlier sections of this book for the carrier and as can be done also for the anti- and sym- porters, one can study the kinetics of the sodium and potassium transport on this primary transporter. One can measure a Michaelis constant Km for each of the cationic substrates, sodium and potassium, and for the phosphate derivatives, ATP, ADP and P_i, where binding is studied for each of the two forms E_1 and E_2. (Note that ATP, ADP, and P_i can bind to each form, E_1 and E_2, although their binding sites exist only at the intracellular face of the membrane). The Na, K-ATPase is quite clearly asymmetric. E_1 has a high affinity for sodium and for ATP and ADP, but a low affinity for potassium and P_i. In contrast, E_2 has a high affinity for potassium and P_i, but a low affinity for sodium, ATP and ADP. E_1's high affinity for sodium and low affinity for potassium enable the pump to bind sodium in preference to potassium at the intracellular face (where the binding-sites of E_1 are accessible) and this maximises the rate of pumping of sodium out of the cell. Similarly, E_2's high affinity for potassium and low affinity for sodium enables the pump to bind potassium preferentially at the extracellular face and maximises the rate of inward pumping of potassium.

How does the splitting of ATP drive the active transport of the ions? It does so as a result of the strict coupling that occurs between the conformational interchanges of E_1 and E_2 and the splitting of ATP. Coupling ensures that a complete cycle of ion transport and ATP splitting takes place only when: (i) three sodium ions from the intracellular face become bound to E_1, (ii) the conformation alters to E_2, (iii) at the extracellular face potassium exchanges for sodium ions, and finally (iv) the conformation returns to E_1, bringing potassium to the intracellular face. Each time that a net breakdown of ATP into ADP and P_i occurs, the exchange of sodium and potassium has to occur. The ATP is a "high-energy" compound in that the equilibrium between ATP and ADP + P_i is well over on the side of the splitting of ATP at the usual levels of ATP, ADP, and P_i in the cell. Thus, a reaction leading to hydrolysis of ATP tends to occur. If this reaction occurs on the pump, the exchange of sodium with potassium will, under physiological conditions, be coupled with it, in the direction of sodium efflux and potassium influx (Fig. 1). But this overall reaction, like any chemical reaction, is in principle reversible. If the cell's resources of ATP are low in comparison with the concentration of ADP and P_i, the reaction of ATP hydrolysis might not be so favourable, energetically. If, in addition, extracellular levels of sodium are high and those of potassium intracellularly are low, a reversal of pumping will occur. Sodium will enter the cell and potassium will leave it, on the pump. The direction of travel around the scheme of Fig. 1 will be reversed and ATP synthesis will occur. Thus it is ATP splitting, driving the interconversion of E_1 and E_2 that brings about pumping. The change of affinity for the cations during this interconversion speeds up pumping but does not drive it.

IT IS NOT PUMP ASYMMETRY THAT DRIVES ACTIVE TRANSPORT

The textbooks teach us that an ion pump works by loading its substrate at a high affinity site at the side from which pumping occurs, and then discharging the substrate from a low affinity site at the other face of the membrane. The conventional view is that the energy from hydrolysis of ATP is used to bring about this change of the site affinities, and that this is what drives pumping. But Hill and Eisenberg (1981) showed that this cannot be the whole story or even the major part of it (see also Hill, 1989). The release of the ATP's free energy can occur anywhere around the pumping cycle and yet pumping will still occur. Consider the conventional scheme for the Na/K ATPase (Fig. 1).

The upper half of the figure shows the pumping of sodium linked to the phosphorylation of the enzyme, the lower half, potassium pumping linked to dephosphorylation. Nowhere in the scheme as written is it necessary to state whether there is a change of pump affinity for either ion. Yet, whatever the details of the scheme, such a pump will pump sodium and potassium in opposite directions against their respective concentration gradients, if the splitting of ATP is tightly coupled to the ion movements. It is this tight coupling, Hill and Eisenberg emphasise, that drives the pump. We have measured directly how tight this coupling is in the case of the Na-K ATPase reconstituted into artificial phospholipid vesicles (Karlish and Stein, 1982a). Take the turnover number of the overall reaction of Na efflux and potassium influx coupled to ATP hydrolysis (some 40 per second at room temperature) as 100%. The turnover number for the potassium movements in the absence of sodium is less than 10% of this. Turnover numbers in the absence of ATP are 10 to 100 times lower again. Jencks (1983) has shown us that it is the erection of appropriate energy barriers and directed dismantling that ensures tight coupling.

A few moments' reflection will convince one that pumping will occur, in a strictly coupled pump, even if the pump's affinities for ions are quite contrary to those ordinarily found. Take the case where there is a low affinity site at the side from which pumping is occurring (the "whence" side in what follows). The pump will bind the ion, at this site, with a small probability and will discharge it from a high affinity site at the side to which pumping is occurring (the "whither" side), again to a small extent, provided only that the cycle of Fig. 1 continues to take place, that is, provided that the overall equilibrium between ATP, ADP and P_i, and the prevailing ion gradients are such as to allow ATP splitting to continue. It is thermodynamics that determines whether or not pumping will occur, not kinetics.

A SIMPLE CARRIER CAN BE ASYMMETRIC AND YET IT CANNOT PUMP

One can, indeed, easily demonstrate the falsity of the assumption that an asymmetry of affinities must bring about pumping. We know that many simple carrier systems, which cannot bring about the active transport of their substrates, are asymmetric in their kinetics as well as in their structure. Two well-studied examples are the glucose transport system (Baker and Widdas, 1973) and the uridine transport system of the human red blood cell (Cabantchik and Ginsburg, 1977). Both of these are facilitated diffusion systems. They catalyse the movement of their substrates, glucose or uridine respectively, across the membrane but only move them down their respective concentration gradients. Their kinetics are, in many modes of transport, simple. A maximum velocity, V_{max}, is reached and one can define a Michaelis parameter, K_m, being that substrate concentration at which one-half of this maximum velocity is achieved.

Interestingly, and revealingly, both these systems develop their asymmetry on cold-storage of the cells, and they do this without developing any capacity for pumping their substrates (Weiser et al., 1983; Plagemann and Wohlhueter, 1984). There is good evidence that it is the accumulation within the cells of some competitive inhibitor of transport that induces these asymmetries. Consider what will happen if a non-penetrating competitive inhibitor of transport is present at cytoplasmic face of a carrier system. Within the cell, substrate and the inhibitor will complete. The effect of such a competitive inhibitor on the efflux of the substrate is to raise the effective K_m, leaving V_{max} unchanged. For influx, on the other hand, substrate and inhibitor cannot compete - they are on opposite sides of the membrane. The K_m for influx is unchanged and the system is now clearly asymmetric in its kinetics. It retains the original high affinity at the extracellular face, but demonstrates a low affinity at the cytoplasmic surface! (V_{max} for influx is, however, reduced). The competitor binds some of the carrier at the cytoplasmic face and holds it there, trapping it, and taking it out of the pool of carriers that are available for substrate transport, thereby reducing the V_{max}. Even at the highest concentrations of substrate, the effect of this inhibition at the cytoplasmic face cannot be overcome. Thermodynamics requires that for these non-pumping systems, in all circumstances, the ratio of V_{max} to K_m must be the same for efflux as it is for influx, and this occurs by the raising of the K_m for efflux, while V_{max} for influx is lowered).

The kinetic asymmetry of a carrier system depends on the difference in free energy between the carrier conformations that take part in transport. It is easy to show (Stein, 1986, p.257) that, for a carrier in which the loaded forms reorient across the membrane faster than do the unloaded forms (and this is true for the majority of the carrier systems), the asymmetry depends on the difference in free energy between the two conformations of the unloaded carrier. I proceed to show that this is exactly true for the important exchange-only carriers.

Consider Fig. 2. This depicts an exchange-only carrier which exists in two forms, E_1 and E_2, of different energy levels (stabilities). In the figure each horizontal line depicts the energy level (in the standard state) of the components written just above the line. The arrows with deltas are the free energy changes of the steps indicated and give the overall binding constants for the ion or (for later figures) for ATP at each face of the membrane.

ES_1 and ES_2 are the two forms of the carrier-substrate complex which interconvert during the conformation transformation which is the transport event. The figure depicts, in addition to the free energy levels of the various forms, the rate constants of the various transitions and the intrinsic binding affinity for substrate at side 1 and side 2, namely $K_1(\mathrm{int})$ and $K_2(\mathrm{int})$. The rate of exchange of S from side 1 to side 2 is given by very simple equation (Stein, 1986):

$$v12 = \frac{S_1 S_2/Ree}{S_1 + S_2 + S_1 S_2}$$

where $S_1 = [S_1]/K_1$ and $S_2 = [S_2]/K_2$, and K_1 and K_2 being the apparent dissociation constants which govern the transport parameters, and the square brackets, [ ], represent concentrations. Ree is the resistance for this exchange reaction (the reciprocal of its maximum velocity) and is given by

$$(1+g1)(1+g2/b2) + (1+g2)(1+g1/b1) + 1/b1 + 1/b2$$

in terms of the rate constants of the model, and is limited by whichever of the four parameters, b1, b2, g1, or g2 is the smallest. The apparent dissociation constants are given by:

$$K_1 = (1/f1)(b1g2)/(g1 + g2)$$
$$= K_1(\mathrm{int})\, g2/(g1 + g2)$$

and

$$K_2 = (1/f2)(b2g1)/(g1 + g2)$$

$$= K_2(\text{int})\ g1/(g1 + g2)$$

To check that this equation is indeed valid for exchange consider first, what happens when $[S_1]$ and $[S_2]$ go to infinity. Then v12 goes to 1/Ree and, correctly, reaches the maximum velocity. Now take $[S_2]$ as infinity but put $[S_1] = K_1$. Then the velocity is one-half of V_{max}. Thus a substitute concentration of $[S_1] = K_1$ determines, correctly, one-half of maximum velocity.

No steady-state transport experiment reveals directly the intrinsic dissociation constants, K_1(int) and K_2(int). Rather, it is the apparent dissociation constants which are revealed as the half-saturation concentrations, K_1 and K_2, in transport measurements from side 1 or side 2 of the system, respectively. Now the ratio of these, which can be usefully considered as apparent affinities, is the kinetic asymmetry,

$$K_1/K_2 = b2g1f1/b1g2f2$$

which is itself given by

$$\exp(\text{Delta } G(E_1 - E_2)/RT)$$

and hence given exactly by the difference in steady-state free energy between the two forms of the unloaded carrier, E_1 and E_2.

Since for all substrates that ride on the same carrier, the unloaded form is the same, the kinetic asymmetry for all substrates that ride this carrier is identical, a strong result. This means that whatever are the intrinsic affinities for carrier and substrate at any face of the membrane, the ratio of apparent affinities, and it these that a steady-state kinetic experiment will measure, will be determined only by the free energy difference between the two forms of the free carrier. (In the case of a carrier plus competitive inhibitor system, the free energy difference takes into account the overall free energy change for the binding of competitive inhibitor to the carrier at the prevailing concentration of the inhibitor, as well as the free energy difference between the unloaded forms of the carrier). Later we shall use an important corollary of the main result: if a carrier system behaves symmetrically in its kinetics, the two forms of its unloaded carrier do not differ in free energy i.e. they are iso-energetic.

WHAT ASYMMETRY DOES IS TO RENDER A PUMP EFFECTIVE

What does all this have to do with ion pumping? We have just shown that asymmetry does not in itself lead to pumping. Yet we know that ion pumps are asymmetric. Is this fact of significance for pumping? The answer is: Yes! Asymmetry increases the effectiveness of an ion pump. While it is true that a pump will pump with ion affinities low at the whence side and high at the whither, it will operate very slowly under these conditions. Much of the pump enzyme will be trapped unloaded at the whence side, unable to take part in the pumping process. Much of the time the pump will return loaded with its substrate from the whither side, having shuttled substrate fruitlessly across the membrane. (Both these situations waste pumping time not pumping energy (Honig and Stein, 1978). In a well-coupled pump, an ion that returns along the path from the whither to the whence side will resynthesise ATP by reversal of the splitting process. No energy is lost, only time).

Time is of significance since our ion pumps are not isolated systems. They are embedded in natural membranes which are leaky to the pumped substrates. At the steady-state, the pumped substrate will leak back down the built up concentration gradient just as fast as pumping occurs. It will leak back without hydrolysis of ATP. This wastes the energy that was invested into ion pumping. A pump will be effective if it is designed to operate rapidly, and for this it must have that high affinity at the whence side and a low affinity at the either side that we have always known was important for pumping. We now see that this aspect of pump design is required to render ion pumping effective by speading up the pump cycle, a kinetic problem not a thermodynamic one, and neglected in the Hill-Eisenberg analysis.

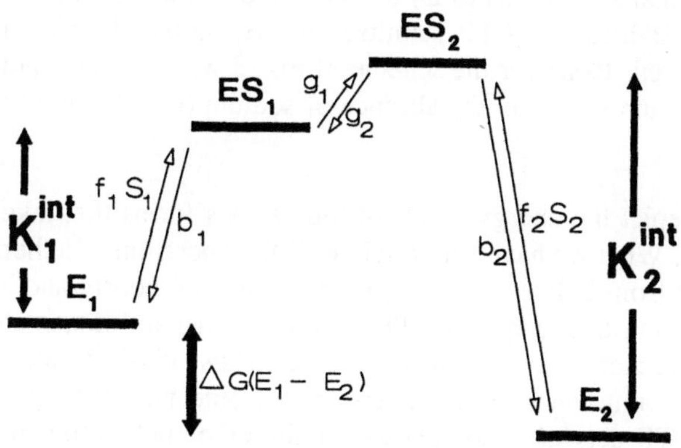

Fig. 2. Energy levels for an exchange-only transport system. Each horizontal line depicts the energy level (in standard state) of the components written just above the line. The arrows with deltas are the free energy changes of the steps indicated and give the overall binding constants for the ion or (for later figures) for ATP at each face of the membrane.

HOW IS ASYMMETRY MAINTAINED IN THE POTASSIUM HALF OF THE PUMP CYCLE?

Consider, first, the simpler, potassium half of sodium-potassium pump's cycle. Sachs (1981) has shown and we have confirmed (Karlish and Stein, 1982 a,b) that, in the absence of sodium and hence of ATP splitting, the system behaves as a simple carrier for potassium ions. Net transport is slower than exchange transport so that the unloaded carrier reorients more slowly than does the potassium - loaded form. In the absence of ATP and P_i, the system behaves almost symmetrically, kinetically, with high affinity sites for rubidium, the analogue of potassium, being demonstrable at both faces of the membrane. K_m is 0.6 mM for transport from the cytoplasmic face, 0.2 mM from the extracellular face. This must mean (as we have just argued above, and see also Stein, 1986, p.257) that the two carrier conformations E_1 and E_2 must be more-or-less isoenergetic. But the intrinsic affinities of E_1 and E_2 for potassium are very different, being high at the extracellular face and low at the cytoplasmic face, by about two orders of magnitude. These intrinsic affinities are just in the ratio that effective pump design would require (high at the whence side for potassium - the extracellular face; low at the whither side - the cytoplasmic face). Notice from the formulae developed in the previous section that the absolute values of the apparent affinities are always less than the corresponding values of the intrinsic dissociation constants, K(int), being determined by the fractions g1/(g1 + g2) and g2/(g1 + g2). they can be very much less than the values of the respective K(int) term if the equilibrium is far from the particular side (of the membrane) under consideration. How are these intrinsic affinities realised during potassium pumping, when both ATP and P_i are present?

First note that the forms E_2 and E_2P are also not very different in free energy, the phosphate bond here being a low energy bond. The effective affinity of K for the system at side 2 is much the same whether or not P_i is present (Stein, 1986, p.560). When ATP is present, however, the asymmetry of potassium binding at its whence and whither sides does not depend on the difference of free energy between E_1 and the phosphorylated form E_2P, but rather on the differential binding of ATP to these two forms. It has been shown (see Stein, 1986, p.560) that ATP binds to E_2 forms with a low affinity, to E_1 forms with a high affinity. Hence the addition of ATP stabilises the E_1 form, allowing its low affinity for potassium to be realised. Consider the scheme of Fig. 3 which depicts potassium movements on the sodium-potassium pump, in the absence of sodium (i.e., the lower half of the scheme of Fig. 1).

Again, we depict the energy levels of the various forms that take part in potassium-potassium exchange. What we have, in principle, is the operation of a thermodynamic "box". Whether one travels from E_2P to (E_1 + P_i) by binding ATP poorly and K well, or from (E_1 + P_i) to E_2P by binding K poorly and ATP well, the initial and final states are the same, so the free energy difference is independent of route. Thus, if ATP binds poorly to E_2P and well to (E_1 + P_i), K must in like but reverse ratio, bind poorly to (E_1 + P_i) and well to E_2P. Essentially, ATP acts as a competitive inhibitor of potassium movement, its action being localised at the cytoplasmic face, just as if it was one of the non-penetrating competitive inhibitors that we discussed above. (Note that the ATP remains at the cytoplasmic face (side 1) although it is binding to forms E_2P and (E_1 + P_i) that bind K at, respectively, side 2 and side 1).

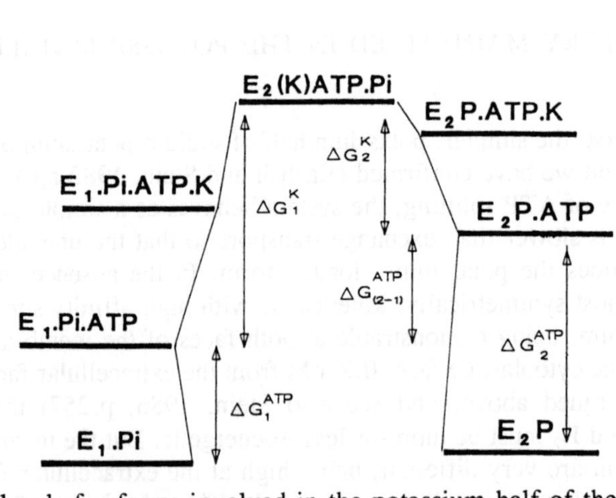

Fig. 3 Energy levels for forms involved in the potassium half of the Na - K pump. Conventions as in Fig. 2.

Comparing Figs 2 and 3 we see that the effective exchange-carrier for potassium is (E_1ATP + P_i) at side 1 and (E_2P.ATP) at side 2. These differ in free energy by Delta G $ATP_{2\text{-}1}$, the differential free energy for ATP-binding. Hence, from our argument around Fig. 2, it is this free energy difference that determines the kinetic asymmetry of the system for potassium, i.e. the high affinity at side 2 compared with that at side 1. Whatever the intrinsic affinities for K at side 1 and side 2, the effective affinity ratio will be determined only by the differential binding of ATP at the two faces of the membrane.

This role for ATP is merely a regulatory one, and it is not split in the process. (If ATP is not split, the energy of its hydrolysis cannot be used for pumping, thus the change of affinity - important for effective pumping - is not what brings about pumping!) The asymmetry in binding of potassium is given by the free energy difference for the binding of ATP to the two forms E_1 and E_2P of the potassium carrier (here the Na/K ATPase).

HOW IS ASYMMETRY MAINTAINED IN THE SODIUM HALF OF THE PUMP CYCLE?

The cytoplasmic face of the pump is the whence side for sodium but the whither side for potassium. Thus we cannot use ATP binding to ensure that sodium binds more avidly at its whence face i.e. to the (E_1 + P_i) form. We used that trick already to stabilise effectively the E_1 form and reveal its low affinity for potassium. The differential affinity of ATP for the E_1 form determines a lower affinity for ALL co-substrates to this form, sodium as well as potassium. Some other design principle is needed to generate a high affinity for sodium at the cytoplasmic face. The solution used is a kinetic one The complex Na.ATP.E_1 takes a different route to get to E_2 than does the complex K.ATP.E_1. The ion which is bound to E_1.ATP determines the fate of the E_1+ATP complex - whether the ATP is split or whether not. Ion binding is the fate-determining step!

K.ATP.E_1 merely undergoes a conformation change to yield the occluded form ATP.E_2(K). In contrast, Na.ATP.E_1 undergoes, of course, a chemical change with eventual phosphorylation of the enzyme and its reorientation to yield eventually Na.E_2.P_i + ADP. The reaction of phosphorylation followed by reorientation is heavily over to the side of Na.E_2.P_i, so that the E_2 form is stabilised over the E_1 form, or E_1 destabilised over E_2 (Fig 4). The binding of Na to E_1 is thus followed kinetically by a change which removes the Na.E_1 complex as fast as it is formed.

The binding of sodium to the carrier is heavily stabilised and the system behaves with a high affinity for sodium. The standard state free energy of E_1 + ATP is, of course, far higher than that of E_2P + ADP (Fig 4). The effective exchange-only carrier for sodium is (E_1 + E_2) at side 1 and (E_2P + ADP) at side 2. Since E_1 and E_2P are iso-energetic (see the previous section), the free energy difference that is effective in determining the kinetic asymmetry of the system is exactly that for the hydrolysis of ATP to ADP + P_i. It is because, as we have seen, E_1 and E_2P are more-or-less isoenergetic, that ALL of the chemical energy in splitting ATP to ADP + (P_i bound to E_2) can be used to stabilise the binding of sodium to the pump from side 1. Sodium can thus bind with the required high affinity at its whence side.

Splitting of ATP drives the whole pump cycle thermodynamically, but effective pump design ensures that the splitting of ATP takes place in a particular step which ensures the satisfaction of the kinetic requirement for a high affinity for sodium at the cytoplasmic face, the whence side for that ion. In contrast, binding of ATP without splitting reveals the high affinity for potassium at the extracellular face, the whence side for potassium.

It is important to note that in all the above discussion, the electrogenic aspect of ion has been neglected. As has been pointed out, the sodium pump exchanges three sodium, for two potassium ions, so that a cycle of pumping moves a positive charge out of the cell, producing an inward electric current. The effects of electric forces on the pump's activity are clearly discussed in Laeuger, (1984).

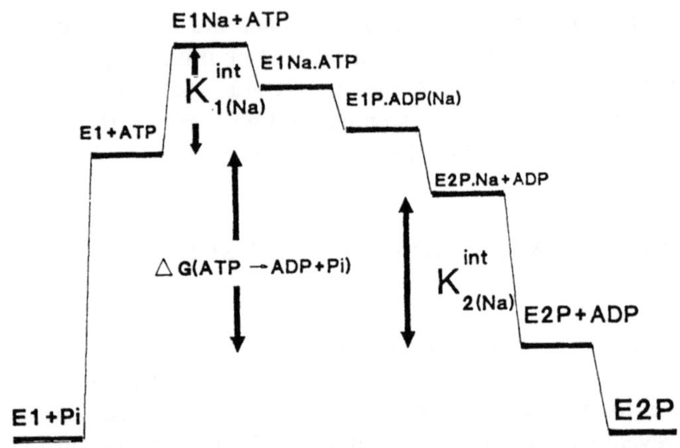

Fig. 4. Energy levels in the sodium half of the Na-K pump cycle. Conventions as in Fig. 2.

A COMMENT ON THE ATP-SYNTHETASES

It is of interest to extend these considerations by including also the F_0F_1 ATPases or ATP synthetases. Detailed studies have uncovered a substantial amount of information about these systems (see, for instance, Amzel and Pedersen, 1983, and many of the articles collected in the book "The Ion Pumps" (Stein, 1988)). The sequences of all the sub-units that comprise this multi-subunit enzyme are known and the overall physical structure of the molecule is clear from electron microscope studies. The kinetic analysis of the transport events are in their infancy. We do not even know if the system behaves towards protons as a carrier or as a channel (using the strict definitions in Stein, 1986). If the former, then we are entitled to attempt to model the system in terms of a carrier with its two (or more) conformations. If the latter, then we must consider some quite different type of model using a theoretical framework which is not clear to me. In the absence of firm knowledge, I shall work on the assumption that we are dealing with a proton carrier, transforming between two different conformations that allow access to proton at the two sides of the membrane alternately, and try make testable predictions from such a model.

The system, of course, works in reverse from the sodium pump that we have just considered in that, physiologically, its role is to harness an existing electrochemical gradient of protons to produce ATP. One might, however, set up a model similar in outline to that of Fig. 1 but in which now the synthetase alternates between two conformations, one of which (say, conformation 1) is accessible at side 1 of the membrane to protons and binds ATP, the other (conformation 2) being accessible to protons only from side 2, and binding ADP and P_i. Such a model describes a system with the capacity to harness the electrochemical gradient of protons to synthesise ATP, if the conformation charges are strictly coupled to the proton translocations and to the chemical reaction of ATP synthesis or its breakdown. Strict coupling will ensure the progress of the chemical reaction and its linkage to the transport event. All one needs to postulate are a conventional active centre for ATP hydrolysis and synthesis and effective energy barriers to slippage events. To ensure, however, that the system operates effectively against the existing leaks or protons and the spontaneous or metabolism-linked hydrolysis of ATP, the mitochondrion, chloroplast, or bacterium will have to ensure that the cycle of transport and chemical reaction takes places fast in the direction of ATP synthesis. This means that conformation 2 must bind ADP + P_i tightly, to ensure that movements leading to synthesis are favoured, while conformation 1 must release its ATP readily, that is, bind it poorly. In contrast with the sodium pump, for the ATP synthetase there is little kinetic advantage to be gained from binding the transported substrate (here protons) differentially at the side of the membrane. A high affinity at the whence side only results in under-usage of the electrochemical gradient, while a low affinity at the whither side can be achieved as well by a low absolute affinity as by a differential affinity for protons at one side or the other. Indeed, if there is no differential binding of protons, all the energy in the electrochemical gradient of protons can be used for ATP synthesis.

There are two general schemes for the design of such an ATP synthetase, based, respectively, on the symport (co-transport) and uniport (simple carrier) paradigms. In the former, ADP + P_i binds, and ATP unbinds, from a carrier which has already bound protons, and only the form bound to both proton and nucleotide can undergo the conformation change which brings about the translocation of protons. In the uniport scheme, proton can bind and unbind in the absence of nucleotide but only forms in which F_0 is bound to proton while F_1 is splitting ATP can translocate. In both cases, translocation of proton and the chemical reaction of ATP synthesis are be tightly linked. The two schemes make subtly different predictions, however, and should therefore be distinguishable experimentally. (Stein and Laeuger, 1990).

These kinetic schemes do not account for the structural details of the ATP synthetase (that it has a basic three-fold symmetry in its structure) and the fact that it has a multiplicity of binding sites for the nucleotides, of differing affinities, sites that interact in negatively co-operative fashion between themselves. Extensive studies by Boyer and his colleagues (Boyer

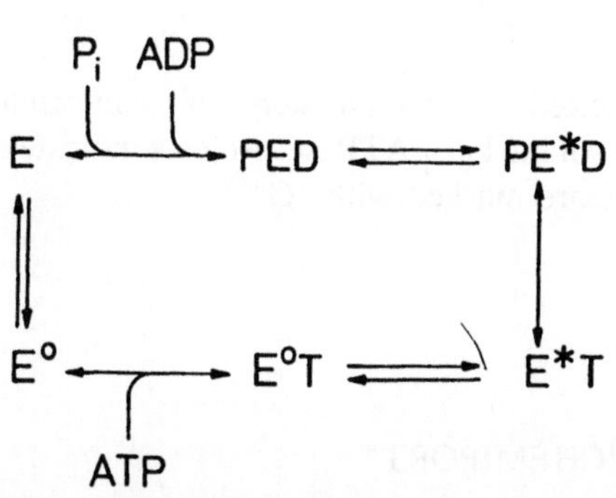

Fig 5. Reaction cycle of a single catalytic sub-unit of the F_0F_1 ATPase, after Boyer and Kohlbrenner (1981). E, E^* and E^0 are different conformational states of the catalytic subunit with affinities for ligands as defined in the text.

and Kohlbrenner, 1981) have led, indeed, to the view that the F_0F_1 ATPase exists in three forms which differ in their affinity for the ligands ATP and ADP ($+P_i$). We follow Boyer et al. and retain the designation of the three conformations of the F_1 subunits of the F_0F_1 ATPase as E^0 for the form which has a preferentially higher affinity for ATP, E for the form with a preferentially higher affinity for ADP ($+P_i$), and E^* for the form which binds both ATP and ADP + P_i very strongly while they undergo the chemical transformation on bound ADP + P_i into bound ATP (Penefsky, 1985). (See Fig 5).

We assume that the whole F_0F_1 ATPase molecule, composed of three subunits of alpha/beta dimers, acts in a co-ordinated fashion so that conformation changes are allowed only when the F_0F_1 ATPase is in one of two forms competent to undergo the conformational change that transfers substrate across the membrane, namely E^0.PED.E^*T or E^0.E.PE^*D (Fig. 6). H can bind to or debind from the F_0F_1 ATPase during any or all off the steps during which the ligands ATP, ADP and P_i bind or debind, or during the steps in which the chemical reaction of ATP bound to E^* breaks down to ADP + P_i bound to E^*, or during the corresponding synthesis of ATP on E^*.

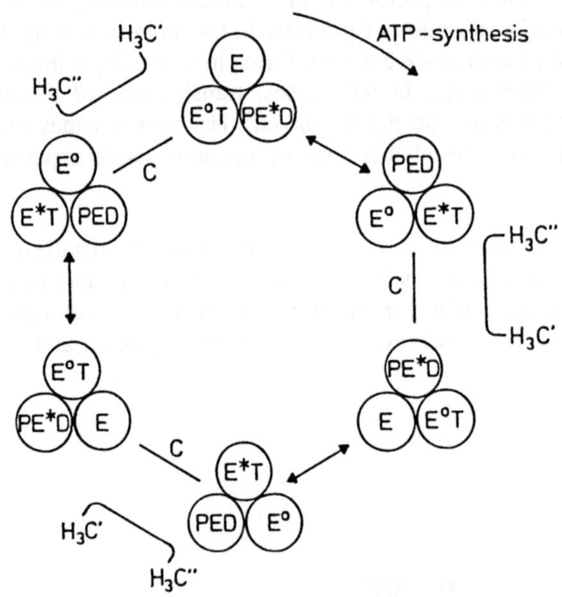

Fig 6. Cyclic, concerted interconversion of conformational states of the (alpha)3(beta)3-complex of the F_0F_1 ATPase. Arrows indicate equilibrium transitions. Conformational changes are marked with "C".

F_0F_1 AS A CLASSICAL "CHEMIPORT"

Figure 7 depicts some possible cases of the "chemiport" model for F_0F_1 ATPase. The ligands, H2, H1, ATP, ADP and P_i bind and debind to F_0F_1 ATPase at various steps around the chemiport cycle. On the models drawn, binding/debinding is ordered but one can also take into account the possibility that these events may be random. The double arrows in the diagram represent the conformation changes or rotation events which interconvert the states of the subunits of F_1 and, on the "chemiport" model, bring about that conformation change which carries the substrate across the membrane.

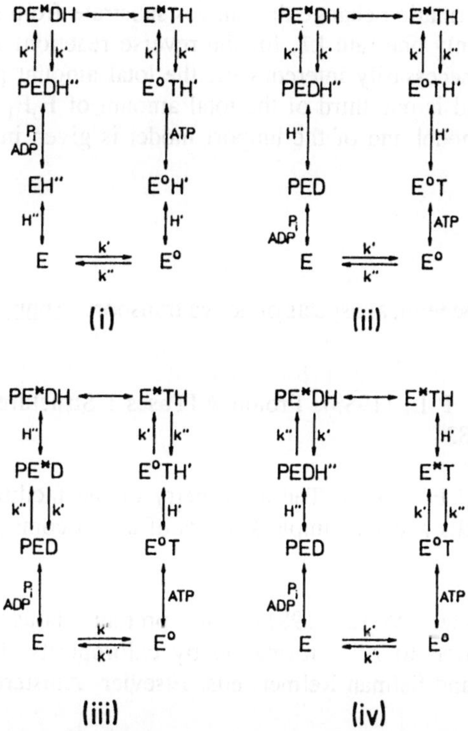

Fig. 7. Four forms of the "chemiport" model for the action of the F_0F_1 ATPase. In each model E, E^0, and E^* are states of the F_0F_1 ATPase as defined in the text. The order of binding of the transportable substrate H and of the phospho-ligands, ATP, ADP and P_i varies. In (i) and (ii), all ligands bind and debind in different order but between the same pair of conformation changes. In (iii) and (iv) the ligands bind/debind after different conformational changes.

Note an important feature of the "chemiport" model: The diagrams of Figure 7 are all isomorphous with the diagram of Figure 1 and are thus perfectly isomorphous with the conventional model for the action of the sodium pump and the calcium pump of animal cell membranes (see Stein, 1986). To see this, write Nacyt for H1, Naex for H2, in the figure and E_1 for E_0, E_2 for E, with finally, Eocc for E^* (where Nacyt is the sodium concentration with the cell, Naex that at the cell exterior and E_1, E_2, and Eocc are, respectively, the forms of the sodium pump with sodium-binding sites that face the cytoplasm, that face the exterior solution, or are the "occluded" form of the pump, in which sodium ions are unavailable at either face of the membrane).

It is important, then, that both the E_1E_2 and F_0F_1 models can be written in the same form, emphasising the essential unity of the coupling of transport with chemical reactions in these cases. Perhaps all these systems should be considered, indeed, as $E_1E_2E_3$ systems! (in the sodium and calcium pumps, of course, the pump-enzyme is phosphorylated during the process in which the cation is carried across the membrane whereas in the F_0F_1 ATPase, no phosphorylation of the enzyme itself takes place rather the ATP (and the ADP) is very tightly bound).

Since the F_0F_1 ATPase is here considered to act in concerted fashion there is only one rate of reaction for conformation change that moves substrate from side 2 to side 1, namely k2 as in Figure 7, and only one rate k1, for the reverse reaction. In addition, since forms such as E^0, E, and E^*, necessarily interconvert, the total amount present of each of those forms is always equal and is one-third of the total amount of F_0F_1 ATPase present. A full kinetic analysis of this model and of the uniport model is given in the paper by Stein and Laeuger (1990).

REFERENCES

Albers, R.W. (1967). Biochemical aspects of active transport. Annu. Rev. Biochem. 36: 727-756.

Amzel, L.M., Pederson, P.L. (1983). Proton ATPases : Structure and Mechanism. Ann. Rev. Biochem. 52: 802-824.

Baker, G.F., Widdas, W.F. (1973). The asymmetry of the facilitated transfer system for hexoses in human red cells and the simple kinetics of a two component model. J. Physiol. (London) 231: 143-165.

Boyer, P.D., Kohlbrenner, W.E. (1981). The present status of the binding-change mechanism and its relation to ATP formation by chloroplasts. In: Energy Coupling in photosynthesis. Selman and Selman-Reimer, eds. Elsevier, Amsterdam . pp.231-240.

Cabantchik, Z.I., Ginsburg, H. (1977). Transport of uridine in human red cells. Demonstration of a simple carrier. J. Gen. Physiol. 69: 75-96.

Gardos, G. (1954). Accumulation of K ions in human red blood cells. Acta Physiol. Acad. Sci. Hung. 6: 191-199.

Garrahan, P.J., Glynn, I.M. (1967). The sensitivity of the sodium pump to external sodium. J. Physiol. (Lond.) 192: 175-188.

Glynn, I.M., Richards, D.E.(1982). Occlusion of sodium ions by the sodium-potassium pump: Its implications for the mechanism of active transport. J. Physiol. (Lond.) 330: 17-43.

Glynn, I.M., Karlish, S.J.D. (1975). The sodium pump. Annu. Rev. Physiol. 37: 13-55.

Glynn, I.M., Karlish, S.J.D. (1990). Occluded cations in active transport. Annu. Rev. Biochem. 57: 171-205.

Glynn, I.M. (1956). Sodium and potassium movements in human red cells. J. Physiol. (Lond.) 134: 278-310.

Harold, F.M. (1986). "The Vital Force: a Study of Bioenergetics." W.H.freeman and Company, New York.

Hill, T.H. (1989). "Free energy Transduction and Biochemical Cycle Kinetics." Springer-Verlag, New-York.

Hill, T.L., Eisenberg, E. (1981). Can free energy transduction be localised at some crucial part of the enzymatic cycle? Quart. Rev. Biophys. 14: 463-511.

Honig, B., Stein, W.D. (1978). Design principles for active transport systems. J.Theor. Biol. 75: 299-305.

Jencks, W.P. (1983). What is a coupled vectorial process? Curr. Topics. Memb. Trans. 19: 1-19.

Jorgensen, P.L., and Andersen, J.P. (1988). Structural basis for E_1-E_2 conformational transitions in Na,K-pump and Ca-pump proteins. J. Membrane Biol. 103: 95-120.

Karlish, S.J.D. (1980). Characterisation of conformational changes in (Na,K) ATPase labelled with fluorescein at the active site. J. Bioenerg. Biomembr. 12: 111-136.

Karlish, S.J.D., Stein, W.D. (1982a). Passive rubidium fluxes mediated by Na-K-ATPase reconstituted into phospholipid vesicles when ATP- and phosphate-free. J. Physiol (London) 328: 295-316.

Karlish, S.J.D., Stein, W.D. (1982b). Effects of ATP or phosphate on passive rubidium fluxes mediated by Na-K-ATPase reconstituted into phospholipid vesicles when ATP- and phosphate-free. J. Physiol (London) 328: 317-331.

Laeuger, P. (1984). Thermodynamic and kinetic properties of electrogenic ion pumps. Biochim. Biophys. Acta. 779: 307-341.

Penefsky, H.S. (1985). Energy-dependent dissociation of ATP from high-affinity catalytic sites of beef-heart mitochondrial adenosine triphosphatase. J.Biol. Chem. 260: 13735-13741.

Plagemann, P.G.W., Wohlhueter, R.M. (1984). Kinetics of nucleoside transport in human erythrocytes: alterations during blood preservation. Biochim Biophys Acta, 778: 176-184.

Post, R.L., et al. (1969). Flexibility of an active center in sodium-plus-potassium adenosine triphosphatase. J. Gen. Physiol. 54: 306-326.

Post, R.L., Hegyvary, C., Kume, S. (1972). Activation by adenosine triphosphate in the phosphorylation kinetics of sodium and potassium ion transport ATPase. J. Biol. Chem. 247: 6530-6540.

Sachs, J.R. (1981). Mechanistic implications of the potassium-potassium exchange carried out by the sodium-potassium pump. J. Physiol. (Lond.) 316: 263-277.

Schatzmann, H.J. (1953). Cardioactive glycosides as inhibitors for the active transport of potassium and sodium through the erythrocyte membrane (in German). Helv. Physiol. Pharmacol. Acta. 11: 346-354.

Schatzmann, H.J. (1983). The red cell calcium pump. Annu. Rev. Physiol. 45: 303-312.

Skou, J.C. (1957). The influence of some cations on an adenosine triphosphatase from peripheral nerves. Biochim. Biophys. Acta 23: 394-401.

Skou, J.C. (1983). On the mechanism behind the ability of Na, K-ATPase to discriminate between Na^+ and K^+. Curr. Topics Memb. Trans. 19: 323-341.

Stein, W.D. (1986). Transport and Diffusion across Cell Membranes, Academic Press, Orlando, Fla.

Stein, W.D. (1988). Editor: "The Ion Pumps. Structure, Function, and Regulation". Progress in Clinical and Biological Research Vol. 273. A R Liss, Inc. New york.

Stein, W.D., and Laeuger, P. (1990). Kinetic properties of F_0F_1 ATPases: Theoretical predictions from alternating-site models. Biophys. J. 57: 255-267.

Weiser, M.B., Razin, M., Stein, W.D. (1983). Kinetic tests of models for sugar transport in human erythrocytes and a comparison of fresh and cold-stored cells. Biochim. Biophys. Acta 727: 379-388.

3

VOLTAGE DEPENDENT CONDUCTANCES:

GATING CURRENTS AND SINGLE CHANNEL RECORDINGS

Francisco Bezanilla

Department of Physiology
UCLA
Los Angeles, CA 90024 U. S. A.

INTRODUCTION

Since Hodgkin and Huxley published their famous series in 1952, we know that the generation and propagation of the nerve impulse, or action potential, is the result of transient permeability changes of the axon membrane to sodium and potassium. These authors studied the membrane currents of the squid axon as a function of membrane voltage using the "voltage clamp" technique. They found that the membrane permeability is voltage and time dependent and is the result of at least two different conduction systems for sodium and potassium (see Fig. 1). In this classical work, Hodgkin and Huxley introduced the concept of voltage dependent ionic conductances and they even separated the process of conduction from the process of opening and closing of the permeability pathway. Today the permeability pathway is called ionic channel of which a large number of different types have been identified and the primary structure has been determined for several of them (e.g. Noda et al., 1984).

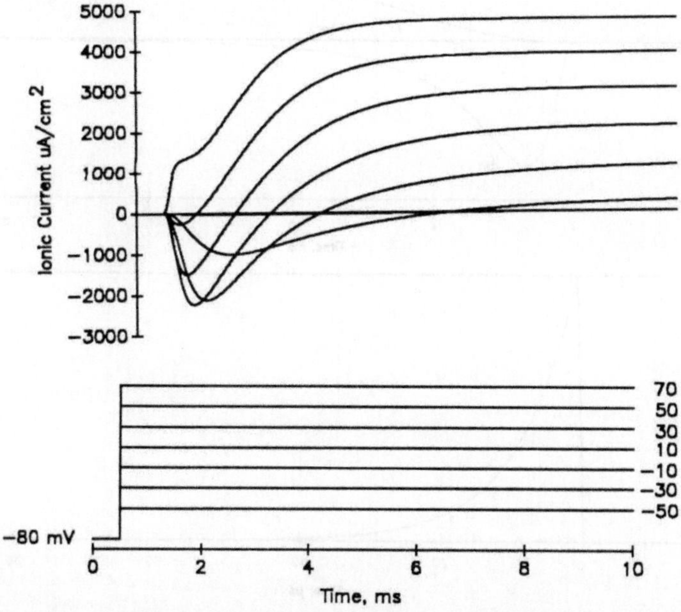

Figure 1. Ionic currents of the giant axon for several depolarizing steps.

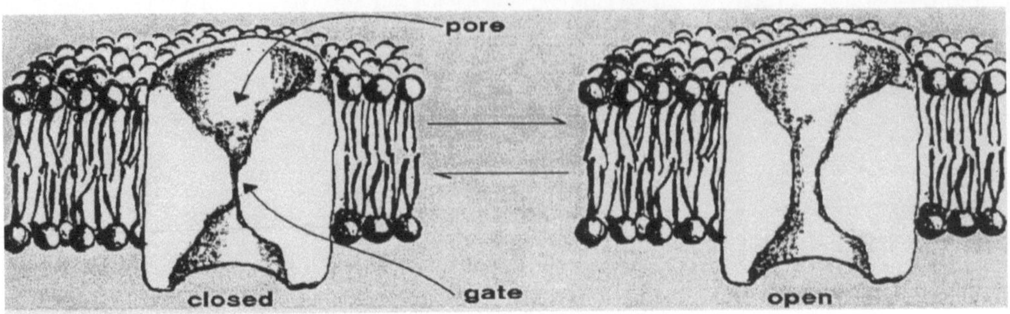

Figure 2. The two main functional components of an ionic channel.

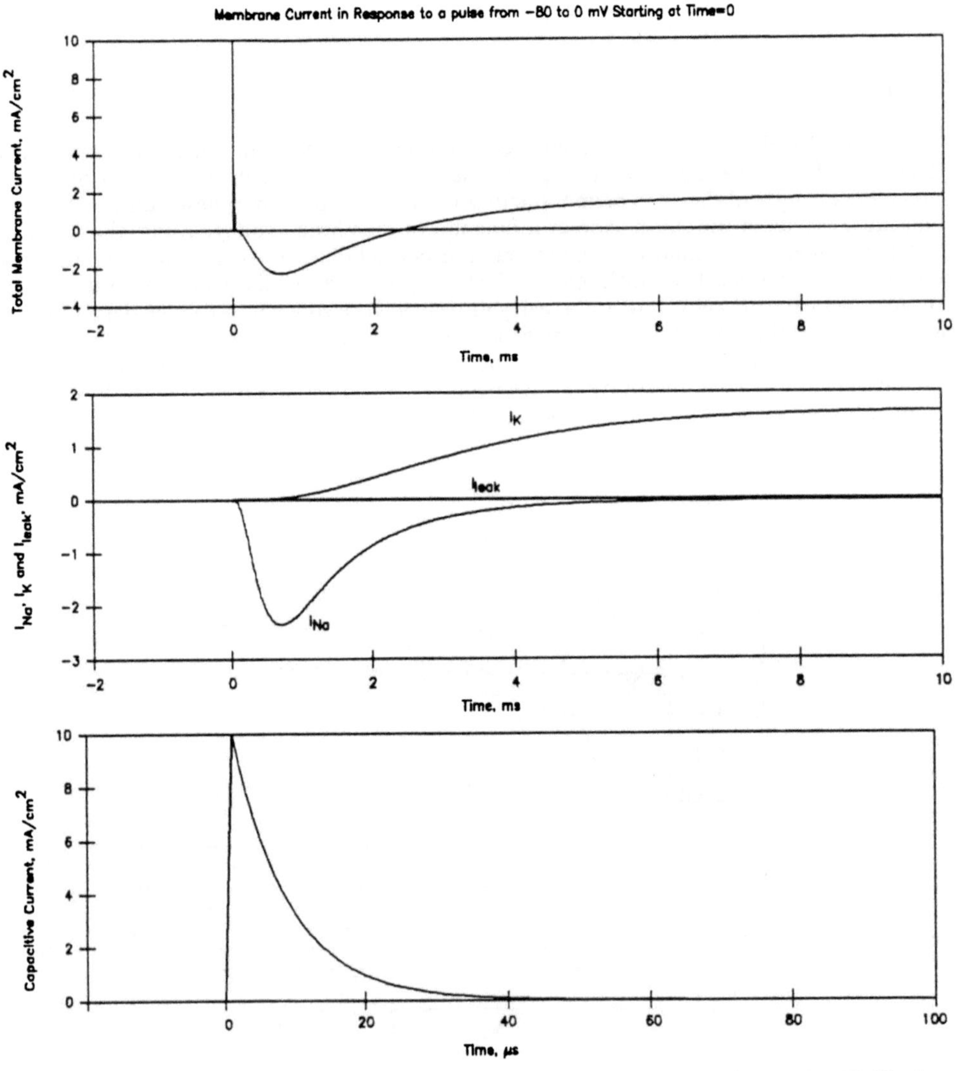

Figure 3. The main components of the membrane current for a pulse to 0 mV. Notice the different time scale for the capacitive current and the change in current scale for the ionic currents.

The "typical" channel is viewed as having a conducting part or "pore" and a system to open and close it called "gate" (Fig. 2).In this chapter we will be concerned with the voltage dependence of the ionic conductances. We will briefly review the three major electrical expressions of these conductances: macroscopic ionic currents, single channel fluctuations and gating currents. We will develop the equations that relate them and a simple model will be used to set up the basic kinetic equations of a voltage dependent channel to illustrate the origin of these electrical measurements.

SEPARATION OF THE CURRENT COMPONENTS IN SQUID AXON

The complex time course and voltage dependence of the currents in squid axon can be dissected in several components for one voltage pulse only, as shown in Fig. 3. The two major components are ionic and capacitive currents. The capacitive current results from the charging of the membrane capacitance but part of that capacitance is the result of molecular

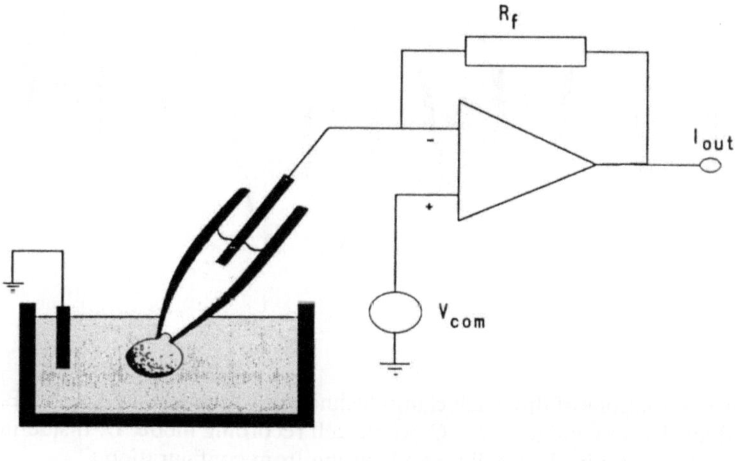

Figure 4. Patch clamp technique. The pipette is connected to a current-to-voltage converter and the output signal (I_{out}) is conditioned by subtracting V_{com} and further amplified and filtered.

rearrangement within the channels and other polarizable membrane dielectrics. From single channel data recording we know that the two main ionic currents in squid axon, the sodium and potassium currents are produced by distinct types of channels: one sodium channel (Bezanilla, 1987) and at least three different types of K channels (Llano et al., 1988).

SINGLE CHANNEL RECORDINGS

The patch clamp technique (Neher and Sackmann, 1983) has allowed the recording of a very small area of membrane and, ideally, it is possible to observe the activity of one channel opening and closing as the membrane potential is controlled.

A glass pipette, usually fire polished, and extremely clean is approached to the surface of a clean membrane. After contact, usually a small amount of suction makes a high resistance seal (several gigaohm) between the membrane and the glass effectively isolating the patch of membrane attached to the pipette from the rest of the cell. The pipette is connected to the summing junction of a current-to-voltage transducer connected to amplifiers and signal conditioning circuits. The membrane voltage of the isolated patch can be modified by changing the voltage of the amplifier summing junction or the voltage of the bath (Fig. 4).

Several variations of the technique are possible because the seal between the glass and the membrane is extremely tight. It is possible to keep the pipette attached to the cell (cell attached configuration), excise the patch (inside-out configuration) or break in the cell (whole cell recording) or pull out after breaking in and allow the reconfiguration of membrane in the outside-out configuration (Fig. 5).

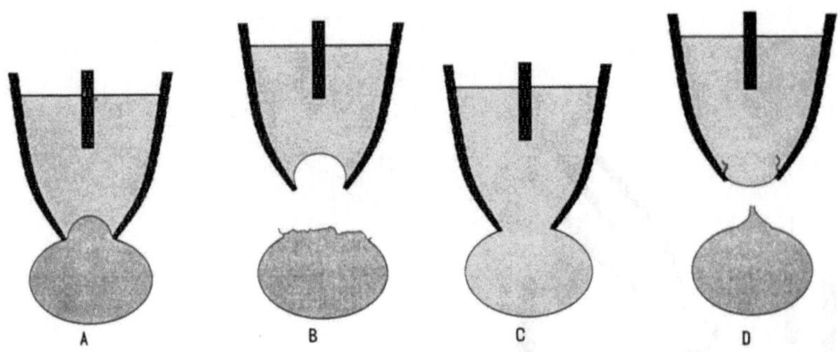

Figure 5. Variations of the patch clamp technique. A, cell attached mode. B, excised inside-out configuration. C, whole cell recording mode. D, Inside-in configuration, obtained by pulling and cutting from configuration C.

In Fig. 6, a pulse of voltage is applied several times to record the activity of one channel in the patch. Notice that each trace is a different trial and, as expected from a random process, every trial produces a different current pattern. If we do n trials and make n very large, the average current will look like the last trace. The current will have the same time course of a macroscopic current produced by a large number of independent, non-interacting identical channels. The macroscopic current reflects the fraction of channels open as a function of time. The question that may be asked is: what is the relationship between the unitary currents and the macroscopic current?

RELATIONSHIP BETWEEN SINGLE CHANNEL AND MACROSCOPIC CURRENTS

The current through a generic single open channel is pictured in the i-V curve of Fig. 7

and the single channel current is given by

$$i = \gamma(V)(V - V_e)$$

where $V-V_e$ is the driving force and $\gamma(V)$ is the single channel conductance. The total number of channels is N, then the maximum current is

$$I_{max} = Ni = N\gamma(V - V_e)$$

But at any given time, only a fraction f of channels is open, then the current is given by

$$I(V,t) = N\gamma(V)\ f(V,t)\ (V - V_e) \qquad (1).$$

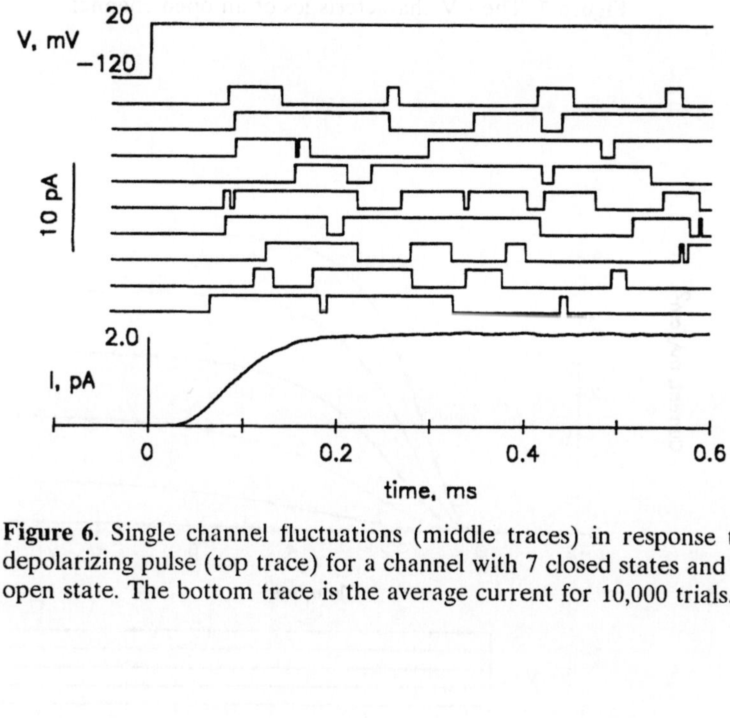

Figure 6. Single channel fluctuations (middle traces) in response to a depolarizing pulse (top trace) for a channel with 7 closed states and one open state. The bottom trace is the average current for 10,000 trials.

We then do the "regular" I-V procedure as pictured in Fig. 8, where we pulse from a holding otential V_h (where most of the channels are closed) to several potentials V (Fig 8., bottom) and we record the currents as a function of time (Fig 8., top). The current measured at a given time (10 ms in this case) plotted againts V gives an isochronal I-V curve which is shown as 'I-V' in Fig. 9.

Then we perform the "instantaneous" I-V procedure as pictured in Fig. 10. We take the voltage to a value V_1 where $f(V_1, t=t_1) = f_1$ and we give pulses to other voltages (Fig 10, bottom) and measure the currents for each pulse (Fig. 10, top).

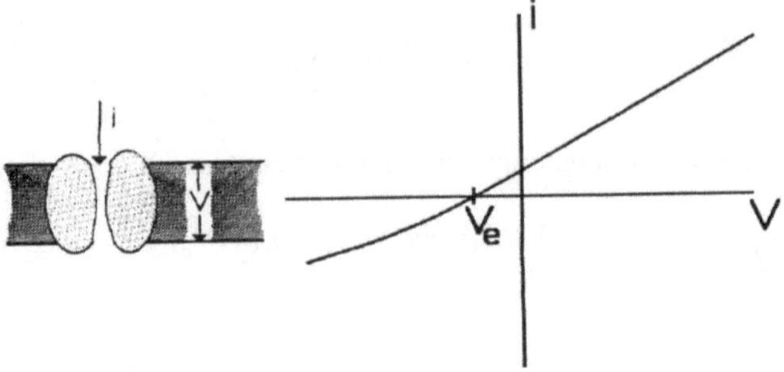

Figure 7. The i-V characteristics of an open channel.

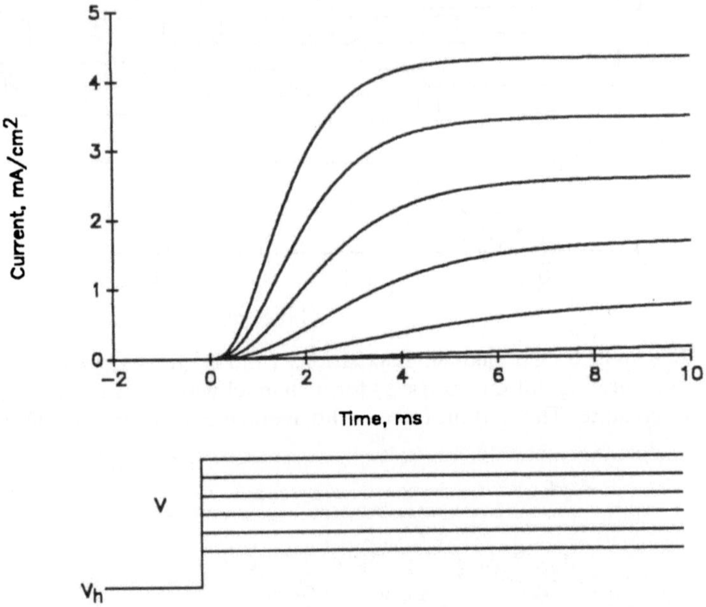

Figure 8. The experiment to obtain a regular I-V curve. The family of pulses is shown on the bottom and the resulting currents on the top traces.

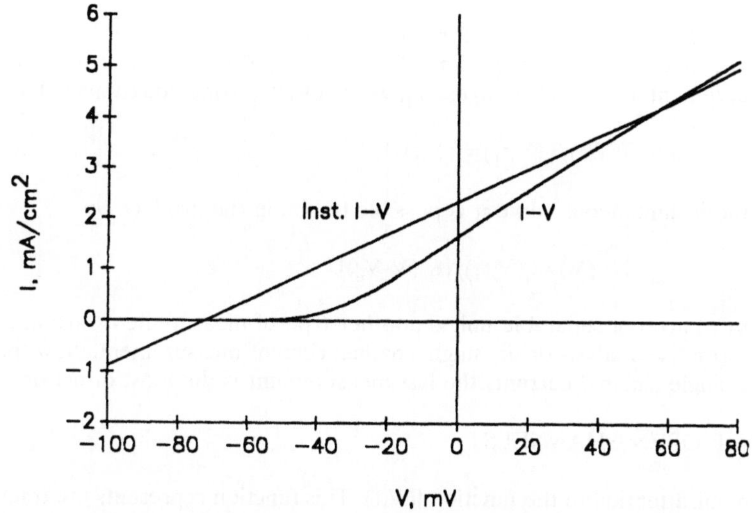

Figure 9. The regular I-V and the instantaneous I-V curves.

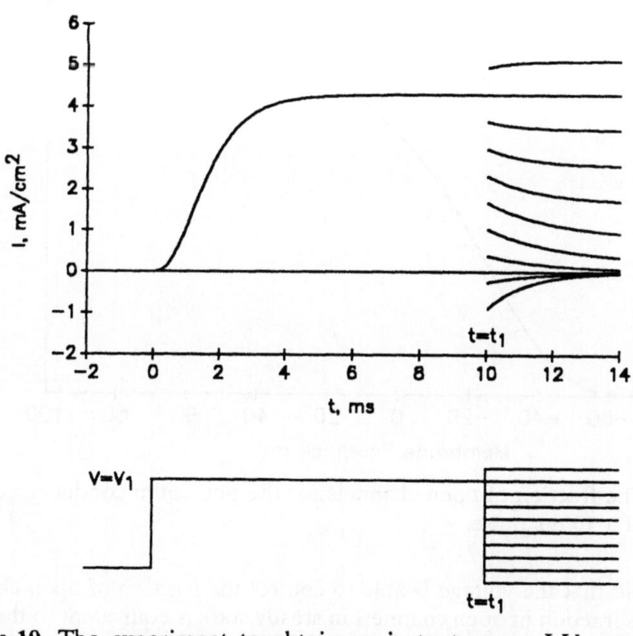

Figure 10. The experiment to obtain an instantaneous I-V curve. The family of pulses is shown on the bottom and the resulting currents on the top traces.

The current measured at the start of the second pulse I_i is plotted as instantaneous I-V in Fig 9 and it is given by

$$I_i(V,t_1)=N\gamma(V)\ f_1\ (V-V_e) \qquad (2).$$

Notice that $f=f_1$ for all V at the start of the second pulse. This is because there is no time between the end of the first pulse to the beginning of the second pulse for the fraction to change. Repeating the measurement at several voltages, V_1, we can write, using equations (1) and (2)

$$I(V,t)/I_i(V,t_1)=f(V,t)/f_1$$

Also, from the instantaneous curve it is possible to obtain the product $N\gamma$

$$N\gamma(V)=I_i(V,t_1)/[f_1\ (V-V_e)].$$

But N and $\gamma(V)$ are not separable unless another type of measurements are used such as: i) toxin binding, ii) noise analysis or iii) single channel current measurement. Now that it is possible to measure single channel currents the last measurement is the most direct one.

THE FRACTION OF OPEN CHANNELS

We will turn our attention to the function f(V,t). This function represents the fraction of open channels as a function of voltage and time. For example for the potassium conductance measured after a 10 ms pulse it is shown in Fig. 11.

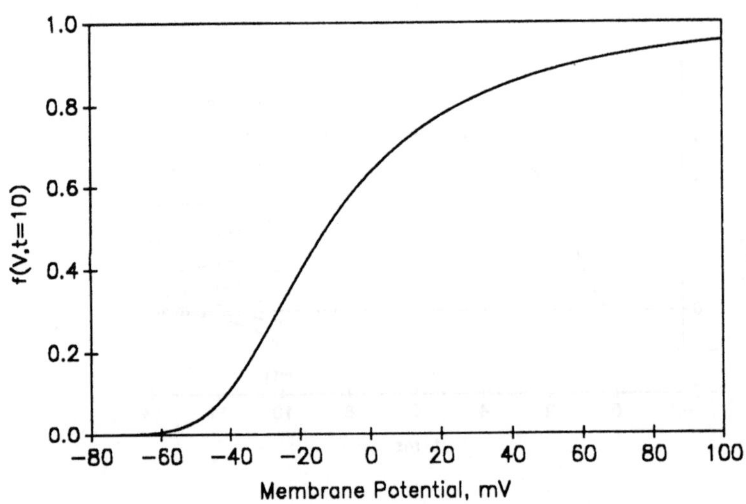

Figure 11. The fraction of open channels for the potassium conductance at the end of a 10 ms pulse.

How is it possible that the voltage is able to control the fraction of open channels? We must remember that the fraction of open channels in steady state is equivalent to the probability of finding the channel open for a long period of time. Then the question is now, how is it possible that the membrane voltage controls the probability of finding the channel open?

For this to occur, it is necessary that an electric charge or dipole could be displaced by the change in membrane potential. The distinction between a charge and dipole is not possible with electrical measurements only. It is easy to imagine that the displacement of the charge or dipole will change the conformation of the channel and as a consequence it will go from a closed to an open conformation. The charge or dipole movement will produce a small current which have been called "gating current" because it is an expression of the movement of a component of the channel gate.

We will review briefly the methods to detect gating currents and then we will develop a very simple model that can account for the voltage dependence of the fraction of open channels and that it predicts *macroscopic* current, gating currents and single channel behavior. This model has been found to be too simple to account for the real data but it gives an insight of how voltage gating may occur in voltage dependent channels.

RECORDING OF GATING CURRENTS

The movement of charge produced by the rearrangement in the channel molecule produces a small current that can be observed experimentally under the appropriate conditions (see Armstrong, 1981; Bezanilla, 1985). When a voltage step is applied to the membrane, the membrane capacitance is charged and large transient capacity current flows that can be easily recorded during voltage clamp experiments. The current produced by the gate rearrangement is also transient in nature but normally orders of magnitude smaller than the capacity current (see the section on gating current prediction in the two state model).

To detect the gating current, it is necessary to separate it from i) the regular capacity current and ii) the ionic current. The separation from the ionic current is done by effectively eliminating the ionic current by non-permeant ion substitution and channel blockers. The separation from the regular capacity current is based on the linear properties of this capacity current and the non-linear characteristics of the gating current. As we have seen, the fraction of open channels is a non-linear function of voltage and the orientation of the dipole or charges must also have a non-linear dependence with membrane potential as will be shown in the section of the Two State Model. This non-linearity implies that the current observed depends on the region of potential they are recorded. For example if we step from the potential where all the channels are closed to a potential where they open, we expect to see reorientation of charges and a consequent gating current, but if we step in the opposite direction, no movement will occur because all the particles are in the closed position and they cannot move any more and no current will be observed.

This asymmetric property is used to record gating current and a typical pulse protocol is shown in Fig. 12. In this case, we assume that we are recording gating currents from a channel that is closed in the negative region and opens in the positive region. The steady (holding) potential is maintained negative to keep the majority of the channels closed, then a test pulse to a positive potential is delivered and the current is recorded. We then jump to a more negative region (subtracting holding potential) where practically all the channels are closed and give the test pulse divided by four in amplitude but delivered four times and we record the currents for each pulse and add them synchronized with the start of the pulses. The sum of all the currents produced by the four subtracting pulses is subtracted from the current produced by the test pulse and the resultant current will contain only the components which are not voltage independent, eliminating all the linear capacitance and linear remanent ionic current. This procedure is called p/4 but there are many variations such as p/-4 (where the subtracting pulses have the opposite polarity of the test pulse) and P/n (where *n* subtracting pulses are used). The main reason of using several small pulses instead of one subtracting pulse is to maintain the subtraction within the linear region and to prevent membrane breakdown when the subtracting pulses take the membrane to extreme potentials.

The major limitation in recording gating current is the signal-to-noise ratio. To record a relatively clean signal it is important that the number of channels per unit area of membrane be very high because as the membrane area is made larger the linear capacitance increases with the corresponding increase in the linear capacitive current and background noise. Among all the preparations used to record gating currents, the squid axon has provided one of the best signal-

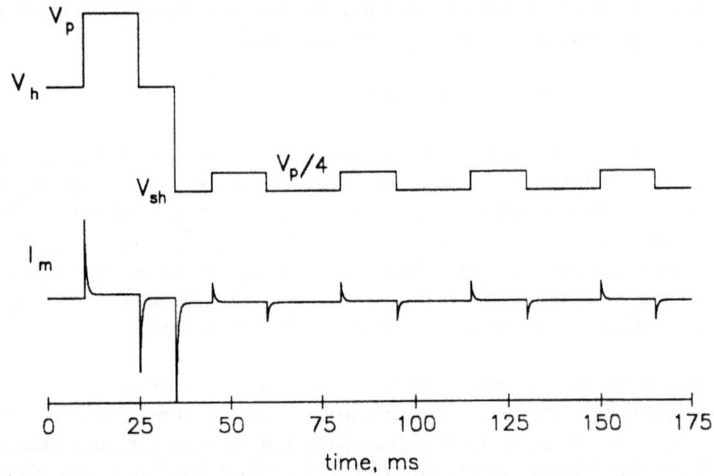

Figure 12. The test pulse V_p is applied at the holding potential V_h and four times with one fourth of its amplitude at the holding potential V_{sh}. The currents produced by the four small pulses are subtracted from the current of the test pulse.

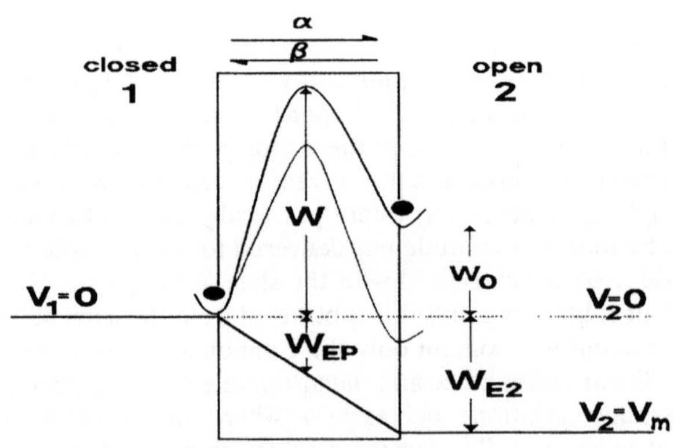

Figure 13. Energy diagram for a two state model. For details see text.

to-noise ratio because it has a very high channel density and the potential can be controlled with a low noise voltage clamp.

THE TWO STATE MODEL

Assume that an electric charge can only exist in two states. There is an energy barrier that separates two energy wells which represent the two stable positions of the charge.

The energy diagram for this situation is pictured in Fig. 13. c is the fraction of particles in side 1 (the closed position) and a is the fraction of particles in side 2 (the open position).

Then

$$c+a=1$$

The rate constants for the transitions, according to Eyring (e.g. Bezanilla, 1985), will be

$$\alpha=(kT/h)\exp(-W_T/kT)$$
$$\beta=(kT/h)\exp(-[W_T-W_o])$$

where k and h are the Boltzmann and Plank constants respectively and T is the absolute temperature and W_T is th energy barrier.

In the absence of voltage

$$W_T=W.$$

In the presence of voltage, we have to add an electric energy term which can be derived as follows. The electric work is

$$W_E=-\int_0^x q\,E\,dx'$$

Assuming a constant field, we get

$$W_E=-q\,E\,x.$$

In our case $V_1=0$ and we are interested in the electric work at the maximum (W_{EP}) and minimum (W_{E2}) of the barrier to calculate its contribution to the total energy. To simplify, assume that the barrier is in the middle of the membrane, then

$$W_{EP}=-qEx/2$$

and

$$W_{E2}=-qEx$$

Therefore the total energy is

$$W_T=W-qEx/2$$

But $Ex=-V_m$, therefore

$$W_T=W+qV_m/2$$

and we write the rates α and β as

$$\alpha = (kT/h) \exp\{-[(qV_m/2kT) + (W/kT)]\}$$

$$\beta = (kT/h) \exp\{-[(-qV_m/2kT) + (W-W_o)/kT)]\}$$

The equation for the movement of particles is

$$da/dt = -\beta a + \alpha c$$

but

$$c + a = 1$$

then

$$da/dt = -\beta a + \alpha(1-a) \qquad (3).$$

Steady State Solution of the Particle Movement.

The solution of eq. (3) in steady state gives us the distribution of the fraction of particles in the open state as a function of the membrane potential, which is equivalent to the fraction of open channels for a membrane with more than one channel

$$a(V,t=\infty) = a_\infty = \alpha/(\alpha + \beta)$$

$$= 1/[1 + \exp\{(qV_m + W_o)/kT\}].$$

This equation can be written in a more familiar form by defining $V_o = -W_o/q$

$$a_\infty = 1/[1 + \exp\{q(V_m - V_o)/kT\}].$$

The distribution of the transferred charge with potential can be calculated from the previous equation knowing the effective charge translocated and the total number of channels. The effective charge is

$$q = ze\delta$$

where z is the valence of the particle and e is the electronic charge ($1.6 \cdot 10^{19}$ coulombs). In the case we have been discussing, the charge moves across the whole membrane field. If it moves less, say a fraction δ, the effective translocated charge decreases proportionally. If the charge is of the opposite polarity but it moves in the other direction the result is the same. Therefore, the sign cannot be distinguished and the magnitude is actually the product of charge times distance. Assuming a total number of channels N, the charge vs. potential distribution (Q-V curve) is given by

$$Q = Nze\delta/[1 + \exp\{(ze\delta(V_m + V_o)/kT\}].$$

The above equations show that in a 2-state model the voltage dependence of the fraction of open channels is identical to the voltage dependence of the charge distribution. This is a direct consequence of the fact that every time a particle is translocated from position 1 to 2, the channel will open. For the model we have been developing, assuming a valence of -2 for the particle, the curve of fraction of open channels vs V is shown in Figure 14 and it is similar to the example presented in Fig. 11, which corresponds to the case of the potassium conductance in squid axon. For both the sodium and potassium channels the f-V curve is displaced to the right with respect to the Q-V curve, which is an indication of a process with multiple closed states.

An important parameter in the voltage dependent channel is its valence and an estimate of it can be obtained from the fraction of open channels in the voltage region where the fraction is small. In our example, when $V = -\infty$,

$$a_{\infty}(V=-\infty) = \exp\{ze(V-V_o)/kT\}$$

In the case of the sodium channel, the conductance curve increases e-fold in about 6 mV, then using this equation evaluated at V and at V+6,

$$a_{\infty}(V+6)/a_{\infty}(V) = \exp(6ze/kT) = e^1$$

which implies that $6ze/kT = 1$ and knowing that e/kT at room temperature is about 25 mV

$$z = 25/6 \approx 4,$$

that is, four electronic charges moving across the entire field will explain the voltage dependence of the sodium channel.

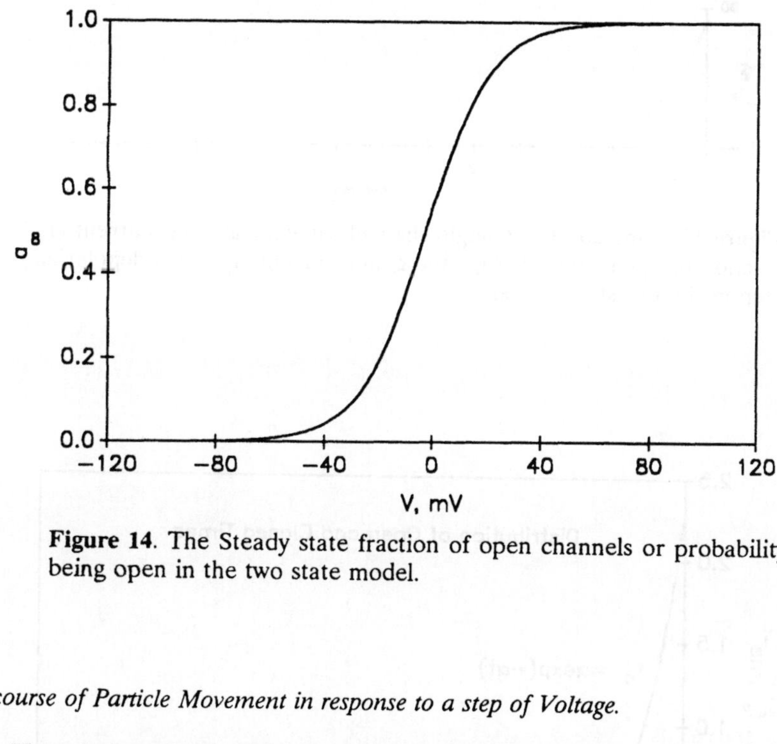

Figure 14. The Steady state fraction of open channels or probability of being open in the two state model.

Time course of Particle Movement in response to a step of Voltage.

The solution of the equation (3) for the fraction of open channels a(V,t) for a step of potential from a voltage V_i to a voltage V is

$$a(V,t) = a_{\infty} - (a_{\infty} - a_i)\exp[-(\alpha + \beta)t]$$

with

$$a_i = \alpha(V_i)/[\alpha(V_i) + \beta(V_i)],$$

and the curve is plotted in Fig 15.

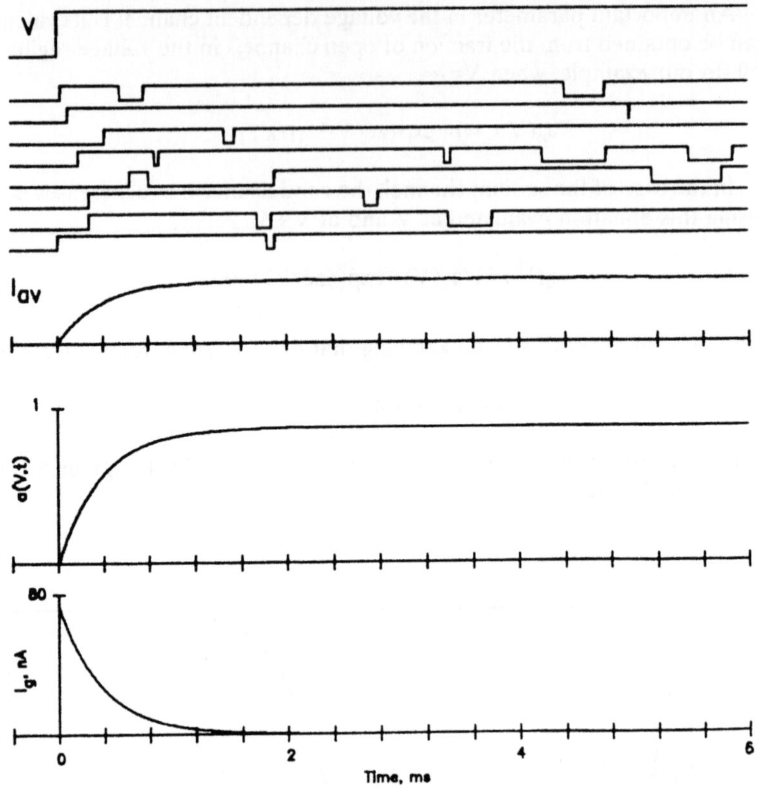

Figure 15. Time course of single channel currents, average current (I_{av}), fraction of open channels (a_∞) and gating current (I_g) for a depolarizing step in the two state model.

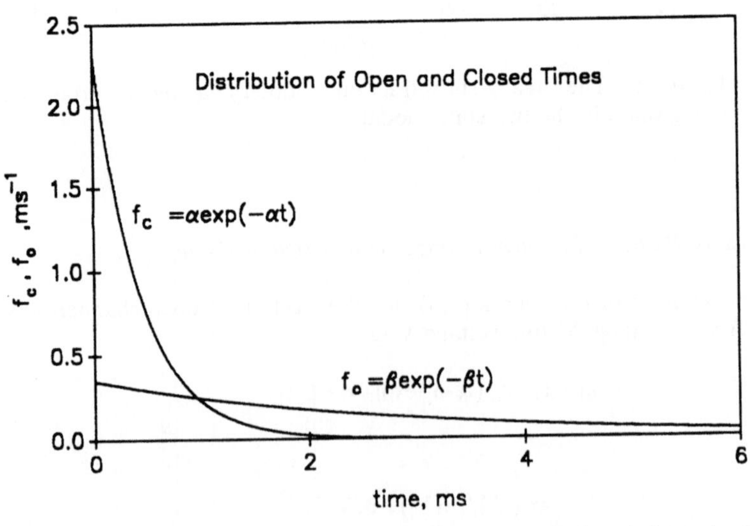

Figure 16. Probability density function for open and closed times in the two state model.

The current produced by the charge movement I_g can be calculated as the product of the effective charge times the net movement of particles

$$I_g = Nze\delta(c\alpha - a\beta)$$

where $Nze\delta$ is total effective charge displaced by N channels and $c\alpha - a\beta$ is the net movement from side 1 to side 2. Replacing the expression for a(V,t) in the equation for I_g we find

$$I_g = Nze\delta(\alpha + \beta)(a_\infty - a_i)\exp[-(\alpha + \beta)t]$$

This is the equation for the "gating current" (Fig 15, bottom trace), which is the movement of the charges responsible for channel opening.

The Distribution of Open and Closed Times

The last prediction we will make with the Two State Model is the distribution of open and closed times in single channel records (See Colquhoun, 1971). Lets define P_l as the probability of leaving state 1 in the time interval t to $t + \Delta t$. This probability is given by

$$P_l = \alpha\Delta t + o(\Delta t)$$

where $o(\Delta t)$ obeys the following limit

$$\lim_{\Delta t \to 0} \frac{o(\Delta t)}{\Delta t} = 0.$$

Then the probability of not leaving in the same time interval P_n will be

$$P_n = 1 - \alpha\Delta t - o(\Delta t).$$

Define further $P_N(t)$ as the probability of not leaving in the time interval 0 to t and $P_N(t+\Delta t)$ as the probability of not leaving in the interval between 0 and $t+\Delta t$. Then we can write

$$P_N(t+\Delta t) = P_N(t)P_n$$

because they are independent events. Replacing the value of P_n in the above expression for P_N we get

$$P_N(t+\Delta t) = P_N(t)[1 - \alpha\Delta t - o(\Delta t)]$$

which can be rearranged as

$$\frac{P_N(t+\Delta t) - P_N(t)}{\Delta t} = -[\alpha - o(\Delta t)]P_N$$

As we approach the limit when $\Delta t = 0$,

$$\lim_{\Delta t > 0} \frac{P_N(t+\Delta t) - P_N(t)}{\Delta t} = \frac{dP_N(t)}{dt}$$

we then can write a differential equation for $P_N(t)$

$$\frac{dP_N}{dt} = -\alpha P_N$$

This equation has the solution

$$P_N(t) = P_N(0)\exp(-\alpha t)$$

But $P_N(0) = 1$ because at time zero the particle is, by definition, in its initial position. $P_N(t)$ is the probability that the particle is still in position 1 (closed) at time t and it decays exponentially with time, that is, at very long time the probability will be zero. The probability that the particle has left the closed state is $1-P_N$ and this is equivalent to the probability that the lifetime in state 1 is less or equal than t which we will define as F(t)

$$F(t) = 1-P_N = 1-\exp(-\alpha t)$$

The density function is calculated as

$$f_c = \frac{dF}{dt} = \alpha\exp(-\alpha t)$$

which is the closed time distribution. In the same manner, the open time distribution is given by

$$f_o = \beta\exp(-\beta t).$$

Both density functions, which are plotted in Fig 16, are exponential distributions with mean α and β, respectively. The reciprocals of the means are the mean closed and open times.

These equations tell us that in a two state model the measurement of the lifetime distribution of the open and closed periods of the single channel events give exponential curves. From these exponential distributions it is possible to extract α and β which are the elementary rate constants of the transition between the two states.

MORE THAN TWO STATES

The two state model has shown us that it is possible to obtain the elementary rates α and β using macroscopic currents, gating currents or, more directly, from single channel analysis. There is no need to perform all the experiments to extract the rate constants because any of them is sufficient.

The situation is more complicated when the channel is known to have more than two states. For example, both sodium and potassium conductances develop with a lag in response to a depolarizing pulse. This lag is the result of the evolution of the channel through several closed states before it reaches the open state (see Fig. 6). If we model the channel as we did with the two state model but we extend the number of closed states, we find that in this case the equation that relates the probability of populating any of the states is of a higher order and the expression of the probability of being in the open state will be the sum of several exponential terms. The time constants, or the reciprocals of the eigenvalues, will be complicated functions of the elementary rate constants. It is interesting to note that the eigenvalues are the same for the gating currents and ionic currents but the coefficients of the exponential terms are quite different.

Our aim is to ultimately obtain the elementary rate constants of all the transitions of the multistate model. The measurement of macroscopic ionic currents alone will only give poor estimates of the rate based on the eigenvalues fitted from the currents. By measuring gating currents, a better estimate can be obtained because the coefficients of several of the exponential components are going to be different from the coefficients of the ionic currents. In a model with several closed states and one open state, the gating currents will give better estimates of the components involved in between closed states, provided there is enough charge movement between those states. Finally, single channel recordings will provide even more information because the coefficients of the exponential components are also different. Single channel recordings will be particularly useful to estimate the transitions which are in proximity to the open state because that is what the single channel fluctuations are showing. In fact, if there is only one open state, the mean open time will be a direct measurement of the reciprocal of the sum of the rates leaving the open state. In particular, if the open state is connected only to one closed state, the mean open time *is* the reciprocal of the elementary rate connecting the open with the close state. Unfortunately the mean close time does not represent, in this case, the rate of opening but again it contains the eigenvalues of the solution of the differential equation of the multistate model.

It is then clear that in determining the rates of a multistate channel, it is necessary to use the information provided by ionic currents, gating currents and single channel recordings. It is still true that all the information is in any of the three type of measurements but to extract it becomes a numerical challenge that only becomes easier as all the measurements are used in combination. Even, when all these measurements are available and the information is extracted by simultaneous fitting of all the data, the rates estimated will depend on the model selected in the first place. Therefore, it is necessary to test different models to obtain the best estimates and the most appropriate will normally be selected using a statistical criterion such as to maximize the likelihood (Horn and Vandenberg, 1984).

The actual description of the molecular events occurring during channel gating will be possible by combining the description of kinetic models, such as the ones we have discussed, with a knowledge of the chemical structure of the molecule and the positions of the responsible groups within the molecule. The hope of elucidating the physical description of voltage dependent gating at the atomic level resides in the combination of electrical measurements with the structure emerging from Molecular Biology techniques (Noda et al., 1984), including site directed mutagenesis (Stuhmer et al., 1989), together with structural techniques such as crystallography and spectroscopic analysis.

Aknowledgements: Many thanks to Dr. A. M. Correa and E. Perozo for reading and commenting on the manuscript. Supported by USPHS grant GM30376. Some of the computations were done using SCoP, developed by the National Biomedical Simulation Resource of Duke University Medical Center.

REFERENCES

Armstrong, C. M., 1981, Sodium channels and gating currents. Physiol. Rev. 61, 644-683.

Bezanilla, F., 1985, Gating of Sodium and Potassium Channels. J. Membrane Biol. 88, 97-111.

Bezanilla, F., 1987, Single sodium channels from the squid giant axon. Biophys. J. 52, 1087-1090.

Colquhoun, D., 1971, "Lectures on Biostatistics". Clarendon Press, Oxford.

Hodgkin, A. L. and Huxley, A. F., 1952, A quantitative description of Membrane current and its application to conduction and excitation in nerve. J. Physiol (Lond). 117, 500-544.

Horn, R. and C. Vandenberg, 1984, Statistical Properties of Single Sodium Channels. J. Gen. Physiol. 84, 505-534.

Llano, I, C. K. Webb and F. Bezanilla, 1988, Potassium Conductance of the squid giant axon: single channel studies. J. Gen. Physiol. 92, 179-196.

Noda, M., Shimizu, S., Tanabe, T., Takai, T., Kayano, T., Ikeda, T., Takahashi, H., Nakayama, H., Kanaoka, Y., Minamino, M., Kangawa, K., Matsuo, H.,Raftery, M. A., Hirose, T., Inayama, S., Hayashida, H., Miyata, T. and Numa, S., 1984, Primary structure of the *Electrophorus electricus* sodium channel deduced from cDNA sequence. Nature 312, 121-127.

Sackmann, B. and E. Neher., 1983, "Single-Channel Recording". Plenum Press, New York.

Stuhmer, W., F. Conti, H. Suzuki & X. Wang, M. Noda, N. Yahagi, H. Kubo, & S. Numa., 1989, Structural parts involved in activation and inactivation of the sodium channel. Nature 339, 597-603.

4

METHODS OF DETERMINING THE ROUTES OF INTESTINAL WATER TRANSPORT

R.J. Naftalin
Department of Physiology, King's College London
The Strand London WC2R 2LS

INTRODUCTION

The main reasons for measuring net water movements across epithelia are to determine the forces which generate the flows and to assess the routes by which water traverses the tissue. Determining water flows by indirect methods is a compromise between expediency and accuracy and is often based on unjustified assumptions with regard to tissue behaviour. For this reason direct measurement of water flow is preferable.

<u>Two forces generate net water flow Jv</u>

a) Two kinds of force generate net water movement J_v ($cm^3\ s^{-1}$ per cm^2 tissue or channel area, or $\mu l\ cm^{-2} h^{-1}$) across layers of intestinal cells, the first is hydraulic pressure, P [units: cm H_2O; mmHg; atm.; kPa (1 atm. = 100kPa = 1000cm H_2O = 760mmHg)] which causes filtration flow through <u>all</u> pores within the membrane in proportion to their fluid conductance, L_p.

b) The second force is osmotic pressure $d\pi$, this induces water flow in proportion to the concentration difference of impermeant solute across a barrier permeable to water. This pressure is applied <u>only</u> across the pores which retard solute flow: (where 1 mOsm= 26cm H_2O at 37^oC) across the water conducting pathway.

The common parameter used to relate both hydraulic and osmotic pressure induced water flows, J_v to their conjugate forces is the hydraulic conductance L_p. (cm s^{-1} cm H_2O^{-1}).

Hence $J_v = L_p(dP - \sigma_i RT\Delta C)$; where R is the gas constant and T temperature °K at 37°C RT = 26cmH_2O/mOsm; ΔC is the concentration difference in mOsm across the osmotic barrier. The reflexion coefficient, σ_i is an index of the mobility of solute within the membrane pores relative to that of the solvent. If the mobility, ϖ_i of the solute is equal to the mobility of solvent, ϖ_s then no restraint is placed by the membrane on the

Cell Membrane Transport, Edited by D. L. Yudilevich *et al.*
Plenum Press, New York, 1991

solute relative to the solvent, and $\sigma_i = 1-\varpi_i/\varpi_s = 0$; in this case the solute exerts zero osmotic pressure. On the other hand, if the solute had zero mobility within the pore, then accordingly $\sigma_i = 1$ and the solute exerts its full osmotic potential = ΔCRT (Katchalsky and Curran, 1965). Solute flow across membranes may be retarded by membranes as a result of steric factors, or charge factors, or both. Charge attraction or repulsion between the ionised solute and the charge within the pore wall can be an important factor in determining solute mobility across the membrane. The most important solute exerting osmotic pressure in transporting epithelia is NaCl, hence pores which do not reflect NaCl cannot be used to generate a significant osmotic pressure and hence do not contribute to osmotically driven flow across the epithelium.

Problems relating to parallel and series resistances to water flow

Even when net water transport is measured unambiguously, a lumped estimate of the force required to generate net flow across the epithelium may be misleading as flow may not be uniform across the epithelial surface. In many epithelia both hydraulic pressure and osmotic pressure- induced flows of water occur simultaneously, leading to fluid circulation between the small and large channels in parallel. Thus net flows cannot be interpreted unless the proportion and direction of flow via each type of channel is known. This requires a knowledge of the pore size characteristics as well as the design of the hydraulic network within the epithelium.

An example of mixed water flows induced by both hydraulic pressure and osmotic pressure occurs in the small intestine both *in vitro* and *in vivo* (see below)

The problem of parallel channels with wide differences in hydraulic conductance and reflection coefficient σ_s toward small solutes is recognised (Durbin et al.,1956; Diamond 1979) is compounded by the problem of series barriers to water movement within a tissue.

In small intestine or gall-bladder in addition to the epithelial layer which generally consists of a single layer of cells, there are many layers of supporting tissue consisting of basement membranes, connective tissue, both tightly and loosely packed and smooth muscle layers. These series barriers to water and solute movement may exert a large and variable hydraulic resistance which can have a profound effect on transepithelial, as well as on trans-tissue water and ion movement. The small intestine and gall-bladder have a tightly adherent submucosal layer consisting of basement membrane, muscularis mucosa, - capillaries and lymphatics which may provide drainage routes *in vitro* perfused tissue and smooth muscle layers. Stripping this tissue without incurring damage to the epithelial layers can only be done at the plane of scission between the longitudinal and circular muscle layer. Removing the outer longitudinal muscle removes some, but by no means all of the hydraulic resistance (Ahsan et al; 1987,1988; Naftalin and Tripathi 1985, 1986). Submucosal layers also contain nerve plexuses which innervate the epithelia, vascular and muscle cells and thereby affect

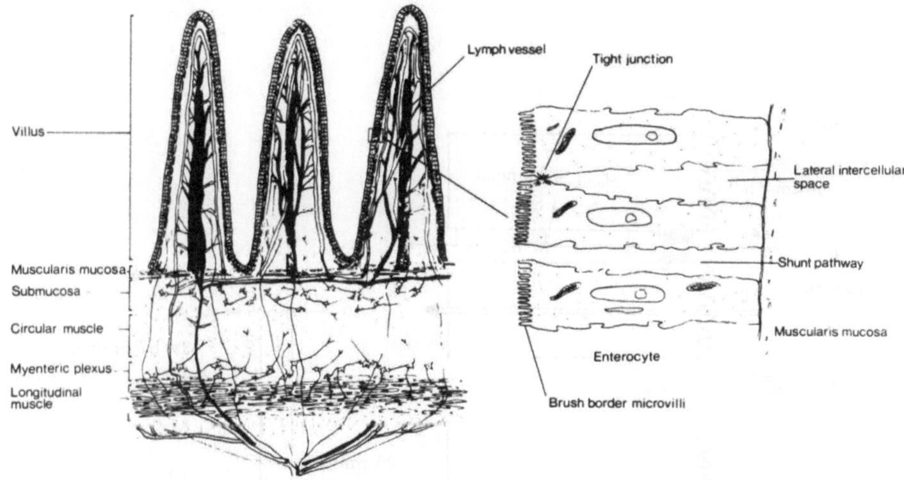

Fig 1. Diagram showing the mucosal and submucosal structures in small intestine and the parallel channels across the mucosal layers.

the mechanical tone of the tissue. This has a significant effect on the overall hydraulic conductance. Until recently the problem of the influence of series resistance on the net water flow and tissue compliance has been ignored. However, with new methods of determining both transepithelial water and tissue volume flow simultaneously, it is clear that these factors have considerable significance both to interpretation of *in vitro* flows and also on the role of autonomic control factors on *in vivo* flows. The series resistance in small intestine has a low conductance, but a negligible reflexion coefficient towards small molecular weight solutes like NaCl and sucrose, although it reflects solutes with a molecular weight above 4kDa.

Distensibility of the interstitial compartment. Series capacitance.

The series array of hydraulic resistors at the mucosal and serosal surfaces of the tissue provides a distensible interstitial compartment . The distensibility of the interstitial compartment creates two problems; firstly, volume expansion of the interstitial compartment implies that fluid crosses the second barrier only at a fraction of the rate that it crosses the first barrier: the remainder accumulates within the interstitial space. In the case of the small intestine the interstitium can swell by 100% of the initial volume, i.e by approximately 50µl /cm^2 in 2-3 h. Thus net trans-tissue flux is not necessarily an accurate index of transmucosal flux

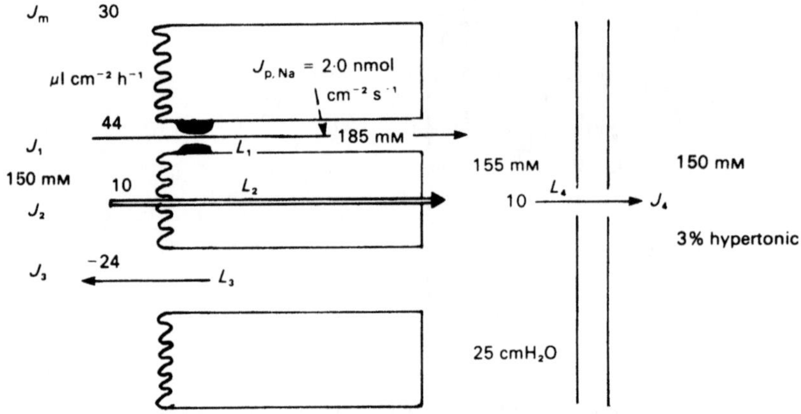

Fig 2. Effect of series resistance to transepithelial flow on flows across the mucosal routes. The flows are obtained from direct measurements of the component L_p's of the transcellular, tight and wide paracellular and transserosal routes. The pressures and concentrations are estimates from the flows and L_p's.

A second problem relating to the interstitial compliance is that the volume increase leads to a rise in interstitial pressure. This substantial pressure rise (20-40cm H_2O) is required to generate the force necessary to induce mass flow across the serosal resistance but also causes fluid reflux via the wide shunt pathway within the mucosa. Backflux of fluid is a high proportion of net flux across the intestinal mucosa; it varies from 20% of net influx in steady state to ca 200% in the initial transient following addition of a fast acting secretagogue e.g. prostaglandins, theophylline or carbachol. (Ahsan et al., 1987, 1988) there is a net reflux of fluid via the wide shunt channels due to the fall in sub-mucosal distensibility and the consequent rise interstitial pressure.

These problems only become evident when a method for high resolution monitoring of volume flows simultaneously across both mucosal and serosal surfaces.

METHODS OF DETERMINING WATER FLOW

a) Streaming potential:

Osmotic or hydraulic pressure induced water flow via ion-selective channels can cause a charge separation of ions entrained within the water stream within the membrane pores. This generates a potential difference across the pore (for review Diamond 1979; Tripathi and Boulpaep 1989) The basis of this potential is the separation of charge due to the frictional interactions between the pore wall and the counter-ions within the pore, hence if the pores are too small to permit passage of any ions e.g the red cell membrane water channels; or alternatively, the channels are so large that they do not discriminate between co-ion or counterions, no streaming potential is developed across the pore. Thus, neither water flows across the wide shunt channels within epithelia, nor the water channels across cell membranes are detectable using electrokinetic methods.

Recently this electrokinetic method of detecting water flows has fallen into disfavour because of the view that much of the streaming potential was generated by unstirred layer effects which lead to either concentration polarization of ions at the *cis* side of the membrane or to concentration depletion at the *trans* side (Diamond 1979).

The view that unstirred layer effects are responsible for the streaming potential across renal tubular epithelia has been questioned by Tripathi and Boulpaep (1988, 1989) who have carried out equivalent circuit analysis of the flows across *Ambystoma* proximal tubule and shown that streaming potentials are generated by paracellular flows and not by transcellular flows or by concentration polarization of solutes within unstirred layers on either side of the tight-junctions. Because of the small dimensions of the tubule lumen and the concentration clamping effect of tubular perfusion, the contribution of unstirred layers in this preparation is negligible. It was also shown earlier that the assumption of Van Os et al.,(1979) that there are large unstirred layer effects which interfere with determination of streaming potential measurements in small intestine and gall-bladder is an error due to failure to recognize tissue volume change resulting from hydraulic resistance of the submucosal layers (Naftalin and Tripathi, 1985, 1986).

Isotope clearance

In vivo isotope clearance measurements have been use to monitor intestinal blood flow with some success. However, there is currently some controversy of the role of countercurrent flows within intestinal villi (review Lundgren 1982).

Fluorescence monitoring of isotope exchange

An interesting new method of monitoring the rate of D_2O exchange across perfused renal tubules has recently been described using the fluorophore aminonaphthalene trisulphonic acid (ANTS) (Kuwahara and Verkman, 1988) . The principle of the method is that the fluorescence of ANTS is 3.2-fold lower in H_2O than in D_2O. Isolated cortical and outer medullary collecting tubules from rabbit were perfused with an isosmotic D_2O buffer and bathed in either an isosmotic H_2O or D_2O buffer at lumen flow rates. Lumen fluorescence was monitored by quantitative epifluorescence microscopy. P_d was determined from tubule geometry, lumen flow, ANTS fluorescence at different wavelengths. This new method represents a considerable advance on the previous methods and could be usefully exploited in investigating tissues with complex structures.

Volumetric methods

a) Change in concentration of non-absorbable markers

One frequently used method of determining net fluid movement across a membrane or epithelium is to monitor the change in the concentration of impermeant marker on one side as water is extracted from or added to the solution containing the marker. High optical absorbance, high molecular weight markers such as haemoglobin, (Curran, 1960) can be detected using optical spectroscopy . Alternatively, it is possible to detect volume change by using the change in radioactivity per unit volume of a labelled volume marker, Powell and Malawer(1968); Pappenheimer and Reiss (1987). The advantage of this method is its simplicity. However, there are problems relating both to its accuracy and efficacy. The accuracy of the method is limited by the resolution of the marker detection system; use of a constant perfusion flow device with a constant volume cell considerably increases accuracy so that the method can be applied to monitoring nanolitre changes in flow in renal collecting tubules (Kuwahara and Verkman, 1988). However, if samples have to be obtained by pipetting this limits accuracy to. It is also important to ensure that the volume marker is detected in a completely homogeneous solution. Uneven mixing of marker within the bathing solution can easily occur if there is mucous present or if the tissue surface is unevenly perfused. The presence of high concentrations of non absorbable marker e.g. PEG, can produce artefacts. There are two reasons for this, firstly colloids can exert osmotic pressure across wide shunt channels which have a very high Lp, so the flow induced by even low concentrations can be considerable, secondly the osmotic coefficient of many colloids, in particular, PEG is higher than 1 hence 2mM PEG will exert an osmotic pressure of approximately 100cm H_2O (Ahsan et al., 1987)

Determination of water flows in renal collecting tubules by fluorescence quenching of the non absorbable marker, fluorescein sulphonate

The intensity of fluorescent emission from renal tubular perfusion fluid containing a fluorophore is proportional to the number of fluorophore molecules within the tubule segment and to the brightness (quantum yield and molar absorptivity) of the fluorophore. Thus fluorescence is dependent on the path length or geometry of the perfusion system as well as the concentration of fluorophore molecules. However, concentration dependent change in fluorescence intensity can be used to monitor the concentration changes of fluorophore within the perfusion fluid. Autoquenching of fluorescence by high concentrations > 5mM of fluorescein is a useful method of obtaining a signal which responds to concentration rather than the amount of fluorophore in the light path and hence is useful for determining concentration changes in solutions with a variable geometry.

An alternative approach to autoquenching by a single fluorophore is to use a ratiometric method of determining concentration change of impermeant fluorophores with a double fluorophore system e.g. pyrenetetrasulphonic acid(PTSA) and fluorescein sulphonate(FS). The principle of this method is that PTSA fluorescence is quenched by low concentrations of FS. Hence, with proportionately increasing FS and PTSA concentrations, FS fluorescence increases, whereas PTSA fluorescence is unchanged as it is quenched by the raised concentration of FS. Thus the ratio of PTSA/FS fluorescence decreases with raised concentrations of both fluorophores. The advantage of this technique applied to monitoring changes in renal tubular luminal concentration of non absorbable markers is that it provides an intensity and geometry independent measure of transepithelial volume flow. This technique has been applied to measuring the time dependence of changes in osmotic water permeability of cortical collecting duct following addition and removal of vasopressin (Kuwahara et al .,1988).

These new methods of adapting fluorescence microscopy and video-enhanced imaging are still being developed. The scope and applicability of these microscopic methods will increase as instrumentation and the software to run it is made generally available.

Volumetric displacement as determined by volumetric titration

Mass flow of solvent into or out of the mucosal bathing solution will alter the volume of the bathing solution and so alter the meniscus height. Deviations of the meniscus height can be detected by circuit breakers which activate fluid pumps which replace the fluid to the original level. The rate of fluid loss can then be determined from the amount of current required to generate the calibrated pump flux (Bourguet and Jard 1964; Fischbarg, Warshawsky and Lim, 1976). The advantage of this system is that it ensures constant volume in the bathing solution, is readily calibrated and automated. The disadvantages of the method are that surface tension will tend to cause an erratic response of the pump. However, this can be smoothed electronically. A more serious disadvantage of the method is that if the tissue is present in the same bath as that where the outflow is detected, changes in tissue volume may mask the changes in the volume of the bathing solution

Capacitance probe method of detection of movement of the solution meniscus

Volumetric methods

Adoption of a capacitance probe to monitor changes in volume of bathing solution within an Ussing chamber by Wiedner (1976) was a major advance methodology. The high sensitivity of this device together with avoidance of the surface tension problem makes this a very useful way of monitoring small changes in the height of the meniscus of a fluid chamber. Van Os et al., (1979) used a capacitance probe to monitor change in fluid volume of mucosal and serosal bathing solutions.

Unfortunately the technique had some associated problems. The main problem being that without some form of external support, a unilateral change in pressure within one of the bathing solution chambers caused major movement artefact of the tissue. When a hydrostatic pressure was applied unilaterally, the tissue bent in the direction of the pressure gradient so that the volume of the chamber from which pressure was applied was increased in size and the lower pressure chamber decreased. This effect was erroneously interpreted to be an absorptive flux the the estimated L_p of the tissue was too high. It is only possible to apply external hydrostatic pressure providing there is a rigid permeable support for the tissue which prevents any translation. Another problem with this method is it takes no account of tissue swelling during absorption, or of tissue shrinkage during following exposure to a hypertonic mucosal bathing solution.

Optical lever in addition to capacitance probe

The problems associated with the capacitance probe as outlined above have been overcome by simultaneous measurement of the rate of tissue swelling and transepithelial water flow. Fig. 3 shows a diagram of the capacitance probe and optical lever (Naftalin and Tripathi, 1985). The optical lever consists of a lightweight mirror which rests against the mucosal surface of the tissue. As the tissue expands the mirror rotates forward pivoting at the hinge attached to the lower edge. A light beam is reflected by the mirror through the solution and its movement is detected, amplified and recorded with photodetectors and a chart recorder. As the tissue gains volume due to the excess rate of mucosal inflow over serosal outflow the rate of net mucosal water flow J_m can be obtained from the algebraic sum of the serosal outflow, J_s and the rate of tissue volume change, J_t.

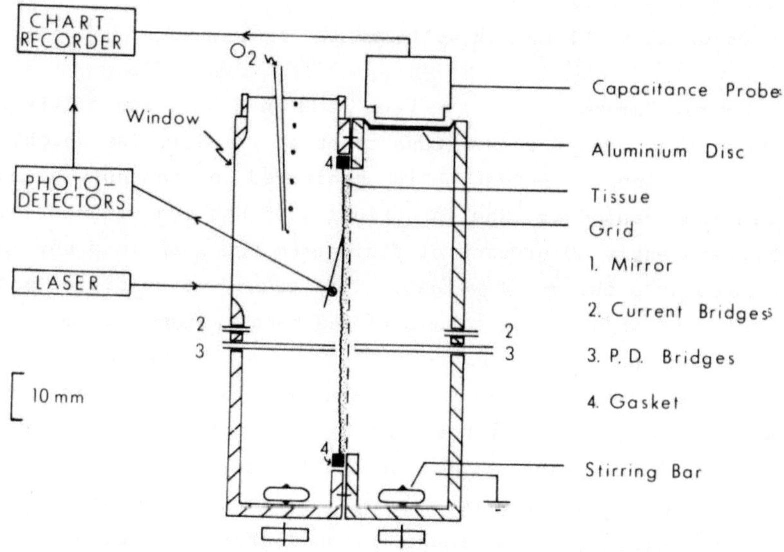

Fig 3. The flux chamber with capacitance probe and optical lever

It has been shown that mucosal and serosal flows are independent. Both the rate and direction of net flow across the mucosal and serosal surfaces may differ. Furthermore the sensitivity to osmotic probes of differing size is dissimilar on the mucosal and serosal surface demonstrating that the tissue behaves as a series array of hydraulic resistors bounding a large hydraulic capacitance (Fig. 2)

Gravimetric methods.Sacs

Either everted sacs (Smyth and Taylor, 1957) or isolated lengths of intestine in which the lumen is perfused with oxygenated recycled solution (Parsons and Paterson, 1964) can be used to monitor water flow. The advantage of this technique is that it can measure both water and solute transport simultaneously; it is also simple, only very modest equipment is required. However, the sac method is associated with several problems which limit both its accuracy and applicability.

These are:

a) The method has aa very low time resolution, (only one or two readings per sac) the tissue must be removed from the bathing solution and dried before weighing; this reduces the accuracy and gives rise to tissue deterioration due to stretching.

b) Sub-mucosal hydraulic resistance of intestinal tissue leads to interstitial oedema with a consequent rise in interstitial pressure. This causes ion and fluid re- flux across stretch enlarged shunt channels in the mucosal surface.

Method of monitoring continuously perfused loops *in vivo*

Considerable improvements have been made on resolution of the sac method of monitoring water flow, (Cassuto et al., 1983). The weight of a luminally perfused loop of intestine is monitored continuously using a sensitive tension transducer. The intestinal loop has its vascular supply intact and consequently absorption of fluid from the loop into the animal can be monitored as a change of weight. This method is sensitive and gives a continuous record of both the change in the total weight of the perfusion solution and the tissue weight.

The critical problems which are related to this preparation are:

1) No account is taken of the effects of changes in volume of the capacitance vessels on tissue weight, so an increase in the volume of the venous capillary bed of the perfused intestinal loop may be interpreted as a secretory flux; similarly a decrease in capillary volume alone could be misinterpreted as enhanced absorption.

2) No account is taken of changes in interstitial volume so that development of interstitial oedema is interpreted as secretion. For these

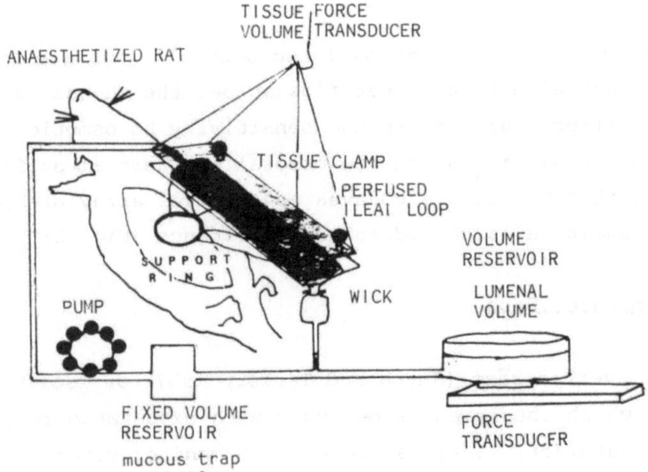

Fig 4 Method of monitoring the change in luminal perfusion fluid and tissue volume change simultaneously.

reasons the results of drug and other treatment induced changes in fluid absorption reported using this method must be treated with caution as at least part of these changes could be due to other causes.

For these reasons the method of Cassuto et al., (1983) has been modified so that the change in luminal volume and tissue volume can be monitored simultaneously(see below) (Jodal, Lundgren and Naftalin, 1987)

Optical methods

Microscopy

Monitoring the changes in the volume of transporting epithelia has been attempted by measurement of the separation distance between adjacent ridges on the surface of the goldfish intestinal mucosa (Siegenbeek Van Heukelom et al., 1981). This method has the disadvantage that it does not accurately measure volume changes in the vertical axis. Expansions within the submucosa which elevate the mucosa are not readily monitored using optical microscopy, so one of the dimensions of the volume change is not monitored. The resolution of straightforward optical microscopy, without computer averaging of the image, is probably insufficient to monitor volume change accurately see Carpi-Medina et al., (1984). For this reason, together with its laboriousness the method has not been extensively used.

A similar approach to that of Siegenbeek van Heukelom et al., (1981) is that of Spring and Hope (1979); Persson and Spring (1982). Using a microscope with very small depth of field the cell volume of *Necturus* gall bladder is derived from measurements of the area enclosed by the light microscopic image of the cell boundaries. Spring (1979). This is aided by computerized planimetry and stepper control of the microscope. By stepping the plane of focus through the entire thickness of the tissue the volume changes of a single epithelial cell can be estimated by summing the cross sectional areas of cell. Although it is possible to measure the transient changes in cell volume occurring following unilateral osmotic perturbation and thereby monitor the permeability of the apical and baso-lateral membranes independently there are several methodological drawbacks to this technique:

a) the resolution of the change in cell depth is very poor, so the upper and lower membrane surfaces are poorly delineated;

b) the monitored change in volume is confined to a single cell and takes little account of compressive and bending forces which may affect individual cell shape change occurring independently within the rest of the tissue;

c) the method measures only changes in volume following unilateral osmolality changes in the bathing solution. The steady state fluxes were measured with reference to transepithelial fluxes and not across the single cell; so large discrepancies may exist between flow across the single cell and the lumped estimate of transcellular flow;

d) the convolutions of the baso-lateral membrane foldings are not easily monitored by light-microscopy, hence no accurate estimate can be obtained of the true membrane area.

Because of these problems serious internal inconsistencies have arisen. The estimates of single transcellular hydraulic permeability exceed the transepithelial hydraulic permeability (Persson and Spring 1982).

A considerable advance on this optical method has been made by Carpi-Medina et al., (1984); Carpi-Medina and Whittembury (1988). The wall volume of rabbit straight proximal tubule is computed by an image processor that measures several luminal and outer diameters along a segment of isolated proximal straight tubule with high precision and at a high rate. Because of its simple geometry the tubular cell volume change can be readily estimated from the observed position changes in tubular and luminal surface. This method has a sufficiently high resolution to measure the transients in tubular cell volume with accuracy and reproducibility (Carpi-Medina et al., 1988).

Volume changes in single cells using changes in intracellular ion-probe concentration to monitor the hydraulic permeability of the membranes

Zeuthen(1982) used intracellular [Na^+] as a probe for rapid cell volume changes. Intracellular Na^+ is monitored with a Na^+- sensitive electrode and the changes in intracellular ion concentration following exposure of the cell to anisotonic solution can be used to monitor membrane hydraulic conductance. The same principle has been used by Reuss and coworkers (see chapter by Reuss and Cotton). However they used a quaternary ammonium compound loaded into the cells using an ionophore. Providing the measurements are matched with morphometric determinations of the membrane surface area of the cells, then estimates of the hydraulic conductance of the apical and baso-lateral hydraulic conductance can be obtained both from the osmotic transients and steady- state flows.

DETERMINING THE ROUTES AND MODE OF WATER TRANSPORT IN SMALL AND LARGE INTESTINE *in vitro*

It is possible to discriminate between channels across which pumped Na^+ can generate an osmotic pressure driven flow and channels which do not restrict electrolyte movement and therefore cannot sustain an osmotic flow. This is done by monitoring the osmotic flow per osmole generated by osmotic probes of varying size.

Figure 5 shows the effect of varying size of a series of uncharged polyhydric solutes on flow across the mucosal and serosal surfaces of rabbit ileum. The relationship between flow and probe size is fitted to the Renkin equation (Renkin and Gilmore (1973). This equation models the relationship between the reflexion coefficient of a solute to the ratio of the solute radius : pore radius.

The best fit between flows and solute radius of the large solutes was obtained with a pore radius of approximately 6nm for both the mucosal and serosal flows. With this large pore radius, the osmotic flow induced by small molecular weight solutes like NaCl across the serosal surface is negligible as $\sigma_{NaCl} = 0$. However, small molecular weight solutes are observed to induce flow and generate a streaming potential across the mucosal surface. Additionally low molecular weight solutes like ethylene glycol and glycerol induce water flows across the mucosal surface which does not generate any streaming potential. These findings are consistent with three types of water channel across the mucosal solution: a wide paracellular route which has a radius of approximately 6-7nm and an L_p of $2x10^{-9}$ cm s^{-1} cm H_2O^{-1}; an intermediate channel which permits a cation selective streaming potential and hence is also paracellular with a radius approximately 0.7nm with and L_p of approximately $2x10^{-8}$ cm s^{-1} cm H_2O^{-1} and a smaller transcellular channel which permits only water flow with an L_p of approximately $2x10^{-9}$cm s^{-1} cm H_2O^{-1}. Fig. 5 shows that the osmotic pressure induced flow across the mucosal surface is consistent with at least two sets of paracellular pores.

Evidence for a submucosal compartment

Following a unilateral decrease in the tonicity of the mucosal bathing solution by 50 or 100mosm kg^{-1} there is a transient rise in fluid inflow across the mucosal surface, however as the tissue expands the rate of inflow decreases until it returns to control levels (Fig. 6A). Addition of a hypertonic solution to the mucosal side reduces the rate of fluid inflow across the mucosal surface below control levels. Flow may become negative i.e. net fluid outflow is observed across the mucosal surface (Fig. 6B). However this fluid outflow is not accompanied by any concurrent reversal of flow across the serosal surface. The absence of any correlation in the extent or direction of net flow across the mucosal and serosal surfaces indicates that there is a central compartment bounded by dissimilar membranes. Direct evidence for a wide mucosal shunt pathway with a high L_p is obtained from the effect of low concentrations of PEG added to the mucosal solution (Fig. 7). 3mosm PEG in the mucosal solution causes a large sustained reduction of net mucosal flow which is not reversed until the PEG is removed from the bathing solution. This contrasts with the effect of addition of 50mosm NaCl or sucrose where the decrease in fluid inflow is only transient . These data indicate that low molecular weight solutes can equilibrate between the mucosal bathing solution and the submucosal fluid compartment thereby reducing the transcellular osmotic gradient. The absence of delay or relaxation in the effect of PEG 4000 indicates that this solute does not equilibrate across the wide mucosal paracellular shunt.

Direct determination of the L_p across the serosal surface by raising the hydrostatic pressure in the mucosal bathing solution and maintaining the rigid support at the serosal surface permits an accurate estimate of the interstitial pressure required to sustain the observed fluid movements across the serosal surface. The substantial interstitial pressure observed in control tissue 20-40cm H_2O could give a large reflux of fluid via the wide paracellular shunt pathways across the mucosal surface.

Evidence that this is the route of fluid reflux during induction of fluid secretion by secretagogues like theophylline or choleragen is that there is increased efflux of the paracellular marker probe ^{14}C-labelled mannitol which moves exclusively via paracellular pathways and a substantial secretagogue dependent decrease in the L_p of the serosal surface is observed following exposure to theophylline or choleragen. This effect can be reversed by muscle relaxant drugs like clonidine or dopamine. Thus it is evident that the direction of net flow across the whole epithelium is not the same as the direction of transmucosal flow and that many secretagogues exert their effects by extracmucosal effects i.e. by increasing the hydraulic resistance via the submucosal tissues (Ahsan et al., 1987, 1988).

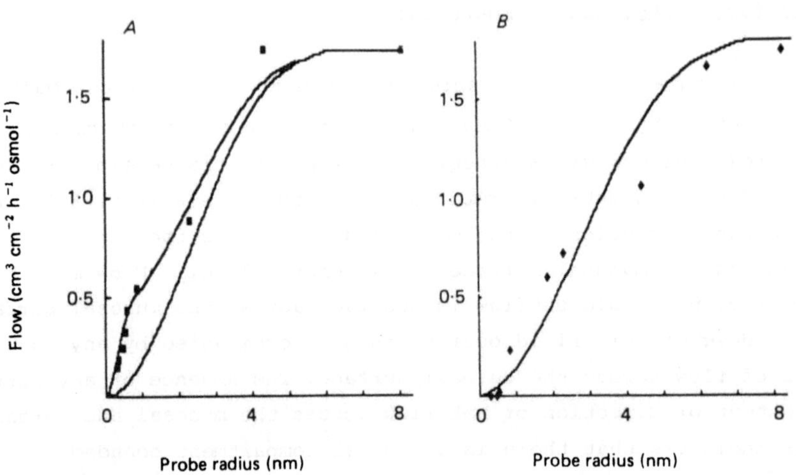

Fig. 5 Panel A shows the effect of variation of probe radius on flow across the mucosal surface and panel B shows the effect of varying probes of the same size on serosal flows.

These findings indicate the problems of interpreting indirect effects of drugs on the electrical parameters in terms of changes in water flow. It is also clear that interpretations of data limited only to short circuit current and isotope fluxes, which disregard the submucosal compartment and the possibility of independent fluxes across the mucosal and serosal surfaces can be seriously in error.

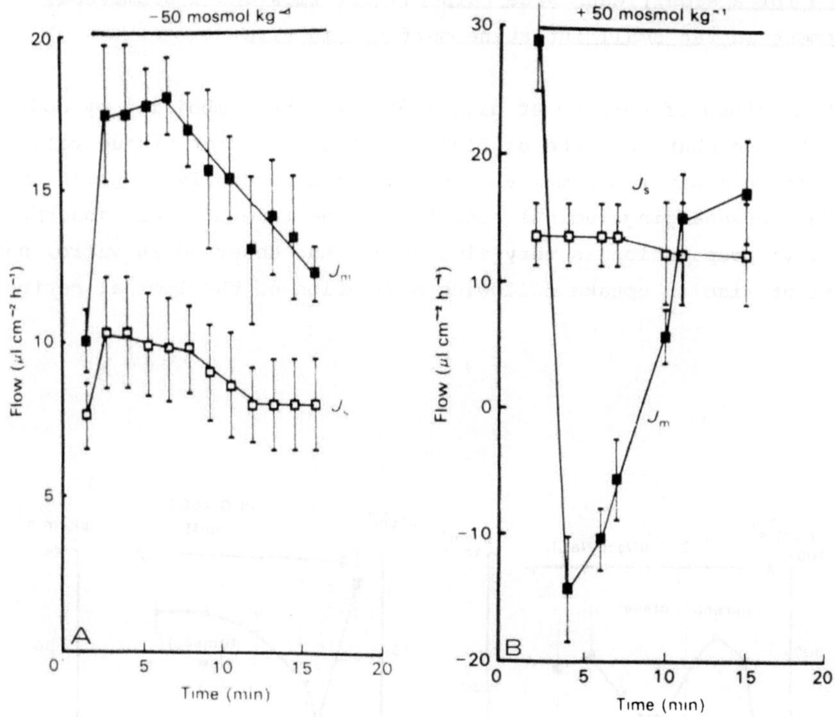

Fig. 6 Effects of reducing (Panel A) or raising (Panel B) the tonicity of the mucosal bathing fluid by 50mosmol/kg on flow across the mucosal surface (filled squares) and across the serosal surface (open squares).

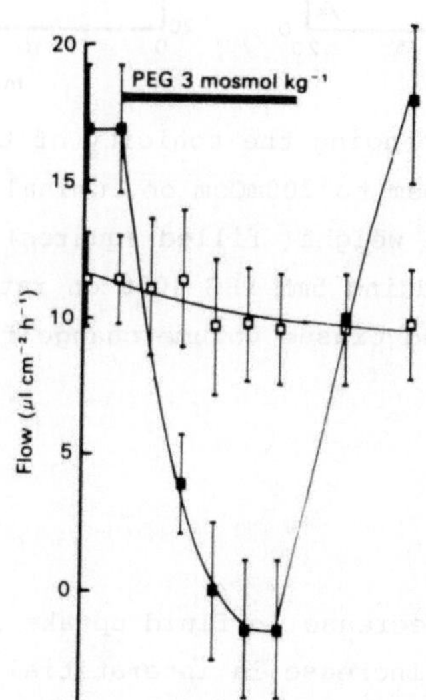

Fig. 7 Effect of addition of 3mosmol/kg PEG 4000 on flow across the mucosal (filled squares) and serosal surface (open squares). Tissue is exposed to PEG only during the time indicated by the bar.

Evidence for a significant wide paracellular flow and a submucosal compartment in rat small intestine perfused *in vivo*

The method of Cassuto et al., (1983) has been modified by Jodal, et al., (1987) so that the rate of fluid absorption and of tissue volume change can be monitored simultaneously and independently (figs.8A and B). The effect of changing luminal tonicity on the rate of fluid absorption in the *in vivo* preparation is very similar to that observed *in vitro*, namely a transient rise in uptake following a dilution of the luminal perfusion

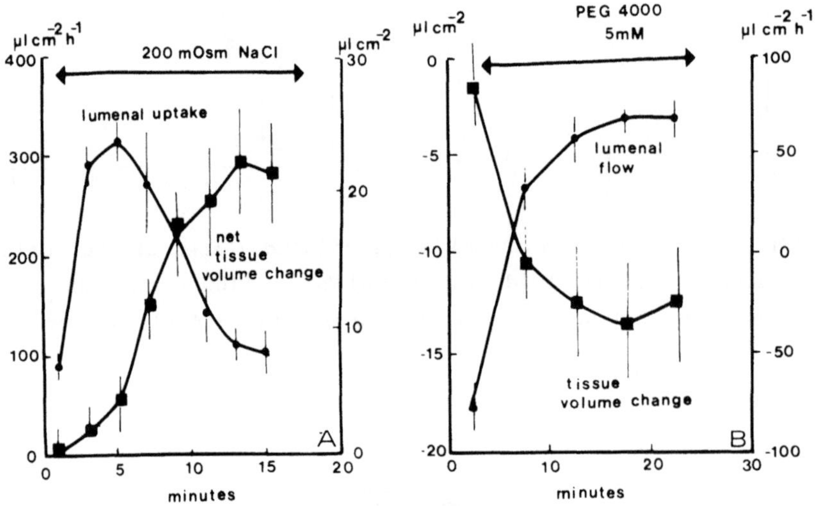

Fig. 8A Effect of changing the tonicity of the luminal bathing solution from 300mOsm to 200mOsm on luminal uptake rate(filled circles) and tissue weight(filled squares).

8B Effect of adding 5mM PEG 4000 on rate of luminal uptake (filled circles) and tissue volume change(filled squares).

solution, followed by a decrease in fluid uptake as the tissue volume increases due both to an increase in interstitial fluid volume and also perhaps to an increase in the volume of the capillary bed within the perfused loop. Similarly the effect of adding a low concentration of a macromolecule(PEG 4000) to the luminal perfusion solution indicates that there is 10-fold larger change in flow/osmole of macromolecule than found with NaCl.

These findings indicate that a wide shunt channel and the large submucosal space are a properties of intact tissue in the *in vivo* perfused animal and not simply an artefact resulting from a large increase in the submucosal hydraulic resistance resulting from cessation of capillary perfusion as is the case in the *in vitro* perfusion preparation. They also indicate that there is a substantial submucosal fluid compartment which equilibrates osmotically with the luminal solution and which modifies osmotic flow across the mucosal layers of the tissue.

Hypertonic fluid absorption by rabbit and rat descending colon

Descending colon from rat, rabbit and man is known to produce a very hypertonic absorbate *ca.* 500mosm kg^{-1}. (Parsons and Paterson 1964; Powell and Malawer 1968; Duthie et al., 1964; Harris and Shield 1973). The tonicity of the absorbate gives useful information regarding coupling between salt and water movement.

The mechanism of production of the hypertonic absorbate is of interest. The absorbate tonicity may be estimated from the slope of the line relating net Na flux to net water flux dJ_{Na}/dJ_{H_2O} (Curran and Schwartz, 1960). Powell and Malawer (1968) using an *in vivo* rat intestinal preparation found that the relationship between salt and water varies. In jejunum transport was isotonic, but more distal regions of the gut transport hypertonically. Ileum 398±11 μosmoles/ml; colon 544±24 μosmoles/ml. The electrical resistance of the intestine increases distally- the jejunum has a low resistance *ca.* 25Ω cm^2 whereas the colon has a resistance ca. 250-500Ωcm^2 (Frizzell, Koch and Schultz, 1976; Fromm and Schultz, 1981). As the increase in resistance is due to the decreasing ionic conductance of the tight-junctions, it may be inferred that the tonicity of the absorbate is inversely related to the ion permeability of the tight-junctions.

An alternative method of measuring absorbate tonicity is to monitor the osmolarity of the absorbate directly as it emerges from the serosal surface of the tissue (Bleakman and Naftalin, 1990). By monitoring the change in osmolarity, $[Na^+]$, $[K^+]$ and volume of a serosal perfusion solution by monitoring vapour pressure osmometry, net Na^+ with a flame photometer and net volume flow the change in concentration of an impermeant volume marker ^{14}C-labelled polyethylene glycol (m.w.4000) it is possible to estimate the net solute flux across the tissue from this data it is possible to estimate the tonicity of the absorbate (Figs. 9 and 10).

As the absorbate tonicity is hypertonic to the mucosal solution and raises the tonicity of the the serosal perfusion solution, it generates a secondary osmotic flow as a result of the transepithelial osmotic gradient acting on cells which are not normally exposed to the local hypertonicity generated by the active transport process which may not be uniformly distributed within the tissue . Additionally because the Na^+ concentration in the serosal solution is higher than in the mucosal bathing solution there will a passive leakage of NaCl from the serosal to the mucosal bathing solution. Both these errors tend to cause an underestimation of the absorbate tonicity. When allowance is made for the passive leakage components, it is estimated that the tonicity of the "active" component of colonic absorption is ca 1000mOsm kg^{-1}.

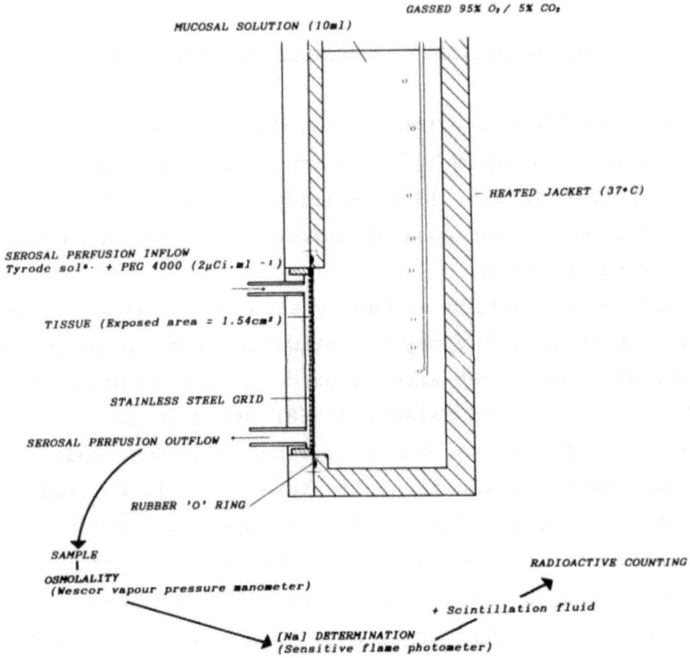

Fig.9 Diagram of chamber used to monitor net water and solute transport across descending colonic mucosa.

The passive hydraulic conductance and NaCl permeability across whole tissue are obtained from the ouabain-insensitive water and NaCl movements obtained when the osmolarity of the mucosal solution is reduced from 298mOsm kg^{-1} to 198mOsm kg^{-1}. The passive L_p of rabbit colonic mucosa is $1.3 \pm 0.1 \times 10^{-9}$ cm s^{-1} H_2O^{-1} whereas the Lp of the active route is $5 \pm 1 \times 10^{-11}$ cm s^{-1} cm H_2O^{-1}. This wide discrepancy between the "active" and "passive" L_p's confirms that there must indeed be separate routes for "actively" induced water flow from osmotic gradients generated within the tissue and passively induced water flow generated by transepithelial osmotic gradients.

Correlation of function with colonic structure

According to the following considerations a hypertonic absorbate requires a transepithelial route with a very low L_p.

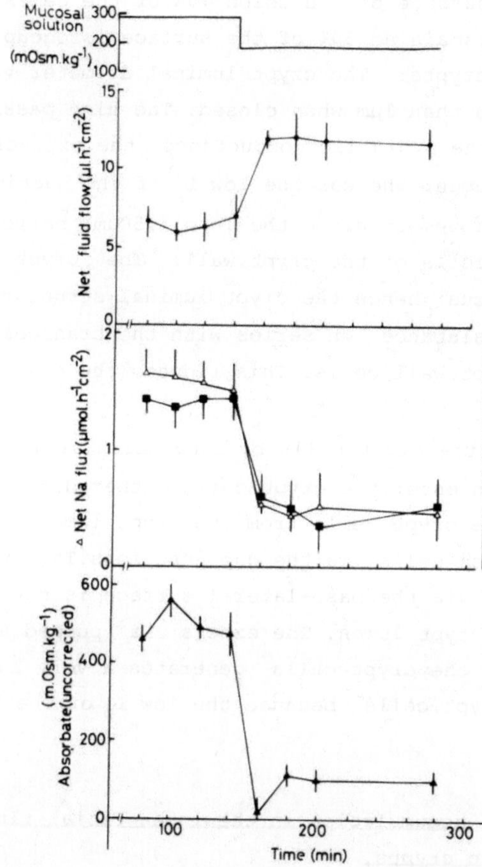

Fig.10 Fluxes of water (upper panel); net Na^+ and solute (middle panel) and the tonicity of absorbate (lower panel) during a period of perfusion with isotonic then 100mosmol/kg hypotonicity.

Net water flux = $J_v = L_p.\Delta P$.

The concentration of the absorbate $C^* = \delta J_s / \delta J_v$

where J_s = net solute flux

i.e. $C^* = \delta J_s / L_p \delta P$.

The mucosa of descending colon consists of closely packed crypts which are blind ended tubules opening at the luminal surface of the colon. In rabbit there are approximately 10000 crypts/cm^2. In rabbit colon *ca.* 70% of all the epithelial cells lie below the luminal surface within the crypt walls. On the luminal surface of the colon 70% of the cells are a monolayer of surface cells, the remaining 30% of the surface is occupied by the external cells of the crypts. The crypt luminal diameter varies from 40μm when fully open to less than 1μm when closed. The high passive transepithelial L_p is the hydraulic conductance the extracryptal surface cells of the colonic lumen; whereas the low L_p of the active route is consistent with fluid movement along the long (150μm) narrow axis of the crypts and out across cells of the crypt wall. The crypt lumen is filled with high viscosity mucus; hence the crypt luminal structure exerts a large extrinsic hydraulic resistance in series with the transcellular hydraulic conductance of the crypt wall cells. This reduces the overall L_p of the active route (Fig. 11).

Na^+ can access the crypt cells by intercellular movement via gap junctions. Thus Na^+ can enter the crypt cells either directly across the apical membranes of the crypt cells from the crypt lumen, or by percolating from the colonic surface cells via the gap junctions.Thus the Na^+ pumped out of the crypt cells via the baso-lateral surface is not limited by the slow delivery via the crypt lumen. The excess Na^+ pumped across the basolateral surface of the crypt cells generates a very large osmotic pressure across the crypt cells because the low L_p of the active route (Fig. 11).

Direct evidence of Na^+ accumulation in the interstitial fluid surrounding the colonic crypts.

Extracellular Na^+ has been monitored previously in epithelial tissues using ion-sensitive microelectrodes, Zeuthen (1983) and X-Ray microprobe microanalysis of frozen hydrated sections (Gupta et al., 1978). Use of the impermeant acid form of SBFI, a fluorescent probe for Na^+, along with high resolution video-enhanced microscopy to detect the fluorescence emission of SBFI is a more sensitive and convenient tool for detecting extracellular Na^+ than these other methods (Tsien, 1989 and Minta et al., 1987) and its use provides an opportunity to test a new hypothesis relating to colonic absorptive function.

If crypts have an absorptive function, then the crypt lumen should be depleted of Na^+ and the interstitial fluid contain a higher

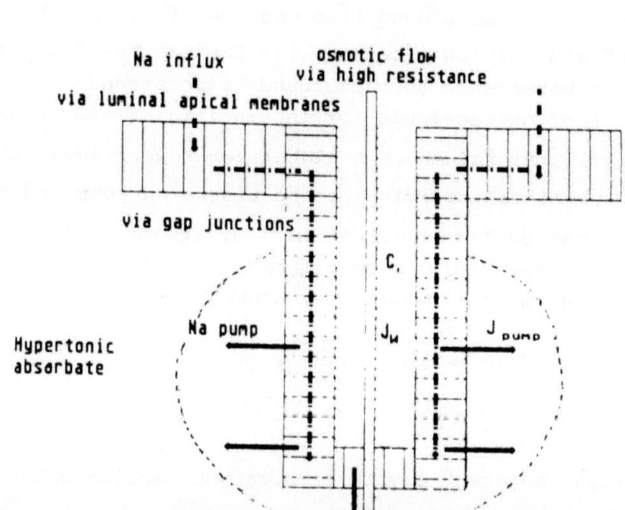

Fig. 11 Diagram showing the suction function of the crypt as it generates a hypertonic pericryptal solution.

concentration of Na salts than in the bulk solution; whereas if the crypts have a secretory function, then the crypt lumen should contain fluid hypertonic to the interstitial fluid i.e. a high concentration of Na salts. The conventional view of colonic function is that colonic crypts are routes of fluid and electrolyte secretion Welsh et al., (1982). Hence, accurate determination of the concentration of Na^+ within the crypt lumen and interstitial fluid using the extracellular fluorescent marker, SBFI should discriminate between these two views.

The local concentrations of Na^+ were quantified by determining regional fluorescence intensity using the computer averaged fluorescence excitation of $SBFI.Na^+$ complex (340-360nm) after background fluorescence from the cells and interstitial fluid was subtracted. This was done using computer enhanced video-microscopy. Comparison of the local emission is obtained with respect to background fluorescence of either the whole field or a neighbouring field of low intensity. Estimation of the $[Na^+]$ from the ratio of local fluorescence to background fluorescence. If there are local regions $(_l)$ of extracellular or interstitial fluid in which the steady state $[Na^+_l]$ is maintained at a higher (or lower) level than the bulk solution $[Na^+_b]$ by the local activity of the tissue Na-pump and the concentration of free SBFI (unbound to Na^+) at steady-state is uniformly distributed in all extracellular compartments, i.e. at steady-state $[SBFI_b] = [SBFI_b]$ then the following relationship exists:

$$SBFI.Na^+_l]/[SBFI.Na^+_b] = [Na^+_l]/[Na^+_b]$$

Thus $[SBFI.Na^+_l]$ is linearly related to $[Na^+_l]$

A group of eight adjacent crypts from rat descending colonic mucosa perifused with low Na^+ Tyrode (4.8mM Na^+) is shown in Plate 1a with bright-field illumination. The crypts appear as rosettes of illuminated cells surrounding a central dark lumen; the crypts themselves are surrounded by dark interstitial fluid. The fluorescence emission of the same field with background fluorescence from the cells and interstitial fluid subtracted is in pseudocolour in Plate 1b. The maximal fluorescent intensity is situated in the interstitial fluid surrounding the crypts. There is virtually no emission from the crypt cells or from the crypt lumen. This indicates $[Na^+_{(l)}]$ =6-10 fold above the concentration in the bulk solution. This degree of hypertonicity can be sustained for periods of 3-4h. During this time there is no evidence of any Na^+ accumulation above background levels within the crypt lumens. Similar accumulations of Na.SBFI within the interstitial fluid of both rabbit and rat colon can be observed in tissues incubated with 150mM NaCl Tyrode. This accumulation is ouabain sensitive in rabbit colon.

These data confirm that the colon is capable of generating very large ouabain-sensitive Na^+ concentration gradients across the crypt -wall. The absence of Na^+ accumulation within the crypt indicates that the crypts are not the site of Na^+ secretion during the absorptive mode.

The dehydrating function of crypts

One of the important functions of crypts is to dehydrate faeces.The fluid leaving the terminal ileum is approximately 90% water whereas hard faeces in rabbit and man contains between 60% and 70% water. Removal of water results in consolidation of faeces. This consolidation reduces the water permeability from $4x10^{-2}$ µm/s for faeces with a water content >65% to $3x10^{-5}$ µm/s for faeces with a water content of <60%. The pressure required to consolidate faeces from 70% water to < 60% water is between 0.5-1MPa (5-10 atmospheres)(McKie et al., 1990).

The maximal luminal pressure which can be exerted by human colonic smooth muscle is around 15kPa and this only for very short intervals the resting pressure is much lower <2kPa. Thus is certain that there is insufficient wall pressure to squeeze water out of faeces. The alternative is that water is sucked out of faeces via the crypts which convert the osmotic pressure gradient across the wall to a high fluid tension within the crypt lumen. An osmotic pressure of 1M can exert a fluid tension of 22.4 atm or 2MPa. This tension is sufficient to dehydrate faeces to the observed levels within the short time available (1-3h). The long narrow crypt lumen permits large fluid tensions to be developed and sustained. The importance of the crypt structure is twofold first to act as a hydraulic resistance and thereby reduce the L_p of the transcryptal route and secondly as a device for matching the load resistance within the liquid content of the colonic lumen to the internal resistance of the hydraulic pump. As the load resistance increases the crypt wall collapses inward thereby increasing the hydraulic resistance of the transcryptal route.

The subatmospheric pressure is confined to the crypt lumen and hence it provides a parallel hydraulic pathway across the mucosal surface. In contrast to the small intestine, this parallel network cannot be detected by probe analysis alone but requires the different kind discrimination provided by measurement of the tonicity of the active and passive routes of absorption.

CONCLUDING REMARKS

Several different methods have been described for monitoring fluid movement across epithelia. The important distinction has been drawn between wide channels capable of conducting fluid but incapable of supporting an osmotic gradient using osmotic probes of different size. Another distinction has been drawn between channels capable of supporting a large osmotic gradient which have a very low L_p and channels which support near isotonic water flow. It is apparent that new optical methods of measuring water and ion movements in discrete microscopic regions of tissue are now becoming available. Applications of these methods to the study of tissue substructures e.g colonic crypts or intestinal villi should extend our knowledge and insight of epithelial function.

A

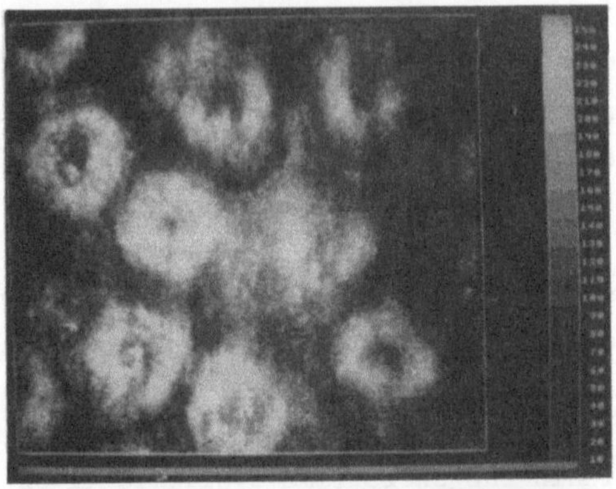

B

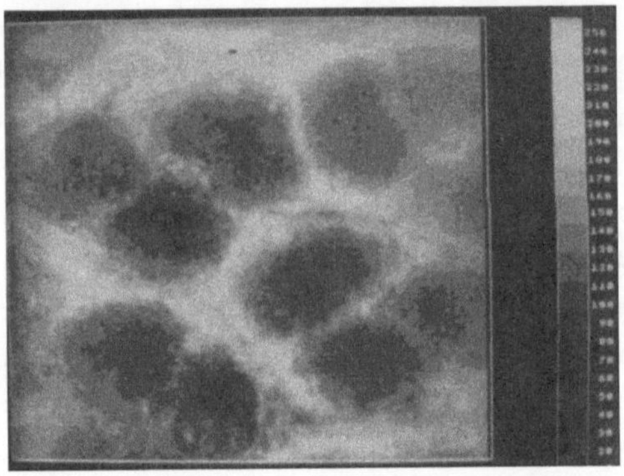

Plate 1A Bright field illuminated video-enhanced micrograph of colonic crypts. The crypt cells are shown as bright rosettes surrounding the dark lumen.

Plate 1B Shows the fluorescent image from Na.SBFI within the interstital fluid. The high intensity fluorescence within the interstitial spaces indicates that there is a large [Na^+] gradient across the crypt wall.

References

Ahsan M. A., Ilundain A., Naftalin R. J., Sandhu B. and Smith P. M. 1987 Effects of theophylline, choleragen and loperamide on rabbit ileal fluid and electrolyte transport *in vitro*. Br. J. Pharmac. 92, 743-754.

Ahsan, M. A., Naftalin R. J. and Smith P. M. 1988 A submucosal mechanism for catecholamine-induced increases in fluid absorption in rabbit ileum *in vitro*. J. Physiol. (Lond.). 404, 385-405.

Bleakman D. and Naftalin R.J. 1990 Hypertonic fluid absorption from rabbit descending colon *in vitro*. Am. J. Physiol. 258, 21 G377-G390

Bourguet, J. and Jard, S. 1964 Un dispositif automatique de mesure et dénregistrement du flux net d'eau a travers la peau et l vessie des amphibiens. Biochim. Biophys. Acta 88. 442-444

Carpi-Medina P. Lindemann B., Gonzalez E. and Whittembury G. 1984 The continuous measurement of tubular volume changes in response to step changes in contraluminal osmolality. Pfluger's Archiv. 400 343-348.

Carpi-Medina P. and Whittembury G. 1988 Comparison of transcellular and transepithelial water osmotic permeabilities (P_{os}) in the isolated proximal tubule (PST) of the rabbit kidney. Pfluger's Archiv. 412 66-74.

Cassuto J. J., Siewert A., Jodal M. and Lundgren O. 1983 The involvement of intramural nerves in cholera toxin induced intestinal secretion. Acta Physiol. Scand. 117, 195-202.

Curran P.F. 1960 Na, Cl and water transport by rat ileum *in vitro* J.Gen. Physiol. 43 1137-1148.

Curran P.F. and Schwartz. G.F. 1960 Na, Cl and water transport by rat colon J. Gen. Physiol. 43; 555-571.

Diamond, J. M. (1979) Osmotic water flow in leak epithelia. J. Membr. Biol. 51 195-216.

Durbin R.P., Frank H. and Solomon A.K. 1956. Water flow through frog gastric mucosa. J. Gen. Physiol. 39 535-551.

Duthie H.L., Watts. J.M., De Dombal F.T. and Galligher J.C. 1964 Serum electrolyte and colonic transfer of water and electrolytes in chronic ulcerative colitis. Gastroenterology. 47 525-530.

Fischbarg J. C., Warshawsky R.E. and Lim J.J. 1976 Pathways for hydraulically and osmotically induced water fluxes across epithelia. Nature 266 71-73.

Frizzell, R. A., Koch, M.J. and Schultz S.G. 1976 Ion transport by rabbit colon. I Active and passive components. J. Membr. Biol. 27 297-316 1976

Fromm, M. and Schultz S.G. 1981 Potassium transport across rabbit descending colon *in vitro* evidence for single file diffusion through a paracellular pathway. J. Membr. Biol. 63 93-98.

Gupta, B.J., Hall, T.A. and Naftalin R.J. 1978 Microprobe measurement of Na, K and Cl concentration profiles in epithelial cells and intercellular spaces of rabbit ileum. Nature 272 70-73

Harris, H.L and Shield, R. 1973 Absorption and secretion of water and electrolytes in chronic ulcerative colitis. Gut 14 784-789.

Jodal, M., Lundgren. O. and Naftalin, R.J. 1987 J. Physiol.Lond 391 29P.

Katchalsky, A. and Curran, P.F. 1965 in Nonequilibrium thermodynamics in Biophysics. Harvard University Press Cambridge Mass.

Kuwahara M. and Verkman A.S. 1988 Direct fluorescence measurement of diffusional water permeability in the vasopressin-sensitive kidney collecting tubule. Biophys. J. 54 587-593.

Kuwahara, M., Berry, C.A. and Verkman.A.S. 1988 Rapid development of vasopressin-induced hydroosmosis in kidney collecting tubules measured by a new fluorescence technique. Biophys. J. 54: 595-602.

Lundgren, O. 1982 Handbook of Physiology The cardiovascular system 4: Microcirculation of the gastrointestinal tract and pancreas.

McKie, A.T., Powrie W. and Naftalin R.J. 1990 Mechanical aspects of rabbit fecal dehydration. Am. J. Physiol 258, 21 G391-G394

Minta A., Harootunian A.T., Koo J.P.Y and Tsien R.Y.1987 J.Cell Biol. 105 89a.

Naftalin R.J. and Pedley K.C. 1990 Video enhanced imaging of the fluorescent Na^+ probe SBFI indicates that colonic crypts absorb fluid by generating a hypertonic interstitial fluid. FEBS Letts 294 187-194.

Naftalin R. J. and Tripathi S. 1985 Passive water flows driven across the isolated rabbit ileum by osmotic, hydrostatic and electrical gradients. J. Physiol. (Lond.). 360, 27-50.

Naftalin R. J. and Tripathi S. 1986 The roles of the paracellular and transcellular pathways and submucosal space in isotonic water absorption by rabbit ileum. J Physiol. (Lond). 370, 409-432.

Pappenheimer J. R. and Reiss K. Z. 1987 Contribution of solvent drag through intercellular junctions to absorption of nutrients by the small intestine of the rat. J. Membr. Biol. 100, 123-136.

Parsons D.S. and Paterson C. R. 1964 Fluid and solute transport across rat colonic mucosa Quart. J. Physiol. 220-230.

Persson B.E. and Spring K.R. 1982 Gallbladder epithelial cell hydraulic water permeability and volume regulation. J. Gen Physiol. 481-505.

Powell, D.W. and Malawer S.J. 1968 The relationship between water and solute transport from isosmotic solutions by rat intestine. Am. J. Physiol. 227 1436-1444.

Powell D.W. 1987 *in* Physiology of the gastrointestinal tract. 2nd Vol 2nd edition. Ed in Chief L.R.Johnson Raven Press Intestinal Water and Electrolyte transport. 1267-1306.

Renkin E. and Gilmore J.E. 1973 Handbook of Physiology, Renal Physiology Chap. 9 pp 185-348.

Siegenbeek Van Heukelom J., Van Den Ham M.D., Albus H. and Groot J.A. 1981 Microscopical determination of the filtration permeability of the mucosal surface of goldfish intestinal epithelium. J.Membr. Biol. 63 31-39.

Smyth D.H.and Taylor C.B. 1957 Coupled transport of water and solutes by an *in vitro* intestinal preparation. J.Physiol. 136 632-648.

Spring K.R. 1979 Optical techniques for the evaluation of epithelial transport processes. Am. J. Physiol. 237 F167-F174.

Spring K.R. and Hope A. 1979 Fluid transport and the dimensions of cells and interspaces of living *Necturus* gallbladder. J. Gen. Physiol. 73 287-305.

Tripathi S. and Boulpaep E.L.1988 Cell membrane water permeabilities and streaming currents in *Ambystoma* proximal tubule. Am.J. Physiol. 255 F188-F203.

Tripathi S. and Boulpaep E.L. 1989 Mechanisms of water transport by epithelial cells. Quart. J. of Exp Physiol. 74 385-417.

Tsien, R.Y. 1989 Annu. Rev Neursci. Chapter 12 227-253

Van Os C.H., Wiedner G. and Wright E.M. 1979 Volume flows across gallbladder epithelium induced by small hydrostatic and osmotic gradients J. Membr. Biol. 49 1-20

Welsh M.J. Smith P.L., Fromm M. and Frizzell R.A. 1982. Crypts are the site of intestinal fluid and electrolyte secretion. Science Wash. DC 218 1219-1221.

Wiedner G. 1976 Method to detect volume flows in the nanoliter range. Rev of Sci. Instr. 47, 775-776

Zeuthen T. 1981 Relations between intracellular ion activities and extracellular osmolality in Necturus gallbladder epithelium. J. Membr. U 66 109-121.

Zeuthen T. 1983 Ion activities in the lateral intercellular spaces of gallbladder epithelium transporting at low wxternal osmolarities. J. Membr. Biol. 76 113-122.

Part 2: ISOTOPIC METHODS FOR FLUX DETERMINATION IN VARIOUS MEMBRANE PREPARATIONS

Chapter 5

Characterization of transport systems at the blood-tissue interface by single circulation paired-tracer dilution in perfused organs: A review of recent studies.
David L. Yudilevich, Luis A. Sobrevía and L. Felipe Barros

Chapter 6

Some inexact methods for the analysis of transport kinetics using dual tracers.
I. Christopher H. Smith

Chapter 7

Transport studies in isolated enterocytes
Ana Ilundain

Chapter 8

Cell culture in the study of epithelial development and function
Francisco Sepúlveda

Chapter 9

New approach to the study of transport of biomolecules by microcarrier cell cultures perfused in a column applying a high resolution paired-tracer technique
Salvador Perán, Manuel Muñoz and María Saiz

Chapter 10

Methodological guide for studying epithelial transport with isolated membrane vesicles
Jurg Biber and Heini Murer

5

CHARACTERIZATION OF TRANSPORT SYSTEMS FOR SOLUTES AT THE BLOOD SIDE OF ENDOTHELIAL AND PARENCHYMAL CELLS BY SINGLE CIRCULATION PAIRED-TRACER DILUTION: A REVIEW OF RECENT STUDIES

David L. Yudilevich, Luis A. Sobrevía and L. Felipe Barros

Physiology, Biomedical Sciences Division, King's College, University of London, Campden Hill Road, London W8 7AH, U.K. and Departamento de Fisiología y Biofísica, Facultad de Medicina, Universidad de Chile, Casilla 70055 Santiago, Chile

INTRODUCTION

Carrier-mediated transport of blood-borne substrates at the blood-tissue interface in the microcirculation is a very rapid process and influx of labelled molecules can often be measured within the time of a single circulation through an organ. The plasma membranes involved in the tracer uptake may be those of (*i*) the endothelial cell, (*ii*) parenchymal cell and (*iii*) neuronal terminals.

The multiple-tracer single circulation technique may be used to measure the passive permeability of intercellular junctions in endothelium and also the saturable facilitated transport mechanisms present in the plasma membrane of cells. Rapid metabolic transformation of certain labelled substrates may influence the tracer dilution profile thus complicating interpretation of transport studies, but in these cases the technique has become a powerful tool for assessing metabolic transformations at the blood-tissue interface. It was introduced in the 1950's and since then many workers have shown it to be a valuable methodology for studies in organs *in vivo* or isolated perfused tissues. This topic was the subject of a Symposium in which the application of these tracer techniques to many organs and a wide range of substrates was illustrated (see book edited by Yudilevich and Mann, 1985). We here review the literature which has appeared since that comprehensive publication with special emphasis on the so called "paired-tracer method" (see Yudilevich and Mann, 1982) which has enabled successful characterization of specific transport systems and their hormonal regulation in perfused organs. In this relatively physiological condition membrane transporters have been characterized with a degree of accuracy resembling that obtained with isolated membrane vesicles (e.g. for placental amino acid transport see Yudilevich and Sweiry, 1985; Yudilevich and Barros, 1990).

The far reaching scope of the single passage tracer methodology is reflected by its recent application to the study of saturable transport kinetics in perfused cells cultured on

Cell Membrane Transport, Edited by D. L. Yudilevich *et al.*
Plenum Press, New York, 1991

microcarrier beads (Perán and McGee, 1986; Mann et al. 1989; see also in this volume Chapters by Perán et al. and Mann et al.). Moreover, the passive permeability of endothelial cell junctions in the microcarrier cultures has recently been studied using the multiple tracer dilution technique (Haselton et al. 1989; Toothill et al. in press).

METHODS AND INTERPRETATIONS

The multiple tracer dilution method consists of a rapid intra-arterial injection of a bolus containing two (or more) radioactively labelled molecules, which is followed immediately by rapid sequential collection of samples of venous blood. Samples are collected at a fast rate (seconds or even 3-4 per s) depending on the venous outflow rate and the particular aim of the experiment. Further samples may be collected for a few minutes for assessing tracer backflux and cellular retention of the labelled substrate. The dilution profile of a test labelled compound is compared with either an intravascular reference tracer or an extracellular tracer of similar size and free-diffusion coefficient. We here follow a mathematical analysis for single-circulation multiple tracer dilution data which was reviewed by Yudilevich and Mann (1982) and by Yudilevich (1985). A summarized version is given below. Most valuable information, including the kinetics of unidirectional influx across the cell membrane, can be obtained with a simple interpretation of the tracer data.

I. MULTIPLE TRACER DILUTION. The reference tracer is confined to the intravascular space (see Figure 1). The lower concentration of the test tracer, with respect to the reference, indicates that the test molecules have diffused into a space larger than the vascular compartment, and results are quantified in terms of a capillary extraction and an extravascular space (Martín de Julián and Yudilevich, 1964; see Table 1). See Table 1.

Measurement of extraction, capillary permeability and extravascular space of distribution. The very early tracer dilution data shown in Fig. 1 is used to measure the unidirectional transfer of the molecules across the capillary endothelium. It can be seen that for sucrose and ouabain, the first few samples show relatively constant "extraction". This is the parameter *E(O)*, from which capillary permeability-surface products (*PS*) are estimated (Table 1). Both molecules have similar free-diffusion constants, and therefore this result suggests that the extraction for both molecules is related to their passive movement across the intercellular clefts. The latter part of the dilution curves is related to the events occurring beyond the capillary wall. For example, in Fig. 1 the interstitial [^{14}C]sucrose space was 0.08 ml/g whereas the "virtual" *EVV* for [^{3}H]ouabain was 0.47 ml/g. One of the advantages of Martín de Julián and Yudilevich analysis is the striking representation of the serial barriers to diffusion which are represented in the tracer dilution profiles. For example in Fig. 1, sucrose and ouabain (and dopamine and mannitol) have similar zero-time extractions (i.e. similar capillary permeability) whereas the slope of the lines in Fig. 1C and D are different. In fact, ouabain binds to Na-K-ATPase which is known to be an ectoenzyme in the sarcolemmal membrane and dopamine is taken up by nerve terminals. The Martín de Julián and Yudilevich analysis has also been applied to the study of capillary permeability and uptake for electrolytes, vitamin B-12, insulin, ouabain and catecholamines in the salivary gland (see Yudilevich and Mann, 1982).

II. PAIRED-TRACER DILUTION. The single circulation dilution technique was modified by Yudilevich and colleagues (see review by Yudilevich and Mann, 1982) to measure cell uptake in any organ. In this case an intravascular marker is not used and the reference tracer is a molecule which is confined to the extracellular compartment. Here the reference and test molecules should have equal capillary permeability and interstitial space distribution. Cellular uptake is estimated from the tracer concentrations in each venous sample (Table 1). It is interesting to point out that in organs such as the brain (tight junctions of the blood-brain barrier) it is the capillary endothelium that is the site of this saturable uptake, whereas in other tissues parenchymal cells membranes are the site of transport or binding represented in such data. Fig. 2 illustrates a paired-tracer experiment in the placenta.

Unidirectional influx and Michaelis-Menten kinetics. It can be seen in Fig. 2 that a plateau (or a maximal value) is observed in the uptake curves and this is used to estimate the maximal uptake, U_{max}. This value is used to estimate unidirectional influx, v (Table 1, Fig. 3). The logarithmic transformation of the uptake corrects for the exponential profile from the

TABLE 1.Equations used to estimate various parameters by means of single circulation tracer dilution techniques. The original multiple tracer method compares the concentrations of the test tracer with that of an intravascular reference tracer and graphs of i) extraction vs time, ii) *PS* vs flow, iii) Martín de Julián and Yudilevich equation (to obtain *k*) are fused to further analyze tracer dilution curves (Fig. 1). In what we call "paired-tracer dilution", the test tracer concentration is compared with an extracellular reference tracer (Fig. 2). A graph of uptake vs time, or vs the accumulated reference tracer, is essential to obtain a maximal uptake U_{max} and the backflux, *b*. A graph of *v* versus *C* leads to a Michaelis-Menten analysis (Fig. 3).

MULTIPLE TRACER DILUTION

a) EXTRACTION	$E = 1 - [test]/[intravascular]$
b) CAPILLARY PERMEABILITY *	$PS = -F \cdot ln[1 - E(0)]$
c) TURNOVER RATE OF EXTRAVASCULAR SPACE	k
d) EXTRACELLULAR VOLUME OF DISTRIBUTION	$EVV = E(0) \cdot F / k$

PAIRED-TRACER DILUTION

a) CELL UPTAKE	$U = 1 - [test]/[extracellular]$
b) UNIDIRECTIONAL INFLUX **	$v = -F \cdot ln(1-U_{max}) \cdot [C]$
c) MICHAELIS-MENTEN KINETICS	$v = V_{max} \cdot [C] / (K_m + [C])$
d) CELLULAR BACKFLUX ***	$b = 1 - U_{tot} / U_{max}$
e) TISSUE RETENTION	R = *total extracellular - total test*
f) LEAKAGE ****	L' = *100 - % extracellular recovered*
	L'' = *% extracellular other side cell*

* *P* is the permeability constant (cm/sec), *S* is the capillary surface area per gram of tissue and *F* is the rate of plasma flow (ml/min·g of tissue).

** *[C]* is the substrate concentration in plasma or artificial perfusate.

*** U_{tot} is the overall uptake calculated from the integrated recoveries of both tracers.

**** *L'* is based only on the injection side sampling. *L''* is estimated from sampling fluid across the cell barrier.

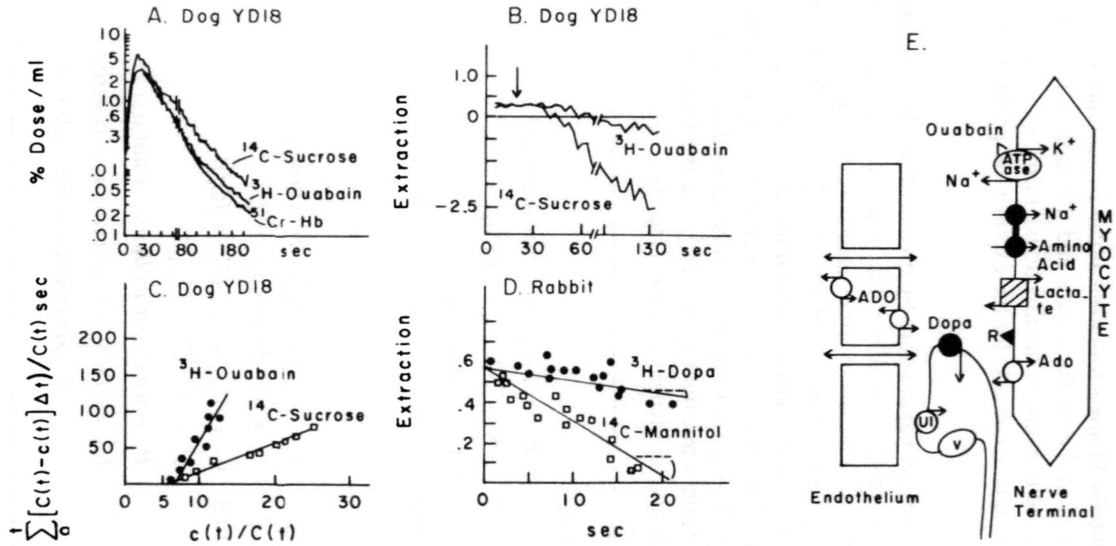

Fig. 1. Multiple tracer dilution curves in the perfused heart of the dog (A, B, C) and rabbit (D). A mixture containing three tracers was injected intra-arterially. **A**: Normalized venous concentrations of the intravascular (^{51}Cr-haemoglobin), extracellular (^{14}C-sucrose) and test (^{3}H-ouabain) tracers. **B**: Capillary extraction (*E*): $E = 1 - c(t)/C(t)$ where $C(t)$ is the extravascular marker, ^{51}Cr-Hb, and $c(t)$ is either of the other tracers; the arrow indicates the peak of the dilution curves. **C**: Martín de Julián and Yudilevich analysis of the same tracer data: the extravascular "virtual" space (*EVV*) is estimated from the turnover constant, *k* (reciprocal of the slope of the fitted line for the initial points); *E(0)*, is zero-time extrapolated extraction. **D**: Perfused rabbit heart. Following an intra-arterial bolus injection of [^{99}Tc(m)]albumin (vascular marker), D-[1-^{14}C]mannitol and [1,8-^{3}H]dopamine, coronary sinus effluent samples 2-3 per second) were collected. The plots is equivalent to panel C (Smith, 1985), and the dotted line gives the slope of the curve (*k*). **E**: Hypothetical model for carriers receptors and enzymes in the heart microcirculation (see text for other symbols). Panels A, B, and C taken from Yudilevich and Wheeler, 1988; panel D from Norman, 1988; panel E from Yudilevich, 1989.

arterial to venous end of the capillary and it is important only when cellular uptake is high. For lower values a simpler expression may be used: $v = F \cdot U_{max} \cdot C$. Influx measured at different concentrations of substrate in the perfusate appears to follow Michaelis-Menten saturation kinetics from which a half-maximal saturation, K_m, and a maximal influx rate, V_{max}, can be estimated (Fig. 3).

Tracer efflux. The test tracer may exit rapidly from the cell (into the blood) either as the unchanged substrate, or a labelled metabolite, and this is apparent as a downslope in the uptake curve (Fig. 2). Therefore, the comparison between U_{max} and U_{tot} (obtained from the total tracer recovery up to 6 min) is used as an index of backflux (Table 1). Very characteristic patterns of cellular efflux are observed for substrates in various experimental conditions and efflux can be stimulated or depressed. As a result of influx and efflux there is an amount of tracer "retained" in the tissue (Table 1).

An overview of the results in various organs is given below with the aim of illustrating the notable range of carrier systems which have been identified and characterized by the paired-tracer dilution method. Table 2 summarizes kinetic parameters for various substrates.

RESULTS

HEART (Lactate, noradrenaline). Carrier-mediated tissue uptake of many blood-borne substrates is measurable within the time of a single circulation through the heart (see review Yudilevich 1989). Figure 1, illustrates a multiple tracer experiment and a hypothetical model. Tracer molecules diffuse across the intercellular junctions of the endothelium (arrow) and this "passive capillary permeability" is represented by the tracer extraction. In the interstitium substrates such as noradrenaline may be secreted by nerve terminal vesicles (v) and rapid reuptake occurs via system U1. Ouabain binds to Na-K-ATPase. Amino acids (some co-transported with Na^+), lactate and other compounds have specific protein carrier molecules in the sarcolemma. Receptors (R) such as muscarinic cholinergic receptors have also been kinetically characterized by paired-tracer dilution (Chaumet-Riffaud et al., 1984). Adenosine (ADO) is taken up by the endothelial cell and the myocyte.

In Figure 1B the fact that [^{3}H]ouabain (MW 585) and [^{14}C]sucrose (MW 342) exhibit similar initial *E* values, about 0.35, suggests that both molecules are passively transported across the capillary endothelium, i.e. that there is no binding of ouabain to the endothelium. Norman (1988) showed that dopamine is not transported by the endothelium since its capillary extraction E(0) was similar to that of mannitol (Fig. 1D) and neuronal uptake was demonstrated by inhibition of [^{3}H]dopamine uptake induced by specific blockers (in experiments such as Fig. 1D the slope of [^{3}H]dopamine extraction curve approached that of [^{14}C]mannitol). Wangler et al. (1989), in the guinea-pig heart, found an initial extraction of [^{3}H]adenosine higher than that for a reference extracellular molecule and this suggested the presence of a transport mechanism for adenosine at the endothelium.

In studies with noradrenaline in the rabbit heart endothelial cell uptake appeared to be negligible whilst the kinetic (Table 2) and specificity of inhibitors demonstrated the presence of neuronal UPTAKE 1 system (Mann and Yudilevich, 1984b). In further paired-tracer experiments an specific system for monocarboxylate substrates was suggested (see lactate kinetics in Table 2) (Mann et al., 1985). Glucagon inhibited the sarcolemmal uptake of lactate, pyruvate and acetate.

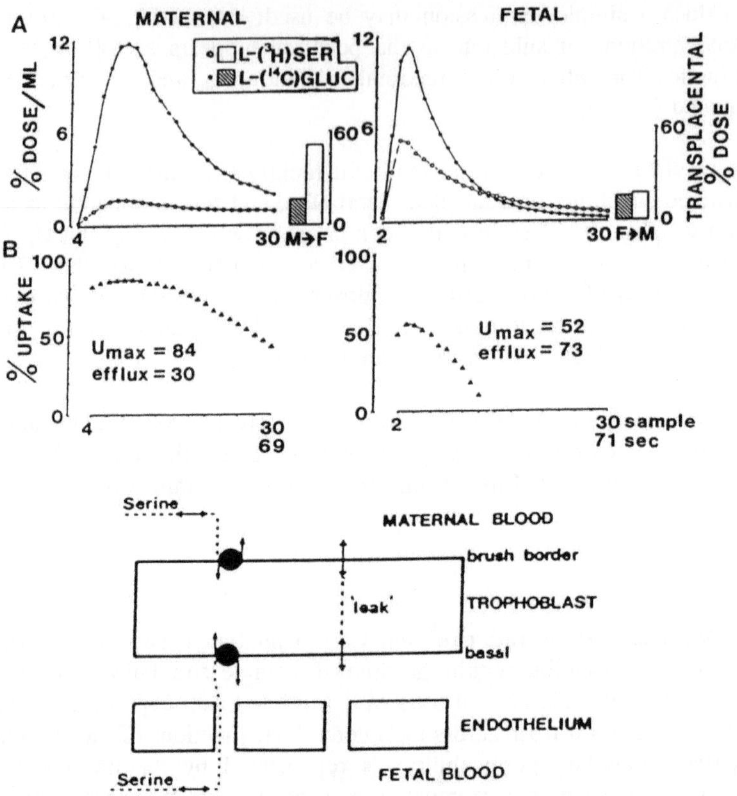

Fig. 2. The paired-tracer dilution technique for the characterization of transport at the maternal (brush border) and fetal (basal) sides of the placenta. A mixture of L-[^{3}H]serine and an extracellular marker (L-[^{14}C]glucose) was injected into either the maternal or fetal arterial inflow of the isolated dually-perfused guinea-pig placenta. A bolus injection (100 µl in 1-2 sec) was immediately followed by rapid sampling of the ipsilateral side venous effluent: 30 samples in about 70 s plus a final 4 min accumulative collection. **Upper panels**: Venous dilution profiles for L-[^{3}H]serine and L-[^{14}C]glucose; tracer concentrations are normalized with respect to the injected dose. **Lower panels**: L-[^{3}H]serine uptake patterns. Uptake of L-serine is calculated for each sample: U = {1 - L-[^{3}H]serine/L-[^{14}C]glucose} x 100. Maximal plateau values (U_{max}), shown by the joined points, were followed by a downfall related to tracer backflux ("efflux", %, in the Figure). The unidirectional influx, v, is estimated from this parameter (see text). Transplacental transfer (% dose) of both tracers was assessed in a single 6 min sample collected from the contralateral circulation (columns in A panels). Movement of L-[^{14}C]glucose represents a non-specific pathway and is termed "leakage". Placental retention (*R*) of label may be calculated from the total tracers recoveries in both circulations (See Table 1). **Lower Figure.** A diagrammatic representation of the term hemomonochorial placental barrier (human and guinea pig). The figure illustrates the ability of the paired-tracer method to characterize transport systems (filled circles with arrows) at either plasma membrane (brush border or basal) of the trophoblast. The endothelium is here assumed to be a passive barrier (for both tracers equally) with negligible uptake of serine in a single circulation. Data taken from Yudilevich and Wheeler (1988).

PLACENTA (Amino acids, glucose, calcium, choline, folate, adenosine, nucleosides). In order to understand the mechanisms involved in the transfer of substrates from the maternal circulation to the fetal circulation we have studied both blood-tissue interfaces by means of the paired-tracer method (Fig. 2; Yudilevich et al. 1979; see review Yudilevich and Wheeler, 1988). The haemomonochorial placenta (human and guinea-pig) has a single syncytial trophoblast separating maternal from fetal blood (Fig. 2). This is a particularly interesting organ for such studies and they have been performed in the guinea-pig placenta either *in situ* (fetal side perfusion) or in the isolated dually-perfused placenta. The results show that both sides of the trophoblast exhibit specific transport mechanisms (Figs. 2-5) and saturation kinetics for unidirectional influx of various substrates is summarized in Table 2.

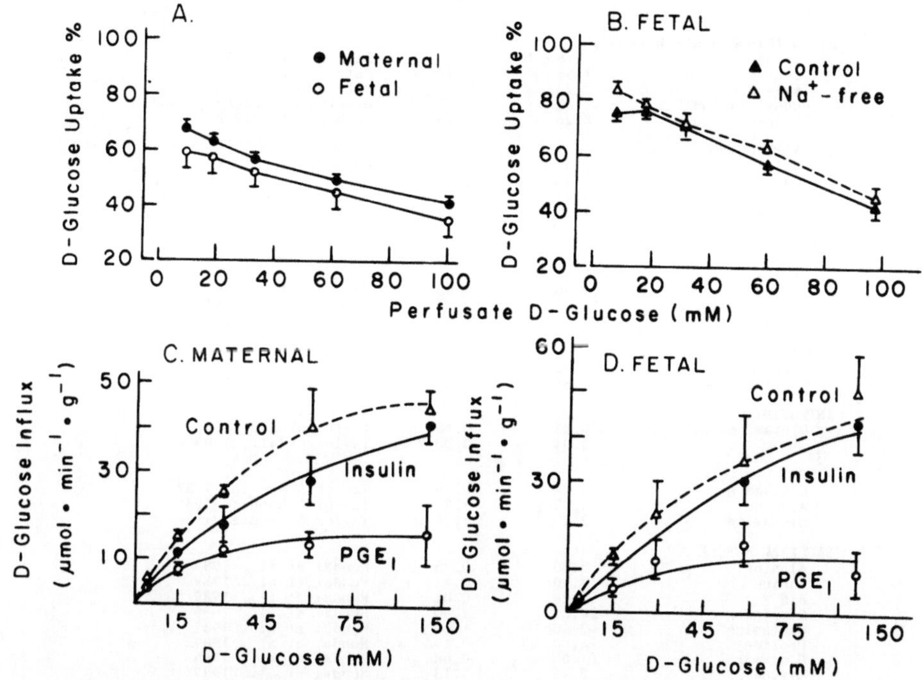

Fig. 3. Concentration dependence of unidirectional D-glucose uptake. A: Dually-perfused placentas (DPP) were perfused successively with various concentrations of unlabelled D-glucose (5.5-100 mM) and a tracer bolus was injected into either the maternal or fetal artery. **B**: The effect of extracellular sodium removal (Na-Free perfusate) on D-glucose uptake was tested in the in situ singly perfused placenta (SPP). **C,D.** The effect of insulin and prostaglandin E_1 (PGE_1) on unidirectional influx across the brush-border (maternal) and basal (fetal) membranes. Kinetic experiments were performed in DPP with perfusates containing either 150 µU/ml porcine insulin or 1 µM PGE_1. The uppermost curve in each panel represents the control influx data derived from the uptake values presented in A. Data from Wheeler and Yudilevich (1988)

TABLE 2. Kinetic parameters measured with the paired-tracer method for a range of substrates in various organs.

	Km (mM)	Vmax (µmol/min·g)	Reference
HEART			
Lactate	19.00	8.40	Mann et al., 1985
Noradrenaline	0.002	0.002	Mann & Yudilevich, 1984
PLACENTA			
Brush-border			
Alanine	10.30	14.30	Eaton et al., 1982
Phenylalanine	3.10	3.70	Eaton et al., 1982
Lysine	1.70	1.90	Wheeler & Yudilevich, 1989b
Glucose	70.00	53.00	Wheeler & Yudilevich, 1989a
Calcium	0.18-1.15	0.12-0.59	Sweiry & Yudilevich, 1984
Choline	0.12	0.08	Sweiry & Yudilevich, 1985
Basal			
Alanine	8.40	4.40	Eaton et al., 1982
Phenylalanine	17.00	8.20	Yudilevich & Eaton, 1980
Lysine	0.90	0.90	Wheeler & Yudilevich, 1989b
Glucose	87.00	82.00	Wheeler & Yudilevich, 1989a
Calcium	1.80	1.66	Sweiry & Yudilevich, 1984
Choline	0.13	0.07	Sweiry & Yudilevich, 1985
BRAIN (Blood-brain barrier)			
Valine	1.76		Betz et al., 1975
Leucine	1.58	0.32	Betz et al., 1975
Isoleucine	0.73		Betz et al., 1975
Phenylalanine	0.85	0.13	Brenton & Gardiner, 1988
Glucose	8.26	1.75	Betz et al., 1973
SALIVARY GLAND			
Lysine	0.49	0.231	Mann et al., 1984
Alanine	0.83	0.655	Mann & Yudilevich, 1984a
Phenylalanine	6.40	1.71	Bustamante et al., 1981
Serine	1.20	0.99	Mann & Yudilevich, 1987
PANCREAS			
MeAIB	1.70	0.49	Norman & Mann, 1986
Serine	14.40	25.90	Norman & Mann, 1989
Glutamine	14.00	13.40	Mann et al., 1986
Phenylalanine	16.00	12.30	Mann & Peran, 1986
Lysine	2.40	2.44	Muñoz et al., 1988
STOMACH			
Alanine	3.92	0.29	Bravo et al., 1988
Leucine	52.86	4.60	Bravo et al., 1988
INTESTINE			
Glutamine (a)	0.88	0.45	Taylor et al., 1989
(b)	0.01	0.001	Taylor et al., 1989
LIVER			
Glutamate	0.29	0.18	Taylor & Rennie, 1987
Alanine	4.00	3.29	Taylor & Rennie, 1988
Glutamine	2.25	3.38	Taylor & Rennie, 1988
SKELETAL MUSCLE			
Alanine	3.90	0.33	Hundal et al., 1989
Serine	3.40	0.41	Hundal et al., 1989
AIB	1.26	0.01	Hundal et al., 1989
Leucine	20.00	2.80	Hundal et al., 1989
Glutamine	7.10	1.02	Hundal et al., 1987
Proline	4.10	0.196	Hundal et al., 1989
Glutamate	1.05	0.086	Hundal et al., 1989
Lysine	2.13	0.13	Hundal et al., 1989
Lactate	21.00	0.84	Watt et al., 1988
Glucose	0.07	0.064	Rennie et al., 1983
LUNG			
PGE_2	0.002	0.012	Syrota et al., 1982
$PGF_{2\alpha}$	0.005	0.004	Hellewell & Pearson, 1983
Adenosine	0.03	0.018	Hellewell & Pearson, 1983
TESTIS			
Leucine	0.06	0.019	Bustamante & Setchell, 1982
ISOLATED CELLS			
VERO CELLS			
Phenylalanine	3.80		Perán & McGee, 1986
HUVEC (c)			
Leucine	0.24	0.027	Mann et al., 1989
BAEC (d)			
Arginine	0.20	0.008	Mann et al., 1990

(a) first component; (b) second component; (c) HUVEC: human umbilical vein endothelial cells; (d) BAEC: bovine aortic endothelial cells. AIB, aminoisobutyric acid; MeAIB, 2-(methyl-amino)isobutyric acid. PG, prostaglandin. All natural amino acids are of the L-configuration.

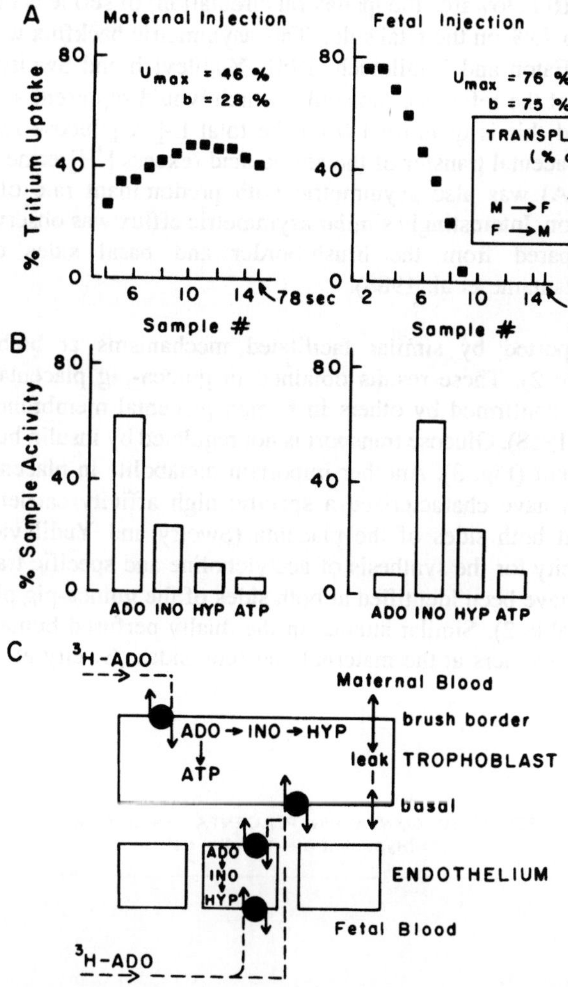

Fig. 4. Adenosine uptake and metabolism at both maternal and fetal sides of the dually-perfused guinea-pig placenta. Perfusates contained 100 μM unlabelled adenosine. **A:** Uptake of [^{3}H]adenosine at maternal and fetal blood-tissue interfaces following an arterial bolus injection of [^{3}H]adenosine and [^{14}C]sucrose (extracellular marker). **B:** A sample comprised of aliquots taken from each of the ipsilateral venous outflow samples collected over a 70-80 s period, was subjected to thin-layer chromatographic analysis and tritiated compounds were identified (ADE, adenosine; INO, inosine; HYP, hypoxantine; ATP, this and other phosphorylated products). Transplacental ^{14}C refers to the [^{14}C]sucrose "leak" and ^{3}H to tritiated compounds. **C:** A model for placental transport and metabolism of adenosine in the guinea pig placenta. At the brush border adenosine carriers mediate the influx and efflux of [^{3}H]adenosine. Following intracellular metabolism to inosine (adenosine deaminase), hypoxanthine (xanthine oxidase) and ATP (adenosine kinase), labelled metabolites also exit the trophoblast at either surface. At the fetal blood-tissue interface similar events may also occur in the endothelium. An additional route for movement of molecules is represented by the leak pathway (see Fig. 2). Data taken from Wheeler and Yudilevich (1988).

Tracer efflux from the cell into the circulation (Table 1) which is general for blood-tissue transport studies, is well illustrated by the results in the placenta. Fig.2B shows that the backflux (tracer efflux towards the ipsilateral circulation) of serine is only 30% on the maternal side compared to 73% on the fetal side. This asymmetric backflux was observed for most other amino acids (Eaton and Yudilevich, 1981; Yudilevich and Sweiry, 1985). There was also a rapid transfer of the reference molecule which would represent a "leak" pathway across the trophoblast and this is quantified from the total L-[^{14}C]glucose recovered in 5-6 min). The specific transplacental transfer of the amino acid (excess [^{3}H]serine with respect to L-[^{14}C]glucose in Fig. 2A) was also asymmetric with predominant rate of transfer from maternal-to-fetal circulation. Interestingly similar asymmetric efflux was observed in placental membrane vesicles prepared from the brush-border and basal sides of the human syncytiotrophoblast (Bustamante et al. 1988).

Glucose is transported by similar facilitated mechanisms at both sides of the trophoblast (Fig. 3, Table 2). These results obtained in guinea-pig placenta (Wheeler and Yudilevich, 1989a) were confirmed by others in human placental membranes vesicles (see Yudilevich and Wheeler, 1988). Glucose transport is not regulated by insulin but prostaglandin E_1 inhibits glucose transport (Fig. 3). Another important metabolite in placenta is folate and paired-tracer experiments have characterized a specific high affinity carrier system which appeared to be similar at both sides of the placenta (Sweiry and Yudilevich, 1988). The placenta has a high capacity for the synthesis of acetylcholine and specific transport systems for choline, a precursor, have been identified at both sides of the guinea-pig placenta (Sweiry and Yudilevich, 1985; Table 2). Similar studies in the dually-perfused human placenta also demonstrated choline transporters at the maternal and fetal sides (Sweiry et al. 1986).

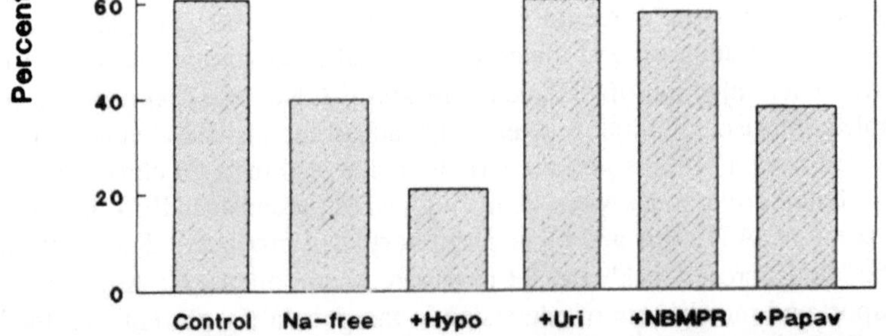

Fig. 5. Uptake of hypoxanthine at the fetal side of the guinea-pig placenta. [^{3}H]hypoxanthine and D-[^{14}C]sucrose were injected as a single bolus into the umbilical artery and uptake was measured from the umbilical venous outflow samples. The data shows significant Na^+-dependence, self-inhibition (4mM) and inhibition by papaverine (papav, 19mM) added to the perfusate; in contrast, uridine (uri, 10mM) and NBMPR (10μM) were ineffective inhibitors. Unpublished results, Barros and Yudilevich.

Adenosine and other nucleosides appeared to be transported at both maternal and fetal poles of the trophoblast (high tritium uptake in Fig. 4A) and the presence of a nucleoside transporter was demonstrated by means of specific inhibitors (Wheeler and Yudilevich, 1988). Nucleoside transporters have just been characterized at both sides of the human placenta (Barros et al., 1990). Tritium backflux (Fig. 4A) was much higher at the fetal (75%) than at the maternal (28%) side. Rapid metabolism of adenosine was demonstrated by the presence of labelled metabolites in the venous effluent (Fig. 4B). The higher rate of metabolism seen at the fetal blood-tissue interface (90 % of the tritium was associated with metabolites) led to the suggestion that the endothelium could play an important role in the regulation of adenosine levels in the fetal blood (Wheeler and Yudilevich, 1988). Recently we have shown in the perfused guinea-pig placenta that a separate transporter appears to exist which is specific for hypoxanthine and purine bases (Fig. 5, unpublished results Barros and Yudilevich).

BLOOD-BRAIN BARRIER (Amino acids, glucose, insulin). The brain microcirculation is the simplest for studies with the multiple paired-tracer technique, insofar as, the blood-tissue interface at which carrier mediated exchange occurs is the endothelium, characterized by tight interendothelial junctions. This was demonstrated by the early observations of Crone (1963, 1965) and Yudilevich et al. (1972) for glucose and amino acid respectively. In the anaesthetized dog, an intracarotid tracer injection was followed by sample collection from the saggital sinus. In further studies, in the isolated perfused head, the kinetic of glucose (Betz et al., 1973) and amino acid transport (Betz et al., 1975) was investigated. Betz et al. (1973) found that insulin had no effect on glucose transport across the blood-brain barrier. Even though these measurements pioneered the paired-tracer method (as described in Table 1) to estimate kinetics in perfused organs, the data in the brain is scarce compared to that in other organs (see Table 2).

SALIVARY GLAND (Amino acids). Unidirectional uptake of 21 amino acids, dopamine, noradrenaline and serotonin has been measured in the perfused cat submandibular gland (Figure 6, see review Mann and Yudilevich, 1987). Kinetic parameters for lysine, alanine, serine and phenylalanine are given in Table 2. The poor cross-inhibition among these amino acids and the pattern of Na^+-dependence suggested the presence of Systems y^+, ASC and L at the basolateral side of the epithelium (Bustamante et al., 1981; Mann and Yudilevich, 1984a; Mann et al., 1984). The absence of uptake for α-(methylamino)isobutyric acid (MeAIB), a model analogue for the insulin-sensitive System A suggested that this broadly distributed amino acid transport system is not present in this epithelium. Perfusion of glands with the metabolic inhibitor 2,4-dinitrophenol selectively inhibited unidirectional influx and stimulated the efflux of large neutral amino acids and lysine.

PANCREAS (Amino acids, insulin). In the perfused rat pancreas paired-tracer experiments allowed characterization of amino acid transport systems at the basolateral side of the exocrine epithelium (Fig. 6; Table 2). Contrary to the salivary gland MeAIB uptake was high, Na^+-dependent and cross-inhibition experiments with a large number of amino acids suggested the presence of System A. Systems ASC, asc and L were also demonstrated. When the regulatory effects of insulin, somatostatin and cholecystokinin on amino acid transport were studied (Mann and Norman, 1984; Norman et al. 1989), somatostatin was found to modulate the stimulatory action of insulin. As illustrated in Fig. 7, the kinetic studies of L-serine influx show that insulin modulates the activity of System asc and this effect was shown to be calcium dependent (Norman and Mann, 1987). Interestingly influx of serine was increased in pancreata of diabetic rats and this effect was abolished by the addition of insulin to the perfusate (Norman and Mann, 1984). Influx of system A amino acids appeared to be inhibited by the ionophore monensin but was unaffected by low potassium or calcium

(Norman and Mann, 1987). Efflux of labelled L-phenylalanine, or L-alanine, was reduced in the presence of a competitor (in the injectate) and this was interpreted as cis-inhibition. Moreover tracer efflux was shown to be accelerated in a concentration dependent manner, suggesting trans-stimulation (Mann and Perán, 1986).

STOMACH (Amino acids). The kinetic of unidirectional uptake of L-alanine and L-leucine was measured in the perfused dog stomach by means of the paired-tracer technique (Table 2; Bravo et al., 1988). High uptake values were reported for aspartic acid, but lysine was taken up to a very small extent. The maximal cellular uptake and the unidirectional influx

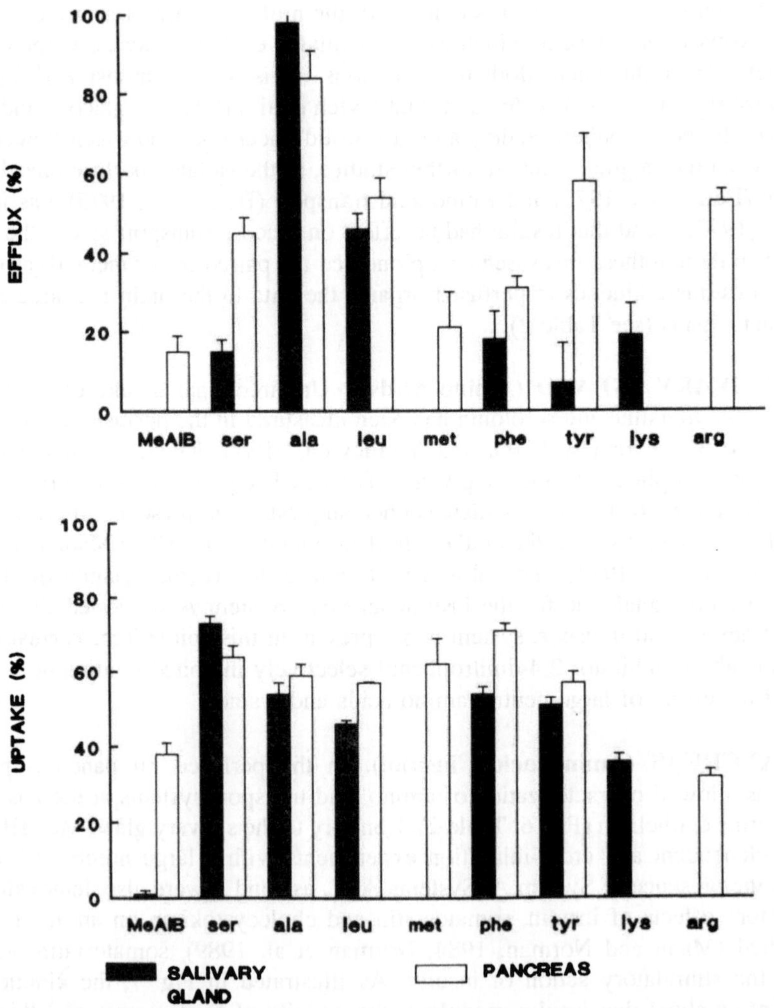

Fig. 6. Comparison of unidirectional amino acid influx and tracer efflux in the isolated perfused cat salivary gland and rat exocrine pancreas. Measurements by paired-tracer dilution. Values are mean ± SEM. Data taken from Mann & Yudilevich (1984a), for salivary epithelium, and Mann & Perán (1986), for the pancreas.

of the amino acids were increased when gastric secretion was stimulated with histamine. These results suggest that the common transport systems available to amino acids may operate at the basolateral membrane of the gastric epithelium and they have recently been compared with observations made in rabbit isolated oxyntic glands (Bravo and Sobrevía, in press).

SKELETAL MUSCLE (Amino acids, glucose, insulin). The paired-tracer isotope dilution technique has been employed in a perfused rat hindlimb preparation to obtain detailed kinetic characterization of acidic, neutral and basic amino acids transport across the sarcolemmal membrane (Table 2; Hundal et al., 1989). Competitive and non-

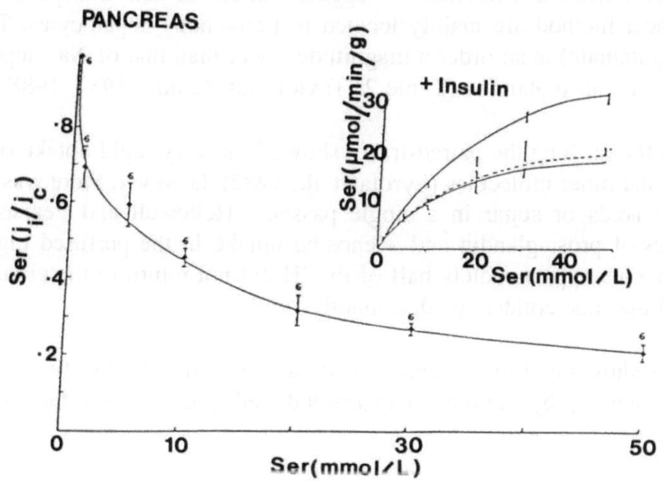

Fig. 7. Stimulation of L-serine transport by exogenous insulin in the isolated rat exocrine pancreas. L-serine influx was measured at increasing perfusate concentrations of unlabelled L-serine in the absence or presence of 100 μU/ml bovine insulin. The inset shows kinetics of serine transport in the absence or presence of 16.8 mM D-glucose (dotted curve) or 100 μU/ml insulin (solid upper curve). Transport was paradoxically stimulated only by exogenous insulin. Taken from Norman et al. (1989).

competitive interactions between substrates suggested that alanine entry appeared to be mediated via the ASC- and L-systems with the A-system playing a minor role. Glutamate and aspartate appear to be transported exclusively by a dicarboxylate amino acid carrier (pH dependent). Insulin had not detectable effect on amino acids substrates for ASC, L, y^+, dicarboxylic or glycine transport systems (Hundal et al., 1989). Transport of glutamine in skeletal muscle appeared to occur via a variant of System N (Hundal et al., 1987). This system is shared by histidine and asparagine, and is both Na^+-dependent and insulin sensitive. A thorough discussion of these results in skeletal muscle in relation to metabolism was made by Christensen (1990).

In kinetic studies on unidirectional glucose influx (Table 2) Rennie et al.(1983) showed that V_{max} was increased almost threefold when insulin was added to the hindlimb perfusate.

Watt et al. (1988) suggested that at least two processes are involved in lactate influx into the sarcolemma in muscle of the perfused skinned rat hindlimb (Table 2). Reducing pH increased the unidirectional lactate transport.

OTHER ORGANS

Intestine. Transport kinetics at the basolateral side of the intestinal epithelium of glutamine and other neutral amino acids was studied in the isolated dually-perfused rat jejunum (Table 2). Glutamine transport was Na^+-independent and appeared to be transported by System L, insensitive to insulin and glucagon, and unaffected by perfusate pH (Taylor et al., 1989).

Liver. Evidence was presented to suggest that amino acid transporters characterized by the paired-tracer method are mainly located in perivenous hepatocytes. The capacity of System X^- (for glutamate) is an order a magnitude lower than that of Na^+-dependent systems transporting alanine and glutamine (Table 2; Taylor and Rennie, 1987, 1988).

Lung. In the rat lung the paired-tracer showed the very rapid uptake of prostaglandin PGE_1, serotonin and other molecules (Syrota et al., 1982), however, there was no measurable uptake of amino acids or sugar in a single passage. Hellewell and Pearson (1983) have measured kinetics of prostaglandin and adenosine uptake in the perfused piglet lung. After injection of adenosine approximately half of the ^3H did not return to the circulation and was found in intracellular nucleotides, predominantly ATP.

Testis. As shown in Table 2, experiments in the perfused rat testis have demonstrated that L-leucine is taken up by saturable, carrier-mediated processes (Bustamante and Setchell, 1982).

Mammary gland. A brief communication reported uptake of phenylalanine and threonine in isolated-perfused mammary gland of the guinea-pig (Mepham et al., 1984).

Choroid plexus. The permeability of the blood-cerebrospinal fluid barrier to [^3H]thyrotropin-releasing hormone (TRH) was studied in the isolated perfused choroid plexus of the sheep. [^3H]TRH uptake was markedly reduced by proline and by the inhibitor of amidase and aminopeptidase activity, bacitracin (Zlokovic et al., 1985).

CULTURED CELLS (Vero, endothelial, fibroblasts, carcinoma: amino acids). Microcarrier beads with confluent cells were perfused in small columns, and uptake and efflux of a labelled substrates were assessed during a single transit through the column. In the original application of the paired-tracer dilution technique to perfused cultures (Vero cells) unidirectional influx of phenylalanine was studied (Table 2; Perán and McGee, 1986). More recently the kinetics of leucine (Mann et al., 1989) and arginine (Mann et al., 1990) transport were studied in human umbilical vein and bovine aortic endothelial cells (Table 2). It can be seen in Fig. 8 that the pattern of dilution curves and uptake are identical to those in perfused organ (eg. placenta in Fig. 2). At tracer concentration in the perfusate (and injectate), significant U_{max} values were found for a range of amino acids, whereas uptake for the system A analogue MeAIB uptake was negligible (Fig. 8). L-serine influx was markedly dependent upon extracellular Na^+ (low Na^+ in perfusate) and it was transported via system ASC and not a Na^+-independent system asc as in the exocrine pancreas (Mann and Perán, 1986). The uptake of glucose in the endothelial cells was low and unaffected by 30 min perfusion with insulin

(Mann et al., 1989). Influx and efflux of serine and phenylalanine were high in human fibroblasts and human colon tumour cells (Perán et al., 1990).

CONCLUSIONS

We have here reviewed kinetic studies made with the "paired-tracer" technique (Yudilevich and Mann, 1982) and summarize the data available in the literature in Table 2. It can be seen that the main group of substrates studied are amino acids and that there is great scope for further research on these as well as many other important substrates, in different organs. Christensen's reviews (Christensen 1982, 1990) on amino acid transport have

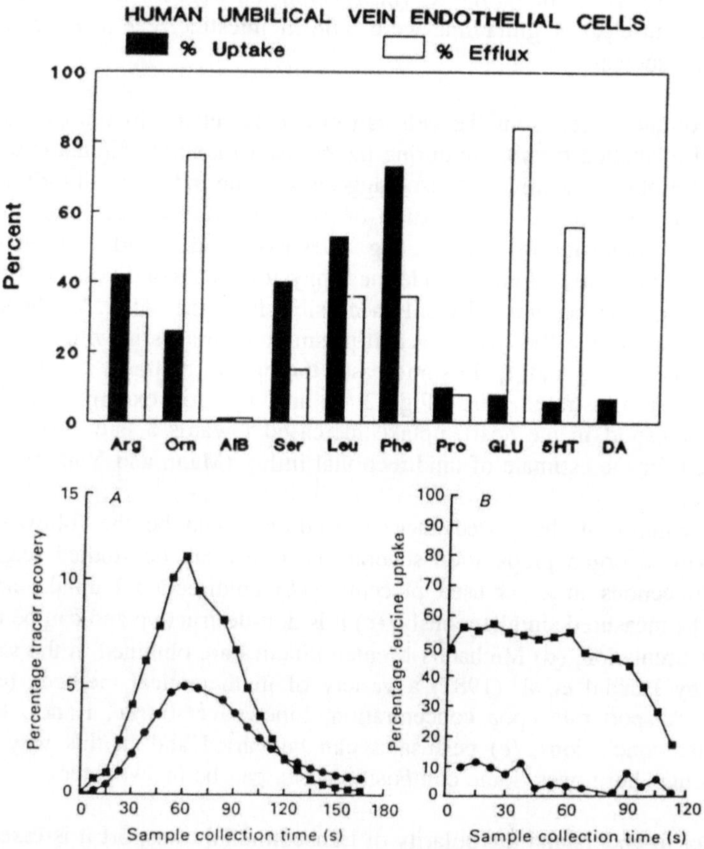

Fig. 8. Unidirectional amino acid uptake and cellular efflux in perfused microcarrier cultures of vascular endothelial cells. Endothelial cell microcarrier columns were perfused at 0.3 ml/min with a Krebs-Henseleit bicarbonate medium. The lower panel illustrates a paired-tracer dilution profile for L-[^{3}H]leucine and D-[^{14}C]mannitol in one experiment. The inhibitory effect of 1mM L-phenylalanine on L-^{3}H]leucine uptake is illustrated in B. The upper panel shows the uptake and efflux of various amino acids, 5-hydroxytryptamine (5-HT) and dopamine (DA). Data taken from Fig. 1 and Table 1 in Mann et al. (1989)

highlighted the importance of interorgan substrate transfer which involves influx and efflux across the plasma membrane of cells facing the blood. He then emphasized the value of in vivo procedures and discussed results obtained with the paired-tracer procedure.

There are generally striking differences in individual organs with respect to the specificity and control mechanisms for transport systems. For example, the luminal side of the brain endothelium does not have the widely distributed system A or ASC for short-chain neutral amino acids (Yudilevich et al., 1972; Christensen, 1982). In the pancreas there is a high uptake of MeAIB (specific for system A) which on the contrary is not transported in the salivary gland (Fig. 6). In the blood-brain barrier insulin was not effective on glucose transport (Betz et al., 1973), whereas experiments in the perfused hindlimb have confirmed that insulin stimulates glucose transport in perfused muscle (Rennie et al., 1983). Insulin was ineffective in modulating glucose transport in the placenta (Fig. 3). Insulin regulation of serine uptake in the perfused exocrine pancreas has been demonstrated in both normal and diabetic animals (Mann and Norman, 1984). Table 2 shows that the affinity constant of certain substrates may vary in different organs. In the rat the K_m (mM) estimated were for leucine 0.06 in testis and 20.0 in skeletal muscle; for glutamine were 0.88 in intestine, 2.2 in liver, 7.1 in skeletal muscle, 14.0 in pancreas.

Efflux of the tracer from the cell can either be related to the exit of the original substance or of a labelled metabolite during the course of a single circulation. This can be a methodological problem, however, it also suggests that the indicator dilution technique may provide a unique tool for the investigation of rapid metabolism by intra- or extracellular enzymes (e.g. for adenosine in placenta, Fig. 4; Barros et al., 1990). The venous sampling technique enables the study transport under near physiological condition, confining the tracer exchange to barrier surfaces normally only accessible from the blood. We have discussed in the Introduction and in the Results the actual plasma membranes in which these phenomena may be located (Fig. 1, 2 and 4). In some cases the pattern of tracer uptake is not as "well behaved" as those illustrated here (Fig. 2, 4 and 8). For example in the studies of noradrenaline transport in the heart, uptake increased towards a late maximal value, U_{max}, which was used for the estimate of unidirectional influx (Mann and Yudilevich, 1984b).

Other features of the paired-tracer procedure would be the following: (*a*) in the artificially perfused organ preparation several substrates can be studied (e.g. up to 10-15 paired-tracer injections in a perfused placenta), (*b*) unidirectional uptake and also tracer backflux, can be measured simultaneously, (*c*) it is non-destructive and can be used in organs with an intact circulation, **(d)** Michaelis-Menten kinetics are obtained in the same organ and as discussed by Hundal et al. (1987) a variety of mathematical methods (e.g. non-lineal regression of transport rate upon concentration, Lineweaver-Burke, Hanes, Eadie-Hofstee) produce similar conclusions, (*e*) perfusates can be varied and in this way the effect on transport of drugs, hormones, ionic composition, etc. can be investigated.

In order to understand the polarity of transepithelial transport it is essential to gather information on the characteristics and regulation of influx and efflux across both poles of epithelia (Boyd and Yudilevich, 1987). The paired-tracer method allows characterization of transport phenomena at the basal side of epithelia which generally has been less studied than the apical or brush-border side. This approach has been the most powerful means for investigating carrier mediated transport processes at the blood-tissue interface of many organs in a manner comparable to other in vitro methods such as tissue slices, isolated cells, membrane vesicles. An advantage is that in the perfused organ normal cell polarity is maintained, and therefore the blood-side of the cells can be investigated selectively.

Furthermore, the cytoplasmic machinery is intact and this should be important for the study of the regulatory mechanisms involved in the transport processes. In addition it has been demonstrated that transport systems may be greatly modified or even disappear during the preparation of isolated membranes or in cultured cells (Christensen, 1982, 1990). This last point must be kept in mind when using the new application of the paired-tracer method to the study of transport in perfused columns of cultured cells (Fig 8; see Chapter 9 by Peran et al. and Chapter 23 by Mann et al.).

ACKNOWLEDGEMENTS

Our present research is generously supported by research Grants to DLY from the British Diabetic Association and from the Wellcome Trust, U.K. We thank Dr G.E. Mann for his discussions of this paper and Mr Richard Bucknall for his help in preparing the manuscript.

REFERENCES

Barros, L.F., Bustamante, J.C., Yudilevich, D.L. & Jarvis, S.M. 1990. Adenosine transport and nitrobenzylthioinosine binding in human placental membrane vesicles from brush-border and basal sides of the trophoblast. J. Membr. Biol. 117, in press.

Betz, A.L., Gilboe, D.D., Yudilevich, D.L. & Drewes, L.R. 1973. Kinetics of unidirectional glucose transport into the isolated dog brain. Am. J. Physiol. 225: 586-592.

Betz, A.L., Gilboe, D.D. & Drewes, L.R. 1975. Kinetics of unidirectional leucine transport into brain: effects of isoleucine, valine, and anoxia. Am. J. Physiol. 228: 895-900.

Boyd, C.A.R. & Yudilevich, D.L. 1987. Blood-tissue movement of amino acids in various organs: the basolateral membrane and its importance in the polarity of transepithelial transport. In: *Amino Acid Transport in Animal Cells.* Eds. D.L. Yudilevich and C.A.R. Boyd, pp. 120-156, Manchester University Press, Manchester

Bravo, I. & Sobrevía, L.. Kinetics of the L-alanine transport through the basolateral membrane in isolated oxyntic glands. Biochim. Biophys. Acta (in press).

Bravo, I., Fuentes, O., Pozo, M. & Rojas, S. 1988. Kinetics of L-alanine and L-leucine transport at the blood-tissue interface in the perfused stomach. Cell. Mol. Biol. 34: 323-336.

Brenton, D.P. & Gardiner, R.M. 1988. Transport of L-phenylalanine and related amino acids at the ovine blood-brain barrier. J. Physiol. 402: 497-508.

Bustamante, J.C. & Setchell, B.P. 1982. Kinetics of the facilitated diffusion of leucine into the perfused testis of the rat. J. Physiol. 324: 49P.

Bustamante, J.C., Mann, G.E. & Yudilevich, D.L. 1981. Specificity of amino acid uptake at the basolateral side of the epithelium in the cat salivary gland *in situ*. J.Physiol 313: 65-79

Bustamante, J.C., Yudilevich, D.L. & Boyd, C.A.R. 1988. A new form of asymmetry in epithelia: Kinetics of apical and basal sulphate transport in human placenta. Q. J. Exper. Physiol. 73: 1013-1016

Chaumet-Riffaud, Ph., Girault, M. & Syrota, A. (1984) Characterization of muscarinic cholinergic receptors in the isolated perfused rat heart. J. Physiol. 348: 11P

Crone, C. 1963. The permeability of capillaries in various organs as determined by use of the indicator diffusion method'. Acta physiol. scand. 58: 292-305.

Crone, C. 1965. Facilitated transfer of glucose from blood into brain tissue. J. Physiol. 181: 103-113.

Christensen, H.N. 1990. Role of amino acid transport and countertransport in nutrition and metabolism. Physiol. Rev. 70: 43-76

Christensen, H.N. 1982. Interorgan amino acid nutrition. Physiol. Rev. 62: 1193-1233
Eaton, B.M. & Yudilevich, D.L. 1981. Uptake and asymmetric efflux of amino acids at maternal and fetal sides of placenta. Am. J. Physiol. 241: C106-112.
Eaton, B.M., Mann, G.E. & Yudilevich, D.L. 1982. Transport specificity for neutral and basic amino acids at maternal and fetal interfaces of the guinea-pig placenta. J.Physiol. 328: 245-258.
Haselton, F.R., Mueller, S.N., Howell, E.M., Levine, E.M. & Fishman, A.P. 1989. Chromatographic demonstration of reversible changes in endothelial permeability. J. Appl. Physiol. 67: 2032-2048.
Hellewell, P.G. & Pearson, J.D. 1983 Kinetic analysis of prostaglandin F2 alpha and adenosine uptake in the piglet pulmonary circulation. J. Physiol. 343: 48P.
Hundal, H.S., Rennie, M.J. & Watt, P.W. 1987. Characteristics of L-glutamine transport in perfused rat skeletal muscle. J. Physiol. 393: 283-305.
Hundal, H.S., Rennie, M.J. & Watt, P.W. 1989. Characteristics of acidic, basic and neutral amino acid transport in the perfused rat hindlimb. J. Physiol. 408: 93-114.
Mann, G.E. & Norman, P.S.R. 1984. Regulatory effects of insulin and experimental diabetes on neutral amino acid transport in the perfused rat exocrine pancreas: Kinetics of unidirectional L-serine influx and efflux at the basolateral plasma membrane. Biochim. Biophys. Acta 778: 618-622.
Mann, G.E. & Perán, S. 1986. Basolateral amino acid transport systems in the perfused exocrine pancreas: sodium-dependency and kinetic interactions between influx and efflux mechanisms. Biochim. Biophys. Acta. 858: 263-274.
Mann, G.E. & Yudilevich, D.L. 1984a. Discrimination of parallel neutral amino acid transport systems in the basolateral membrane of cat salivary epithelium. J. Physiol. 347: 111-127.
Mann, G.E. & Yudilevich, D.L. 1984b. Rapid transcapillary exchange and unidirectional neuronal uptake of noradrenaline in the perfused rabbit heart. J. Physiol. 348: 589-600.
Mann, G.E. and Yudilevich, D.L. 1987. Regulation of amino acid influx and efflux at the basolateral plasma membrane of the salivary epithelium: effects of parasympathetic nerve stimulation. J. Dental Res. 66: 569-575.
Mann, G.E., Pearson, J.D., Sheriff, C.-J. & Toothill, V.J. 1989. Expression of amino acid transport systems in cultured human umbilical vein endothelial cells. J. Physiol. 410: 325-339.
Mann, G.E., Sheriff, C-J. & Pearson, J.D. 1990. N^G-monomethyl-L-arginine and other cationic amino acids inhibit L-arginine uptake in perfused microcarrier cultures of bovine aortic endothelial cells. In: *Nitric Oxide from L-Arginine: A Bioregulatory System*, Eds. Moncada, S. and Higgs, E.A., pp. 331-339. Elsevier, Amsterdam.
Mann, G.E., Wilson, S.M. & Yudilevich, D.L. 1984. Characteristics of a cationic amino acid transport system in the basolateral membrane of the cat salivary epithelium. J. Physiol. 351: 123-134.
Mann, G.E., Zlokovic, B.V. & Yudilevich D.L. 1985. Evidence for a lactate transport system in the sarcolemmal membrane in the perfused rabbit heart: Kinetics of unidirectional influx, carrier specificity and effect of glucagon. Biochim. Biophys. Acta 819: 241-248.
Martín de Julián, P. & Yudilevich, D.L. 1964. A theory for the quantification of transcapillary exchange by tracer dilution curves. Am. J. Physiol. **207**, 162-168.
Mepham, T.B., Overthrow, J.I., Short, A.H., & Warne, D.E. 1984. Amino acid transport interactions in lactating mammary gland of guinea-pig. J. Physiol.357: 73P
Norman, P.S.R. 1988. Capillary extraction of [7,8-^{3}H]dopamine in the isolated perfused rabbit heart. J. Physiol. 407: 16P.

Norman, P.S.R. & Mann, G.E. 1986. Transport characteristics of System A in the rat exocrine pancreatic epithelium analyzed using the specific non-metabolized amino acid analogue α-methylaminoisobutyric acid. Biochim. Biophys. Acta 861: 389-394.

Norman, P.S.R. & Mann, G.E. 1987. Ionic dependence of amino-acid transport in the exocrine pancreatic epithelium: calcium dependence of insulin action. J. Membr. Biol. 96: 153-163.

Norman, P.S.R., Habara, Y. & Mann, G.E. 1989. Paradoxical effects of endogenous and exogenous insulin on amino acid transport activity in the isolated rat pancreas: somatostatin-14 inhibits insulin action. Diabetologia 32: 177-184.

Perán, S. & McGee, M.P. 1986. Unidirectional flux of phenylalanine into Vero cells. Measurement using paired tracers in perfused cultures. Biochim. Biophys. Acta 856: 231-236.

Perán, S., Muñoz, M. & Saiz, M.T. 1990. Rapid and steady-state amino acid transport in perfused human fibroblasts and colon adenocarcinoma cells: effects of methotrexate. Biochim. Biophys. Acta 1024: 233-240.

Rennie, M.J., Idstrom, J-P., Mann, G.E., Schersten, T. & Bylund-Fellinius, A-C. 1983. A paired-tracer dilution method for characterizing membrane transport in the perfused rat hindlimb. Biochem. J. 214: 737-743.

Smith, I.C.H. 1985. An interpretation of single-pass uptake curves for measuring cellular transport. In: *Carrier-Mediated Transport of Solutes from Blood to Tissue*. Eds. Yudilevich, D.L. and Mann, G.E., pp. 397-402. Longman, London.

Sweiry, J.H. & Yudilevich, D.L. 1984. Asymmetric calcium influx and efflux at maternal and fetal sides of the guinea-pig placenta: kinetics and specificity. J. Physiol. 355: 295-311.

Sweiry, J.H. & Yudilevich, D.L. 1985. Characterization of choline transport at maternal and fetal interfaces of the perfused guinea-pig placenta. J. Physiol. 366: 251-266.

Sweiry, J.H. & Yudilevich, D.L. 1988. Characterization of folate uptake in guinea pig placenta. Am. J. Physiol. 254: C735-C743.

Sweiry, J.H., Page, K.R., Dacke, C.G., Abramovich, D.R. & Yudilevich, D.L. 1986. Evidence of saturable uptake mechanisms at maternal and fetal sides of the perfused human placenta by rapid paired-tracer dilution: studies with calcium and choline. J. Dev. Physiol. 8: 435-445

Syrota, A., Girault, M., Pocidalo, J-J. & Yudilevich, D.L. 1982. Endothelial uptake of amino acids, sugars, lipids, and prostaglandins in rat lung. Am. J. Physiol. 243: C20-C26.

Taylor, P.M. & Rennie, M.J. 1987. Perivenous localisation of Na-dependent glutamate transport in perfused rat liver. FEBS Lett. 221: 370-374.

Taylor, P.M. & Rennie, M.J. 1988. Amino acid fluxes across sinusoidal membranes of the perfused rat liver: relationships with portal ammonia concentration. In: *Advances in Ammonia Metabolism and Hepatic Encephalopathy*. Eds. Soelers, P.B., Wilson, J.H.P, Meyer, A.J., Holm, E. Amsterdam: Elsevier, pp, 45-52.

Taylor, P.M., Egan, C.J. & Rennie, M.J. 1989. Transport of glutamine across blood-facing membranes of perfused rat jejunum. Am. J. Physiol. 256: E550-E558.

Toothill, V.J., Eaton, B.M., Davis, H., Pearson, J.D. & Mann, G.E. Solute permeability of endothelial cell junctions in perfused microcarrier cultures of human umbilical vein endothelial cells. Int. J. Microcir. Clin. Exp. (abstract in press).

Wangler, R.D., Gorman, M.W., Wang, C.Y., DeWitt, D.F., Chan, I.S., Bassingthwaighte, J.B. and Sparks, H.V. 1989. Transcapillary adenosine transport and interstitial adenosine concentration in guinea pig hearts. Am. J. Physiol. 257: H89-106.

Watt, P.W., MacLennan, P.A., Hundal, H.S., Kuret, C.M. & Rennie, M.J. 1988. L(+)-Lactate transport in perfused rat skeletal muscle: kinetic characteristics and sensitivity to pH

and transport inhibitors. Biochim. Biophys. Acta 944: 213-222.
Wheeler, C.P.D. & Yudilevich, D.L. 1988. Transport and metabolism of adenosine in the perfused guinea-pig placenta. J. Physiol. 405: 511-526.
Wheeler, C.P.D. & Yudilevich, D.L. 1989a. Effect of insulin, prostaglandin E1 and uptake inhibitors on glucose transport in the perfused guinea-pig placenta. J. Dev. Physiol. 11: 159-169.
Wheeler, C.P.D. & Yudilevich, D.L. 1989b. Lysine and alanine transport in the perfused guinea-pig placenta. Biochim. Biophys. Acta 978: 257-266.
Yudilevich, D.L. 1985. Characterisation of membrane carriers and receptors at the blood-tissue interface by single circulation paired-tracer dilution. In: Carrier mediated transport of solutes from Blood to Tissue. Eds. Yudilevich, D.L. and Mann, G.E., pp. 1-31. Longman, London.
Yudilevich, D.L. 1989. Blood-tissue transport of substrates in the heart: studies by single circulation tracer dilution. Int. J. Microcirc. Clin.Exp. 8: 397-409.
Yudilevich, D.L. & Barros, L.F. 1990. Transport of amino acids and nucleosides in the placenta. Biochem. Soc. Transactions 14: 1146-1150
Yudilevich, D.L. & Eaton, B.M. 1980. Amino acid carriers at the maternal and fetal surfaces of placenta by single circulation paired-tracer dilution. Kinetics of phenylalanine transport. Biochim. Biophys. Acta 596: 315-319
Yudilevich, D.L. & Mann, G.E. 1982. Unidirectional uptake of substrates at the blood-side of secretory epithelia: stomach, salivary gland, pancreas. Fed. Proc. 41: 3045-3053.
Yudilevich, D.L. & Mann, G.E. (eds.) 1985. Carrier-mediated Transport of Solutes from Blood to Tissue. Longman, London.
Yudilevich, D.L. & Sweiry, J.H. 1985. Transport of amino acids in the placenta. Biochim. Biophys. Acta 822: 169-201.
Yudilevich, D.L. & Wheeler, C.D.P. 1988. Placental transport: the brush border and the basal sides of the trophoblast, and a role for the fetal endothelium?. In: *Fetal and Neonatal Development.* Ed. Jones, C.T., pp. 47-57. Perinatology Press, New York.
Yudilevich, D.L., de Rose, N. & Sepulveda, F.V. 1972. Facilitated transport of amino acid through the blood-brain barrier of the dog studied in a single capillary circulation. Brain Res. 44: 569-578.
Yudilevich, D.L., Eaton, B.M., Short, A.H. & Leichtweiss, H.-P. 1979. Glucose carriers at maternal and fetal sides of the trophoblast in guinea-pig placenta. Am. J. Physiol. C205-212.
Zlokovic, B.V., Segal, M.B., Begley, D.J., Davson, H. & Rakic, L. 1985. Permeability of the blood-cerebrospinal fluid and blood-brain barriers to thyrotropin-releasing hormone. Brain Research 358: 191-199.

6

SOME INEXACT METHODS FOR THE ANALYSIS OF TRANSPORT KINETICS USING DUAL TRACERS

I. Christopher H. Smith

Biomedical Sciences Division
King's College London
London U.K.

INTRODUCTION

Paired tracer dilution studies have provided a revolution in understanding the cellular transport of metabolites and other biologically active substances. The potency of the method lies in the twin features of selectivity and speed. Selectivity is obtained by the test tracer being paired with a reference tracer which, ideally, differs only in the features under investigation. Speed is obtained by perfusion against or close to the active site of transport. Over the last forty years the techniques have been developed by many laboratories using whole organ perfusion of the capillary bed to give rapid access to the transporting cells (see *Yudilevich*, this book). More recently the perfusion of cultured cells on carrier beads (see *Mann, Sherrif, Toothill and Pearson*, this book) have proved to allow a similar access time. It is thus rather surprising to discover that there is no consensus of how to interpret the results obtained by these methods. Old equations have been applied to new techniques with scant regard for the fundamentals of the method. New analytical methods have come, and gone, without apparently being able to fully exploit the obvious experimental power of the dilution technique. This article briefly describes the analytical problems and considers the options available to overcome them.

THE BOLUS TECHNIQUE

The classic experimental approach in tracer dilution studies has been the upstream (arterial) rapid injection of a mixture of two or more tracers and the subsequent downstream (venous) sampling of the transient. Typically the injection lasts one second and sampling occurs for a minute or more at intervals of one second or more. The tracers injected may be classified as:

- flow (vascular) markers - assumed restricted to the flow compartment
- space (extravascular) markers - assumed to exit and return to the bulk flow
- transport marker - moves further according to transport studied

Cell Membrane Transport, Edited by D. L. Yudilevich *et al.*
Plenum Press, New York, 1991

Uptake Analysis

To measure transport sensitively the most obvious idea would be to use as reference and test tracers the space and transport markers respectively. Movement of the test molecule relative to the reference would thus indicate that, at least somewhere, specific transport or binding of the test substance had occurred. For example, in looking at carrier-mediated transport of sugar one might use L-glucose and D-glucose as reference and test molecules. This is called the paired-tracer technique and is the approach used by Yudilevich and associated workers (Yudilevich, 1985) employing the time-dependent parameter:

$$\text{Uptake,} \quad U(t) = (\text{ref} - \text{test}) / \text{ref}$$

However before one can interpret the movement of the transport marker relative to the space marker one must understand what the transient of the space marker represents. For example it is ingenuous to use the Crone - Renkin logarithmic transform for a vascular reference:

$$\text{Permeability,} \quad PS = -\text{flow} \times \ln(\text{test} / \text{ref})$$

and extrapolate this to measure cellular transport (**v**) :

$$\text{Transport,} \quad v = -\text{concentration} \times \text{flow} \times \ln(1 - U_{max})$$

with the argument that this 'corrects for the exponential profile from the arterial to venous end of the capillary' (Yudilevich, 1985). That depends on what the space marker venous concentration means in terms of the extravascular concentration.

Exchange Models

A second approach would be to use (at least) all three types of markers and try to provide a quantitative recreation of the physical processes involved in the dispersion of the markers (Bassingthwaighte and Goresky, 1984). There are firm benefits to this strategy. (a) It forces a direct statement of the physical model used. (b) It is not necessary to assume that, for example, the space marker provides an exact passive reference for the test molecule; it is sufficient that it marks processes adequately to extract its parameters. (c) It maximises use of the entire transient to derive single parameters. [In uptake analysis selection of times of interest, such as 'early uptake' or 'maximal uptake' are used]. Some of the physical processes considered are:

- dispersion due to the technique of bolus injection
- dispersion due to variable path length and flow speed in the non-exchanging regions
- a similar dispersion in the exchange regions (capillaries)
- dispersion due to diffusion into non-flowing (extravascular) regions
- movement into cells facing flow (endothelial) and non-flow (parenchymal) regions
- return transport (backflux) from these cells

These processes are hard to model even individually. For example the diffusion into the extravascular regions requires an efficient description of the geometry of the initial restriction barrier and the spaces beyond. The limiting nature of this problem is reinforced when one tries just to identify where the physical locations of the classically-measured extravascular space really exist. Overall it is clear that the complexity of detail, combined with lack of our knowledge, means that the physical models must remain just that - models.

Despite these problems it has become clear that the physical models are (a) an effective tool in the practical analysis of dilution transients and (b) are an essential part of any attempt to interpret dilution transients (Bassingthwaighte and Sparks, 1986). There is, however, a real problem that the models are - by necessity - a deep specialisation. Whilst these models must continue to be developed and used, there is also a clear need for simplified models and rules which can be used by researcher's whose primary interest lies in specific cell mechanisms rather than vascular dynamics.

Inexact Methods

One of the worst problems to interpret is the effects of capillary heterogeneity and how this is related to heterogeneity of transits in the non-exchanging large vessels. Do either long-path or slow-flowing large vessels, for example, tend to link to long or to slow-flowing capillaries (Bassingthwaighte, King and Roger, 1989). The combinations are multiple. An oversimplified but uncomplicated assumption is that all variance in the flow is due to the large vessels and it is on this basis that I have described some simple models of transport kinetics (Smith, 1985). It has the merit, when acceptably valid, of forcing the data to be presented in terms of the idealised 'average' capillary. Quantitative interpretation of the data must be restricted to the central region of the transient (which is dominated by the average capillary) and to the time integral of the transient. However both early and late transients can be used as a qualitative indicator of what transport mechanisms are operating. In particular they can be used for comparing transports of differing substances, or differing concentrations of a single test substance, in the same preparation, i.e. where the capillary dynamics remain constant.

The main observations to be accounted for in paired-tracer transport studies are a) a relative removal of the test substance and b) its partial return. This implies there are at a minimum three transport pathways:

- unidirectional uptake
- return flux
- sequestration

The shape of the uptake curve, U(t), can be initially low and progressively increasing, steady, or initially high then falling. These differences must be ascribable to specific transport sites, and not to vascular dynamics, since they occur even in the same preparation but with different test molecules (Mann and Yudilevich, 1985).

Low Early Uptakes The most surprising outcome of the simple model, compared to experimental results, is the case of an uptake which is initially zero and progressively rises. This is seen with a traditional concept of cellular transport in which the test substance must first move into the non-flowing extravascular compartment before specific transport into the parenchymal cell. In this case the relative removal of the test substance can only be apparent at the venous outflow in proportion to the amount of backflux from the extravascular space - i.e. tending to zero at early stages of the transient and then progressively rising. However experimental examples of a zero then rising uptake is relatively rarely seen (but see uptake of L-phenylalanine or ouabain in the heart, Mann and Yudilevich, 1985; Yudilevich, 1989). There are two sorts of explanation for the infrequent observation of what might reasonably be expected. Firstly that transport only infrequently takes place from the depths of the interstitial spaces, i.e. that cell transport is maximised close to the capillary interface (see below). Secondly the backflux from the interstitial space is faster than the central distribution of transient times. The latter is rather unlikely in view of the variation of uptake shape with the molecule transported and in view of the timescale for extravascular washout as evidenced by extraction curves relative to a vascular marker.

The direct quantitative assessment of these types of uptake curves is essentially impossible. The best approach is perhaps to form the integral net uptake:

$$U_{TOT}(t) = \Sigma^{t} (test - ref) / \Sigma^{\infty} (ref)$$

This term, if there is no return from the cellular compartment, moves progressively to the required average uptake, U_o. However a rise then fall is more likely, the fall showing - at least in part - return flux from the cellular compartment. This interpretation becomes more sure the later one looks. Indeed if the curve of $U_{TOT}(t)$ can be followed after the reference transient is essentially zero, then the problem is simply that of a cellular washout and can be analyzed as described below. This analysis will give the average uptake, U_o, a fractional sequestration and the return flux rate constant. It requires, of course, the collection of data well beyond the tail of the reference transient. Note that the term U_o is not to be taken as meaning the zero time uptake but the uptake by an average capillary under the imaginary conditions that no backflux takes place during the time-scale investigated. Note also that application of the Crone - Renkin equation does not give the apparent permeability of the cell but the effective permeability of the transport unit (capillary + parenchymal cells).

Initially Steady Uptakes. I next consider the case where uptake starts at a finite value and continues at this level at least until the central region of the transient. Essentially transport must be taking place at a site immediately close to the flow compartment. The steady uptake can be treated as a good estimate of U_o and can be interpreted as a permeability by the Crone - Renkin equation. What 'immediately' means depends, of course, on the resolution time of the experiment. An obvious candidate would be transport at the endothelial cell. However dispersion, or lack of high frequency sampling, may force 'immediate' events to include transport from the extravascular space. Control experiments, with vascular and space-marking tracers, would indicate if the extravascular washout is clearly resolvable. If, as often seems to be the case, washout is resolvable and - on the basis of the total quantity of transport - it appears that the transport is not occurring just at the endothelial cell (e.g. pancreatic uptake of amino acid, see Mann et al., 1989), there is a problem in identifying the site of transport. It behaves as if it were neither the endothelium nor the underlying cell - which would require a delayed uptake. One solution is that the cellular uptake does not take place uniformly over the cell. The cell membrane may be polarised such that transport takes place preferentially at sites close to the capillary .

Late decay of uptake and return flux from the cell. There is clear evidence, based on procedures that measure or interfere with cell metabolism, that return flux from cells can take place within the immediate time-scale of the single-pass transient (Gorman et al., 1986, Norman and Mann, 1986). This can be analyzed simply only if the role of the extravascular space is insignificant and transport essentially taking place from near the flow compartment. A modified Martin de Julian analysis (Smith, 1985) is then appropriate. This three-compartment model (here flow compartment, cellular compartment and metabolic sink) gives:

$$E(t) = E_o - k \times (1 + \beta) \times (R \times \Sigma ref - \Sigma test) / ref \qquad (1)$$

where R is the net recovery of the test tracer (obtained directly from the totals of the transient), k is the return rate constant and β is the relative chance of an intracellular test molecule moving to the sink compared to its return flux to the circulation. The sequestration ratio can be estimated from the steady state equation:

$$1 - R = E_o \times \beta / (1 + \beta) \qquad (2)$$

provided that a reliable measure of E_o is obtained from the straight-line plot of E(t). The equations have changed to using E (extraction) since the implicit hypothesis is now that the reference tracer is a vascular marker.

SQUARE-WAVE LOADING AND STEADY STATE ANALYSIS

If the experimental problem is to measure a unidirectional flux, when there is clear evidence of return flux, the analysis is at least simplified by having a firm estimate of the steady state net fluxes. Examples above using the bolus technique used the total transient integral to estimate the recovery - which is a measure of the metabolic sink. This situation is however far more directly seen by using a square-wave loading and wash-out procedure (Boyd and Parsons, 1979, Mann et al., 1989). As seen in Fig. 1 the uptake (relative difference between test and reference curves) can be seen to start at a large value (U_o) and move towards a steady state value (1 - R). Following this there is a reasonably exponential washout of the test tracer. The time-scale in these experiments (20 min loading) is such that heterogeneity of the circulation is of little importance. The analysis essentially follows equations (1) and (2) above noting that the rate constant of exit after the end of loading is $k \times (1 + \beta)$, i.e. membrane transport rate appears to be speeded by the relative sequestration rate, β . This rate can be estimated either using the overall integrals and equation (2) or fitting an exponential to the uptake during loading which moves from an initial value of U_o to an infinite time value of R.

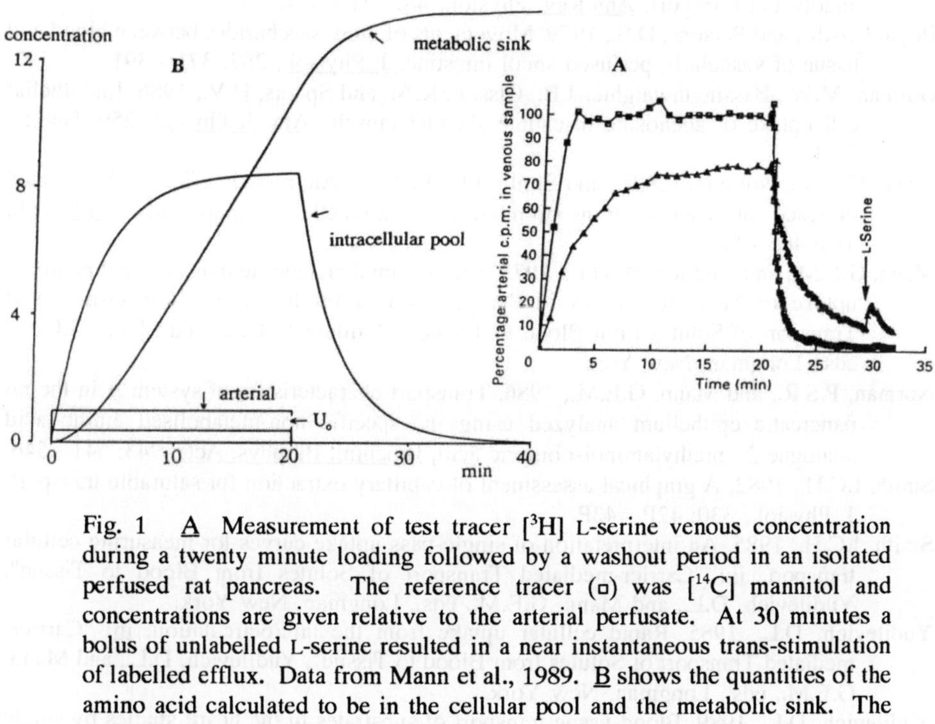

Fig. 1 A Measurement of test tracer [^{3}H] L-serine venous concentration during a twenty minute loading followed by a washout period in an isolated perfused rat pancreas. The reference tracer (□) was [^{14}C] mannitol and concentrations are given relative to the arterial perfusate. At 30 minutes a bolus of unlabelled L-serine resulted in a near instantaneous trans-stimulation of labelled efflux. Data from Mann et al., 1989. B shows the quantities of the amino acid calculated to be in the cellular pool and the metabolic sink. The sequestration ratio, β, was 0.4 relative to the return flux. The carrier asymmetry, given by the influx to efflux ratio, was 20.1 .

CONCENTRATION - DEPENDENT UPTAKE

Transport of substrates such as serine are saturable and are conventionally described by the Michaelis-Menton equation, $V / V_{max} = C / (C + K)$ where C is concentration and K the concentration to achieve half-saturation. Under these conditions the Crone - Renkin equation does not apply. Interpretation of the tracer uptake now depends on whether this tracer uptake matches the actual ('cold') uptake. For simplification the two extremes are given by cellular uptake followed by sequestration only (β infinite) or by uptake with return flux only ($\beta = 0$). In the latter case the unlabelled concentration is constant along the length of the capillary and the Crone - Renkin equation can be used to give an apparent permeability. However with sequestration the true concentration also drops along the capillary length. Integration of the Michaelis - Menton equation gives the linear Smith plot (Smith, 1982):

$$-F \times \ln(1 - U_o) = V_{max} / K - C \times U_o \times F / K$$

where F is the perfusate flow rate.

REFERENCES

Bassingthwaighte, J.B., and Goresky, C.A., 1984, Modelling in the analysis of solute and water exchange in the microvasculature, in: Handbook of Physiology, Sec. 2, eds. Renkin, E.M., and Michel, C.C., Am.Physiol. Soc., Bethesda.

Bassingthwaighte, J.B., King, R.B., and Roger, S.A., 1989, Fractal nature of regional myocardial flow heterogeneity, Circ. Res., 65: 578 - 590.

Bassingthwaighte, J.B., and Sparks, H.V., 1986, Indicator dilution estimation of capillary endothelial transport, Ann.Rev. Physiol., 48: 321 - 334.

Boyd, C.A.R., and Parsons, D.S., 1979, Movements of monosaccharides between blood and tissue of vascularly perfused small intestine, J. Physiol., 287: 371 - 391.

Gorman, M.W., Bassingthwaighte, J.B., Olsson, R.A., and Sparks, H.V., 1986, Endothelial cell uptake of adenosine in canine skeletal muscle, Am. J. Physiol. 250: H482 - H489.

Mann, G.E.M., Norman, P.S.R., and Smith, I.C.H., 1989, Amino acid efflux in he isolated perfused rat pancreas: trans-stimulation by extracellular amino acids, J. Physiol., 416: 485 - 502.

Mann, G.E.M., and Yudilevich, D.L., 1975, Sarcolemmal metabolite transport and neuronal uptake of biogenic amines in the perfused rabbit heart, in: "Carrier-mediated Transport of Solutes from Blood to Tissue", Yudilevich, D.L., and Mann, G.E.M.. eds., Longman, New York.

Norman, P.S.R., and Mann, G.E.M., 1986, Transport characteristics of system A in the rat pancreatic epithelium analyzed using the specific non-metabolised amino acid analogue 2 - methylaminoisobutyric acid, Biochim. Biophys. Acta, 943: 541 - 546.

Smith, I.C.H., 1982, A graphical assessment of capillary extraction for saturable transport, J. Physiol., 330: 42P - 43P.

Smith, I.C.H., 1985, An interpretation of single-pass uptake curves for measuring cellular transport, in: "Carrier-mediated Transport of Solutes from Blood to Tissue", Yudilevich, D.L., and Mann, G.E.M.. eds., Longman, New York.

Yudilevich, D.L., 1985, Rapid cellular uptake from the microcirculation, in: "Carrier-mediated Transport of Solutes from Blood to Tissue", Yudilevich, D.L., and Mann, G.E.M.. eds., Longman, New York.

Yudilevich, D.L., 1989, Blood-tissue transport of substrates in the heart: studies by single circulation tracer dilution, Int. J. Microcirc. Clin. Exp., 8: 397 - 409.

7

TRANSPORT STUDIES IN ISOLATED ENTEROCYTES

A. Ilundáin

Depto. Fisiología y Biología Animal
Universidad de Sevilla, Sevilla, Spain

INTRODUCTION

The purpose of this paper is to illustrate experimental strategies that may be adopted in the laboratory to approach transport studies in isolated enterocytes with isotopic methods. Since the work carried out with this biological preparation is vast I shall focus on methodologies set up to: a) investigate the energetics of organic solute active transport (e.g.sugars and dipeptides), b) reveal the separate pathways involved in the transfer of a given solute across the cell membrane (e.g. K^+ transport) and c) study intracellular transport events.

TECHNICAL PROPERTIES OF ISOLATED ENTEROCYTE PREPARATIONS

Advantages

1. Isolated epithelial cells represent the smallest functional transport unit of the tissue and an experimental model which is considerably less complex than intact tissue preparations.

2. Cell suspension allows a high degree of homogeneity in cell populations, that can be easily handled with micro-pipettes and readily subjected to changes in bathing medium and rapid sampling techniques.

3. Intracellular volume of the epithelial cell population was impossible to determine accurately in intact tissue preparations. Therefore the concentration gradients of substrate established by the epithelial cells were poorly estimated. In isolated enterocytes the correction for non-specific, extracellular binding of the substrate and evaluation of the intracellular volume, and hence substrate concentration gradients, can be more precisely defined.

4. The use of intact cells circumvents some of the questions

Cell Membrane Transport, Edited by D. L. Yudilevich *et al.*
Plenum Press, New York, 1991

regarding possible alterations of inherent membrane properties which accompany the process of cell lysis, divalent cation precipitation and resealing of membrane vesicles.

5. The cells are able to maintain steady-state solute gradients for longer intervals of time than membrane vesicles, hence they can be used to monitor formation and maintenance of steady- state solute gradients, as well as unidirectional solute influxes. Therefore they are a suitable experimental model for kinetic studies.

6. Because all cell boundaries are equally exposed to suspending medium they are ideally suited: a) for evaluation of the types of transport systems present in the entire plasma membrane, b) for studying the effects and binding of agonists acting at any locus of the membrane and c) to discern whether an agonist directly affects transport events in enterocytes or indirectly causes the transport effects via neural structures.

7. Cells can be depleted of their ATP and thus permit to have control over the ion gradients and electrical membrane potential across the cell membrane. Therefore, systemic evaluation of the effects of individual driving forces (chemical and electrical gradients), operative on a given transport system, can be assessed independently.

8. ATP-depleted isolated enterocytes behave as vesicles but offer a technical advantage of larger intracellular volume than classic membrane vesicle preparations, so that imposed chemical and/or electrical gradients persist for a much longer interval of time before being dissipated by diffusion.

9. They can be rendered leaky and intracellular transport events can then be monitored.

Drawbacks

1. Isolated cells lose their normal fixed orientation. As a result transport events occurring at a given locus can only be precisely defined as long as there are means available for distinguishing between the various operative systems involved in the transfer of a metabolite. Work with some intact tissue preparations and membrane vesicle preparations, enriched in membranes from either the mucosal or serosal border of the cell, provide information regarding the cellular locus of a given transport system.

CELL ISOLATION

The best conditions obtained so far for preparing enterocytes are those described by Kimmich (1975) for the chick small intestine. The method provides suspensions of enterocytes in high yield and with good retention of morphological, metabolic and transport properties, and has been used, with minor modifications, to isolate epithelial cells from other animal species such as rat (Bronner et al., 1983) and rabbit (Brown & Sepúlveda,1985a) or other intestinal regions (Ferrer et al., 1986; Ahnen et al., 1988; Kaunitz, 1988). All experiments described in the present work have been carried out in enterocytes isolated from chicken (4-6 week-old) small intestine.

The animal is killed by decapitation and a section of the intestine is removed. The tissue is rinsed clean with ice-cold buffer containing 0.1 mM DL-dithiothreitol (to reduce the mucus content of the preparation), then everted and opened longitudinally. Pieces of 3-4 cm are then transferred to the isolation buffer containing 1 mg/ml hyaluronidase and incubated at 37ºC in a shaking water bath (70 cycles/min) for 30 min. Unless otherwise stated, the composition of the buffer used in this and subsequent incubations is (in mM): NaCl, 80; $CaCl_2$ 1; mannitol, 100; K_2HPO_4, 3; $MgCl_2$, 1; Tris-HCl (pH=7.4), 20; B-hydroxybutyrate, 0.5 and 1 mg/ml bovine serum albumin. B-hydroxybutyrate is present as a non-actively accumulated nutrient for the suspended cells. After incubation, intestinal segments are transferred to hyaluronidase-free buffer where cells are released by gentle shaking with a plastic pipette tip. Cells are filtered through a cotton gauze and then through a 200 µm nylon mesh before being washed twice with standard buffer by resuspension and centrifugation. The final concentration of the cell suspension varies depending on the requirements of the particular experiment.

Ca^{2+}-chelators, such as citrate (Ferrer et al.,1986) or EDTA (Velasco et al., 1986a), have also been employed instead of hyaluronidase. However cells isolated by EDTA became highly permeable (Velasco et al., 1986a) to small molecules, such as ions, and they are not suitable for transport studies.

It is worth pointing out that cells are always exposed only to plastic containers and pipette tips.

Cell viability is asessed by determining the fraction of the cell population able to exclude 0.2% trypan blue (Girardi et al., 1956) which usually ranges from 60% to 75%.

TRANSPORT MEASUREMENTS

The isotopic method consists of measuring at timed intervals the radiolabelled substrate in one of the two membrane-enclosed compartments. Using this method, two different experimental approaches can be performed to investigate solute transport across the cell plasma membrane: uptake and efflux measurement experiments.

<u>Uptake Experiments</u>

The uptake of the substrate is measured at 37ºC, using cells at a final concentration of 15-25 mg cell protein/ml. Incubations are started by adding 1 ml of cell suspension to 3 ml of buffer containing the substrate (S) under study, together with a trace of radiolabelled substrate (*S) and 0,1 μCi/ml labelled-PEG 4,000 (*PEG), as an extracellular space marker (radiolabelled inulin or any other non-permeant molecule can also be used). The cell suspension and incubation buffer (both prewarmed at 37ºC) are mixed in a 20 ml polyethylene beaker held in a thermostated bath and shaken at 70 cycles/min to maintain adequate stirring and homogeneity of the cell suspension. Uptake is terminated as follows: at timed intervals 200 μl aliquots are transferred to microfuge tubes containing 500 μl ice-cold buffer and 250 μl of the oil mixture di-n-butyl phthalate: dinolyl phthalate 3:2 (v/v) (1.03 density). Cells are immediately separated by centrifugation (10,000 g, 20 s) and, unless otherwise stated, cell pellets are extracted in 0.2 ml of perchloric acid (3%) by vortexing vigorously. Aliquots are taken from the extract supernatant after centrifugation and the *S and *PEG in incubation fluids and cell pellet extracts are measured in a liquid scintillation counter. The S uptake is calculated taking into account the trapped extracellular volume, estimated from the amount of extracellular marker associated with the pellet. Protein content of the pellet can be determined by the Lowry method (1951).

Intracellular concentration of S is obtained by dividing the intracellular content of S by the intracellular water content. This value is calculated by subtracting the trapped extracellular water from the total water space. Total water space is estimated by measuring the equilibrium uptake (~30 min) of radiolabelled 3-O-methylglucose (a nonmetabolized substrate) in the presence of 250 μM phloridzin (an inhibitor of concentrative sugar uptake). Under these experimental conditions, at the steady-state the intracellular sugar concentration is equal to that in the incubation buffer. Chicken enterocytes water intracellular space is ~ 3μl/mg cell protein.

Efflux Experiments

A concentrated cell suspension (40-60 mg cell protein/ml) is first labelled by incubating at 37ºC in a shaking water bath for 30 min in the presence of the radiolabelled substrate (*S) under study. Cells are then washed in radioisotope-free, ice-cold buffer and resuspended in buffer to a final concentration of 20-30 mg cell protein/ml. The rate of *S loss from the cells is then measured by dilution of aliquots of this suspension with appropiate radioisotope-free buffer, followed by incubation at 37 ºC in a shaking water bath. Efflux is terminated as in the uptake measurement experiments.

*S efflux is expressed as an apparent efflux rate constant that has units of min^{-1} and that is calculated as follows. If the loss of substrate from cells is from a single intracellular compartment the *S remaining in the cells should be a decreasing monoexponential function of time of the form:

$$C_t = (C_o - C_\infty) \cdot e^{-k \cdot t} + C_\infty$$

where t is time, C_t, C_o, and C_∞ are radioactivity in the cells at times t, 0 and infinity respectively, and k is an exponential constant for efflux. Assuming C_∞ is negligible, the equation simplifies to:

$$C_t = C_o . e^{-k.t} \quad \text{and then:} \quad k = (\ln C_o - \ln C_t) / t$$

ENERGETICS OF EPITHELIAL TRANSPORT OF ORGANIC SOLUTES

Several organic solutes are actively transported across the plasma membrane of intestinal epithelial cells, that is at steady-state their concentration in the intracellular media is greater than in the extracellular media.

It is now known that concentrative transfer of several organic solutes is driven by the Na^+ chemical gradient and by the cell membrane potential (see Kimmich, 1981; Eddy, 1987 for reviews). The Na^+ gradient can be also used by cells to create other ion chemical gradients, that in turn can act as a driving force for the active uptake of certain organic solutes. Thus, recent work has shown that the intestinal transport of dipeptides appears to be coupled to the electrochemical gradient of H^+, which is maintained by the Na^+/H^+ exchanger system. In this case there would be an indirect coupling of peptide transport to the Na^+-gradient.

Evaluation of Organic Solute Active Accumulation

Concentrative transfer is defined as the net movement of a substrate between two compartments against a concentration gradient. Figures 1 and 2 shows that at steady-state the concentration of 3-O-methylglucose (3-OMG) or glycyl-L-sarcosine (gly-sar) within cells is greater than in the incubation buffer. This would indicate that both substrates have been actively transported across the cell membrane.

By using radiolabelled substrates to follow the movement of a given substrate into the cell, only the amount of radioactivity present inside the cell can be measured. However, once inside the cell the substrate can be metabolized or bound to intracellular components. Futhermore, substrates such as dipeptides can be hydrolized by membrane and cytosolic dipeptidases, giving amino-acids that could also be actively transported. In either case, the total amount of label in the intracellular medium may be higher than outside but the substrate might not be accumulated.

Therefore to obtain an accurate value of intracellular concentration of the substrate it is necessary to know the amount of substrate that enters the cell intact and remains intact and free in the cytosol. This can be achieved by interrupting the active accumulation of substrate after a period of substrate accumulation has been allowed. Following the addition of an inhibitor, active accumulation of substrate is precluded and only diffusion down a concentration gradient is possible. As shown in Figures 1 and 2, 3-OMG and gly-sar are released from the cells after addition of the inhibitor of the active transport process.

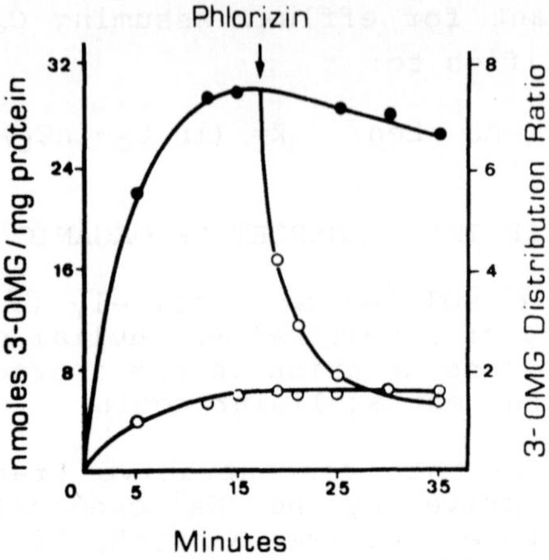

Fig. 1. Effect of phloridzin on 3-OMG accumulation by isolated chick intestinal epithelial cells. Cells were allowed to accumulate 3-OMG (1mM) for 17 min at which time phloridzin (200 μM) was added. (●) Control, (O) phloridzin. Distribution ratio is the ratio of cellular to media sugar concentration at a given point of time.(From Kimmich et al., 1977).

Since phloridzin binds to the sugar carrier and thus prevents the binding of sugar, this test is sufficient to demonstrate that a concentration gradient of 3-OMG has been established prior to inhibitor addition. However, as the proton ionophore FCCP can also affect metabolic activity, and dipeptides can be hydrolysed, the above observations do not allow us to discern whether the labelled substrate has been accumulated by the cells in the form of free dipeptide or free

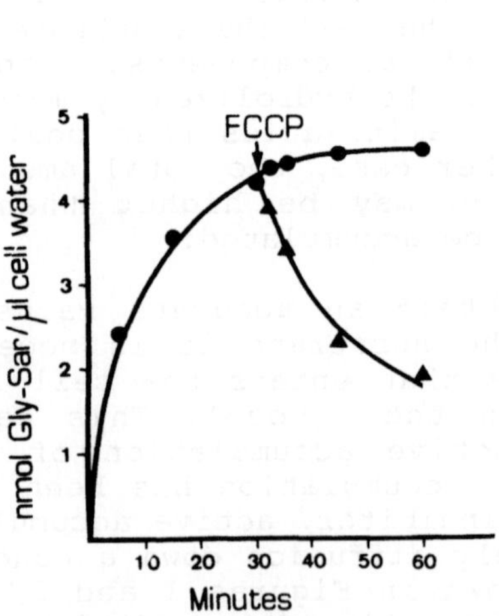

Fig. 2. Effect of FCCP on Gly-Sar uptake by enterocytes. Cells were allowed to take up Gly-Sar (1 mM) for 30 min. At this time 25 μM FCCP (carbonyl cyanide p-trifluoromethoxyphenylhydrazone) was added. (●) control, (▲) FCCP. (From Calonge et al., 1989).

amino-acids. Chromatografic studies of the incubation buffer and of the extracts of cell pellets, after a period of incubation in the presence of dipeptide, allow the evaluation of the degree of dipeptide hydrolysis by membrane and cytosolic dipeptidases respectively. These studies revealed that gly-sar is highly resistant to the attack of either brush-border or cytosolic dipeptidases (Calonge et al.,1989).

Na-Chemical Gradient and Intestinal Transport of Organic Solutes

Experimental tests of the Na^+-gradient, as a source of energy for organic solute active accumulation, center on the following questions:

a) Does transport of organic solutes exhibit an absolute dependence on Na^+?

b) Do inhibitors of the Na^+-pump also inhibit accumulation of organic solutes?

c) Would organic solute stimulate ouabain-sensitive Na^+-efflux?

d) Would the presence of a Na^+ gradient across the cell membrane lead to a parallel formation of a concentration gradient of the substrate without the intervention of ATP?

The following are experimental approaches that may be adopted to answer each of these questions:

a) Figure 3 shows that Na^+ removal from bathing solutions prevents sugar accumulation. Under these conditions the

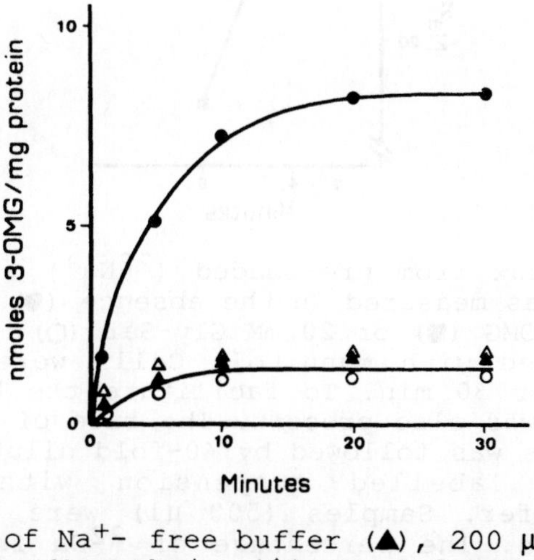

Fig. 3. Effects of Na^+- free buffer (▲), 200 μM phloridzin (O) and 0.1 mM ouabain (Δ) on 1.5 mM 3-OMG uptake by isolated chicken enterocytes. (●) Control. (From Bolufer et al., 1982).

intra-to extracellular sugar concentration ratio is 1 and it is not different from that obtained in the presence of phloridzin. Decreasing external Na^+ concentration also reduced gly-sar uptake. However, even in the absence of Na^+, the intra-to extracellular gly-sar concentration ratio was greater than one (Calonge et al.,1989).

b) Figure 3 shows that ouabain (inhibitor of the Na^+ pump) prevented 3-OMG accumulation, however gly-sar active accumulation was not abolished by ouabain.

c) The Na^+,K^+,ATPase activity and hence, the ouabain-sensitive Na^+-efflux is expected to be increased by those substrates that are cotransported with Na^+. Figure 4 shows that 3-OMG, but not gly-sar, stimulated Na^+ efflux.

These findings indicate that whilst sugar active uptake is entirely Na^+-dependent, the active uptake of gly-sar presents only a partial dependence on external Na^+.

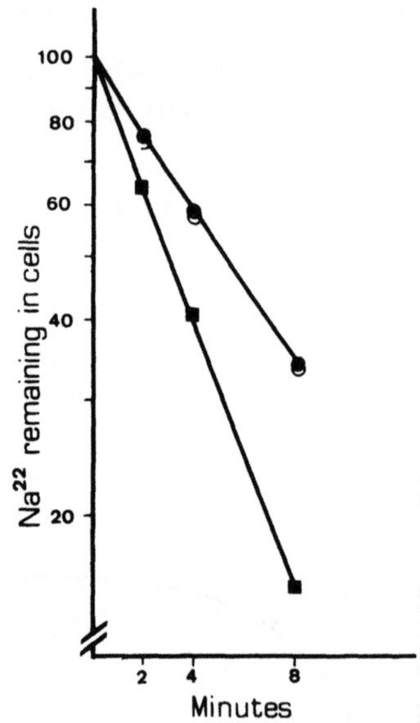

Fig. 4. Na^+ efflux from pre-loaded ($^{22}Na^+$) enterocytes. Na^+ efflux was measured in the absence (●) and presence of 20 mM 3-OMG (■) or 20 mM Gly-Sar (○). Isotonicity was maintained with mannitol. Cells were labelled with $^{22}Na^+$ for 30 min. To facilitate the Na^+-loading 1mM ouabain was also present. The rate of $^{22}Na^+$ loss from the cells was followed by 40-fold dilution of aliquots of the labelled suspension with radioisotope free-buffer. Samples (500 µl) were taken at time intervals. The percentage of ^{22}Na remaining in the cells is plotted on a logarithmic scale against time. 100% is the radioactivity present in the samples taken at 0 min. (unpublished observations)

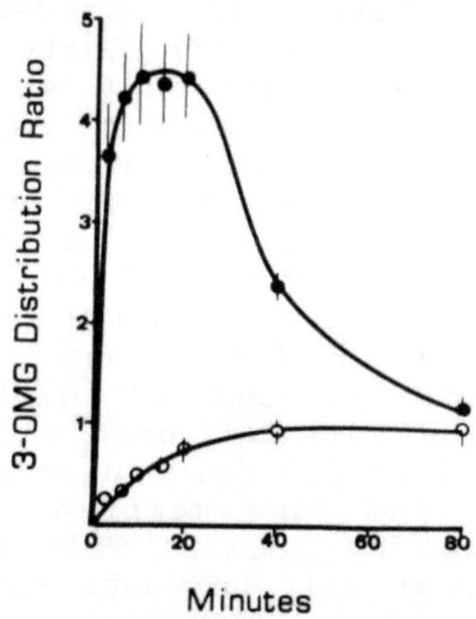

Fig. 5. Accumulation of 3-OMG by ATP depleted enterocytes with an imposed transmembrane Na^+ gradient. Cells were depleted of cellular ATP by pre-incubation for 10 min with 20 μM rotenone, in Na^+-free buffer, and suddenly introduced to a medium containing 90 mM Na^+ and 1 mM [^{14}C]-3-OMG. (●) control, (O) 0.1 mM phloridzin (From Kimmich et al.,1977).

d) The ion-dependence of the transport of a given solute can be further investigated in ATP-depleted cells. According to the ion-gradient hypothesis (Crane,1968), as long as the cell can maintain a concentration gradient for the ion involved, it should have the capability to establish and sustain an intra-to extracellular gradient of the substrate cotransported with the ion. This implies that ion gradients imposed experimentally on ATP-depleted cells should induce formation of solute gradient until the imposed ion gradient dissipates. Figure 5 shows that in ATP-depleted cells and in the presence of an inward Na^+-gradient, sugar uptake briefly "overshoots" the final steady-state equilibrium value to a degree that indicates that a momentary four-fold concentration gradient was achieved. In the absence of Na^+ gradient no overshoot is observed. Phlorizin prevents the overshoot even with the Na^+ gradient. On the other hand, an experimentally imposed Na^+-gradient in ATP-depleted cells does not induce transient formation of gly-sar gradient (Figure 6). Therefore it can be concluded that whereas 3-OMG is cotransported with Na^+, gly-sar is not.

Membrane Electrical Potential and Transport of Organic Solutes

If the complex ion-carrier-substrate is charged then its movement across the cell membrane will be affected by the membrane electrical field.

ATP-depleted enterocytes can also be employed to investigate whether the membrane electrical potential can provide a driving force for concentrative substrate accumulation (Kimmich et al., 1977; Carter-Su & Kimmich, 1979, 1980; Kimmich et al. 1985; Kimmich & Randles,1988). A plasma membrane potential is established by imposing gradients of permeant ions across the cell membrane boundary to create diffusion potentials. The commonest method to impose an inside negative membrane potential is to create an intra- to extracellular K^+ gradient and to add valinomycin to the incubation buffer to increase the membrane K^+ permeability and the magnitude of the K^+ diffusion potential. Figure 7 shows that when a valinomycin- induced K^+ diffusion potential was imposed in conjunction with Na^+ gradient, the metabolically inhibited cells established 8-fold sugar gradient (3-OMG). In the absence of Na^+ the same membrane potential caused no change in sugar uptake by cells. This indicates that the effect of membrane potential on sugar accumulation capability

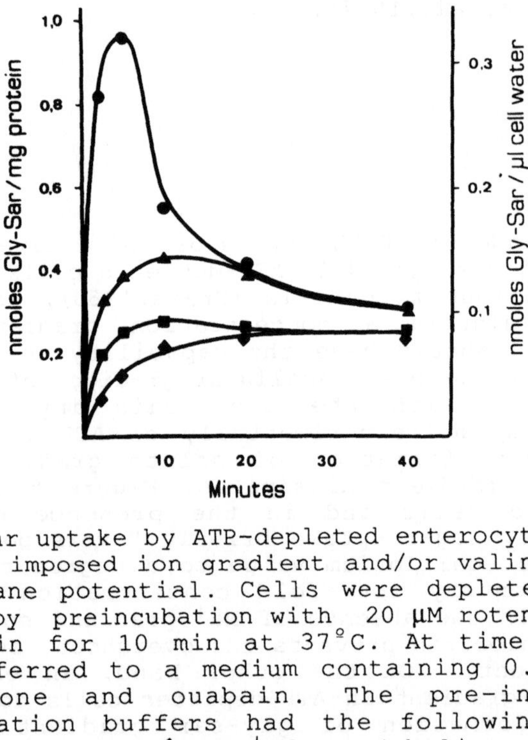

Fig. 6. Gly-Sar uptake by ATP-depleted enterocytes in response to an imposed ion gradient and/or valinomycin-induced membrane potential. Cells were depleted of cellular ATP by preincubation with 20 µM rotenone and 200 µM ouabain for 10 min at 37°C. At time 0 cells were transferred to a medium containing 0.1 mM gly-sar, rotenone and ouabain. The pre-incubation and incubation buffers had the following composition respectively: (◆) Na^+-free (choline substitution), Tris-HCl,buffer, pH 7.4 and 80 mM Na^+, Tris-HCl buffer, pH 7.4; (■) Na^+-free, Tris-HCl buffer, pH 7.4 and Na^+-free, Tris citric acic buffer, pH 5.5; (▲) 120 mM Kgluconate, 20 mM Nagluconate, Tris-HCl buffer, pH 7.4 and K^+-free, 20 mM Nagluconate, Tris-HCl buffer, pH 7.4, 20 µg/ml valinomycin; (●) 120 mM Kgluconate, Na^+-free, Tris-HCl buffer, pH 7.4 and K^+-free, Na^+-free, Tris citric acid buffer, pH 5.5, 20 µg/ml valinomycin. Osmolality was maintained with mannitol (unpublished observations).

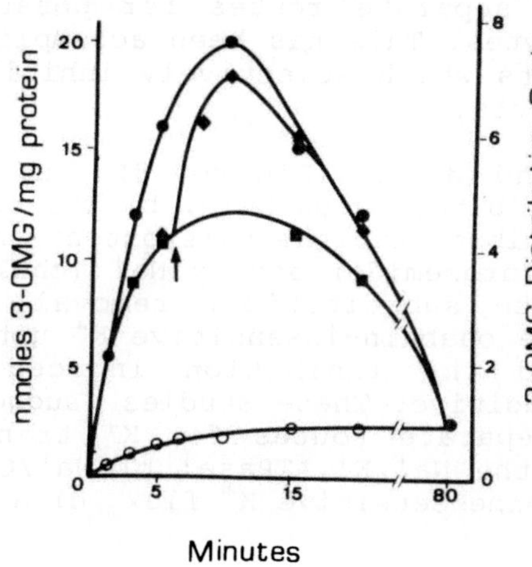

Fig. 7. Accumulation of 3-OMG (1 mM) by ATP-depleted intestinal cells in response to an imposed Na^+ gradient and to a valinomycin-induced membrane potential. Cells were preincubated in Na^+ -free buffer which contained 120 mM KCl and rotenone. At time zero they were transferred to a buffer in which the KCl was replaced by 45 mM Na_2SO_4 and sufficient mannitol to maintain osmolality and which included 1 mM [^{14}C]-3-OMG. In one case 10 µg/ml valinomycin was present from outset (●) and, in another (◆), was added as indicated by the arrow after 7 min incubation. (■) absence of valinomycin, (○) 0.1 mM phloridzin. (From Kimmich et al.,1977).

is absolutely dependent on the presence of Na^+. These findings clearly demonstrated that the Na^+-dependent sugar transport system is driven by cell membrane potential as well as by the chemical Na^+-gradient.

Results obtained with gly-sar are shown in Figure 6. Valinomycin-induced K^+ diffusion potential caused a small transient accumulation. The magnitude of the overshoot was independent of the presence of an inward directed Na^+ gradient. However, when an electrical potential gradient was imposed in conjunction with H^+-gradient, gly-sar was transiently accumulated by a factor of about 3, suggesting that gly-sar uptake is linked to the movement of H^+ down its electrochemical gradient.

POTASSIUM TRANSPORT IN ISOLATED ENTEROCYTES

Because all cell boundaries are equally exposed to the bathing medium, isolated enterocytes allow the study of all separate pathways involved in transfer of a given solute across the cell membrane, and evaluation of the contribution of each transfer route to the overall uptake of the solute. The following are results of experiments set up to investigate the presence of separate routes for potassium transfer in isolated enterocytes. This has been accomplished with the aid of chemical agents which selectively inhibit the function of each transport route.

It was found that (Figure 8): a) K^+ uptake was significantly inhibited by ouabain; b) the ouabain-insensitive K^+ uptake was inhibited to the same extent by loop diuretics (bumetanide and furosemide) and by Na^+ (choline substitution) or Cl^- (gluconate substitution) removal from the bathing solutions; c) the ouabain-insensitive K^+ uptake was inhibited by quinine, and the inhibition induced by quinine and bumetanide was additive. These studies suggest that there are at least four separate routes for K^+ transport in chicken enterocytes: a) the Na^+,K^+,ATPase; b) $Na^+/Cl^-/K^+$ cotransport system; c) quinine-sensitive K^+ flux d) a residual K^+ flux.

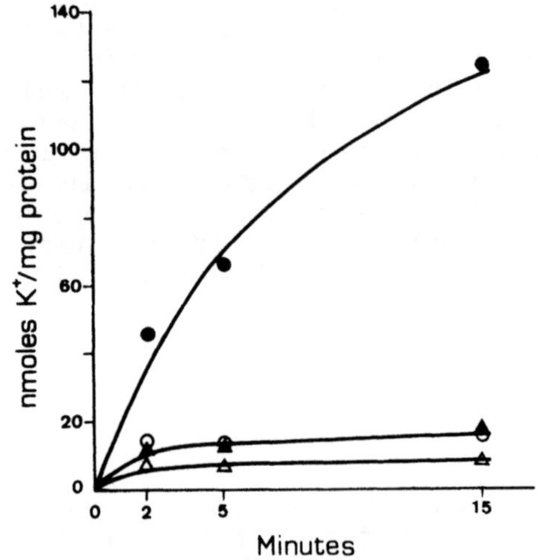

Fig. 8. Routes for K^+ uptake in chicken enterocytes. (●) ouabain-sensitive (Na^+,K^+, ATPase); (▲)ouabain-resistant, bumetanide-sensitive ($Na^+/Cl^-/K^+$ cotransport system); (Δ) ouabain- and bumetanide-resistant, quinine sensitive; (O) residual flux. In these studies $^{86}Rb^+$ was used as a marker for K^+. Once the uptake was terminated as indicated before, the cell pellets were lysed in 1 ml distilled water, and the content of $^{86}Rb^+$ in the incubation buffer and solubilized pellets was estimated by measuring its Cerenkov radiation in a ß-counter. The ouabain-sensitive component was calculated by subtracting in each experiment the 1 mM ouabain-resistant portion from the total ^{86}Rb uptake. The bumetanide-sensitive component is the difference in the uptake of K^+ in the presence and absence of 0.1 mM bumetanide. The quinine-sensitive component is the difference in the uptake in the presence and absence of 125 µM quinine. Residual flux is that portion of K^+ uptake obtained in the presence of ouabain, bumetanide and quinine.(From Montero et al.,1988; Montero & Ilundáin, 1989).

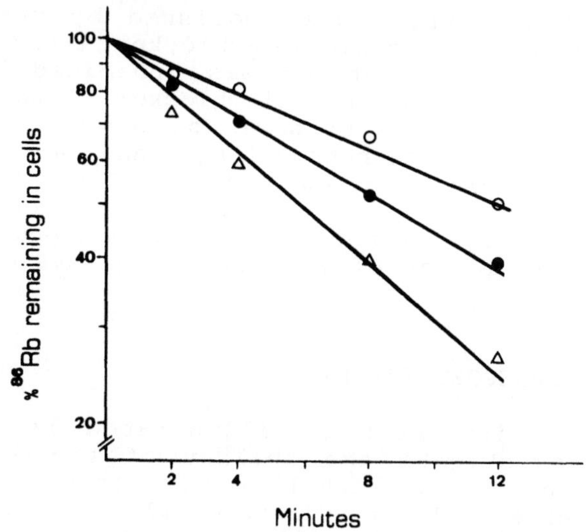

Fig. 9. K^+ efflux from labelled ($^{86}Rb^+$) chicken enterocytes. The rate of $^{86}Rb^+$ loss from the cells was followed by 10-fold dilution of aliquots of labelled suspension with radioisotope free-buffer. Aliquots (200 µl) were taken at time intervals and the content of $^{86}Rb^+$ in the pellets was estimated as indicated in Figure 8. Other details as in Figure 4. (●) control; (△) 5 µg/ml A23187; or (○) 125 µM quinine. (From Montero et al.,1988).

Since K^+ concentration is higher inside cells than outside, mechanism(s) that mediate transport of K^+ down its electrochemical gradient are better characterized by measuring K^+ efflux from the enterocytes. Figure 9 shows that within the time interval studied, K^+ efflux follows first order kinetics with an apparent rate constant of efflux of 0.073 ±0.004 /min.

The exit of K^+ from cells can be mediated either by cotransport systems (Geck & Heinz,1986) and/or by K^+ channels. K^+ channels can be divided into two main classes: those activated by variations in the membrane potential and others activated by Ca^{2+} and modulated by the membrane potential (K_{Ca}) (Schwarz & Passow, 1983; Latorre et al., 1989). Radioisotope flux and patch-clamp studies have revealed the presence of K_{Ca} channels in enterocytes of rabbit (Brown & Sepúlveda, 1985a, b; Sepúlveda & Mason, 1985), rat (Morris et al., 1986) and *Necturus* (Sheppard et al., 1988).

Using isotopic methods inhibitors are useful experimental tools to explore whether K^+ exit from enterocytes is mediated either by cotransport systems and/or by K^+ channels (Brown & Sepúlveda, 1985a,b; Montero et al., 1988; Montero & Ilundáin, 1989). The following observations indicate that K_{Ca} are present in chicken enterocytes and suggest that basal K^+ efflux occurs mainly through these channels: 1) The efflux of K^+ was stimulated by the calcium ionophore A23187. 2) the

A23187-induced K^+ efflux was abolished by the bee venom neurotoxin apamin, a specific K_{Ca} blocker (Lazdunski,1983). 3) The basal rate of K^+ efflux was inhibited by quinine, described as a calcium channel blocker (Schwarz & Passow, 1983), although it appears to have non-specific actions, and by Ba^{2+}, reported to block both voltage- and Ca^{2+}-activated K^+ conductances (Schwarz & Passow, 1983); it was unaffected by 3,4 diaminopyridine, agent thought to selectively block voltage-dependent K^+ channels (Schwarz & Passow, 1983) or by bumetanide, a blocker of K^+-cotransport systems (Geck & Heinz, 1986).

INTRACELLULAR TRANSPORT EVENTS

This last point will be illustrated by experiments designed to investigate the ability of the intracellular organelles of intestinal epithelial cells to store and release Ca^{2+} (Burton et al., 1986; Velasco et al., 1986b; Sepúlveda & Smith, 1987; Ilundáin et al., 1987; Verbost et al., 1987).

In these experiments the plasma membrane is rendered permeable and hence we are dealing with a "suspension of cytosolic components". Therefore the leaky enterocytes are placed in a buffer mimicking the intracellular composition.

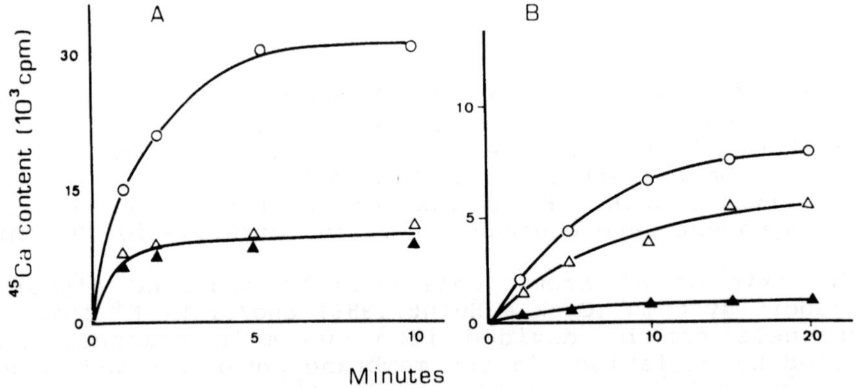

Fig. 10. ^{45}Ca uptake by leaky chicken enterocytes. Cells previously permeabilised with 50 μg/ml saponin were incubated at 37ºC and pH 7.2 in a buffer containing, in mM : 120 KCl; 10 $MgCl_2$; 1.2 KH_2PO_4; 5 Tris ATP; 5 sodium phosphocreatine; 9.5 units/ml creatine phosphokinase; 0.5 mg/ml bovine serum albumin, 25 Tris-HEPES and 5 μCi/ml ^{45}Ca. Initial free calcium concentration was either $2x10^{-5}$ M(A) or 6.7×10^{-7} M (B) obtained by HEDTA and EGTA buffering, respectively. Incubation solutions contained either 3 μg/ml oligomycin and 0.5 μg/ml rotenone (Δ); 3 μg/ml oligomycin, 0.5 μg/ml rotenone and 5 mM vanadate (▲). Control solution (O). (From Ilundáin et al., 1987)

Figure 10 shows that at high Ca^{2+} concentrations Ca^{2+} uptake was rapid and sensitive to a mixture of the mitochondria inhibitors oligomycin and rotenone. Addition of the Ca^{2+}-ATPase inhibitor, vanadate, (that will prevent non-mitochondrial Ca^{2+} sequestration) caused a very small further decrease in Ca^{2+} uptake. When the Ca^{2+} free concentration was reduced to 6.7×10^{-7}M the inhibition by mitochondrial inhibitors was smaller. Addition of vanadate, hovewer, caused a further inhibition. These results suggest that permeable enterocytes transport Ca^{2+} into mitochondrial and non-mitochondrial intracellular stores. Uptake into mitochondrial stores was predominant at high free calcium concentration, while non-mitochondrial stores became dominant at lower Ca^{2+}concentration.

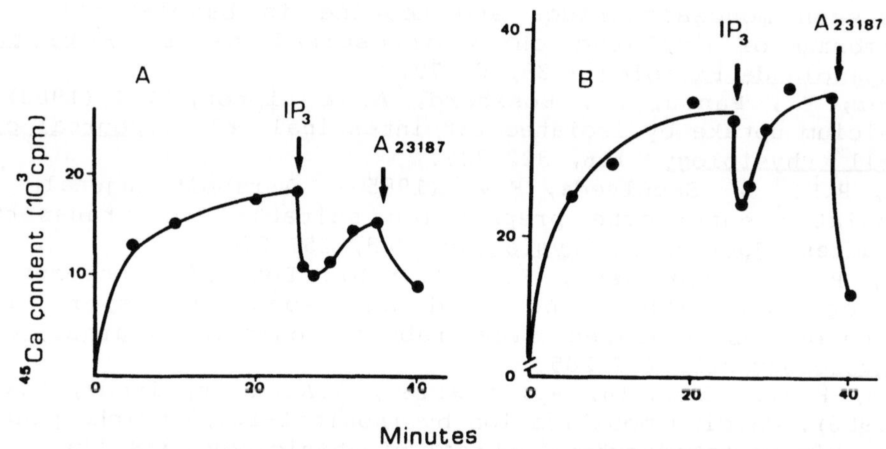

Fig. 11. Effect of IP_3 on the calcium content of leaky chicken-enterocytes. Cells were incubated in the same buffer as that used in figure 10. The initial free calcium concentration was 500 nM as measured with a Ca^{2+}-selective electrode. Incubations were performed in the absence (B) and presence (A) of 3 µg/ml oligomycin and 0.5 µg/ml rotenone. Additions: 5 µM IP_3, 10^{-5} M A23187 (From Ilundáin et al.,1987).

Leaky enterocytes can also be used to explore whether a given agent releases or prevents the release of Ca^{2+} from intracellular stores (Burton et al., 1986; Velasco et al., 1986b; Ilundáin et al., 1987). In these experiments the cells are allowed to take up $^{45}Ca^{2+}$ until steady-state is reached, then the agent under study is added to the incubation buffer and aliquots are taken at short intervals of time. Figure 11 shows that inositol 1,4,5-triphosphate (IP_3), which in a variety of cells is produced in response to external stimulii, causes a rapid and transient release of accumulated Ca^{2+}, both in the presence and in the absence of oligomycin and rotenone. The magnitude of the release was similar in both situations but the re-uptake process was slower in the presence of mitochondrial inhibitors. These findings suggest that Ca^{2+} is released from non-mitochondrial compartments and that, as indicated before, mitochondrial stores are important in sequestering Ca^{2+} when the cytosolic Ca^{2+} concentration is increased.

In summary, isolated enterocytes allow the application of various techniques to study transport mechanisms involved in the transfer of solutes across the cell membranes.

REFERENCES

Ahnen, D.J., Reed, T.A. & Bozdech, J.M. (1988). Isolation and characterization of populations of mature and immature rat colonocytes. American Journal of Physiology 254, G610-G621.

Bolufer, J., Santos, F.J. & Vila, A. (1982). Interactions between monosaccharides and Leucine in basolateral membrane of isolated chick intestinal cells. Revista española de Fisiología 38, 65-70.

Bronner, F., Pansu, D., Bosshard, A. & Lipton, J.H.(1983). Calcium uptake by isolated rat intestinal cells. Journal of Cell Physiology 116, 322-328.

Brown, P.D. & Sepúlveda, F.V. (1985a). A rabbit jejunal isolated enterocyte preparation suitable for transport studies. Journal of Physiology 363, 257-270.

Brown, P.D. & Sepúlveda, F.V. (1985b). Potassium movements associated with amino acid and sugar transport in enterocytes isolated from rabbit jejunum. Journal of Physiology 363, 271-285.

Burton, K.A., Ilundáin, A., O'Brien, J.A. & Sepúlveda, F.V. (1986). Calcium mobilisation by inositol-1,4,5-triphosphate in chicken enterocytes. Journal of Physiology 378,77p.

Calonge, M.L., Ilundain, A. & Bolufer, J. (1989). Ionic dependence of glycylsarcosine uptake by isolated enterocytes. Journal of Cell Physiology 138, 579-585.

Carter-Su, C. & Kimmich, G.A. (1979). Membrane potentials and sugar transport by ATP-depleted intestinal cells: effect of anion gradients. American Journal of Physiology 237, C67-C74.

Carter-Su, C. & Kimmich, G.A.(1980). Effect of membrane potential on Na^+-dependent sugar transport by ATP-depleted intestinal cells. American Journal of Physiology 238, C73-C80.

Crane, R.K. (1968). Absorption of sugars. In "Handbook of Physiology", C.F. Code, ed., sect. 6, vol III, pp 1323 Americal Journal Society , Washington D.C.

Eddy, A.A. (1987). The sodium gradient hypothesis of organic solute transport with special reference to amino acids. In "Amino acid transport in animal cells", ed.Yudilevich,D.L. and Boyd C.A.R., pp 47-86. Manchester University Press.

Ferrer, A., Planas, J.M. & Moretó, M. (1986). Preparation and properties of isolated epithelial cells from chicken caecum and jejunum. Revista española de Fisiología 42, 341-348.

Geck, P. & Heinz, E. (1986). The Na-K-Cl cotransport system. Journal of Membrane Biology 91, 97-105.

Girardi, A.J., Mc Michael, H. & Henle, W. (1956). The use of HeLa cells in suspension for the quantitative study of virus propagation. Virology 2, 532-544.

Ilundáin, A., O'Brien, J.A., Burton, K.A. & Sepúlveda, F.V. (1987). Inositol triphosphate and calcium mobilisation in permeabilised enterocytes. Biochimica et biophysica acta 896, 113-116.

Kaunitz, J.D. (1988). Preparation and charaterization of viable epithelial cells from rabbit distal colon. American Journal of Physiology 254, G502-G512.

Kimmich, G.A. (1975).Preparation and characterisation of isolated intestinal epithelial cells and their use in studying intestinal transport. In: "Methods in membrane biology", Korth, E.,ed., vol IV. Plenum Press, New York, pp 51-115.

Kimmich, G.A.(1981). Intestinal absorption of sugars. In "Physiology of the Gastrointestinal Tract", chapter 41, Johnson L.R., ed., Raven Press, New York, pp. 1035-1061.

Kimmich, G.A., Carter-Su, C. & Randles, J. (1977). Energetics of Na^+-dependent sugar transport by isolated intestinal cells: evidence for a major role for membrane potential. American Journal of Physiology 233, E357-E362.

Kimmich, G.A., Randles, J., Restrepo, D. & Montrose, M. (1985). The potential dependence of the intestinal Na^+-dependent sugar transporter. Annual New York Academic Sciences 456,63-76.

Kimmich, G.A. & Randles, J. (1988). Na^+-coupled sugar transport: membrane potential-dependent Km and Ki for Na^+. American Journal of Physiology 255, C486-C494.

Lazdunski, M. (1983). Apamin a neurotoxin specific for one class of Ca^{2+}-dependent K^+ channels. Cell Calcium 4, 421-428.

Latorre, R., Oberhauser, A., Labarca, P. & Alvarez, O.(1989). Varieties of calcium-activated potassium channels. Annual Review of Physiology 51, 385-399

Lowry, O.H., Rosebrough, N.J., Farr, A.L. & Randall, R.J. (1951). Protein measurement with the Folin phenol reagent. Journal of Biological Chemistry 193, 265-275.

Montero, M.C., Bolufer,J. & Ilundáin, A. (1988). Potassium transport in epithelial cells isolated from small intestine of the chicken. Pflügers Archiv European Journal of Physiology 412, 422-426.

Montero, M.C. & Ilundáin, A. (1989). Effects of anisosmotic buffers on K^+ transport in isolated chicken enterocytes. Biochimica et Biophysica Acta 979, 269-271.

Morris, A.P., Gallacher, D.U. & Lee, J.A.C. (1986) A large conductance, voltage- and calcium-activated K^+ channel in the basolateral membrane of rat enterocytes. FEBS Letters 206, 87-92.

Schwarz, W. & Passow, H. (1983). Ca^{2+}-activated K^+-channels in erythrocytes and excitable cells. Annual Review of Physiology 45, 359-374.

Sepúlveda, F.V. & Mason, W.T. (1985) Single channel recordings obtained from basolateral membranes of isolated rabbit enterocytes. FEBS Letters 191, 87-91.

Sepúlveda, F.V. & Smith, S.M. (1987). Calcium transport by permeabilised rabbit small intestinal epithelial cells. Pflügers Archiv European Journal of Physiology 408, 231-238.

Sheppard, D.N., Giraldez, F. & Sepúlveda, F.V.(1988). Kinetics of voltage- and Ca^{2+} activation and Ba^{2+} blockade of a large-conductance K^+ channel from *Necturus* enterocytes. Journal of Membrane Biology 105, 65-75.

Velasco, G., Dominguez, P., Shears,S.B. & Lazo, P.S. (1986a). Permeability properties of isolated enterocytes from rat small intestine. Biochimica et Biophysica Acta 889, 361-365.

Velasco, G., Shears, S.B., Michell, R.H. & Lazo, P.S. (1986b). Calcium uptake by intracellular compartments in permeabilised enterocytes. Effect of inositol 1,4,5 triphosphate. Biochemical and Biophysical Research Communications 139, 612-618.

Verbost, P.M., Senden, M.H.M.N. & Van Os, C.H.(1987). Nanomolar concentrations of Cd^{2+} inhibit Ca^{2+} transport systems in plasma membranes and intracellular Ca^{2+} stores in intestinal epithelium. Biochimica et Biophysica Acta 902, 247-252.

8

CELL CULTURE IN THE STUDY OF EPITHELIAL DEVELOPMENT AND FUNCTION

F. V. Sepúlveda

AFRC Institute of Animal Physiology
and Genetics Research, Babraham Hall
Cambridge CB2 4AT England

INTRODUCTION

Certain cell lines derived from epithelial tissues have been demonstrated to retain *in vitro* some of the characteristics of their tissue of origin. Polarity of membrane proteins involved in transepithelial transport has been one of the most interesting characteristics that are preserved in these lines, thus allowing the study of ion and non-electrolyte transport in epithelia of simplified cellular composition. In addition they have become the model of choice in studies of the cell biology of membrane polarity generation.

Although there are numerous established cell lines of epithelial origin with potentially interesting properties only a few will be discussed here either because they are the best studied or simply out of personal bias.

<u>Epithelial cell lines of kidney origin</u>: MDCK cells were derived from canine kidney. This was the first line to be identified as forming polarised monolayers *in vitro*. They have also been shown to respond to adrenergic stimulation by secreting chloride ions and have been widely used in studies of the mechanisms of segregation of proteins to the basolateral or apical membrane domains.

LLC-PK_1 cells are from pig kidney origin. They have conserved *in vitro* the sodium-dependent, phloridzin-sensitive glucose transport system in their apical membranes.

<u>Epithelial cells of intestinal origin</u>: T_{84} and Caco-2 (but see below for a list of other lines) cells were derived from colonic cancers. They have been used in studies of the mechanisms of Cl^- secretion and are also being used in the elucidation of intestinal differentiation mechanisms.

CONSERVATION OF EPITHELIAL FEATURES IN MDCK CELLS

Epithelial layers cover the surfaces of body cavities separating a luminal from blood compartments and contributing actively to the differential composition of the two milieux. The expression of this differentiated phenotype is dependent on the morphological and biochemical characteristics of plasma membranes. The cells have an apical membrane in contact with the luminal side, and a basolateral membrane, in contact with

Cell Membrane Transport, Edited by D. L. Yudilevich *et al.*
Plenum Press, New York, 1991

the blood side. Occluding junctions between the cells occur at the apical-basolateral boundary and are thought to be responsible for maintaining the segregation of different membrane protein components. The MDCK cell line was derived from a normal dog kidney. When these cells are grown on standard cell culture glass or plasticware, they form monolayers of closely packed cells with a pavement appearance as observed by phase contrast microscopy. Blisters or hemicysts are also visible which have been shown to result from fluid accumulation into spaces where the monolayer has lifted from the support. MDCK monolayers can also be grown on permeable support, such as Millipore filters (see below), in which case no domes are generated since fluid accumulation at the basolateral side is prevented (Misfeldt et al., 1976; Cereijido et al., 1978).

The presence of domes in cultures of confluent monolayers strongly suggests an epithelial character in the cultured cells. In the case of MDCK cells, electron microscopy of monolayers grown on permeable support reveals that a brush border is expressed at the apical side of the monolayers. Tight-junctions are shown to be present between the cells and provide a boundary to segregate apical and the basolateral membrane. The microvillar membrane is always expressed on the side directly in contact with the bathing medium while the basolateral-type membrane is always that in contact with the support, be it permeable or non-permeable.

Further evidence for the separateness of the two membrane domains comes from experiments on the location of Na^+-K^+-ATPase. This protein occurs at the basolateral membrane of most natural epithelia. Lamb et al. (1981) directly visualised Na^+-K^+-ATPase by autoradiographic observation of the binding of a [^{3}H]-labelled inhibitor of the pump, ouabain. In this way they demonstrated that the pump is exclusively located at the basolateral membrane of MDCK cells.

SUGAR AND AMINO ACID TRANSPORT BY LLC-PK_1 CELLS

The proximal tubule of the kidney and the small intestine are the only two epithelia of the body capable of transporting glucose and other hexoses against their concentration gradients. It is now well documented that the first step in this process is the Na^+-dependent unidirectional transport of the sugar across the brush border membrane. Sugar transport occurs coupled to the transport of Na^+ and it is the electrochemical potential difference for this ion which energises the uphill movement of the sugar. The LLC-PK_1 line was derived by and his co-workers in 1958 (Hull et al., 1976). It has been shown to have a near diploid chromosome number and does not produce tumours in immunosuppressed mice. The morphological characteristics of confluent monolayers of LLC-PK_1 cells are very similar to those described above for MDCK cells. The Na^+-dependent D-glucose uptake mechanism was described for monolayers of LLC-PK_1 cells by Mullin et al., (1980) and Rabito and Ausiello (1980). The non-metabolisable glucose analogue α-methyl-D-glucoside was demonstrated to be concentrated intracellularly by a Na^+-dependent, phloridzin-inhibitable process. Misfeldt and Sanders (1985) grew LLC-PK_1 cells on collagen coated membrane filters and mounted them in Ussing chambers. In these chambers the membranes separate two well stirred compartments allowing measurements of electrical potential differences generated by the epithelial cell layers. A current can also be passed in order to annul the potential difference (short-circuit current), the magnitude of which must be equivalent to the net flux of all the ionic species, which can be measured in parallel with radioisotopes. Addition of glucose to the apical side of an LLC-PK_1 monolayer results in a prompt increase in transepithelial potential and short-circuit current. The increased current is carried by Na^+, is inhibited by phloridzin and can be evoked by other hexoses in the following ranking order of affinities (from highest to lowest): β-methyl-D-glucoside, D-glucose, α-methyl-D-glucoside, 6-deoxy-D-glucose, D-galactose, 3-0-methyl-D-glucoside. Direct evidence for

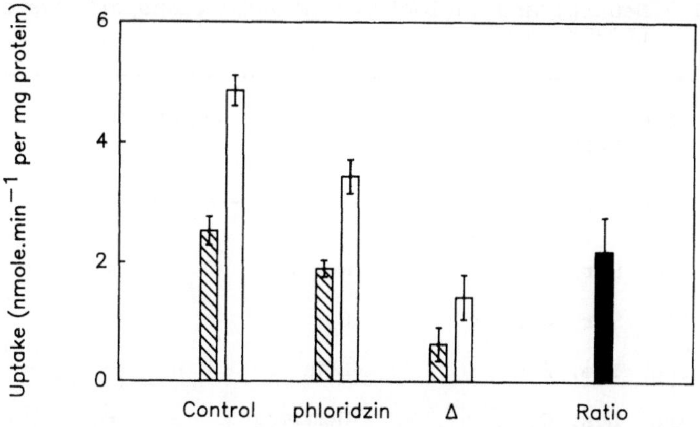

Figure 1. Uptake of Na^+ and glucose across the apical border of an LLC-PK_1 confluent monolayer. Uptakes were measured during 5 min incubations of the apical aspect of a monolayer with ^{22}Na and ^3H-D-glucose. Hatched bars represent glucose uptake and clear bars Na^+ uptake. Glucose and Na^+ concentrations were 30 and 2.5 mM respectively. Δ is the phloridzin (0.1mM)-sensitive uptake from which the "Ratio", uptake of Na^+ to uptake of glucose, was calculated. Values are means ± SEM of eight experiments (modified from Misfeldt & Sanders, 1985).

co-transport of glucose and Na^+ at the apical membrane of LLC-PK_1 cells was also obtained by Misfeldt and Sanders (1985) by measuring unidirectional fluxes of ^{22}Na and [^3H]glucose simultaneously. The results are shown in Fig. 1. The ratio of phloridzin sensitive fluxes of ^{22}Na to that of glucose is 2, suggesting a 2:1 stoichiometry for the co-transport carrier.

The presence of Na^+/glucose co-transport systems in LLC-PK_1 cells suggests that this line might be a useful model for the proximal segment of the nephron. Studies on neutral and cationic amino acid transport by LLC-PK_1 disclosed the presence of the amino acid transport systems listed in Table 1. It was noticed in the course of these experiments that progression of cell growth on plastic lead to a decrease in uptake of neutral amino acids. Autoradiographic studies of the cellular distribution of amino acid taken up revealed marked heterogeneity. Only cells at the periphery of confluent monolayers or at the edges of confluent islands were observed to take up amino acids at high rates (Sepúlveda & Pearson, 1984). A possible explanation for this was that transport systems were located primarily at the basolateral membranes of LLC-PK_1 cells. This was confirmed by comparing the uptake of alanine into cells attached to their impermeable support with that of cells mechanically resuspended or monolayers whose occluding junctions had been permeabilised by exposure to Ca^{2+}-free solutions. These manoeuvres increased amino acid uptake and this was accompanied by an augmentation in Na^+-K^+-pump-mediated potassium uptake. As the pump is located basolaterally in LLC-PK_1 cells these results suggested a similar location for the amino acid transport systems. A similar picture emerges when the uptake of synthetic neutral amino acids is measured into either side of LLC-PK_1 monolayers grown on permeable support (Rabito & Karish, 1982). Studies on lysine transport also suggest that the y^+ system has a basolateral localisation (Sepúlveda & Pearson, 1985). These results suggest that the A, ASC, L and y^+ amino acid transport systems are expressed in LLC-PK_1 cells only in the basolateral membrane domain. This might correlate with their natural localisation. Apical membrane transport systems analogous to those in the proximal tubule cells, do not seem to be expressed by this cell line.

Table 1. Neutral and cationic amino acid transport systems in $LLC\text{-}PK_1$ cells.

System	Preferred Substrates	Specific Substrate	Dependence on Na^+	Li^+	Exchange Capacity
A	Ala, Ser, AIB	MeAIB	+	+	-
ASC	Ala, Ser, Cys	Cys	+	-	+
L	Leu, Ile, Phe	BCH	-	-	+
y^+	Arg, Lys, Orn	Homoarg	-	-	+

Based on data in Sepúlveda and Pearson (1982), Rabito and Karish (1982) and Sepúlveda and Pearson (1985).

CHLORIDE SECRETION AND ITS MODULATION IN MDCK AND COLONIC CARCINOMA CELLS

Colonic carcinoma cells as model small intestinal cells

Recent studies on cell lines established from colon carcinomas have proven interesting for the study of epithelial development and function. Although these cells are of colonic and cancer origin they can mimic *in vitro* some of the characteristics of enterocytic differentiation normally associated with small intestinal cells. This feature has made them a useful tool in studies of the mechanisms associated with the differentiation of the small intestinal epithelium and also in studies of regulation of epithelial transport. More than 50 colonic carcinoma cell lines have been established and published. Table 2, which is taken from a recent review by Zweibaum and Chantret (1989), lists those that have been characterised. They can be classified into cells that show some differentiated characteristics spontaneously in culture, those that can be induced to express these characteristics and finally those which are not differentiated. In addition cells showing vectorial transport are also listed. The list does not include 24 other published lines which have not yet been characterised. According to Zweibaum and Chantret (1989) only Caco-2 cells are spontaneously differentiated, in the sense that, whatever the culture conditions, upon reaching confluency the cells form a polarised monolayer with a fine structure resembling small intestinal cells of tight-junctions and well developed brush border microvilli (see Fig. 2). Enzymes typical of the small intestinal brush-border, but which are absent in the colon, also appear in Caco-2 cells once confluency is reached. These include sucrase-isomaltase, lactase, aminopeptidase N, dipeptidyl-peptidase-N and alkaline phosphatase. Fig. 3, illustrates the presence of sucrase-isomaltase revealed by immunocytochemistry in Caco-2 brush border microvilli.

Caco-2 cells are also able to mediate transepithelial transport as evidenced by the formation of fluid-filled domes past confluency when grown on impermeable support. A similar type of development *in vitro* is obtained with some cell lines whose differentiation *in vitro* has been termed "modulatable" (see Table 2). Under standard growth conditions these cells form multilayered structures with no polarity. When grown in a medium

Table 2. Some human colon carcinoma cell lines classified according to their differentiation characteristics.

Cells showing Transepithelial transport	Enterocyte-like Spontaneous	Enterocyte-like Modulatable	Not differentiated
T_{84}	Caco-2	HT-29	HCT-C
HCT-Gly		HCT-EB	SW-480
HCT-CBS		HCT-Geo	HCT-8R
SW-1116			HCT-R
HCA-7			HCT-Tenn
GCT-ALA			LS-174T
Co-115			HCT-RCA
HCT-Fri			HCT-JVC
HRT-18			HCT-116
HCT-Fet			Moser

Modified from Zweibaum & Chantret (1989)

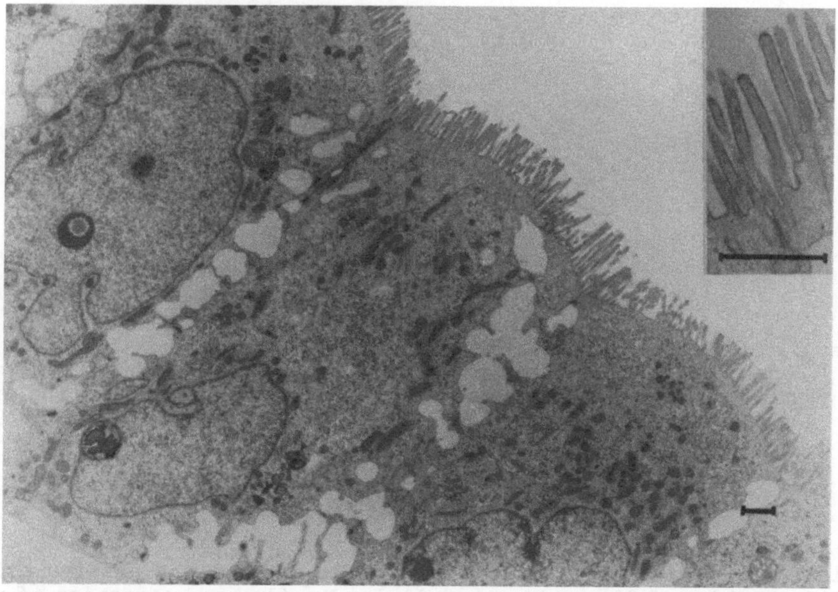

Figure 2. Electron micrograph of a section of Caco-2 cells. Polarity is expressed in this confluent monolayer as evidenced by the presence of apical microvilli and a convoluted basolateral space. The inset shows a higher magnification image of the brush-border microvilli and of a tight-junction. The calibration bars are 1μm long (taken from Zweibaum & Chantret, 1989).

devoid of glucose, however, they undergo a change in morphology and enzyme expression similar to that achieved spontaneously in Caco-2 cells. The differentiation process induced by glucose removal is reversible by subculturing into a medium containing this hexose.

Chloride secretion in MDCK and T_{84} cells

A model for cAMP-dependent Cl^- secretion by the small intestine has postulated the presence of apical membrane Cl^- channels and a mechanism for intracellular Cl^- accumulation across the basolateral membrane. Activation

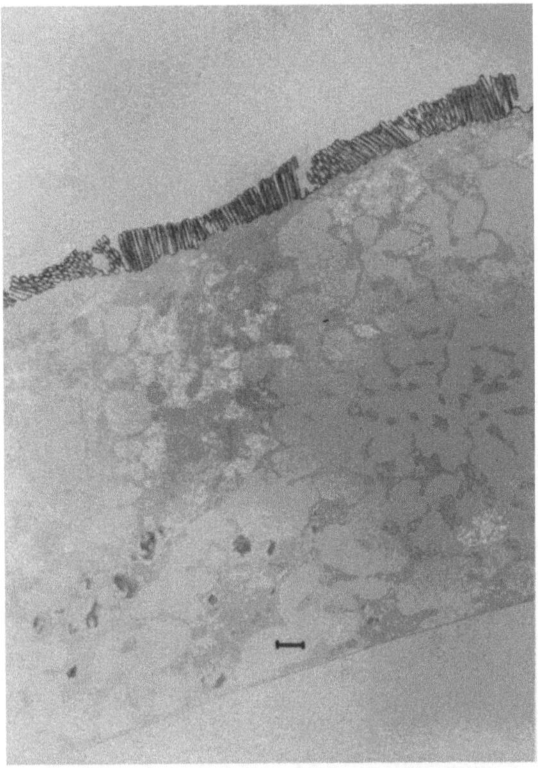

Figure 3. Ultrastructural localisation of sucrase-isomaltase on the brush-border membrane of Caco-2 confluent cells. The electron micrograph shows immunoperoxidase staining as a dark lining of the microvilli. The bar is 1μm long (from Zweibaum & Chantret, 1989).

of apical Cl^- channels by secretagogues would mediate the passive exit of Cl^-. This would be followed by paracellular Na^+ flow and secretion of water would ensue (see Rao & Field, 1983). Analogous Cl^- secretion has been described in cultured epithelial cells that serve as a very useful model in which to study the transport systems and the activation pathways involved in this process. The first cell line shown to effect Cl^- secretion was from kidney origin. Simmons and his coworkers described hormone regulation of transepithelial ion transport in MDCK cells (see Simmons, 1982 for a review). Exogenous ATP, arginine vasopressin, prostaglandin E_1 (PGE_1) and adrenaline all cause increases in short circuit-current. Isotopic flux measurements suggest that increased current

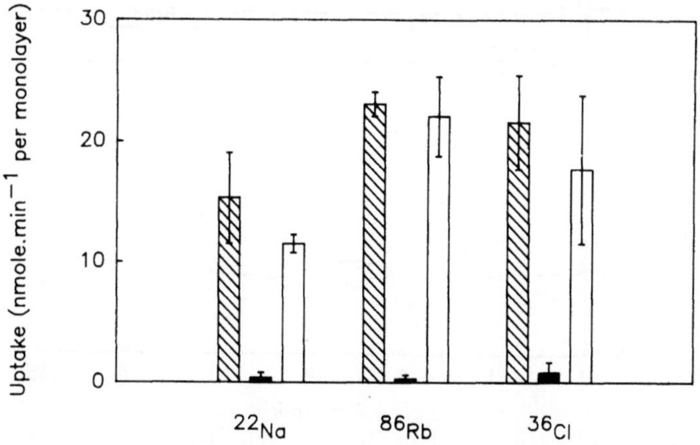

Figure 4. Ion uptake by confluent T_{84} monolayers. Confluent monolayers were grown on permeable support (a collagen-coated polycarbonate filter, 5 µm pore size). Uptake was measured through the basolateral border by exposing only that side of the monolayer to either ^{22}Na, ^{86}Rb (as a tracer for K^+) and ^{36}Cl for a period of 3 min. Measurements were carried out in the absence (hatched columns) or presence (filled columns) of 0.2 mM bumetanide. The bumetanide-sensitive portion of the uptake is also shown (open columns). The monolayers had been previously incubated for 1 h in a buffer containing K^+, Na^+ and Cl^- to deplete the cell of these ions. The uptake buffer contained 0.5 mM ouabain, 70 mM Cl^-, 35 mM Na^+ and 35 mM Rb^+. Equivalent uptake measurements across the apical aspect of the monolayer were not greatly affected by bumetanide and were of the same order of magnitude as those measured in the presence of the inhibitor across the basolateral membrane (filled columns). Values shown are means ± SD of triplicate determinations (After Dharmsathaphorn et al., 1985).

in the presence of adrenaline is caused by chloride ions moving from the basal to the apical bathing solution. The model proposed by Simmons for the MDCK cell is analogous to that proposed for secretagogue-induced secretion of water and electrolytes in the small intestine. Work with T_{84} cells on the other hand, has shown that a similar Cl^- secretion phenomenon can be brought about by a variety of substances. These include vasoactive intestinal peptide, PGE_1, adenosine, cholera toxin, carbachol, histamine and heat-stable toxin of E. coli.

The essential membrane Cl^- transport mechanisms necessary to effect epithelial Cl^- secretion have been described in T_{84} cells (see Dharmsathaphorn & Madara, 1989, for a very comprehensive review of the morphological and transport characteristics of T_{84} cells and for detailed methods for their growth and use in transport experiments). These are a basolateral Cl^- accumulation mechanism, a K^+-recycling mechanism and an apical Cl^- conductance.

Basolateral Na^+, K^+, $2Cl^-$ co-transport mechanism

This cotransporter has been shown to be present in a variety of cells and to be capable of accumulating Cl^- intracellularly above the levels expected for electrochemical equilibrium. A family of drugs, including bumetanide and furosemide, have been shown to be effective inhibitors of this transport system. Evidence for the presence of the co-transporter in

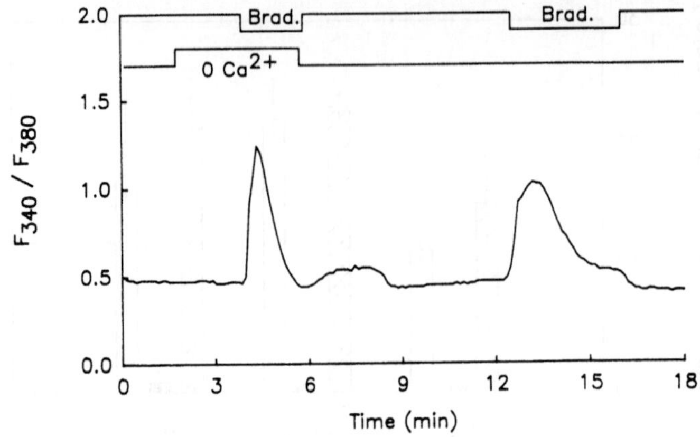

Figure 5. Changes in intracellular Ca^{2+} produced in MDCK cells by bradykinin. Subconfluent MDCK cells grown on glass coverslips were loaded by incubation with the acetoxymethyl ester of Fura-2 (see article by Valdeolmillos and Eisner in this volume). An individual cell was viewed through a fluorescence microscope and illuminated alternately at 340 and 380 nm. The ratio of 500 nm-fluorescence emanating from 340 to that coming from 380 nm excitation was measured; this is proportional to intracellular free Ca^{2+} concentration. At the times indicated the normal extracellular solution was replaced by onedevoid of Ca^{2+} and/or by one containing 10^{-5}M bradykinin (unpublished results from the author's laboratory).

T_{84} monolayers was obtained by measuring the influx of Na^+, K^+ and Cl^- across the basal aspect of confluent monolayers and by testing the effect of bumetanide or the removal of one of the ions on the influx of the others. Fig. 4. shows such an experiment. The radioisotopes Na^{22}, ^{36}Cl and ^{86}Rb were used as tracer substances for Na^+, Cl^+ and K^+ respectively. Influx across the basolateral membranes (measured in the presence of ouabain to abolish the contribution of the Na^+-K^+ pump) was greatly reduced by addition of bumetanide for all three ions. A similar result was obtained by removal of any of the co-ions. Measurements of initial rates, after a 45 s incubation, allowed a calculation of stoichiometry which was $1Na^+:1K^+:2Cl^-$. That this system is an essential step in the secretory process can be demonstrated by the blockade of transepithelial Cl^- secretion by serosal bumetanide.

Recycling of K^+ across the basolateral membrane

Entry of Cl^- across the basolateral membrane can be shown to be accompanied by increased efflux of K^+ to the basal side of the monolayer both in MDCK and T_{84} cells (Brown & Simmons, 1982; McRoberts et al., 1985). This seems to be mediated by Ca^{2+}-dependent K^+ channels and it can often be demonstrated that agonist-induced increased K^+ permeability occurs simultaneously with receptor-induced elevation of cytosolic Ca^{2+}. An example of this is shown in Fig. 5, where it can be seen that bradykinin produces an increase in intracellular Ca^{2+}, measured here as an increase in the ratio of the 340 to 380 nm fluorescence of intracellular Fura-2, in MDCK cells. As with bradykinin action on basolateral K^+ permeability (López-Burillo et al., 1988), the effect on intracellular Ca^{2+} is largely independent of extracellular Ca^{2+} and is probably due to release from intracellular stores. Similar increases in Ca^{2+} are seen in T_{84} cells under the effect of carbachol and histamine. It is not yet clear what the mechanism for increased K^+ recycling is during cAMP- or cGMP-dependent secretion. Differential effects of inhibitors suggests, however, that separate types of K^+ channels might be stimulated by cyclic nucleotides and Ca^{2+} in T_{84} cells.

Presence of Cl^- channels in T_{84} cells

The only regulated Cl^- transport mechanism in the secretory state is an apically located Cl^- channel. Flux measurements with radiolabelled Cl^- through the channel have been difficult to perform due to the low intracellular concentration of the anion and to the low specific activity of $^{36}Cl^-$. It has been possible, nevertheless to study Cl^- influx into T_{84} monolayers taking advantage of the fact that movement through the channel should be bidirectional. In this way it can be demonstrated that vasoactive intestinal peptide (a cAMP-mediated secretagogue) is able to stimulate Cl^- transport which is apically located and insensitive to bumetanide and Na^+ or K^+ removal (Mandel et al., 1986). Direct evidence for the presence of Cl^- channels in T_{84} cells has been obtained recently by the patch-clamp methodology (Halm et al., 1988). The channel is activated by intracellular cAMP and almost identical to that recently described in the natural small intestinal epithelium (Giraldez et al., 1989).

MECHANISM FOR THE SEGREGATION OF APICAL AND BASOLATERAL MEMBRANE COMPONENTS

As it should have become abundantly clear in the previous sections the polarity of epithelial cells is the key to the function of epithelia as vectorial transporters. The mechanisms for generation and maintenance of segregated basolateral and apical domains containing different complements of membrane proteins is one of the most exciting problems in cellular biology. Cultured epithelial cells have been used extensively and most profitably in the study of the generation of polarity. In 1978 Rodríguez-Boulan and Sabatini described that in MDCK cells, influenza, Sendai and Simian virus 5 bud from the apical surface of confluent monolayers while vesicular stomatitis virus and certain retroviruses bud from the basolateral membrane (see Rodríguez-Boulan et al., 1989 for a recent review). Polarised budding is preceded by the preferential segregation of viral glycoproteins. The information for the asymmetrical budding is contained in the proteins themselves as these are directed to the correct membrane domain when expressed from cloned cDNAs. The proteins seem to share a common intracellular biosynthetic pathway up to the point of reaching the Golgi apparatus. They are then sorted in what is termed the trans Golgi network and transported to the correct domain by smooth intracellular membrane vesicles which eventually fuse with the apical or basolateral membrane.

The search for a sorting signal in the proteins which would direct them to their final destination has not yet produced a satisfactory answer. Deletion of the cytoplasmic domain in simian forest virus glycoproteins does not seem to be detrimental for correct basolateral targeting and suggests that whatever signal is involved it must reside in the ectoplasmic domain of these molecules. A firm conclusion in this connection, however, is prevented by some contradictory results and will have to await further work. As for the maintenance of cell polarity, it has long been assumed that actual restriction to protein diffusion on the plane of the membrane is exerted by the tight-junctions. Some experimental evidence, for this has come from monitoring the polarized distribution of fluorescently conjugated lectins in MDCK cells (Dragsten et al., 1981). The probes do not migrate between the apical and basolateral domains under normal conditions. Disruption of the tight-junction structure, i.e. by exposure to low extracellular Ca^{2+}, can be shown, however, to lead to loss of polarity. The absolute necessity of the junctional complex has been challenged by studies showing correct budding of viruses in single cells of sub confluent monolayers (Rodríguez-Boulan et al., 1989). This, and other experiments have led to the proposal that an important role in the generation of polarity must also be played by specific interactions of the proteins with the submembrane cytoskeleton which might get disrupted when the junctions lost.

There are still very large gaps in the understanding of polarity generation in epithelia. The type of approach being followed with cultured epithelia should be able to provide some further answers in the years to come.

A word of caution

Although it is clear that studies with permanent lines of cultured cells have and will be useful in understanding epithelial function and development, they do not replace studies with the naturally occurring epithelia. In that sense they remain models, though very good ones. But regardless of the source, cell lines must be abnormal in some aspects of growth regulation and probably also in other respects. Thus cells grown in culture might express functions not expressed *in vivo* and lack some which are present in the tissue of origin. Despite this they remain a model of choice to establish general principles of epithelial structure and function applicable to the situation *in vivo*.

Acknowledgements

I am very grateful to Dr Alain Zweibaum for providing Figs. 2 and 3, and to Dr K Dharmasathaphorn who kindly made available his review prior to publication.

REFERENCES

Brown, C.D.A. & Simmons, N.L. (1982) K^+ fluxes in tight epithelial monolayers of MDCK cells; evidence for a calcium activated channel. Biochim. Biophys. Acta, 690:95-105.

Cereijido, M., Robbins, E.S., Dolan, W.J., Rotunno, C.A. & Sabatini, D.D. (1978). Polarized monolayers formed by epithelial cells on a permeable and translucent support. J. Cell Biol., 77:853-880.

Dharmsathaphorn, K. & Madara, J.L. (1989) Established intestinal cell lines as model systems for electrolyte transport studies. in: Methods in Enzymol: Biomembranes Biological Transport, ed. S. Fleisher & B. Fleisher. Orlando Florida: Academic Press (in press).

Dharmsathaphorn, K., Mandel, K.G., Masui, H. & McRoberts, J.A. (1985) Vasoactive intestinal polypeptide-induced chloride secretion by a colonic epithelial cell line. Direct participation of a basolaterally localized Na^+, K^+, Cl^- cotransport system. J. Clin. Invest., 75:462-471.

Dragsten, P.M., Blumenthal, R. & Handler, J.S. (1981) Membrane asymmetry in epithelia: Is the tight junction a barrier to diffusion in the plasma membrane? Nature, 294:718-722.

Giraldez, F., Murray, K.J., Sepúlveda, F.V. & Sheppard, D.N. (1989) Characterization of a phosphorylation-activated Cl^--selective channel in isolated Necturus enterocytes. J. Physiol., 416:517-537.

Halm, D.R., Rechkemmer, G.R., Schoumacher, R.A. & Frizzell, R.A. (1988) Apical membrane chloride channels in a colonic cell line activated by secretory agonists. Am. J. Physiol., 254:C505-511.

Hull, R.N., Cherry, W.R. & Weaver, F.W. (1976) The origin and characteristics of a pig kidney cell strain, LLC-PK_1. In Vitro, 12:670-677.

Lamb, J.F., Ogden, P. & Simmons, N.L. (1981) Autoradiographic localisation of [^{3}H]ouabain bound to cultured epithelial cell monolayers of MDCK cells. Biochim. Biophys. Acta, 644:333-340.

López-Burillo, S., O'Brien, J.A., Ilundain, A., Wreggett, K.A. & Sepúlveda, F.V. (1988) Activation of basolateral membrane K^+ permeability by bradykinin in MDCK cells. Biochim. Biophys. Acta, 939:335-342.
Mandel, K.G., Dharmsathaphorn, K. & McRoberts, J.A. (1986) Characterization of a cyclic AMP-activated Cl^- transport pathway in the apical membrane of a human colonic epithelial cell line. J. Biol. Chem., 261:704-712.
McRoberts, J.A., Beuerlein, G. & Dharmsathaphorn, K. (1985) Cyclic AMP and Ca^{2+} activated K^+ transport in a human colonic epithelial cell line. J. Biol. Chem., 260:14163-14172.
Misfeldt, D.S., Hamamoto, S.T. & Pitelka, D.R. (1976) Transepithelial transport in cell culture. Proc. Natl. Acad. Sci., USA, 73:1212-1216.
Misfeldt, D.S. & Sanders, M.J. (1985) Transepithelial transport in cell culture. Mechanisms and bioenergetics of Na^+, D-glucose cotransport. in: Tissue Culture of Epithelial Cells, ed. Taub, M., pp 105-124. New York: Plenum Press.
Mullin, J.M., Weibel, J., Diamond, L. & Kleinzeller, A. (1980) Sugar transport in the LLC-PK_1 renal epithelial cell line, similarity to mammalian kidney and the influence of cell density. J. Cell. Physiol., 104:375-389.
Rabito, C.A. & Ausiello, D.A. (1980) Na^+-dependent sugar transport in a cultured epithelial cell line from pig kidney. J. Membr. Biol., 54:31-38
Rabito, C.A. & Karish, M.V. (1982) Polarized amino acid transport by an epithelial cell line of renal origin (LLC-PK_1) J. Biol. Chem., 257:6802-6808.
Rao, M.C. & Field, M. (1983) Role of calcium and cyclic nucleotides in the regulation of intestinal transport. in: Intestinal Transport, ed. Gilles-Baillien, M. & Gilles, K., pp. 228-239. Berlin:Springer Verlag.
Rodríguez-Boulan, E. & Sabatini, D.D. (1978) Asymmetric budding of viruses in epithelial monolayers: A model system for study of epithelial polarity. Proc. Natl. Acad. Sci., USA, 75:5071-5075.
Rodríguez-Boulan, E., Vega-Salas, D., Sambuy, Y. & Salas, P.J.I. (1989) Mechanisms of segregation of apical and basolateral epithelial domains studied with enveloped viruses. in: Adaptation and Development of Gastrointestinal Function, ed. Smith, M.W. & Sepúlveda, F.V. pp. 115-130. Manchester: Manchester University Press.
Sepúlveda, F.V. & Pearson J.D. (1982) Characterisation of neutral amino-acid transport by cultured epithelial cells from pig kidney. J. Cell. Physiol., 112:182-188.
Sepúlveda, F.V. & Pearson J.D. (1984) Localisation of alanine uptake by ultured renal epithelial cells (LLC-PK_1) to the basolateral membrane. J. Cell. Physiol., 118:211-217.
Sepúlveda, F.V. & Pearson, J.D. (1985) Cationic amino acid transport by two renal epithelial cell lines: LLC-PK_1 and MDCK cells. J. Cell. Physiol., 123:144-150.
Simmons, N.L. (1982) Cultured monolayers of MDCK cells: a novel model system for the study of epithelial development and function. Gen. Pharmacol., 13:287-291.
Zweibaum, A. & Chantret, I. (1989) Human colon carcinoma cell lines as *in vitro* models for the study of cell differentiation. in: Adaptation and Development of Gastrointestinal Function, ed. Smith, M.W. & Sepúlveda, F.V., pp 103-112. Manchester: Manchester University Press.

9

NEW APPROACH TO THE STUDY OF TRANSPORT OF BIOMOLECULES BY MICROCARRIER CELL CULTURES PERFUSED IN A COLUMN APPLYING A HIGH RESOLUTION PAIRED-TRACER TECHNIQUE

S. Perán, M. Muñoz and M.T. Saiz

Membrane Transport Group, Department of Biochemistry and Molecular Biology, Faculty of Medicine, University of Málaga, Málaga, SPAIN

INTRODUCTION

The functional and structural asymmetry of the plasma membrane makes the net uptake of any substrate the result of an interaction between two vectorial components of different magnitudes: one directed to the inside and another one directed to the outside of the cell (Stein, 1986). At the same time, the physiological activity of the cell determines membrane transport not as an isolated process with a finality in itself, but as one step of the metabolic pathways to which is coupled and from which is in many instances dependent. Hence, the mechanisms regulating influx and efflux must be directional. On the one hand, to meet the cellular requirements in response to internal stimuli and, on the other hand, to preserve the functional integrity of the organism responding to extracellular humoral and/or tissue (paracrine) stimuli (Shotwell et al., 1983).

Traditional batch experiments in isolated cells presented major difficulties to discriminate both transport components, mainly due to lack of time resolution, intra-extracellular contamination and/or incomplete washout (Christensen, 1975; Stein, 1986). These problems seriously handicapped the possibility to perform sequential experiments in the same preparation. More than 20 years ago, Yudilevich and Martin de Julian (1965) developed an analysis of double isotope dilution curves which has been applied by other research groups to the study of transport in isolated perfused organs both "in situ" and "in vitro", allowing discrimination between influx and efflux components of net basolateral transport (Boyd & Yudilevich, 1987; Bravo & Yudilevich, 1971; Bustamante et al., 1981; Mann et al., 1986a, b; Mann & Peran, 1986; Muñoz et al., 1988; Yudilevich, 1970; Yudilevich & Mann, 1982; Yudilevich & Swery, 1985) . This technique has the advantage of establishing experimental conditions that resemble the physiological ones and to present a high time resolution in sequential assays in which consecutive stimuli can be applied to the same preparation. However, perfused organs have the disadvantage of

Cell Membrane Transport, Edited by D. L. Yudilevich *et al.*
Plenum Press, New York, 1991

intrinsic non-specificity due to cellular heterogeneity of tissues.

In 1967, Van Wezel developed a technique which improves the yield of cellular cultures by growing cells attached to beaded DEAE-Sephadex . Later on, Smith and Vale (1980) used this technique to study pituitary hormone secretion by packing microcarrier cultures of pituitary cells in a thermostated column and perfusing them with different solutions.

Peran and McGee (1985, 1986) applied the dual tracer dilution technique to the measurement of phenylalanine transport by microcarrier cultures of VERO cells. This new experimental approach overcomes the problems of cellular heterogeneity of perfused organs and it enabled them to analyze, with a simple method, the responses of different transport components to stimuli. Moreover, it has been recently employed to study amino acid transport by perfused microcarrier cultures of endothelial cells (Mann et al., 1989b). At the same time, this method has shown to be valid for the study of hormone transport as well as for the analysis of the assembly of clotting factors on the cellular membrane.

METHODS

Development of microcarrier for tissue cultures

The homogeneous systems (microcarriers), first designed to improve the yield of large scale tissue cultures, can be applied to smaller volumes with good results using either static cultures or suspension cultures.

The original method developed by Van Wezel (1967) used positively charged dextran microcarriers (DEAE-Sephadex 50). This beaded gel was not entirely satisfactory due to its high substitution degree (3.5 meq/g dry weight) which resulted in toxic effects for the cells when utilized at concentrations above 2 mg/ml. Since then, a wide variety of both organic and inorganic microcarriers have been developed which, while they retained all characteristics required for an optimum microcarrier, present specific properties for particular uses (for review see Griffiths, 1986).

Among the wide range of commercially available microcarriers, some of the ones utilized in our laboratory will be discussed in the following paragraphs. The Cytodex series (Pharmacia Fine Chemical, Sweden) consists of three microcarriers types with a common matrix of dextran. Cytodex 1 has similar characteristics to those of DEAE-Sephadex, but still presents a lower substitution degree (1.5 meq/g dry weight) which allows its use at higher concentrations without eliciting any adverse effect on the cells. In contrast to Cytodex 1 which has a positively charged matrix, Cytodex 2 has all charges confined to the bead surface, lowering its substitution degree to 0.6 meq/g dry weight and reducing further toxic effects. Finally, Cytodex 3 presents no charges, having instead a collagen coated surface (60 μg collagen/cm^2 surface).

Another type of microcarrier has a polyacrylamide matrix, positively charged with dimethylaminopropyl groups with a substitution degree of 4 meq/g dry weight (Biocarrier, Bio-Rad, USA). Biosilon microcarriers (Nunc, Demmark) have a polystyrene matrix and a negatively charged surface. Gelibeads microcarriers (KC Biological, USA) are entirely made of gelatine (denatured collagen). Finally, the DE-50 microcarrier (Whatman, USA) presents a cellulose matrix substituted with DEAE (2 meq/g dry weight) and cylindrical shape.

For its utilization in transport studies, the specific adsorptive properties of these gels have to betaken into account since, as we shall show below, they can interfere with the results. Data from our laboratory clearly indicate that Cytodex 3 is an appropriate microcarrier for amino acid transport studies, since its amino acid uptake relative to D-mannitol is only 2-5%. However, Cytodex 3 is not suitable for steroid hormone uptake measurements since more than 80% of the tritiated hormone is retained by nude beads in a single passage. For this kind of transport studies Biosilon microcarriers (≈ 5% retention) may be used instead.

<u>Cell isolation and culture</u>

Human fibroblasts were obtained from a 4 mm ϕ skin biopsy from the subdeltoideal region of the arm. The skin was previously disinfected with an iodinated solution and anaesthetized with mepicain without epinephrine.

Isolation of primary cultures were carried out in a laminar flow hood (Gelaire Twin, Flow Lab., USA) according to the procedure described by Sly and Grubb (1979). Skin biopsies were washed twice in DB(+) balanced salt solution supplemented with glucose, albumin and antibiotics, and then divided into 4-5 fragments. Each of these fragments was transferred to one of the surfaces of a 25 cm culture flask (Corning, USA), covered with fetal calf serum (Flow Lab., USA) and placed in a CO_2 incubator (95% O_2, 5% CO_2) at 37 ºC for 24 hours. After this time, fetal calf serum was replaced by 10 ml supplemented MEM-Glasgow culture medium and the flask left in the CO_2 incubator for another 10 days. When cellular growth surrounding the fragments was observed, the tissue pieces were removed and the medium was changed twice a week until the culture was confluent. Confluent cultures were split into several flasks (2 ml cell solution plus 8 ml culture medium).

Human colonic adenocarcinoma cells were obtained from a surgically removed tumoral piece, according to the procedure described by Bashor (1979), with some modifications. Colonic adenocarcinoma cells were cultured in the same medium as that for fibroblasts and splitting was done when culture reached a high cellular density.

<u>Microcarrier cell culture and perfusion</u>

The perfusions employed the previously described procedure of Perán & McGee (1986). Approximately 8-10 million cells were cultured together with 2 ml of microcarriers beads (Cytodex 3, 25000 beads/ml) in the previously described medium for 24-48 hours in a CO_2 incubator. Once the beads became confluent, cells on beads were packed in a thermostated column (C-10 Pharmacia, Sweden) and perfused with Hanks balanced salt solution supplemented with 5.5 mM D-glucose and 2% BSA. The perfusate was delivered by a peristaltic pump (P3, Pharmacia, Sweden) at constant flow rate of 0.60 ml/min and to a pressure 15-20 mm Hg. At the end of each experiment cell viability was tested by trypan blue staining (>90%).

Uptake of biomolecules by perfused microcarriers cultures

Continuous flux dual-isotope dilution technique

The description and discussion of this technique (Yudilevich, 1970; Yudilevich & Mann, 1982; Yudilevich & Swery, 1985) will be presented in another chapter of this book; only those specially relevant aspects concerning its application to perfused cultures will be discussed here.

The basis of the paired tracer dilution technique to transport studies is the joint administration of a reference tracer molecule (whose diffusion is restricted to the extracellular compartment), together with the test tracer. Both reference and test molecule have to be selected according to their molecular weights and physico-chemical properties to ensure identical passive diffusion coefficients.

The joint administration of these two molecules into the column inflow is followed by a rapid sequential sampling of the effluent, during a single passage through the column. The uptake in each successive sample can be calculated according to the equation 1. Finally, from an effluent sample which the effluent is accumulated for several minutes, cellular efflux of the transported molecule can be estimated according to equation 4.

As shown in Figure 1, the test tracer dilution profile with respect to time, lies under that of the reference tracer, indicating that the diffusible molecule has been diluted into a larger volume. Since this is a two compartmental system, a portion of it has been transferred to the intracellular space, i.e., it has been taken up by the cells. The time-course of this uptake shows a rapidly ascending profile until it reaches a plateau of maximal uptake (Umax), which is followed by a more or less rapid decay due to efflux of the transported molecule and/or its metabolites. The total net uptake (% Ut) obtained from the total recoveries of both tracers is significantly lower than the Umax, suggesting again cellular efflux of the test tracer. This backflux can be calculated from the maximal uptake and the overall uptake from the beginning of plateau, according to equation 4.

In the experiments performed in this work, two experimental models have been utilized: i) pulse administration of radioactive molecules (250 µl/30 s) from which initial influx rates and kinetic parameters of relevant molecules can be estimated (Yudilevich & Mann, 1982), and ii) loading and washout experiments where the isotopes are directly added to perfusates and from which initial and steady-state uptakes as well as cellular efflux for a given molecule can be measured (Boyd & Parsons, 1979).

i) pulse administration experiments: unidirectional uptakes

Rapid uptakes of tritiated molecules (L-^{3}H-serine, 23 Ci/mmol; L-^{3}H-phenylalanine, 37 Ci/mmol; 17β-^{3}H-estradiol, 90

Ci/mmol; ^3H-testosterone, 108 Ci/mmol, and ^3H-Promegestone, 86 Ci/mmol) were measured using a continuous flux high resolution paired tracer dilution technique (Yudilevich & Mann, 1982; Peran & McGee, 1986) during a single passage of the molecules across the cellular network. Briefly, after the administration of a pulse containing both isotopes in the column inflow, 20 samples (60 µl each) were sequentially collected from the column effluent plus a four minute sample which accumulates effluent to asses maximal tracer recovery. The test uptake in each successive sample is measured by directly comparing its dilution profile to that of the reference molecule, D-^{14}C-mannitol, according to the equation:

$$\% U = [1 - (^3H\text{-test}/^{14}C\text{-mannitol})] . 100 \qquad \text{eq. 1}$$

From maximal fractional uptake value (Umax) the corresponding unidirectional influx rate (v, nmol/min.10^6 cells) can be calculated, according to:

$$v = - F . Ca . \ln (1 - Umax) \qquad \text{eq. 2}$$

where F is the perfusion flow (ml/min.10^6 cells) and Ca the perfusate concentration of unlabelled test molecule (mM). In most instances, experimental influx values follow a saturation kinetics that can be fit by a Michaelis-Menten rectangular

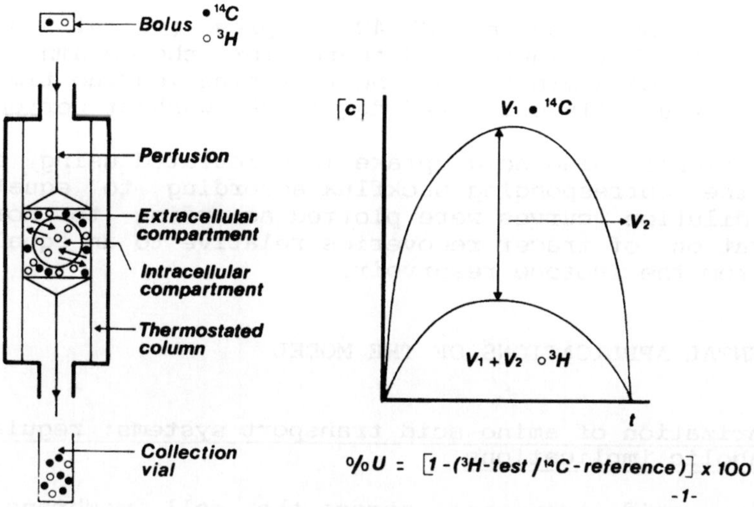

Figure 1. When a bolus containing reference and test tracers is injected in the inlet of a perfusion column in which confluent microcarriers are packed, the non-diffusible molecule (D-^{14}C-mannitol) is only diluted in the extracellular compartment (V1) and the same amount can be recovered as was injected. The diffusible tritiated molecule however crossed the cellular membrane being distributed in both extracellular (V1) and intracellular (V2) compartments. Plotting the amount of recovered ^{14}C and ^3H activities against time, their respective dilution profiles can be attained, and specific uptake of the diffusible molecule can be calculated according to equation 1.

hyperbola and kinetic parameters Kt (mM) and Vmax (µmol/min.10^6 cells) can be estimated. This is the simplest solution, although more complicated models can be applied (Neame & Richards, 1972):

$$v = Vmax \cdot Ca / Kt + Ca \qquad \text{eq. 3}$$

Cellular backflux (%B) is determined in each run from the overall uptake (Ut) from the beginning of plateau of maximal uptake:

$$\%B = [\, 1 - Ut/Umax] \cdot 100 \qquad \text{eq. 4}$$

ii) <u>Loading-washout experiments: initial and steady-state uptakes</u>

This experimental protocol which allows the measurements of initial and steady estate uptakes, is based on the model described by Boyd and Parsons (1979) for the study of sugar transport by the small intestine, and it has been adapted to perfused microcarrier cell cultures. Confluent microcarriers were packed into a small column and perfused with supplemented Hanks solution containing both isotopes, Cells were loaded with tracers for 20-25 minutes before they were switched to an isotope-free perfusate to monitor cellular wash out.

During the wash-in, 30-40 samples (60 µl each) are collected in 3-4 minutes and thereafter the column effluent was sampled once a minute for the remaining loading time. 40-50 samples were usually collected during the washout period.

Fractional amino acid uptake is calculated using equation 1 and the corresponding backflux according to equation 4. Tracer dilution curves were plotted according to normalized concentration of tracer recoveries relative to an equal volume sample from the isotope reservoir.

EXPERIMENTAL APPLICATIONS OF THE MODEL

Characterization of amino acid transport systems: regulation and metabolic implications

Amino acid transport across the cell membrane is a process in which, besides the sequence binding-translocation-dissociation, a full series of control limiting and regulatory factors are involved. These factors modulate transport activities to meet the specific amino acid requirements of the cells (Shotwell, Kilberg & Oxender, 1983). However, even accepting the simplest model, association and dissociation between amino acid and carrier do occur to a velocity that is frequently out of the resolution range of classic batch experiments with isolated cells. Due to this lack of time-resolution, measurement of unidirectional transport rate can be hampered by intracellular metabolism and/or cellular efflux and, consequently, underestimated (Guidotti et al., 1978; Yudilevich & Mann, 1982).

The application of paired tracer dilution technique to the study of amino acid transport in perfused microcarrier cultures during continuous loading overcomes most of these problems. On the one hand, the technique shows a time-resolution close to a second, and, on the other hand, it allows one to maintain the extracellular concentration of substrates, and make sure that metabolic products excreted by the cells are continuously removed, reducing re-uptake to a minimum. Finally, as will be discussed below, this experimental model enables us to study transport regulation by extracellular (hormones, growth factors, drugs, ions, etc) and intracellular (trans-effects, metabolism, etc.) factors.

Here after, we shall discuss the results obtained by the application of this new methodology to the amino acid transport by human colonic adenocarcinoma cells. Panel A in Figure 2 shows dilution profiles of L-[^{3}H]phenylalanine and D-[^{14}C]mannitol obtained by sequentially sampling the column effluent after challenging the preparation with a mix containing both isotopes (300 µl/30s). As it can be seen, the amino acid dilution curve lies below that of D-mannitol. That is, the amino acid has a larger distribution volume than the reference tracer, which in this two compartment model indicates a passage of the amino acid into the intracellular space, i.e., cellular uptake.

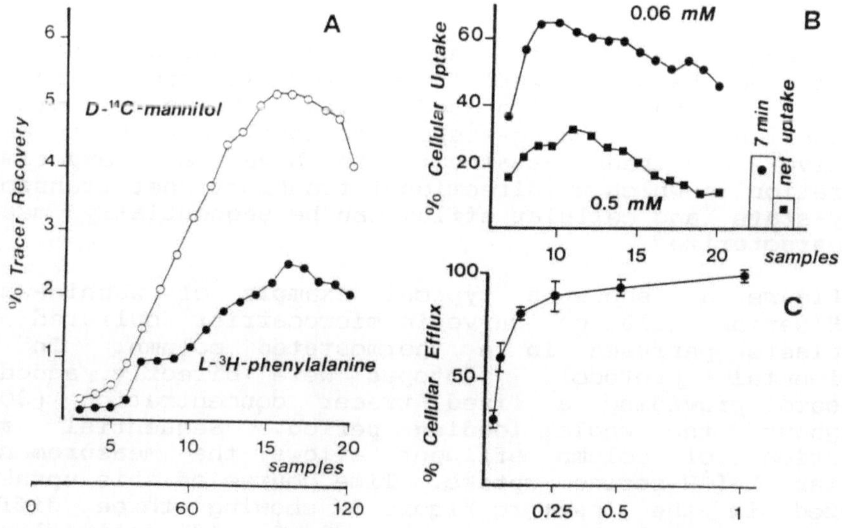

Figure 2. A. Dilution profiles of L-[^{3}H]phenylalanine and D-[^{14}C]mannitol obtained by sequentially sampling effluent of a perfusion column loaded with microcarriers cultures of human colonic adenocarcinoma cells, after the administration of a mix containing both isotopes (300 µl/30 s). B. Cellular uptake of L-[^{3}H]phenylalanine calculated according to equation 1. The uptake is inhibited by increasing concentrations of perfusate unlabelled L-phe. Net 7 min uptakes are also depicted. C. Concentration dependence of L-[^{3}H]phenylalanine efflux, calculated according to equation 4.

Panel B shows the time course of L-[^{3}H]phenylalanine uptake calculated according to eq. 1, which reaches a plateau value within 60 seconds from pulse administration. The maximal uptake value is followed by a decline, suggesting cellular efflux of the transported amino acid. The same panel of Figure 2 shows the specificity of L-[^{3}H]phenylalanine uptake being inhibited by increasing concentrations of unlabelled perfusate L-phenylalanine.

The presence of cellular efflux is also suggested by the fact that 7 min net uptakes have significatively lower values than those obtained for unidirectional uptakes (Fig. 2B).

From maximal initial uptakes measured in the presence of increasing perfusate concentrations of unlabelled L-phe, unidirectional influx of the amino acid can be estimated. Using a Michaelis-Menten analysis, unidirectional influx of L-phenylalanine in colonic adenocarcinoma cells shows a Kt of 0.6 ± 0.2 mM and a Vmax of 172 ± 32 nmol/min.10 cells. This Kt values similar to that reported by Peran & McGee (1986) in VERO cells, where they first described this new application of the paired tracer dilution technique. In the same paper, the authors demonstrated that cell capacity to take up phenylalanine remained constant for at least four hours with Umax values close to 80% and also that phenylalanine uptake can be inhibited in a variable degree by other amino acids such as proline, glycine, threonine, methionine, and phenylalanine itself.

Consequently, this experimental design has been demonstrated to be a powerful tool for exploring amino acid transport systems in isolated cells. Moreover, with the introduction of loading-washout protocols a very remarkable objective has been attained: to have an experimental preparation in which unidirectional transport, net transport at steady-state and cellular efflux can be sequentially measured and characterized.

Figure 3 shows a typical example of washin-washout L-[^{3}H]serine dilution curve in microcarrier cultured human fibroblasts perfused in a thermostated column. In this experimental protocol, isotopes were directly added to perfusate providing a fixed tracer concentration (30 µM) throughout the whole loading period. Sequential sample collection of column effluent allows the measurement of cellular L-[^{3}H]serine uptake. Time course of this uptake is depicted in the inset to Figure 3 showing three different phases. An initial phase in which maximal unidirectional uptake is reached within 30 seconds, is followed by a rapid decline in uptakes values due to efflux of the transported amino acid and/or its metabolites. Finally, a third phase of steady-state between influx and efflux which allows net uptake measurements.

After the 20-25 min loading period, the perfusate is switched to an isotope-free solution to study tracer washout. As it can be seen in Figure 3, once the isotope supply is stopped a rapid inversion on dilution profiles is observed; i.e., less reference than test tracer activity is collected. This is so because the reference tracer is confined to the extracellular space and hence its clearance is faster than that of the tritiated amino acid. Moreover, if at the end of the washout period, when virtually there is no D-[^{14}C]mannitol in the extracellular space, the preparation is pulsed with 10 mM unlabelled L-serine (100 µl/10 s) a prominent peak of H activity is elicited due to trans-stimulation of cellular efflux by homoexchange (Clayman & Scholefield, 1969; Stein, 1986).

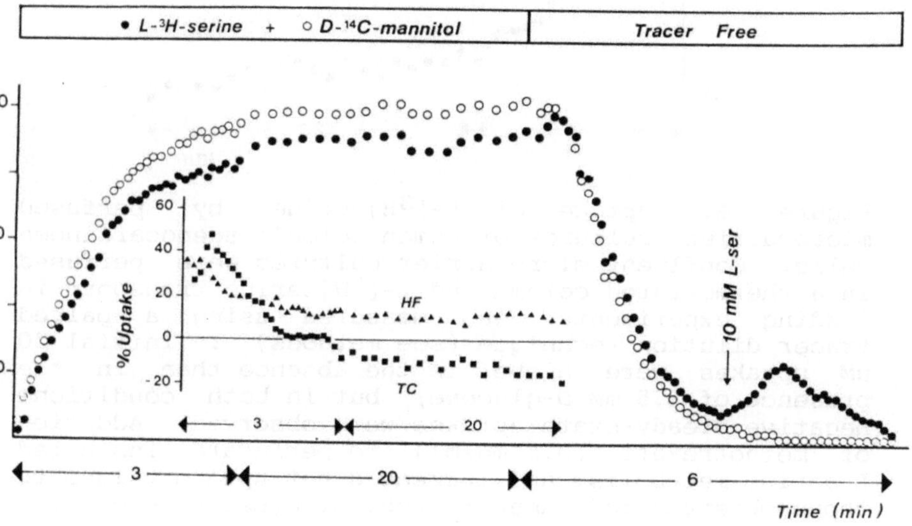

Figure 3. Graphic representation of a L-[^{3}H]serine loading-washout experiment in perfused microcarrier culture of human fibroblasts. Confluent microcarriers were packed into a thermostated column and perfused with a supplemented Hanks solution containing both isotopes to keep a constant tracer supply to the extracellular compartment throughout the loading period. Time course of L-[^{3}H]serine uptakes by human fibroblasts (HF) and human colonic adenocarcinoma cells are depicted in the inset. Once the loading period is completed, the preparation was switched to an isotope-free solution and a rapid inversion of tracer dilution profiles was observed, indicating cellular washout (efflux). Administration of a 10 mM cold serine pulse (100 µl/10 s) elicited a peak of ^{3}H activity in the effluent by trans-stimulation of cellular efflux.

Cellular efflux during the loading period can be estimated by assuming that if efflux were not present, maximal initial uptake would remain constant throughout. Then, the application of equation 4 to each experimental uptake value measured after the achievement of maximal initial uptake, would enable us to follow the time course of both transport components.

The analysis of wash out curves also allows the study of cellular efflux specificity for a given amino acid as well as the characterization of the systems involved. This analysis has been recently published by Mann et al., (1989a) and shall be discussed in another chapter.

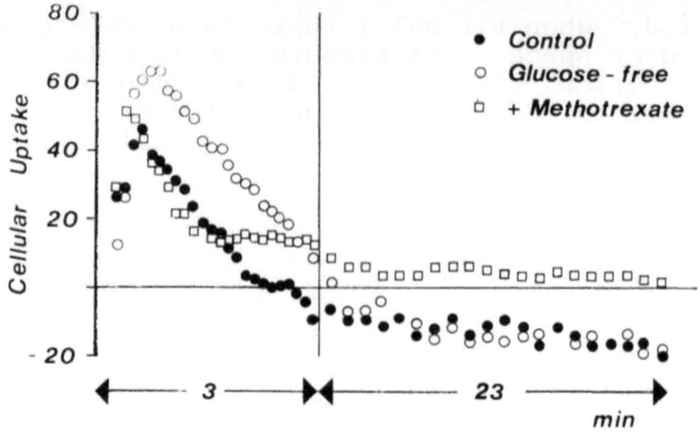

Figure 4. Uptake of L-[^{3}H]serine by perfused microcarrier cultures of human colonic adenocarcinoma cells. Confluent microcarrier cultures were perfused in a thermostated column and L-[^{3}H]serine transport in loading experiments was measured using a paired tracer dilution technique (see methods) . Initial 30 µM uptakes were higher in the absence than in the presence of 5.5 mM D-glucose, but in both conditions negative steady-state uptakes were observed. Addition of methotrexate (0.5 mg/ml) to perfusate inhibited L-serine metabolism and increased net uptake values to those observed for human fibroblasts (see Figure 3).

One of the problems found in the analysis of loading experiments was the efflux of labelled metabolites from cells which is detected as test tracer. So that, when the dilution curves of L-[^{3}H]serine for human colon adenocarcinoma cells were analyzed, it was observed that after reaching a maximal value, L-serine uptake rapidly declined to negative values which remained throughout the loading period (Fig. 4). This negative uptakes values were more likely related to efflux of some labelled metabolite (glycine, H_2O) since it can be prevented by methotrexate (0.5 mg/ml), an inhibitor of TH_4-reductase (Dorr & Fritz, 1980; Dowling et al., 1970).

It is remarkable that treatment of normal human fibroblasts with methotrexate increased initial uptakes of L-serine without having any effect on either net uptake or cellular efflux (Peran et al., 1990). Since L-serine transport by fibroblasts is mostly mediated via a Na-dependent System ASC (Franchi-Gazzola et al., 1982; Gazzola et al, 1980) and since methotrexate inhibits the major metabolic pathway for serine (Snell, 1984, 1986; Thorndike et al., 1979) inducing intracellular accumulation of the amino acid, the observed increased could be due to trans-stimulation of System ASC. On another hand, since this effect is not present in tumoral cells, a direct action of the drug on carrier protein cannot be excluded. Finally, these findings suggest that the cellular efflux observed in L-[^{3}H]serine loading experiments in fibroblasts does actually correspond to that of the original molecule and not to any metabolites, while in tumour cells these might account for a 65% of efflux.

Neoplastic cellular transformation is accompanied by reorientation of metabolism towards increased nucleotide and protein synthesis which requires a continuous supply of substrates for biosynthesis and energy production (Weber, 1983). Glucose and glutamine have been assumed to be the major sources of energy for both processes, although other substrates are extensively taken up by rapidly growing cells (Sauer et al., 1982).

Table I. Effect of glucose deprivation on unidirectional uptake of amino acids by colonic adenocarcinoma cells. In these experiments cells were perfused with a glucose-free Hanks solution for 20 min prior to uptake measurements. Unidirectional uptakes of tritiated phenylalanine, serine, glutamine and lysine were measured in the presence of unlabelled substrate according to the pulse protocol described in Methods. Data were compared by mean of a unpaired Student's t-test; * $p < 0.05$, # $p < 0.005$.

Amino acid	mM	Unidirectional uptake mean ± SE(n)	
		5.5 mM D-glucose	0.0 mM D-glucose
L-phenylalanine	0.1	28 ± 6 (5)	45 ± 5 (3) *
L-serine	0.1	41 ± 2 (7)	52 ± 2 (4) *
L-glutamine	0.5	21 ± 3 (6)	34 ± 5 (6) *
L-lysine	0.1	12 ± 1 (5)	25 ± 5 (3) #

We have applied this methodology to examine whether the glucose levels influence amino acid uptake by tumour cells. As shown in Table I, unidirectional uptake of serine, lysine, phenylalanine and glutamine were increased following a 20 min glucose deprivation. Our findings are consistent with the reciprocal relationship observed between glutamine and glucose oxidation in tumour and fast-growing cells (Kovacevic & Morris, 1972; Zielke et al., 1984).

Transport of steroid hormones across the cytoplasmic membrane of human fibroblast

The effects of steroid hormones are mediated via specific receptors in the nucleus of target cells. It has been proposed and widely accepted that binding occurs in the cytoplasm where the receptor is activated and then translocated to the nucleus (Jensen et al, 1967). However there are several lines of evidence which cast doubt over the proposed mechanisms: King & Greene (1984) using immunochemical techniques with monoclonal antibodies against receptors for estrogen in MCF-7 cells, obtained information which was interpreted as indicating the presence of the cytosolic form of the estrogen receptor in the nucleus. Further, Welshons et al. (1984) have shown non-occupied receptors for estrogen in the nucleus.

In addition, in human and rat pancreas there are certain proteins which bind estrogen but which do not fulfill strict criteria to be considered true receptors (Calandra & Nicola, 1985). Also the presence of steroid receptors on the cytoplasmic membrane (Pietras & Szego, 1977, 1979a) and the non-genomic effects of these hormones (Duval et al., 1983) suggest alternative mechanisms mediating the hormonal effects.

Regardless of the final destination of steroid hormones inside the cell it is clear that the initial step for effect mediation is crossing the cytoplasmic membrane. It has been generally accepted that due to the lipophilic nature of these hormones, passage across the cell membrane is accomplished by simple diffusion (Müller et al., 1977; Müller & Wotiz, 1979). However, evidence accumulated mainly during the last 15 years, questions this thesis (Rao, 1976, 1977a,b; Pietras & Szego, 1979b; Milgrom et al., 1973). Rao (1981) has recently reviewed the subject and emphasized several important characteristics of the steroid hormone uptake which markedly differ from those expected in simple diffusion: saturation kinetics, specificity, high affinity, transport against gradient and energy dependency.

More recently, more complex mechanisms have been suggested to explain the passage of steroid hormones across the cell membranes, including receptor mediated endocytosis (Pietras & Szego, 1985).

Considering the objections against cytoplasmic receptors presented by Szego and Pietras (1984) and more recent data suggesting facilitated membrane transport, it seem possible that steroid hormones could follow two different pathways to mediate effects. One pathway to mediate rapid effects and not requiring genomic mechanisms, i.e., permissive actions. Another pathway, relatively slower would involve nuclear receptors and protein synthesis. The existence of these two pathways would also presuppose mechanisms of internalization other than the widely accepted simple passive diffusion.

There are, however, technical difficulties associated with characterizing mechanisms of rapid transport across the membranes. In the present work we approach the problem of measuring and characterizing transport of steroid hormones across membranes using confluent microcarrier cultures of human fibroblasts perfused in a thermostated column (Peran & McGee,

1986) and applicanting an analysis of dual isotope dilution curves (Yudilevich & Mann, 1982) in loading experiments. This method overcomes most of the problems associated with more traditional approaches and the results obtained provide further evidence for the existence of mediated transport of steroid hormones across the cell membrane.

Figure 5 depicted the uptake curves obtained for estradiol, promegestone and testosterone (TST) by human fibroblasts in perfused microcarrier cultures. For these experiments cells were cultured in the absence of hormone and the perfusion was performed with radiolabelled hormone 100 pM.

In contrast with L-serine (Figure 3), uptake of testosterone increased steadily to reach a maximal value at 60% which was maintained throughout the loading period. Since this stationary phase was reached faster for L-serine and at lower value than that of maximal uptake, these results suggested a different mechanisms mediating TST transport. Estradiol and promegestone uptakes showed a very similar pattern reaching maximal values of 60 and 30%, respectively.

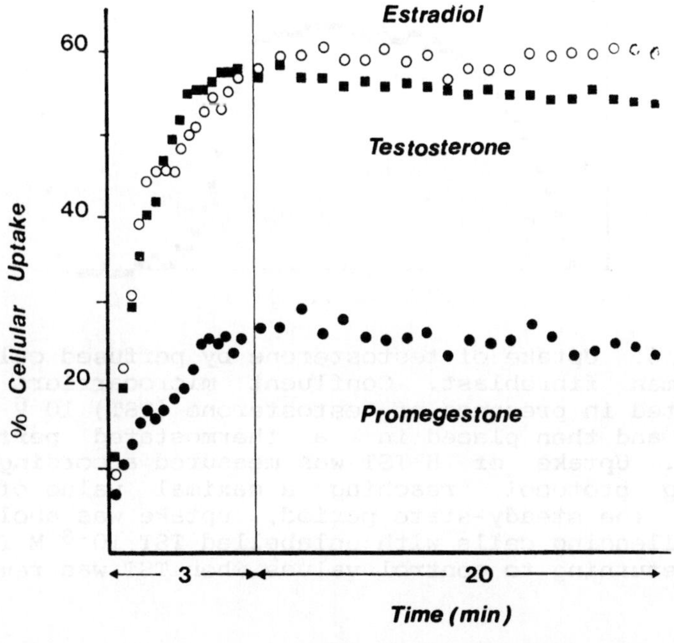

Figure 5. Uptake of steroid hormones by human fibroblasts in perfused microcarrier cultures. Confluent microcarrier cultures of human fibroblasts were placed in a thermostated column and perfused at constant flow rate. After 20 min equilibration period, cells were loaded with tritiated hormones and initial and steady-state uptakes of estradiol (o), testosterone (■) and promegestone (●) were measured, according to the protocol described in Methods.

To avoid possible effects on steroid receptors resulting from hormone deprivation during cell culture, a set of experiments was performed after culturing fibroblasts in the presence of unlabelled TST 10^{-6}M for 24 hours. It is known that steroid homones can modulate the composition and structure of cell receptors. In addition, Jensen et al., (1968) have shown uptake values for estradiol in various tissues which were maximal at 60 min remaining above the basal values after 6 hours, perhaps due to the existence of an excess of receptors.

After incubating human fibroblasts with TST, the steady-state uptake decreased to 30% (Figure 6). As shown in the same figure, this uptake was abolished by addition of TST 10^{-6} M to perfusate, returning to control values after hormone removal. These data support the hypothesis proposing a mechanism of mediated transport for steroid hormones across the cytoplasmic membrane of human fibroblasts.

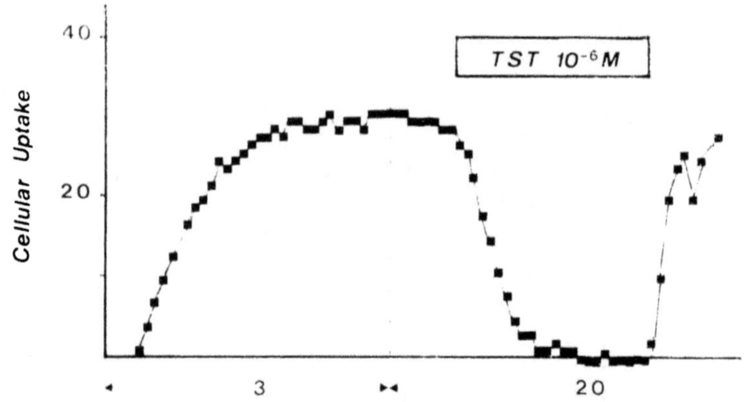

Figure 6. Uptake of testosterone by perfused cultures of human fibroblast. Confluent microcarriers were incubated in presence of testosterone (TST) 10^{-6} M for 24 h and then placed in a thermostated perfusion column. Uptake of H-TST was measured according the loading protocol, reaching a maximal value of 30%. During the steady-state period, uptake was abolished by challenging cells with unlabelled TST 10^{-6} M for 10 min, returning to control values when TST was removed.

Steroid hormones may bind specifically to proteins at the cell membrane and mediate responses with a latency which can vary from seconds to several days (Pietras & Szego, 1977, 1979a; Duval et al., 1983). Mechanisms implicated in the mediation of very rapid responses on neuronal membranes are basically different to those proposed for responses mediated at the nuclear level (Duval et al, 1983). It is possible however that the first step in the uptake of steroids is common, regardless of the pathway of stimulation followed by the hormone and there may be complementary mechanisms for steroid hormone uptake.

Mainly due to the lack of conclusive evidence, it is widely accepted that steroid hormones penetrate the cells by a process of simple diffusion. However more complex transport mechanisms have not been excluded, the time resolution of methods being the most important handicap for their characterization.

To differentiate between the mechanisms, we have applied a high-resolution paired-tracer technique in loading protocols described by Boyd & Parsons (1979), with some modifications to allow for the study of microcarrier cultures maintained in a column. For discussion, results with steroid hormones will be compared to those obtained with L-serine.

Steroid hormone uptakes (exemplified by TST) presented a very different pattern to that of serine, high plateau levels at the steady-state being the main difference. This behaviour does not seem to support the idea of a mechanism of simple diffusion in which a decrease in hormone uptake would be expected as the concentration gradient decreases. In addition, it is known that intracellular steroid concentrations are higher than extracellular ones (Calandra & Nicola, 1985). In our experiments this would be particularly pertinent to uptake measurements effectuated after 24 hours exposure of the cells to TST 1 μM. Cell culture in the presence of unlabelled TST reduced tracer uptake by 50% with respect to standard culture. This is probably due to down-regulation of receptors or saturation of all non-bound receptors. The qualitative characteristics of uptake pattern are the same regardless of culture conditions.

The addition of unlabelled TST during the steady-state phase inhibit uptake in a reversible way, suggesting that the process is specific, saturable and inhibitable.

Finally, the fact that steady-state uptake presents values higher than those obtained during the lineal phase, in the presence of a progressive increase of intracellular steroid concentration, also suggests a mechanism of mediated transport different from either counter-transport or facilitated diffusion.

CONCLUSIONS

The experimental design utilized in these experiments presents 3 major advantages in comparison to traditional methods to measure amino acid transport. Firstly, the model allows a second-to-second follow up of uptake which makes possible a discrimination in the same preparation between unidirectional uptake (linear phase of transport) from net uptake (steady-state phase of transport). Secondly, cellular efflux can be simultaneously measured giving information about the mechanisms controlling substrate exit from the intracellular compartment. Thirdly, it overcomes most of the artifacts derived from intra-extracellular contaminations and avoids the problems derived from stopping the reaction and separation of phases, since the experiments are performed in a sequential way, more closely resembling physiological conditions.

ACKNOWLEDGEMENTS

This work has been financially supported by Grants of FISss 86/1015 and PAI+D 1988. Manuel Muñoz was the recipient of a Beca de Reincorporacion a España, 1988.

REFERENCES

Bashor, M.M. 1979. Dispersion and disruption of tissues. Method. Enzymol., 58:119-131.

Boyd, C.A.R. & Parsons, D.S. 1979. Movements of monosaccharides between blood and tissue of vascularly perfused small intestine. J. Physiol, 287:371-391.

Boyd, C.A.R. & Yudilevich, D.L. 1987. Blood-tissue movement of amino acids in various organs: the basolateral membrane and its importance in the polarity of transepithelial transport. In: "Amino acid transport in animal cells". D.L. Yudilevich & C.A.R. Boyd eds., Manchester University Press, Manchester.

Bravo, I.R. & Yudilevich, D.L. 1971. Liver distribution and uptake of molecules studied by rapid indicator dilution technique. Am. J. Physiol., 221:1449-1455.

Bustamante, J.C., Mann, G.E. & Yudilevich, D.L. 1981. Specificity of neutral amino acid uptake at the basolateral side of the epithelium in the cat salivary gland "in situ". J. Physiol., 313:65-79.

Calandra, R.S. & Nicola, A.F. 1985. "Endocrinología Molecular", 2 ed. El Ateneo, Buenos Aires.

Christensen, H.N. 1975. "Biological Transport", 2d. ed., W.A. Benjamin Inc., Reading, Massachussetts.

Clayman, S. & Scholefield, T. 1969. The uptake of amino acids by mouse pancreas in vitrro. IV. The role of exchange diffusión. Biochim. Biophys. Acta, 173:277-289.

Dorr, R.T. & Fritz, W.L. 1980. "Cancer Chemotherapy Handbook", Elsevier, New York.

Dowling, M.D., Krakoff, I.H. & Karnofsky, D.A. 1970. Mechanism of action of anticancer drugs. In: "Chemotherapy of Cancer", W.H. Cole ed., Lea & Febiger, Philadelphia.

Duval, D., Durant, S. & Delarche, U. 1983. Non-genomic effects of steroids. Interactions of steroids molecules with membrane structures and functions. Biochim. Biophys. Acta, 737:409- 442.

Franchi-Gazzola, R., Gazzola, G.C., Dall'Asta, V. & Guidotti, G.G. 1982. The transport of alanine serine and cysteine in cultured human fibroblasts. J. Biol. Chem., 257:9582- 9587.

Gazzola, G.C., Dall'Asta, V. & Guidotti, G.G. 1980. The transport of neutral amino acids in cultured human fibroblasts. J. Biol. Chem., 255:929-936.

Griffiths, B. 1986. Scaling-up of animal cultures. In: "Animal cell culture. A practical approach", R.I. Freshney ed., IRL Press Ltd., Oxford.

Guidotti, G.G., Borguetti, A.F. & Gazzola, G.C. 1978. The regulation of amino acid transport in animal cells. Biochim. Biophys. Acta, 515:239-266.
Jensen, E.V., Suzuki, T., Kawashima, T. Stumpf, W.E., Jumblut, P.W. & Desombre, E.R. 1967. A two-step mechanism for the interaction of estradiol with rat uterus. Biochemistry, 59: 632-638.
King, W. & Greene, H. 1984. Monoclonal antibodies localize oestrogen receptor in the nuclei of target cells. Nature, 307: 745-747.
Kovacevic, Z. & Morris, H.P. 1972. The role of glutamine in the oxidative metabolism of malignant cells. Cancer Res., 32: 326-333.
Mann, G.E., Habara, Y. & Peran, S. 1986a. Characteristics of L-glutamine transport in the perfused rat exocrine pancreas. Lack of sensitivity to insulin and streptozotocin-induced experimental diabetes. Pancreas, 3:239-235.
Mann, G.E., Muñoz, M. & Peran, S. 1986b. Fasting and refeeding modulates neutral amino acid transport activity in the basolateral membrane of the rat exocrine pancreatic epithelium. Fasting-induced insulin insensitivity. Biochim. Biophys. Acta, 862:119-126.
Mann, G.E. & Perán, S. 1986. Basolateral amino acid transport systems in the perfused exocrine pancreas. Sodium-dependency and kinetic interaction between influx and efflux mechanisms. Biochim. Biophys. Acta, 858:263-274.
Mann, G.E., Norman, P.S.R. & Smith, I.C.H. 1989a. Amino acid efflux in the isolated perfused rat pancreas:trans-stimulation by extracellular amino acids. J. Physiol., 415: 485-502.
Mann, G.E., Pearson, J.D., Sheriff, C-J. Toothill. V.J. 1989b. Expression of amino acid transport systems in cultured human umbilical vein endothelial cells. J. Physiol., 410:325-339.
Martin de Julian, P. & Yudilevich, D.L. 1964. A theory for the quantification of transcapillary exchange by tracer dilution curves. Am. J. Physiol., 207:162-168.
Milgrom, E., Atger, S. & Baulieu, T. 1973. Studies on estrogen entry into uterine cells and on estradiol receptor complex attachment to the nucleus. Is the entry of estrogen into uterine cells a protein-mediated process?. Biochim. Biophys. Acta, 320:267-283.
Müller, R.E., Traish, M. & Wotiz, H. 1977. Interaction of estrogen-receptor complex (R-E) with uterine nuclei. Proc. Natl. Acad. Sci. USA, 43:3464-3468.
Müller, R.E. & Wotiz, H. 1979. Kinetics of estradiol entry into uterine cells. Endocrinology, 105:1107-1114.
Müller, R.E., Traish, M. & Wotiz, H. 1985. Interaction of E2 and estriol with uterine estrogen receptor in vivo and in excised uterior cell suspensions at 37 C: non-cooperative estradiol binding and absence of estriol inhibition of estradiol-induced receptor activation and transformation. Endocrinology, 117:1839-1847.
Muñoz, M., Emery, P.W., Peran, S. & Mann, G.E. 1988. Dietary regulation of amino acid transport activity in the exocrine pancreas. Biochim. Biophys. Acta, 945:273-280.
Neame, K.D. & Richards, T.G. 1972. "Elementary kinetics of membrane carrier transport", Blackwell Scientific Plublications, Oxford.
Peran, S. & Peran-McGee, M. 1985. Transport of L-phenylalanine by isolated VERO cells immobilized in a microcarrier perfusion column: application of a rapid paired-tracer dilution technique. J. Physiol. 360:71P.

Peran, S. & McGee, M.P. 1986. Unidirectional flux of phenylalanine into Vero cells. Measurements using paired tracers in perfused cultures. Biochim. Biophys. Acta, 856:231-236.

Peran, S., Muñoz, M. & Saiz, M.T. 1990. Rapid and steady-state amino acid transport in perfused human fibroblasts and colon adenocarcinoma cells. Effects of methotrexate. Biochim. Biophys. Acta, 1024:233-240..

Pietras, R.J. & Szego, C.M. 1977. Specific binding sites for estrogen at the outer surfaces of isolated endometrial cells. Nature, 265:69-72.

Pietras, R.J. & Szego, C.M. 1979a. Estrogen receptors in uterine plasma membrane. J. Steroid. Biochem., 11:71-83.

Pietras, R.J. & Szego, C.M. 1979b. Metabolic and proliferative responses to estrogen by hepatocytes selected for plasma membrane binding sites specific for estradiol-17β. J. Cell. Physiol., 98:145-160.

Pietras, R.J. & Szego, C.M. 1984. Specific internalization of estrogen and binding to nuclear matrix in isolated uterine cells. Biochem. Biophys. Res. Commun., 123:84-91.

Rao, G., Schulze-Hagen, K., Rao, M. & Brener, H. 1976. Kinetics of steroid transport through cell membranes: comparison of the uptake of cortisol by isolated rat liver cells with binding of cortisol to rat liver cytosol. J. Steroid Biochem., 7:1123-1129.

Rao, M., Rao, G., Eckel, J. & Brener, H. 1977a. Factors involved in the uptake of corticosterone by rat liver cells. Biochim. Biophys. Acta, 500:322-332.

Rao, M., Rao, G. & Brener, H. 1977b. Uptake of estrone, estradiol-17B, and testosterone by isolated rat liver cells. Biochem. Biophys. Res. Commun., 77:566-573.

Rao, G. 1981. Mode of entry of steroid and thyroid hormones into the cells. Mol. Cell. Endocrinol., 21:97-108.

Sauer, L.A., Webster-tayman, J. & Dauchy, R.T. 1982. Amino acid, glucose and lactic acid utilization in vivo by rats tumors. Cancer Res., 42:4090-4097.

Shotwell, M.A., Kilberg, M.S. & Oxender, D.L. 1983. The regulation of neutral amino acid transport in mammalian cells. Biochim. Biophys. Acta, 737:267-284.

Sly, W.S. & Grubb, J. 1979. Isolation of fibroblasts from patients. Method Enzymol., 58:439-443.

Smith, M.A. & Vale, W.W. 1980. Superfusion of rat anterior pituitary cells attached to Cytodex Beads: Validation of a technique. Endocrinology, 102:1425-1431.

Snell, K. 1984. Enzymes of serine metabolism in normal, developing and neoplastic rat tissues. Adv. Enzyme. Regul., 22:325-400.

Snell, K. 1986. Enzymes of serine metabolism in normal and neoplastic rat tissues. Biochim. Biophys. Acta, 843:276-281.

Stein, W.D. 1986. "Transport and diffusion across cell membranes", Academic Press Inc., London.

Szego, C.M. & Pietras, R.J. 1984. Subcellular distribution of oestrogen receptors. Nature, 317:88-89.

Thorndike, J., Pellinemi, F.T. & Beck, W.S. 1979. Serine hydroxymethyltransferase activity and serine incorporation in leukocytes. Cancer Res., 39:3435-3440.

Van Wezel, A.L. 1967. Growth of cell strains and primary cells on microcarriers in homogeneous cultures. Nature, 216:64-65.

Weber, G. 1983. Biochemical strategy of cancer cells and the design of chemotherapy. Cancer Res., 43:3466-3492.

Welshons, W.J., Lieberman, M.A., Gorski, J. 1984. Nuclear localization of unoccupied oestrogen receptors. Nature, 307: 747-749.

Yudilevich, D.L. 1970. Serial barriers to blood-tissue transport studied by single injection indicator diffusion technique. In: "Capillary Permeability", Alfred Benzon Symposium II, C. Crone ed., Munksgaard, Copenhagen.

Yudilevich, D.L. & Mann, G.E. 1982. Unidirectional uptakes of substrates at the blood side of secretory epithelia: stomach, salivary gland and pancreas. Fed. Proc., 41:3045- 3053.

Yudilevich, D.L. & Swery, J.H. 1985. Transport of amino acids in the placenta. Biochim. Biophys. Acta, 822:169-201.

Zielke, H.R., Zielke, C.L. & Ozand, P.T. 1984. Glutamine: a major energy source for cultured mammalian cells. Fed. Proc., 43:121-125.

10

METHODOLOGICAL GUIDE FOR STUDYING EPITHELIAL TRANSPORT WITH ISOLATED MEMBRANE VESICLES

J. Biber and H. Murer

Institute of Physiology, University of Zürich

Winterthurerstr. 190, CH-8057 Zürich

INTRODUCTION

Conceptually, studies with isolated plasma membranes can be placed between traditional physiological approaches and traditional biochemical approaches. In the former, the living cell is regarded as the minimum entity in which processes underlying the function of cells can be investigated; in the latter, cellular components usually have to be purified to homogeneity before studies on the properties of the component can be initiated. Thus studies on isolated membranes can focus either on questions related to cellular physiology, on questions related to the molecular properties of membrane components, or on both.

Transport experiments performed with membrane vesicles can be grouped into three major categories: characterization by biochemical, biophysical and functional methods. Biochemical characterization comprises the kinetic characteristics of a system (e.g. substrate specifity, identification of inhibitors) and also the determination of the molecular weight of the minimum functional unit. Concerning the biophysical characterization, the experiments include, for example, determinations of the driving forces and stoichiometries of transport systems. Functional characterization might involve the interplay of one transport system with other systems in the same membrane or with systems located in the opposite cell membrane, as is the case in an epithelial cell. Isolated membranes have also been used to answer the question of whether or not a particular transport process is subject to an *in vivo* regulation (e.g. by the hormonal status) either through a short term or a long term modification. Finally, isolated membrane vesicles might help to elucidate alterations of membrane transport events associated with disease processes.

For several years, isolated membrane vesicles have been used to investigate the mechanisms of trans-epithelial

Cell Membrane Transport, Edited by D. L. Yudilevich *et al.*
Plenum Press, New York, 1991

transport which is made possible by the existence of quantitative and/or qualitative differences in the transport properties between the luminal (apical, brush border) and the basal-lateral membrane. This article will focus on the use of membrane vesicles isolated from polarized cells and point to some problems inherent to this kind of approach. The general methodological discussion about preparative aspects will be extended by using some practical examples with special emphasis on the isolation of the plasma membranes of the renal proximal tubular cells. Similar procedures have also been applied for the isolation of the microvillus and basal-lateral membranes of other polarized cells, such as enterocytes (Hopfer et al., 1973; Kessler et al., 1978; Mircheff, 1983; Stieger and Murer, 1983), hepatocytes (Meier, 1988) and colonozytes (Biber et al., 1983; Stieger et al., 1986; Dudeja et al., 1989).

ISOLATION OF APICAL AND BASAL-LATERAL PLASMA MEMBRANES

Starting material

The ideal starting material for the purification of plasma membranes is a homogeneous cell population. Unfortunately, this ideal situation is not always fulfilled when working with organs such as the kidneys, the small intestine or the colon. As a compromise, the problem is often partially solved by a crude anatomical dissection of the tissue/organ.

To isolate apical membranes from the proximal tubular cells by the divalent precipitation technique (see below), whole kidney cortex is used as the starting material. Note however, that in the superficial (outer) cortex convoluted parts of the proximal tubules (segments S1 and S2) are dominant, whereas in the inner cortex the straight portions (segments S3) will contribute to the majority of the brush border membranes. Indeed, brush border membranes isolated from either the outer or the inner cortex exhibit different transport kinetics for a variety of substrates (Turner and Moran, 1982; Brunette et al., 1984). In analogy, the small intestine is at first best divided into the duodenum, jejunum and ileum (Lücke et al., 1981).

In contrast to the isolation of the microvillous membranes, a crude anatomical dissection of a tissue does usually not solve the problem of inhomogeneity of basal-lateral membrane preparations, since basal-lateral membranes of different cell types do not differ much with respect to the isolation criteria. Thus, then starting with kidney cortex, basal-lateral membranes of the cells of the distal tubule and of the thick ascending limb of Henle's loop contribute significantly to the purified membrane fraction. To obtain a purer membrane fraction with respect to the basal-lateral membranes of the proximal tubular cells proximal tubules need to be isolated first (see Vinay et al., 1981).

Choice of the method

As a general rule one might say that the procedure used to isolate membrane vesicles to study membrane transport should be fast and should result in high yields of sealed and tight (transport competent) vesicles. Different techniques have been developed and described to isolate luminal and basal-lateral membranes from epithelial cells (Murer and Gmaj, 1986; Mircheff, 1983; Murer et al., 1984; Kinne-Saffran and Kinne, 1989). Most of these methods, such as the divalent precipitation technique or the Percoll density gradient centrifugation (see below) yield reasonably pure brush border or basal-lateral membrane fractions. But note that these methods produce either brush border or basal-lateral membranes separately. Sometimes however, the properties of the apical and basal-lateral membranes need to be compared directly and therefore the two membranes should be exposed during isolation to identical buffer conditions and separation procedures in order to minimize experimental artefacts. As an example, for a simultaneous isolation of the two plasma membranes separation by free flow electrophoresis will be discussed (see below).

Homogenization

With respect to the disruption of a tissue or cells there are no general rules for choosing a certain homogenization method, except that the homogenization should ensure complete physical separation of the different types of membranes and membrane domains and ensure functional integrity of the membranes of interest. Empirically, brush border membrane isolation is often started by homogenizing the tissue under strong conditions (e.g. Warring blender), whereas for preparing basal-lateral membranes the tissue is homogenized rather mildly (e.g. glass-Teflon homogenizer). This different behaviour of the two membranes is most likely due to a different stabilization of the bilayer by the underlying cytoskeletal elements.

In most cases homogenization is performed in a isotonic buffer at a neutral pH-value, which is based on a sugar such as mannitol or sucrose. In order to prevent proteolytic activity (which will be activated by the homogenization) it is recommended that protease inhibitors are included in the homogenization medium.

Isolation of microvillous membranes

The method most frequently used to isolate the luminal membranes of a variety of epithelial tissues is based on the different interaction of the luminal membrane and the other cellular membranes with the divalent ions of magnesium or calcium. This method has been introduced by Schmitz et al.(1973) and subsequently has been used and modified by many others for the isolation of small intestinal (Kessler et al., 1978) and proximal tubular brush border membranes (Booth and Kenny, 1974; Evers et al., 1978; Biber et al.,

SCHEME 1. Isolation of brush border membranes from rat kidney cortex.

Rats of approximately 200 g body weight are killed by a blow on the neck or by decapitation. The kidneys are removed as quickly as possible (prolonged ischemic conditions might influence the final result) and placed in icecold isolation buffer (300 mM D-mannitol, 5 mM EGTA, 12 mM Tris-HCl, 0.1 mM phenyl-mehtyl-sulfonyl-fluoride (PMSF), pH 7.1). The fibrous capsula is removed and thin (<0.5 mm) cortex slices are cut off with a razor blade. Place cortex slices from two kidneys in 15 ml of isolation buffer and homogenize with a Polytron homogenizer (e.g. Kinematica, Switzerland) at high speed (setting 5) for two minutes. Keep suspension cooled during homogenization. Then add 21 ml of chilled water and 430 µl of 1M $MgCl_2$ to yield a final concentration of 12 mM $MgCl_2$. Proceed as follows:

After 15 min. on ice, the sample is centrifuged at 4,500 rpm (1,850 g_{av}) for 15 min.

Pellet$_1$ (P_1) Supernatant$_1$ (S_1)

16,000 rpm (23,600 g_{av}) for 30 min.

S_2 P_2

Resuspend pellet P_2 in 30 ml of 150 mM D-mannitol, 6 mM Tris-HCl(pH 7.4), 2.5 mM EGTA, 0.1 mM PMSF and homogenize with a glass-Teflon potter, add $MgCl_2$ up to 12 mM and after 15 min, centrifuge at 4,500 rpm.

P_3 S_3

16,000 rpm for 30 min.

S_4 P_4

Resuspend P_4 in 30 ml of buffer used for further experiments, homogenize and centrifuge at 16,000 rpm for 30 min.

P_5

Resuspend final brush border membrane fraction in a small volume of the experimental buffer; up to about 10mg of protein/ml. Keep final vesicle suspension on ice until use.

1981). Successful use of this method has also been described for the isolation of the apical membranes of cultured epithelial cells (Moran et al., 1982; Brown et al., 1984; Malmström et al., 1987). The idea behind this technique is that luminal membranes exhibit a high density of negative charges at their surface producing intramembranous saturation of the anionic sites in the presence of divalent cations. In contrast, other membranes, such as basal-lateral membranes will be aggregated through intermembranous cross-linking by divalent cations due to a lower density of negative charges.

Note, that the use of high concentrations of calcium or magnesium (usually 10 mM) can also be problematic. Since calcium for example is known to activate membrane-bound phosholipases and since calcium might directly interact with a transport system, magnesium might be favoured over calcium. In fact as shown by Hauser et al.(1980), intestinal brush border membrane vesicles isolated by the Ca^{++}-precipitation method exhibit high levels of lysophospholipids, whereas in the same vesicles, but isolated by a Mg^{++}/EGTA method, lysophospholipids are almost absent. Furthermore, as demonstrated in Figure 1 (comparing Na^{+}/D-glucose transport in proximal tubular brush border membrane fractions isolated by either the calcium or by the magnesium technique), the choice of the divalent cation might have substantial influence on the leak permeabilities of the membrane vesicles. The differences as illustrated in Figure 1 are most likely due to different non-specific permeabilities of the vesicles for the anions chloride and isothiocyanate. Compared to Fig. 1A, reduced uptake of D-glucose as shown in Fig. 1B is thus most likely due to a high leak permeability for the anions leading to a rapid dissipation of the driving force for the uptake of D-glucose, which is given by the electrochemical gradient of sodium.

A detailed description for the isolation of renal proximal tubular microvillus membranes is given in scheme 1. Note that the precipitation step is performed twice. A second precipitation usually results in a approximately 1.5-fold higher enrichment compared to a single precipitation. When kept on ice, the final vesicles can either be used within the next 10 hours without substantial loss of most of the transport activities and the equilibrium space or can be stored frozen for up to several months in liquid nitrogen.

Isolation of basal-lateral membranes

A convenient and rapid method to isolate the basal-lateral membranes of the small intestine and the kidney cortex is by the use of a density gradient using Percoll as a self orienting medium (Scalera et al., 1980; Scalera et al., 1981). To obtain basal-lateral membranes well separated from the other membranes, a prepurification by differential centrifugation is necessary in order to get rid of most of the nuclei (the presence of nucleic acids usually leads to bad separations), mitochondria and lysosomes. This

Scheme 2. Isolation of basal-lateral membranes from kidney cortex.

Rat kidneys are rapidly removed from the animal, decapsulated and placed into 250 mM sucrose, 10 mM triethanolamine-HCl (pH 7.6) and 0.1 mM phenyl-methyl sulfonylfluorid (always at $4^{o}C$). Cut thin cortex slices (<0.5 mm) and add isolation buffer to make an approx. 10% (w/v) solution. Homogenize the tissue using a glass-teflon homogenizer (first 10 strokes by hand and then 5 strokes at full speed; 1000 rpm).

	Centrifuge homogenate at 2000 rpm for 10 min. Discard pellet 1.
Supernatant 1	
	Recentrifuge for 10 min at 2000 rpm. Discard pellet 2.
Supernatant 2	
	Centrifuge at 14,000 rpm for 20 min.
Pellet 3	
	Resuspend white fluffy top layer of P_3 together with supernatant 3; homogenize gently by three strokes and centrifuge at 14,000 rpm for 20 min. Discard supernatant 4.
Pellet 4	
	Resuspend white fluffy layer in 20 ml of isolation buffer, homogenize gently by three strokes and centrifuge at 14,000 rpm for 20 min. Discard supernatant.
Pellet 5	
	Resuspend white fluffy layer in the kind and amount of buffer required for further experiments:

<u>Free flow electrophoresis</u> <u>Percoll density gradients</u>
(scaled up for starting with five rats)

Free flow electrophoresis: Pellet 5 is resuspended in a small volume (2 ml) of electrophoresis buffer (10 mM triethanolamine, 10 mM acetic acid, 100 mM mannitol, pH 7.4). The buffer for the electrodevessels consists of 100 mM triethanolamine, 100 mM acetic acid, pH 7.4. Electrophoresis is performed at 140 volts/cm and 110 mA (Desaga FF60).

Percoll density gradients: Resuspend pellet 5 in 35 ml of isolation buffer, homogenize by 20 strokes at full speed using a glass-teflon homogenizer. Add Percoll up to 8% (v/v) and centrifuge at 48,000g for 30 min. If possible control centrifugation with an w^2t-integrator.

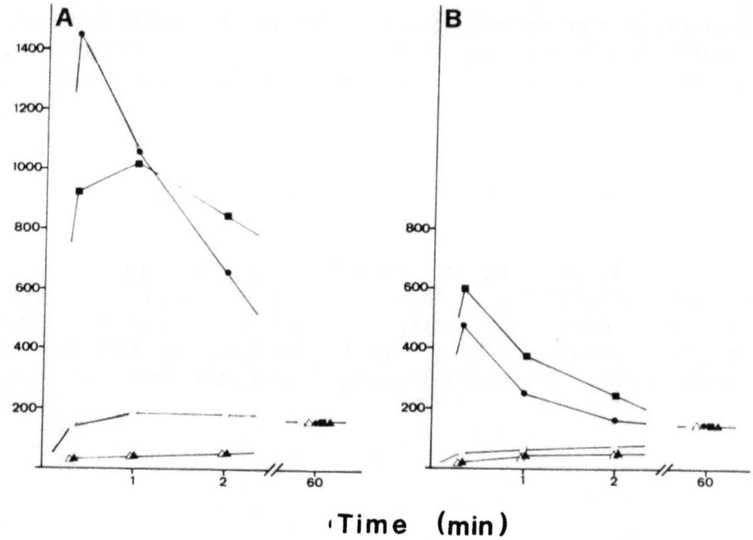

Figure 1. D-glucose uptake by renal brush border vesicles. Vesicles were prepared by either a Mg++/EGTA (A) or by a Ca++ precipitation method (B). Transport was measured either in the presence of inwardly directed gradients of 100 mM NaCl/0.1 mM D-glucose or 100 mM NaSCN/0.1 mM D-glucose (filled symbols) or in the presence of an inwardly directed gradient of 100 mM KCl/0.1 mM D-glucose (open symbols).

prepurification results in the so-called crude plasma membrane fraction. The preparation of such a crude plasma membrane fraction of the kidney cortex is outlined in scheme 2 and a typical separation profile as obtained by a Percoll density gradient centrifugation of such a crude plasma membrane fraction is shown in Figure 2. By pooling of the appropiate fractions it is possible to obtain basal-lateral membranes enriched between 20 and 40-times with respect to the basal-lateral marker enzyme Na/K-ATP'ase. Typically, basal-lateral membranes are recovered from the fractions at a density of 1.038 g/l (in 250 mM sucrose). Since various batches of Percoll might differ slightly, it is recommended to test every batch in order to set up the final Percoll concentration such that marker beads indicating a density of 1.038 g/l are positioned in the upper third of the gradient. To recover the basal-lateral membranes, pooled fractions are suspended in approximately a 10-fold volume of a buffer containing 100mM of NaCl or KCl. This ionic strength allows to centrifuge Percoll at 100000g for 1 hour. Percoll will form a glassy tight pellet and the membranes will form a separate pellet on top of the Percoll pellet.

Although the brush border enzymes are well separated from the basal-lateral marker Na/K-ATP'ase (see Fig.2), brush border membranes recovered from a Percoll gradient are highly cross-contaminated by mitochondria. Thus if brush border membranes recovered from a Percoll gradient would be used for parallel studies, they need to be purified further by e.g. a Mg-precipitation, after Percoll has been removed first.

Simultaneous isolation of apical and basal-lateral membranes

Often it is necessary to purify the two types of plasma membranes of polarized cells under exactly the same conditions in order to compare exactly their different properties. To isolate both, the brush-border and the basal-lateral membranes, simultaneously, separation of the two

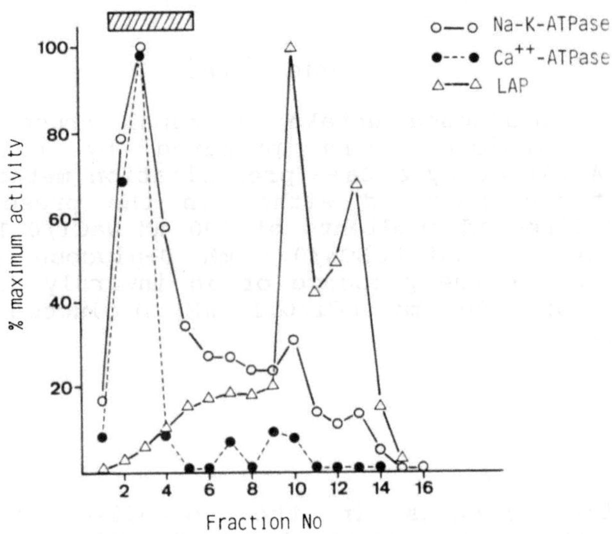

Figure 2. Fractionation of a crude plasma membrane fraction prepared from rat kidney cortex by Percoll gradient centifugation.

membranes by free flow electrophoresis is a convenient way. Since basal-lateral membranes exhibit a higher net negative charge than the apical membranes they are deflected more towards the anode in a electric field (Heidrich et al., 1972; Shlatz et al., 1975; Hannig, 1982; Stieger and Murer, 1983). Although the enrichment factors found for the two membranes are comparable to the methods mentioned above, careful attention has to be given to the problem of cross-contamination: Whereas the basal-lateral fraction often is cross-contaminated by brush border membranes the brush-border membrane fraction often will be contaminated by membranes of the endoplasmic reticulum.

Recently, a simultaneous isolation procedure for the two different epithelial membranes has been introduced, which is based on the different distribution behaviour of the membranes at an interphase formed between two aqueous polymer systems (Mircheff, 1989). Although, theoretically this method has great potential due to the many variables that can be adapted to the specific isolation problem (such as the possibility to modify the net charges of the polymers to influence the distribution of the membranes), the application of the method has been hampered by the low yields of purified membranes suitable for transport experiments.

Storage of vesicles

Storage of isolated vesicles is often an advantage to statistically mix membrane vesicles prepared from different animals and/or prepared on different days. But before freezing membrane vesicles routinely, it is recommended to test carefully first if the transport function(s) of interest will survive such a procedure.

At this point it has to be mentioned that brush border membranes (but not basal-lateral membranes) have also been isolated from frozen (and collected) tissue, such as small intestines or kidney cortex. However notice: Whereas fresh intestines from rats yield excellent vesicles, it has been observed that membranes isolated from frozen rat intestines do not seal properly and therefore are not suitable for transport experiments.

Brush border membranes are normally not critical to freeze and can be frozen directly in an isotonic mannitol buffer. Basal-lateral membranes on the other hand often need the presence of a cryoprotection agent, such as 10% glycerol. To prevent oxidative damage it is recommended to include one to ten millimolar of dithioerythreitol. Freezing is best performed by putting the membranes directly in liquid nitrogen and thawing is normally performed by rapidly warming up the membranes to 37°C.

Purity of the isolated membranes

The purity of isolated mebranes is routinely assessed by the determination of marker enzymes, that is of enzymes which are known to belong to a single cellular component. In polarized cells, aminopeptidases and where present disaccharidases seem to be the most reliable and easy measurable markers for the apical plasma membrane. In some tissues (such as small intestine and proximal tubules) alkaline phosphatase can also be used as a marker for the apical membrane, but in others (e.g. colon) the distribution of this enzyme has to be carefully examined before. For basal-lateral plasma membranes Na/K-ATP'ase is a the most often determined marker enzyme. But note that transport and receptor functions (e.g. adenylate cyclase, Shlatz et al., 1975) of defined localization might also be used. In some cases there might be no enzymatic nor transport function available to follow the purification steps. Such a situation is described by Gorr et al. (1988), who used a monoclonal antibody to follow the purification of the apical membrane of colonocytes.

Often too much emphasis has been placed on the attempt to achieve enrichment factors -that is the ratio of specific activity found in the final membranes over the activity found in the homogenate- as high as possible, overlooking the fact that the enrichment factor in itself is limited by the amount of membrane (relative surface areas) present in the starting homogenate. Also an enrichment factor depends on the degree of the homogeneity of the starting material.

Together with the enrichment factors a complete recovery study should also be presented. Only such data would allow to judge whether an enrichment factor is real and not

Table 1. Specific activities, recoveries and enrichment factors of marker enzymes of rat renal brush border membranes.

Enzyme	spec. activity*) (enrichment)***)	recovery**) (total recovery)
Alkaline phosphatase	11.7 ± 0.18 (17.0± 5.3)	48.3 ± 7.9 (110 ± 11)
Aminopeptidase M	1.2 ± 0.2 (16.7± 3.7)	47.0 ± 9.5 (105 ± 15)
Succinate-cytochrome c oxidoreductase	0.003 ± 0.006 (0.1 ± 0.3)	0.3 ± 0.1 (87 ± 19)
KCN-resistant NADH oxidoreductase	0.006 ± 0.001 (0.010± 0.003)	0.1 ± 0.2 (104± 15)
Na/K-ATP'ase	0.253 ± 0.125 (1.7 ± 1.0)	3.6 ± 1.3 (78 ± 24)
K-phosphatase	0.04 ± 0.01 (2.3 ± 0.8)	6.6 ± 3.6 (104± 14)
Lactate-dehydrogenase	0.048 ± 0.011 (0.06 ± 0.02)	0.2 ± 0.1 (116± 16)
Glucosaminidase	0.025 ± 0.006 (0.3 ± 0.1)	0.7 ± 0.2 (100± 9)

*) Specific activities are given in $\mu mol \cdot min^{-1} \cdot mg^{-1}$.

**) The values represent the percentage of enzyme activity found in the purified membranes compared to the activity of the homogenate which is set to 100%.

***) Ratios of the specific activities (brush border membranes over the homogenate).

related to inactivation or activation of an enzyme activity. Furthermore, recovery data together with morphometric data (if available) would allow estimation of the degree of cross-contamination. As an example: If the total area of a membrane type A in the cell is 10-fold higher than the total area of the membrane to be isolated, a 4% recovery of the marker enzyme of membrane A could mean that there would be a 10% cross-contamination (by area) assuming a 40% recovery of the isolated membrane.

An enzymatic characterization of a rat renal proximal tubular brush border membrane fraction is given in Table I. The following marker enzymes have been used and assayed the same day of preparation: Alkaline phoshatase and aminopeptidase M for the brush border membrane; Na/K-ATP'ase and potassium stimulated phosphatase for the basolateral membrane; glucosaminidase for lysosomes; succinate-cytochrome c oxidoreductase for the mitochondria; KCN-resistant NADH oxidoreducatase for the endoplasmic reticulum and lactate dehydrogenase for cytoplasmic soluble components. As evident from the data, the brush border fraction is reasonably pure with regard to cross-contamination from lysosomes, mitochondria and the endoplasmic reticulum, however the two-fold enrichment of the Na/K-ATP'ase indicates some cross-contamination by basal-lateral membranes, specially if one takes the high relative surface of the basal-lateral membrane into account.

TRANSPORT STUDIES

Transport experiments with radiolabelled isotopes in vesicle populations can provide substantial information about the dynamic properties of transporters. It will not be the aim of this chapter to describe in detail the many experimental conditions used in the past few years, but rather to discuss the basic principles and problems. The various approaches (thermodynamic approach, kinetic approach) and experimental set-up's used to study membrane transport with vesicles are described by Turner (1983) and Hopfer (1989).

The use of radiolabelled isotopes as tracers generally requires low amounts of membrane protein (usually approximately 20 to 50 g per time point). Since at every time point vesicles are separated from the surrounding medium by filtration, transport can be determined only in a non-continuous way. The advantage of continuously monitoring transport processes is offered by spectroscopic techniques, which have been applied for example to analyze electrogenic transports (Schell et al., 1983; Stieger et al., 1983; Eddy, 1989; Freedman and Novak, 1989) and to analyze transports leading to changes in the intravesicular concentration of protons (Warnock et al., 1982; Cassano et al., 1984). More recently, specific dyes have been introduced to study anion transport (such as chloride) in isolated membrane vesicles (Fong et al., 1988). For a detailed discussion of the use of fluorescent dyes to study membrane transport the reader is refered to the article of Eidelman and Cabantchik, 1989.

Basic principles of the rapid filtration technique

A transport experiment using radiolabelled isotopes consists of the following basic steps:

(1) Incubation of an aliquot of the isolated membranes with the labelled substrate for a certain time, at a certain temperature and under certain buffer conditions: To investigate kinetic properties, one has to ensure that transport is analyzed under strictly linear rate conditions. Especially under conditions where transport is determined under the influence of a driving force (such as a sodium-gradient) initial rates are linear for a few seconds only. Incubation times of five seconds and more can be managed manually, but for shorter incubation times it is recommended to use a commercially available rapid mixing device.

(2) Quenching of any solute movement across the membrane: Before separation, the transport process is quenched by diluting the vesicles 50 to 100-fold in a ice-cold isotonic stop solution. However, whereas influx is strongly reduced by dilution, efflux can not be totally blocked to zero during the 10 to 15 seconds needed for filtration and filter-washing. A rough estimation of the substrate loss can be obtained by arbitrarily prolonging the time between the quenching of the reaction and the filtration. If available, inclusion of specific inhibitors might help to reduce loss of the substrate by efflux.

(3) Separation of the substrate in the medium from that associated with the membrane: Immediately after quenching, the membranes are separated by rapid filtration through filters based e.g. on nitro-cellulose of 0.45 to 0.6 μ pore size. Interestingly, membrane vesicles are retained by the filter due to adsorption onto the filter and not due to a true filtration: Note that the mean diameter of apical membrane vesicles is 0.2 m. Adsorption to the filter depends on the buffer composition. Optimal retention is obtained with salt solutions of high ionic strength, such as 250 mM NaCl. Some filters might adsorb enormous amounts of substrate, making it almost impossible to detect the amount specifically taken up by the vesicles. This problem might be overcome by the inclusion of high amounts of the substrate in the stop solution and by presoaking the filters in the stop solution.

(4) Counting the radioactivity: To determine the amount of the isotope specifically taken up, the values have to corrected by an appropiate blank value. Blank values are best obtained by incubating the membranes for zero time (in the cold) with the isotope. The background determined this way will correct for not only for adsorption of the isotope on the filter and for non-specific binding to the membrane but will also account for trapping of the substrate in the filter due to the presence of membrane vesicles.

Membrane orientation

To study kinetic properties of a transport system, a well defined orientation of the vesicular population is required. Concerning brush border membrane vesicles, the vesicle population is of an almost exclusive right-side out orientation (Haase et al., 1978). This has been established by using a variety of techniques, such as by: i) The accessibility of extracellularly localized enzyme acitivities. A measure for the accessibility is to determine the enzyme activity in the presence and absence of a detergent; ii) electron microscopic studies and iii) the use of non-permeating reagents. As demonstrated in Fig.3, the non-permeable reagent ^{32}P-ATP, when added to renal brush border membranes under isotonic conditions (lane A), results in the phosphorylation of a single band only (alkaline phosphatase). However, if ^{32}P-ATP is added together with a detergent (lane B) most of the proteins are phosphorylated. To exclude possible phosphorylation artefacts introduced by the detergent, vesicles have been phosphorylated with ^{32}P-ATP also in hyperosmolar media in order to apply an osmotic shock (lanes C,D,E,for details see Biber et al.(1983a). These results not only demonstrate that most of the brush border membrane vesicles are of a right-side out orientation, but also that the vesicles can be opened transiently by an osmotic shock. Studies on the possibility of incorporating molecules of high molecular weight into the vesicles by an osmotic shock indicated that the maximal molecular weight is in the order of five thousand (unpublished results).

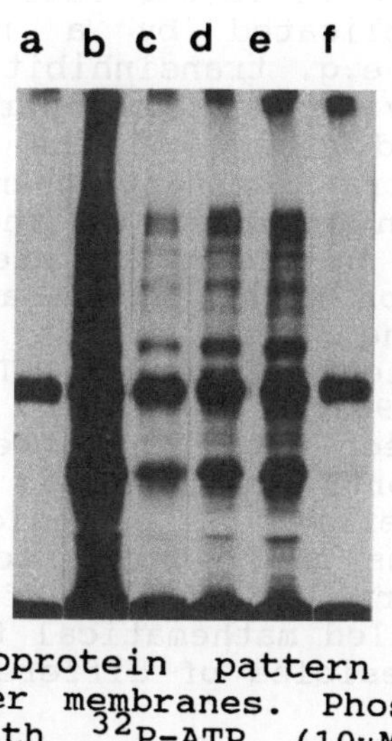

Figure 3. Phosphoprotein pattern of rat kidney cortex brush border membranes. Phosphorylation has been performed with ^{32}P-ATP (10µM, 1 min.,25°C) under isotonic conditions (lane a), in the presence of 0.1% saponin (lane b); under different hypotonic conditions (lane c: 1:4, lane d: 1:8 and lane e: 1:16). Lane f represents phosphorylation of preshocked vesicles.

In contrast to brush border membrane vesicles, random orientation of basal-lateral membrane is what is normally observed (Walter, 1975; Del Castillo and Robinson, 1982; Boumendil-Podevin and Podevin, 1983). As a rough estimation, one has to assume that one third of the membranes form right-side out and that one third form right-side in orientated vesicles. The rest might consist of open, transport incompetent, membrane fragments. A good estimation for the orientation of a basal-lateral membrane fraction can be obtained by the extent of the inhibition of the Na/K-ATP'ase by ouabain when measured either in the absence or in the presence of a detergent (Walter, 1975).

Vesicular size

Special attention has to be given to vesicular size when membrane transport is investigated. Microvillus vesicles isolated from renal proximal tubules or small intestine are of a mean diameter of 0.2 and exhibit an intravesicular volume of one to two microliters/mg protein. On the other hand, the morphological appearance of basal-lateral membranes is highly variable and depends on the homogenization and separation method. Usually the internal volume is stable during the period of the experiment. If necessary, the internal volume can be controlled after preloading the vesicles with an (almost) impermeable tracer by either an osmotic shock or by prolonged incubation (up to several hours). Often, repeated syringing through a 22-gauge needle helps to shorten the time needed to equilibrate the vesicles with the impermeant substrate.

Since vesicles of a diameter in the order of 0.2 exhibit a high surface to volume ratio, kinetic analysis of influx can be complicated by a rapid deviation from linearity (evoked by e.g. transinhibition) due to the rapid change in the intravesicular concentration of the solute. Furthermore the rapid change of the intravesicular buffer composition during transmembrane transport might lead to wrong conclusions in studies on the inhibition of the transport of a solute by another solute. Thus it is possible that mutual inhibition might become apparent for different substrates using the same driving force such as an electrochemical gradient of sodium (Thierry et al., 1982; Barett and Aronson, 1982).

Finally it has been noted that the information obtained by a typical transport experiment is at first information about the mean properties of a vesicle population, since vesicular preparations are heterogeneous in terms of size, shape and/or in terms of the surface concentration of transporters. A detailed mathematical treatment of transport data obtained with vesicles of different sizes is described by Hopfer (1989).

Leak permeabilities

Simple diffusion of a solute across the vesicle membrane is largely due to the intrinsic properties of the membrane and is related to the chemical characteristics of the solute. Thus analysis of the transport characteristics of lipophilic substrates can be very difficult, as influx is mainly governed by the simple diffusion pathway and as efflux during the disequilibrium situation during rapid filtration leads to a significant loss of the solute. Due to the lack of controls for non-mediated uptake, it is almost impossible to analyze specific transport of substrates with a relatively high rate of free diffusion versus a low rate of mediated transport, especially when there is no specific ion requirement or inhibitor sensitivity of the mediated transport.

An additional problem is related to eventually "artificially" introduced permeabilities (such as by preparation artefacts). For example, it has been demonstrated that brush border membranes isolated from kidney cortex and the small intestine exhibit a large (non-specific) proton conductance (Reenstra et al., 1981; Sabolic and Burckhardt, 1984; Cassano et al., 1984). Also, it has been shown that in rabbit jejunal brush border membrane vesicles, the major pathway for sodium is due to a simple diffusion pathway (Gunther and Wright, 1983). Thus, it must be stressed that permeability measurements performed with isolated vesicles can provide information only on the permeability properties of the isolated membrane but might not necessarily be related to the real intrinsic membrane properties as they exist *in vivo*.

Binding problems

In studies with vesicles, the so-called uptake value normally represents the sum of transport into the intravesicular space plus the binding of the solute to the external and internal membrane surface as well as to structures present in the intravesicular space (e.g. cytoskeletal elements). The amount of binding can be analyzed in an equilibrium situation in one of the following three ways: One way would be to analyze the equilibrium uptake of a solute as a function of the osmotic sensitive intravesicular space, which is manipulated by the addition of impermeant solutes (such as mannitol) to the extravesicular space. Ideally, extrapolation of the experimental values to infinite osmolarity (there no intravesicular space is assumed) should reveal zero uptake (Hopfer et al., 1973; Beck and Sacktor, 1975). A second criterion is to compare normalized uptake (solute uptake divided by its concentration in the medium), which should be identical for a variety of different solutes. However this criterion does not necessarily hold for electrolytes, since their equilibrium uptake can also be influenced by Donnan effects, i.e. electrical effects due to fixed charges on the inside of the vesicles. A further test, but measured under non-equilibrium conditions, is performed as follows: Assuming that binding of a solute to the external vesicular surface is a considerably fast process, then linear and initial transmembranal transport can be extrapolated back to a zero incubation time. The amount bound to the external vesicular surface would be indicated by the deviation from zero of the extrapolated line (Biber et al., 1983c).

LOCALIZATION OF TRANSPORT SYSTEMS

Since membrane vesicles can be obtained from either the luminal or the basal-lateral pole of an epithelial cell, the observed transport properties can be attributed to the one or the other membrane (polar distribution of transport systems). Unfortunately, the routine preparations for basal-lateral and brush border membranes are never free of contaminating membrane fractions as indicated by the measurable activities of marker enzymes of different cellular structures (see also Table I and Figure 2). This means that the interpretation of transport data with respect to the localization of transport systems requires careful examination of the contribution of the cross-contaminating membrane.

In the following we will first discuss the problem of cross-contamination of brush border membranes by basal-lateral membranes and vice versa and secondly give a functional approach to determine coexistence of transport systems within the same vesicle. The examples given are by far not a complete list of the possibilities of how to deal with the cross-contamination problem, rather they should stimulate the reader to invent other ways to combine different transports in order to decide whether a given transport system might coexist with another transport system in the same vesicle or not.

Cross-contamination of a membrane preparation

At present the most widely used method for the isolation of the brush border membranes is based on a precipitation technique using the divalent cations Mg^{++} or Ca^{++}. Although this method results in a brush border membrane fraction which is highly enriched (usually in the order of 15-fold) in the corresponding luminally localized marker enzymes, contamination by basolateral membranes will always be a serious problem as is indicated by measurable activities of the Na/K-ATP'ase. An experimental example of how one could deal with this fact/problem is described by Stieger and Murer (1983). When applied onto a discontinuous gradient of 30, 35 and 50% sucrose (w/w) the brush border membrane fraction could be further separated into three membrane populations. Brush border membranes recovered from the 35/50% interphase were found to be further enriched by a factor of 1.5 with respect to the luminal marker enzymes and were found to be 4-fold less contaminated by basal-lateral membranes than the original brush border membranes.

Basal-lateral membrane preparations have to be regarded as being always contaminated to a certain degree by brush border membranes. The following example shall illustrate how one could use different separation criteria in order to correlate a transport activity to one or the other membrane.

As demonstrated by Hagenbuch and Murer (1986), basolateral membranes from rat kidney cortex (isolated by a Percoll gradient) have been further purified by free-flow electrophoresis. This second separation resulted in a clear separation of the activities of Na/K-ATP'ase and leucine-aminopeptidase M. The correlation of the Na/P_i-transport activity with the distribution of the enzyme activities as obtained by the second separation, allowed then to conclude that the Na/P_i-cotransport system has to be attributed exclusively to the apical membrane.

Experimental evidence for the coexistence of transport systems

Usually several transport systems can be identified in a population of membrane vesicles. However, this does not necessarily mean that all of these transport activities coexist in one and the same vesicle, since (as discussed above) a membrane population has to be assumed a priori to be heterogenous.

A first possibility to test for the coexistence of various transport systems would be to further separate a membrane fraction and to correlate transport activities with e.g. enzyme activities as described above. However the possibility always exists that two different types of membranes can not be separated further by the methods available.

The following strategies (see Figure 4) shall illustrate how one might overcome the problem of incomplete fractionation and how to provide evidence for the coexistence of transport systems in the same vesicular structure:

- Example 1: Solute X1 is transported together with sodium via a cotransport mechanism. The resulting increase of the intravesicular concentration of X1 then permits the transport of X2 via an exchange mechanism. Ideally, a concentrative uptake of X2 should be observed. As an example, the anion exchange properties of dog renal brush border membranes (Na/lactate and lactate/urea) are mentioned (Guggino et al., 1983). Another, functionally interesting example is the coexistence of a Na/dicarboxylate cotransport system and a dicarboxylate/ paraaminohippurate exchanger in basolateral membranes from kidney proximal tubules (Shimada et al., 1987).

- Example 2: Two sodium dependent pathways exist. The Na-dependent transport of X1 influences the Na-dependent transport of X2, via competition for the driving force. Typically, this mutual inhibition is observed for potential dependent (electrogenic) transport systems. The inhibition is prevented or reduced by using sodium equilibrated vesicles or by short-circuiting the membrane potential, e.g. by valinomycin and potassium (Thierry et al., 1981; Barett and Aronson, 1982).

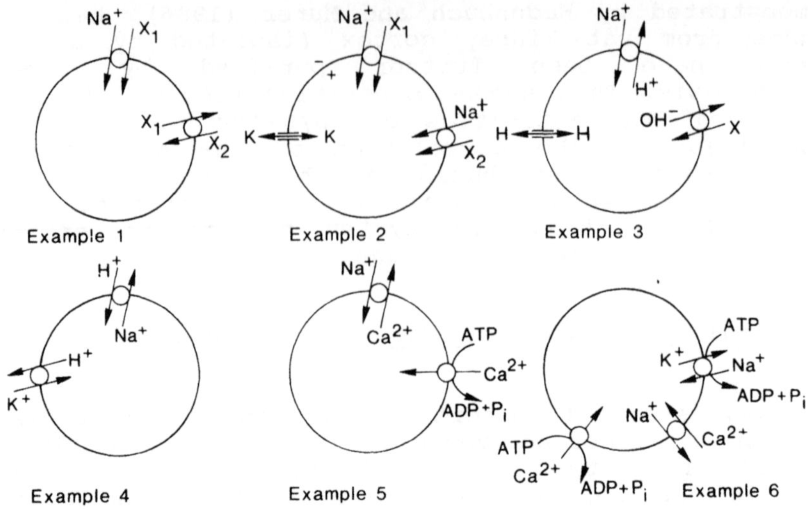

<u>Figure 4</u>. Schematic representations of coupling phenomena of transport mechanisms possibly coexisting in the same vesicular membrane.

- <u>Example 3:</u> An inwardly directed gradient of sodium leads via the operation of the Na/H-exchanger to an alkalinization of the intravesicular space. This results in a increase of the intravesicular pH which could then be used as a driving force for the influx of a solute X via an $OH(HCO_3)/X$ exchanger (Knickelbein et al., 1985).

- <u>Example 4:</u> An outwardly directed gradient of sodium leads to an acidification (measured by the use of acridine orange) of the intravesicular space via the activity of the Na/H-exchanger. This outwardly directed proton gradient can then be used to drive transport of a second solute (Binder and Murer, 1986; Rafizadeh et al., 1986).

- <u>Example 5:</u> In inside out oriented vesicles Ca-uptake could be promoted by the Ca-ATP'ase. Due to the existence of a Na/Ca-exchange mechanism in the same membrane, calcium could be released by the addition of sodium (Gmaj et al., 1979).

- <u>Example 6:</u> This example illustrates the coexistence of the Na/K-ATP'ase, Ca-ATP'ase and the Na/Ca-exchanger. The sodium pump promotes intravesicular, ouabain sensitive accumulation of sodium which might serve as a driving force for the Ca-influx via the Na/Ca-exchanger, which is additional to that mediated by the Ca-pump (Van Heeswijk et al., 1984).

CONCLUSIONS

Studies with vesicles for the elucidation of epithelial transport functions were introduced early in the 1970's (Hopfer et al., 1973; Beck and Sacktor, 1975). Thus they have survived two decades and have certainly not lost the popularity and power to analyze epithelial transport. Such studies have an enormous capacity to give an insight into the properties of the individual transport steps involved in transepithelial transport. In the meantime however we have realized that studies with vesicles also have their limitations as is discussed in part in this article. One major limitation lies in the approach itself, i.e. vesicles are separated from other cellular components/factors which might be required to control the operation/function of the transport steps. Thus the investigator should always keep in mind that in studies with vesicles he is looking only at one small but important piece of the puzzle and that observations at the membrane level need to be integrated into a functional concept at the cellular and organ level.

ACKNOWLEDGMENT

The work mentioned from the author's laboratory was supported by the Swiss National Funds.

REFERENCES

Barett, P.Q. and Aronson, P.S., 1982, Glucose and alanine inhibition of phosphate transport in renal microvillous membrane vesicles. Am. J. Physiol. 242: F126-F131.

Beck, J.C. and Sacktor, B., 1975, Energetics of the Na-dependent transport of D-glucose in renal brush border membrane vesicles. J. Biol. Chem. 250: 8674-8680.

Biber, J., Malmström, K., Scalera, V. and Murer, H., 1983a, Phosphorylation of rat kidney proximal tubular brush border membranes. Role of cAMP dependent protein phosphorylation in the regulation of phosphate transport. Pflügers Arch. 398: 221-226.

Biber, J., Rechkemmer, G., Bodmer, M., Schroeder, P.,Haase, W. and Murer, H.,1983b, Isolation of basolateral membranes from columnar cells of the proximal colon of the guinea pig. Biochim. Biophys. Acta 735: 1-11.

Biber, J.,Stange, G., Stieger, B. and Murer, H., 1983c, Transport of L-cystine by rat renal brush border membrane vesicels. Pflügers Arch. 396: 335-341.

Biber, J.,Stieger, B.,Haase, W. and Murer, H., 1981, A high yield preparation for rat kidney brush border membranes. Different behaviour of lysosomal markers. Biochim. Biophys. Acta 647: 169-176.

Binder, J. and Murer, H., 1986, Potassium/proton exchange in brush border membranes of rat ileum. J. Membrane Biol. 91: 77-84.

Booth, A.G. and Kenny, A.J., 1974, A rapid method for the preparation of microvilli from rabbit kidney. Biochem. J. 142: 575-581.

Boumendil-Podevin, E.F. and Podevin, R.A., 1983, Isolation of basolateral and brush border membranes from the rabbit kidney cortex. Vesicle integrity and membrane sideness of the basolateral fraction. Biochim. Biophys. Acta 735: 86-94.

Brown, C.D.A., Bodmer, M., Biber, J. and Murer, H., 1984, Sodium dependent phosphate transport by apical membrane vesicles from a cultured renal epithelial cell line. Biochim. Biophys. Acta 796: 471-478.

Brunette, M.G., Chan, M., Maag, V. and Beliveau, R., 1984, Phosphate uptake by superficial and deep nephron brush border membranes. Effect of dietary phosphate and parathyroid hormone. Pflügers Arch. 400: 356-362.

Burckhardt, G., 1984, Sodium-dependent dicarboxylate transport in rat renal basolateral membrane vesicles. Pflügers Arch. 401: 254-261.

Cassano, G., Stieger, B. and Murer, H., 1984, Na/H- and Cl/OH-exchange in rat jejunal and rat proximal tubular brush border membrane vesicles: Studies with acridine orange. Pflügers Arch.400: 309-317.

Del Castillo, J.R. and Robinson, J.W.L., 1982, The simultaneous preparation of basolateral and brush-border membrane vesicles from guinea-pig intestinal epithelium, and the determination of the orientation of the basolateral vesicles. Biochim. Biophys.Acta 688: 45-56.

Dudeja, P.K., Foster, E.S. and Brasitus, T.A., 1989, Na/H-antiporter of rat colonic basolateral membrane vesicles. Am. J. Physiol. 257: G624-G632.

Eddy, A.A., 1989, Use of carbocyanine dyes to assay membrane potential of mouse ascites tumor cells. Methods in Enzymology 172: 95-101.

Eidelman, O. and Cabantchik, Z.I., 1989, Continous monitoring of transport by fluorescence on cells and vesicles. Biochim. Biophys. Acta 988: 319-334.

Eveloff, J. and Kinne, R., 1983, Sodium-chloride transport in the medullary thick ascending limb of Henle's loop: Evidence for a sodium-chloride cotransport system in plasma membrane vesicles. J. Membrane Biol. 72: 173-181.

Evers, C., Haase, H. Murer, H. and Kinne, R., 1978, Properties of brush border vesicles isolated from rat kidney cortex by calcium precipitation. Membr. Biochem. 1: 203-219.

Fong, P., Nicholas, P., Widdicombe, J.H. and Verkamn, A.S., 1988, Chloride transport in apical membrane vesicles from bovine tracheal epithelium: Characterization using a fluorescent indicator. J. Membrane Biol. 104: 233-239.

Freedman, J.C. and Novak, T.S., 1989, Optical measurement of membrane potential in cells, organelles, and vesicles. Methods in Enzymology 172: 102-121.

Gmaj, P., Murer, H. and Kinne, R., 1979, Calcium ion transport across plasma membranes isolated from rat kidney cortex. Biochem. J. 178: 549-557.

Gorr, S.-U., Stieger, B., Fransen, J.A., Kedinger, M., Marxer, A. and Hauri, H.-P., 1988, A novel marker glycoprotein for the microvillus membrane of surface colonocytes of rat large intestine and its presence in small-intestinal crypt cells. J. Cell Biol. 106: 1937-1946.

Guggino, S.E., Martin, G.J. and Aronson, P.S., 1983, Specificity and modes of the anion exchanger in dog renal microvillous membranes. Am. J. Physiol. 244: F612-F621.

Gunther, R.D. and Wright, E.M., 1983, Na, Li, and Cl transport by brush border membranes from rabbit jejunum. J. Membrane Biol. 74: 85-94.

Haase, W., Schaefer, A., Murer, H. and Kinne, R., 1978, Studies on the orientation of brush-border membrane vesicles. Biochem J. 172: 57-62.

Hagenbuch, B. and Murer, H., 1986, Phosphate transport across the basolateral membrane from rat kidney cortex: Sodium dependence? Pflügers Arch. 407: 149-155.

Hannig, K., 1982, New aspects in preparative and analytical continous free-flow cell electrophoresis. Electrophoresis 3: 235-243.

Hauser, H., Howell, K., Dawson, R.M.L. and Bowyer, D.E., 1980, Rabbit small intestinal brush border membrane preparation and lipid composition. Biochim. Biophys. Acta 602: 567-577.

Heidrich, H.G., Kinne, R., Kinne-Saffran, E. and Hannig, K., 1972, The polarity of the proximal tubule cell in the rat kidney. Different surface charges for the brush border microvilli and plasma membranes from basal infoldings. J. Cell Biol. 54: 232 - 245.

Hopfer, U., Nelson, K., Perrotto, J. and Isselbacher, K.J., 1973, Glucose transport in isolated brush border membrane from rat small intestine. J. Biol. Chem. 248: 25-32.

Hopfer, U., 1989, Tracer studies with isolated membrane vesicles. Methods in Enzymology 172: 313-330.

Kessler, M., Acuto, O., Storelli, C., Murer, H., Müller, M. and Semenza, G., 1978, A modified procedure for the rapid preparation of efficiently transporting vesicles from small intestinal brush border membranes. Their use in investigating some properties of D-glucose and choline transport systems. Biochim. Biophys. Acta 506: 136-154.

Kinne-Saffran, E. and Kinne, R.K.H., 1989, Membrane isolation: Strategy, Technique, Markers. Methods in Enzymology 172: 3-17.

Knickelbein, R., Aronson, P.S., Schron, C.M., Seifter, J. and Dobbins, J.W., 1985, Sodium and chloride transport across rabbit ileal brush border: II. Evidence for Cl/HCO_3 exchange and mechanism of coupling. Am. J. Physiol. 249: G236-G245.

Lücke, H., Stange, G. and Murer, H., 1981, Sulfate-sodium cotransport by brush-border membrane vesicles isolated from rat ileum. Gastroenterology 80: 22-30.

Malmström, K., Stange, G. and Murer, H., 1987, Identification of proximal tubular transport functions in the established kidney cell line, OK. Biochim. Biophys. Acta 902: 269-277.

Meier, P.J., 1988, Transport polarity of hepatocytes. Seminars in Liver Disease 8: 293 - 307.

Mircheff, A.K., 1983, Empirical strategy for analytical fractionation of epithelial cells. Am.J. Physiol.244: G347-G356.

Mircheff, A.K.,van Os, C.H. and Wright, E.M., 1980, Pathways for alanine transport in intestinal basal lateral membrane vesicles. J. Membrane Biol. 52: 83-92.

Mircheff, A.K., 1989, Isolation of plasma membranes from polar cells and tissues: Apical/basolateral separation, purity, function. Methods in Enzymology 172: 18-33.

Moran, A., Handler, J.S. and Turner, R.J., 1982, Na-dependent hexose transport in vesicles from cultured renal epithelial cell line. Am. J. Physiol. 243: C293-C298.

Murer, H. and Gmaj, P., 1986, Transport studies in plasma membrane vesicles isolated from renal cortex. Kidney Int. 30: 171-186.

Murer, H., Biber, J.,Gmaj, P. and Stieger, B., 1984, Cellular mechanisms in epithelial transport: Advantages and disadvantages of studies with vesicles. Molecular Physiology 6: 55-82.

Rafizadeh, C., Manganel, M., Roch-Ramel, F. and Schaeli, C., 1986, Transport of organic cations in brush border membrane vesicles from rabbit kidney cortex. Pflügers Arch. 407: 404-408.

Reenstra, W.W., Warnock, D.G., Yee, V.J. and Forte, J.G., 1981, Proton gradients in renal cortex brush border membrane vesicles. J. Biol. Chem. 256: 11663-11666.

Sabolic, I. and Burckhardt, G., 1984, Effect of the preparation method on Na/H exchange and ion permeabilities in rat renal brush border membranes. Biochim. Biophys. Acta 772: 140-148.

Sachs, G., Jackson, R.J. and Rabon, E.C., 1980, Use of plasma membrane vesicles. Am. J. Physiol. 238: G151-G164.

Sacktor, B., Rosenbloom, I.L., Liang, C.T. and Cheng, L., 1981, Sodium gradient- and sodium plus potassium gradient-dependent L-glutamate uptake in renal basolateral membrane vesicles. J. Membrane Biol. 60: 63-71.

Scalera, V., Huang, Y.-K., Hildmann, B. and Murer, H., 1981, A simple isolation method for basal-lateral plasma membranes from rat kidney cortex. Membrane Biochem. 4: 49-61.

Scalera, V., Storelli, C., Storelli-Joss, C., Haase, W. and Murer, H., 1980, A simple and fast method for the isolation of basolateral plasma membranes from rat small-intestinal epithelial cells. Biochem. J. 186: 177- 181.

Schell, R.E., Stevens, B.R. and Wright, E.M., 1983, Kinetics of sodium-dependent solute transport by rabbit renal and jejunal brush border vesicles using a fluorescent dye. J. Physiol. (London) 335: 307-318.

Schmitz, J., Preiser, H., Maestracci, D., Ghosh, B.K., Cerda, J.J. and Crane, R.K., 1973, Purification of the human intestinal brush border membrane. Biochim. Biophys. Acta 323: 98-112.

Shimada, H., Moewes, B. and Burckhardt, G., 1987, Indirect coupling to Na^+ of p-aminohippuric acid uptake into rat renal basolateral membrane vesicles. Am. J. Physiol. 253: F795-F801.

Shlatz, L.J., Schwartz, I.L., Kinne-Saffran, E. and Kinne, R., 1975, Distribution of parathyroid hormone-stimulated adenylate cyclase in plasma membranes of cells of the kidney cortex. J. Membrane Biol. 24: 131-144.

Stieger, B., Marxer, A. and Hauri, H.-P., 1986, Isolation of brush border membranes from rat and rabbit colonocytes. J. Membrane Biol. 91: 19-31.

Stieger, B. and Murer, H., 1983, Heterogeneity of brush-border membrane vesicles from rat small intestine prepared by a precipitation method using Mg/EGTA. Eur. J. Biochem. 135: 95-101.

Stieger, B., Burckhardt, G. and Murer, H., 1983, The application of a potential-sensitive cyanine dye to rat small intestinal vesicles. Biochim. Biophys. Acta 732: 324-326.

Thierry, J., Poujeol, Ph. and Ripoche, P., 1981, Interactions between Na-dependent uptake of D-glucose,phosphate and L-alanine in rat renal brush border membrane vesicles. Biochim. Biophys. Acta 647: 203-210.

Turner, R.J., 1983, Quantitative studies of cotransport systems: models and vesicles. J. Membrane Biol. 76: 1-15.

Turner, R.J. and Moran, A., 1982, Further studies of proximal tubular brush border membrane. D-glucose transport heterogeneity. J. Membrane Biol. 70: 37-45.

Van Heeswijk, M.P., Geertsen, J.A. and van Os, C.H., 1984, Kinetic properties of the ATP-dependent Ca-pump and the Na/Ca-exchange system in basolateral membranes from rat kidney cortex. J. Membrane Biol. 79: 19-31

Vinay, P., Gougoux, A. and Lemieux, G., 1981, Isolation of a pure suspension of rat proximal tubules. Am. J. Physiol. 241: F403-F411.

Walter, H., 1975, Tightness and orientation of vesicles from guinea pig kidney estimated from reactions of adenosine triphosphatse dependent on sodium and potassium ions. Eur. J. Biochem. 58: 595-601.

Warnock, D.G., Reenstra, W.W. and Yee, V.J., 1982, Na/H antiporter of brush border vesicles: Studies with acridine orange uptake. Am. J. Physiol. 242: F733-F739.

Wright, E.M.,Harms, V., Mircheff, A.K. and van Os, C., 1981, Transport properties of intestinal basolateral membranes. Ann. N.Y. Acad. Sci. 372: 626-636.

Part 3: ELECTROPHYSIOLOGICAL METHODS

11

THE USE OF ION-SELECTIVE MICROELECTRODES TO STUDY CELLULAR TRANSPORT PROCESSES

J. Fernando García-Díaz and Fernando Giráldez

Department of Physiology, Boston University School of Medicine
Boston, Massachusetts, U.S.A. and Departamento de Bioquímica y
Biología Molecular y Fisiología, Universidad de Valladolid
Valladolid, Spain

INTRODUCTION

Ionic gradients and fluxes are involved in a number of cellular functions including nerve and muscle excitation and transepithelial absorption and secretion. It is of critical importance in these cases to obtain accurate measurements of intracellular ionic activities both at rest and during cell activation. Since the miniaturization of ion-selective glass membranes between 1950 and 1970, ion-selective microelectrodes (ISMs) have been extensively used for this purpose. The method allows not only measurement of steady-state values but also the monitoring of transient intracellular ionic activities. With certain provisos, the estimation of net transmembrane fluxes of ions is also possible. Improvement in the procedures for constructing ISMs and the introduction of liquid membranes (Walker, 1971) have allowed the technique to be used not only in large cells, typically of invertebrates, but also in smaller cells such as smooth muscle fibers, vertebrate neurons or tubular epithelia of the kidney.

Advantages and drawbacks of using ISMs

Advantages and disadvantages of using ISMs have been considered in detail by Thomas (1978) and Ammann (1986). Some papers on the subject, including new applications and developments can be found in a recent symposium (Morris and Krnjevic, 1987). Some of the advantages of using ISMs are :

1) ISMs allow direct and continuous measurement of cytosolic ion activities which are the relevant variables in the description of chemical states (Tsien, 1983). As a consequence of the complexity of the intracellular space, ions are not distributed uniformly within a cell. They can be sequestered into organelles, bound to proteins, or in the form of ion pairs. In any of these conditions they are not free in the cytosolic aqueous phase which directly interacts with the plasma membrane and therefore do not contribute directly to the transmembrane electrochemical potential. Analytical measurements of ion content of cells yield total (free plus bound plus sequestered) concentrations and do not discriminate between bound and free forms of a particular ion.

2) ISMs have a good time resolution for studying a number of ion transport mechanisms. Typical response times are in the order of a few seconds even for microelectrodes having very fine tips and faster response

Cell Membrane Transport, Edited by D. L. Yudilevich *et al.*
Plenum Press, New York, 1991

times have been reported (Coles and Tsacopoulos, 1979; Levy and Fein, 1985).

3) Ion activities, electrical membrane potential and other electrical properties can be simultaneously monitored in resting conditions and during dynamic states.

4) ISMs allow measurements in identified cells *in situ*, which is important in certain tissues, such as the CNS where the properties of the neurons are often dependent on the maintenance of particular connections.

5) ISMs can be used to measure net ion fluxes. This will be discussed in the last section.

6) ISMs can be used to measure cell volume changes. A technique for measuring changes in cell volume using ISMs is described by Reuss in this book.

Some disadvantages and difficulties in using ISMs are:

1) The method is invasive. This is the major problem in using ISMs and usually limits its applicability to relatively large cells. Even in those cases where the cell size permits the penetration of microelectrodes, some electrophysiological expertise is required. Although the technique always carries the danger of a certain degree of cell damage, there are, however, ways of assessing the extent of the inflicted damage and validating the measurements with confidence (see later).

2) Lack of perfect selectivity for the ion of interest. This is a problem that depends on the specific sensor used, but which becomes important when working at low activities, as for example in depletion experiments (Vaughan-Jones, 1979; Giraldez, 1984; Alvarez-Leefmans et al., 1988). Another concern is interference by foreign ions or drugs, for example interference of choline with the K exchanger or loop diuretics (furosemide, bumetanide) with the Cl exchanger.

3) The reading of the ISM is local, reflecting the activity at a particular region in the cytosol. While this may not be a problem for the three major monovalent ions, Na, K and Cl, the activity of Ca in the vicinity of the membrane or intracellular stores need not be identical to that elsewhere in the cytosol. But this property of the ISMs could be an advantage when one is interested in the spatial distribution of Ca inside the cell (Gorman et al., 1984).

THEORETICAL BACKGROUND

Ion activities and ion concentrations

ISMs respond directly to ionic activities. The activity of an ion in solution is its thermodynamically effective concentration and not its analytical concentration. (Effective thermodynamic concentration is that which makes mass action formulae quantitatively exact for describing chemical equilibria). Thus, a 100 mM KCl solution at 25 ^{o}C has an activity of only 77 mM. The inactive fraction represents ions that due to ion-ion and solvent-ion interactions have restricted freedom and, therefore, do not contribute to chemical potential changes (Bockris and Reddy, 1973). In general the more concentrated the solution, the greater the fraction of inactive ions. This is a well-known property of electrolytic solutions, which, although extensively studied, it is not yet completely understood. The classical treatment of the problem is the Debye-Hückel theory (Robinson and Stokes, 1959; Bockris and Reddy, 1973). Free concentrations (c_i) and activities (a_i) are related by an activity coefficient (γ_i), operationally defined as:

$$\gamma_i = a_i/c_i \qquad (1)$$

At a given temperature the activity coefficient of ions in aqueous solutions of low concentration varies with the interionic distance and with the number of charges borne by the ions. Precise values for mean activity coefficients of single-electrolyte solutions are available from tables (Conway, 1952; Parsons, 1959). For very dilute solutions of a single salt $\gamma_i \to 1$ and the value of a_i approaches that of c_i. However for most biological fluids, like the cytosol, $\gamma_i < 1$, and its calculation is not straighforward (Lev and Armstrong, 1975; Tsien, 1983; Ammann, 1986). Since the ISM responds directly to the ion activity, in principle there is no need to be concerned with the value of γ_i in the cytosol. A more relevant activity coefficient is that of the standard solutions used to calibrate the ISMs. For simple solutions this can be calculated or taken from tables. In other cases (as for example with Ca solutions) it may be preferable to use free ion concentrations instead (see later and Thomas, 1978; Tsien, 1983).

<u>Electrical potentials recorded with ISMs. The Nicolsky-Eisenman equation</u>

For a membrane to be useful as an ion-selective electrode it not only has to exclude sample counterions but it must discriminate between ions of the same charge (selectivity). Ion selectivity is governed by both the mobility of the ions in the membrane and by physicochemical factors which influence equilibrium at the membrane-solution interfaces (e.g., partition coefficients, ion-exchange equilibrium constants or, in the case of neutral carriers, the topology of the carrier, the number and properties of binding sites, etc.). A detailed account of the factors responsible for ion-selectivity of electrochemical membranes can be found in Eisenman (1967, 1968), Koryta and Stulik (1983) and Ammann (1986).

How does an ISM work?. As discussed by Tsien (1980), the relevant processes occur at the tip of the microelectrode (see Fig.1). Let us assume that an ionic species i is preferentially taken up by the organic phase of a liquid-type membrane. At steady-state no current flows. That is, the influx of ions into the membrane is balanced by the outflow. A sudden increase of a_i in the sample solution increases the influx of i into the membrane, which, at first, is not balanced by the outflow. As a consequence the electrical potential in the membrane shifts in the direction of restraining any further influx of i while favoring the exit of i from the membrane, until a new steady state is reached. In the presence of various interfering ions j, k, ..., the electrical potential response, V_i, of an ion-selective electrode is usually described by a semi-empirical formalism known as the Nicolsky-Eisenman equation:

$$V_i = E_o + S \log [a_i + \Sigma_j k_{ij} (a_j)^{z_i/z_j}] \quad (2)$$

where E_o is a constant potential difference (for a given temperature) comprising liquid-junction and reference electrode potentials and S is the empirical slope of the ISM response. Its theoretical value is $2.303(RT/z_iF)$. Practically it is defined as $S = nRT/z_iF$ where n is a dimensionless empirical constant chosen so that nRT/z_iF is the <u>observed</u> slope of the line when V_i is plotted as a function of log a_i in the absence of interfering ions. R, T and F have their usual meaning. z_i, z_j, a_i and a_j are the valence and activity (in moles/liter) of i and j, respectively. k_{ij} is the potentiometric selectivity factor, and is also referred to as the selectivity coefficient. It is a measure of the preference by the sensor for the interfering ion j relative to the ion i to be detected (i.e. the primary ion): the smaller the value of k_{ij} the less the interference from ions j experienced by an electrode sensitive to ion i. The selectivity of an electrode is often expressed as $1/k_{ij}$, the reciprocal of the selectivity coefficient. For instance, a selectivity coefficient of 0.05 indicates that the electrode is 20 times more selective to i than to j. Proper assessment of k_{ij} is necessary for accurate measurements of ionic activities inside cells. It follows that for an ideally selective membrane electrode $k_{ij}=0$ and the Nicolsky-Eisenman equation reduces to the Nernst equation:

$$V_i = E_o + S \log a_i \quad (3)$$

DESIGN AND CONSTRUCTION OF ISMs

Types of Membranes

The most common types of ISMs employed for intracellular measurements are made with either glass or liquid membranes. The construction of glass ISMs for the determination of cellular pH, Na and K activities is extensively described by Thomas (1978) and others (Lev and Armstrong, 1975). In the following we will restrict our discussion to liquid membrane ion-selective microelectrodes (LISMs). These are easier to fabricate and there is now a wide range of liquid membranes with appropriate selectivity for several ions of biological interest (see Table I).

The first type of LISM made consisted of the salt of a hydrophobic electrolyte dissolved in a water-immiscible solvent of low dielectric constant (liquid ion-exchanger or LIX; Walker, 1971). Depending on the type and charge of the hydrophobic ion and on the solvent employed, LIX membranes exhibit cation or anion selectivity and discrimination between ions of the same charge (Tsien, 1980). An inorganic ion in the aqueous solution will enter the membrane phase only after it is "trapped" by a hydrophobic ion of the opposite sign, with the two ions forming an uncharged ion pair. Specificity among ions of the same charge follows the order of hydrophobicity. Thus, for instance, the exchanger made by dissolving the lipophilic anion tetraphenylborate in 3-nitro-o-xylene is selective for K against Na (Table I) but is more selective for the more hydrophobic quaternary ammonium ions (Tsien, 1980).

More recently, Simon, Ammann and colleagues (Ammann, 1986; Schefer et al., 1986; Hu et al., 1989) have developed a series of neutral carriers or ligands which contain oxygen atoms arranged in a conformation that enables them to replace the normal hydration shell that surrounds a cation in aqueous solution. Because the complex formed between the carrier and the complexed cation is positively charged it is necessary to add some lipophilic anion, such as tetraphenylborate, to the membrane phase. This has the effect of decreasing the electrical resistance of the microelectrodes, improving their response time and stability. Because these anions behave as exchangers themselves, as discussed before, their concentration in the membrane phase has to be kept low (usually about 1%), in order not to alter the selectivity properties of the mixture. An excess of counterions in the membrane phase, as well as the use of solvents of high dielectric constant, promotes preference for divalent over monovalent ions.

Table I lists the membrane composition of several ready-to-use cocktails that are commercially available. Selectivity against the most important interfering ions is also shown. Note, however, that the values of k_{ij}^{-1} are only indicative. For K and Cl the cocktails are of the ion-exchanger type since they give better overall electrode performance. Two different cocktails are given for H-selective membranes. The second one seems to be faster. Its pH range is 2.0 to 9.0 compared to the 5.5 to 12.0 of the tridodecylamine based cocktail (Fluka Handbook). There are also two Mg sensors in the Table. The one made with ETH-5214 has been recently described (Hu et al., 1989) and probably will be commercially available soon. The pure ligands can be obtained from Fluka, so that those interested can make their own cocktail. For example, using 3-nitro-o-xylene instead of 2-nitrophenyl octyl ether as the solvent for Na-selective membranes improves the response time and selectivity of microelectrodes made with ETH-227 (O'Doherty et al., 1979; Garcia-Diaz et al., 1986). The neutral Ca-selective ligand listed in the table is ETH-129 (Schefer et al., 1986); its detection limit is lower than that of ETH-1001 (Ammann, 1986). For Ca microelectrodes with tip diameters less than 1μm it is best to use a mixture of 88% of the Ca-selective cocktail and 12% polyvinylchloride

TABLE I. LIQUID ION SELECTIVE MEMBRANES.

ION	COMPOSITION (wt-%) LIGAND	SALT	SOLVENT	SELECTIVITY (k_{ij}^{-1})	SOURCE
H	TDDA (10)	KtCB (0.7)	2-NPOE (89.3)	Na,K,Ca>10^{10}	FLUKA (95293)
H	ETH-1907 (6)	KtCB (1.0)	2-NPOE (93)	Na,K>10^{9}	FLUKA (95297)
K	----	KtCB (1.0)	3-NOX (99)	Na:50-80	CORNING (477317)
Na	ETH-227 (10)	NaTPB (0.5)	2-NPOE (89.5)	K:50-200 Ca:0.3-0.6	FLUKA (71176)
Ca	ETH-129 (5)	NaTPB (1.0)	2-NPOE (94)	K,Na,Mg>10^{5}	FLUKA (21196)
Cl	----	QACl	1-DECANOL	HCO_3:10-20	CORNING (477913)
Mg	ETH-1117 (20)	NaTPB (1.0)	PC (79)	Ca:0.08 K:25, Na:12.5	FLUKA (63048)
Mg	ETH-5214 (10)	KtCB (3.0)	2-NPOE (87)	Ca:0.25 K:250, Na:200	Hu et al. (1989)
Li	ETH-149 (9.7)	NaTPB (4.8)	TEHP (85.5)	Na:20, Ca:4 K:200	FLUKA (62380)

The cocktails included here are those currently preferred for intracellular measurements. The amount of each component (in wt-%) is listed in parentheses under its name. Catalogue number is listed under the source name, when available.

Abbreviations:

TDDA : Tridodecylamine.
NaTPB : Sodium tetraphenylborate.
2-NPOE: 2-Nitrophenyl octyl ether.
KtCB : Potassium tetrakis (p-chlorophenyl)borate.
3-NOX : 3-Nitro-o-xylene.
QACl : Chloride salt of quaternary ammonium. (Concentration unknown)
TEHP : Tris (2-ethylhexyl) phosphate.
PC : Propylene carbonate.

dissolved in tetrahydrofuran (Tsien and Rink, 1981; Lanter et al., 1982). The mixture is sucked though the tip and THF is allowed to evaporate. The resulting solid membrane is reported to give microelectrodes with better overall performance. However sucking of the mixture through a submicron tip is quite difficult.

Response time of LISMs depends on the type of sensor and cocktail used, tip size and whether shielding and capacitance neutralization (see later) are used. It ranges from tens of milliseconds to several seconds. Longer response time (minutes) is considered undesirable drift. Detailed information about neutral carriers and related topics can be found in Ammann's (1986) book. A handbook with information on neutral carriers, cocktails and other components used in the fabrication of LISMs is available from Fluka ("Selectophore: Ionophores for Ion-Selective Electrodes", 1988).

Construction of LISMs

Steps in the fabrication of LISMs differ for single and double-barrelled micropipettes. Since the ISMs measure the cell membrane potential plus the activity-dependent potential, it is necessary to subtract the first in order to determine intracellular activities. In a steady state it is possible to measure the cell potential first with a conventional (non-selective) microelectrode and then subtract its mean value from the ISM potential. For more accurate measurement or when one is interested in following the kinetics of ion activity changes following some experimental manoeuvre, it is necessary to measure simultaneously the membrane potential and the ISM potential. In principle this is easily done with a double-barrelled micropipette, where one barrel is filled with an electrolyte solution to measure the membrane potential and the other with the appropriate ion-sensitive cocktail. In practice these electrodes are more difficult to fabricate and the limitations in size and electrode resistance may preclude their use in small cells. For large cells or in highly coupled cell systems, such as epithelia (Garcia-Diaz et al., 1985, 1986) and sheep Purkinje fibers (Eisner et al., 1981), it is possible to use simultaneously two separate single barrelled microelectrodes.

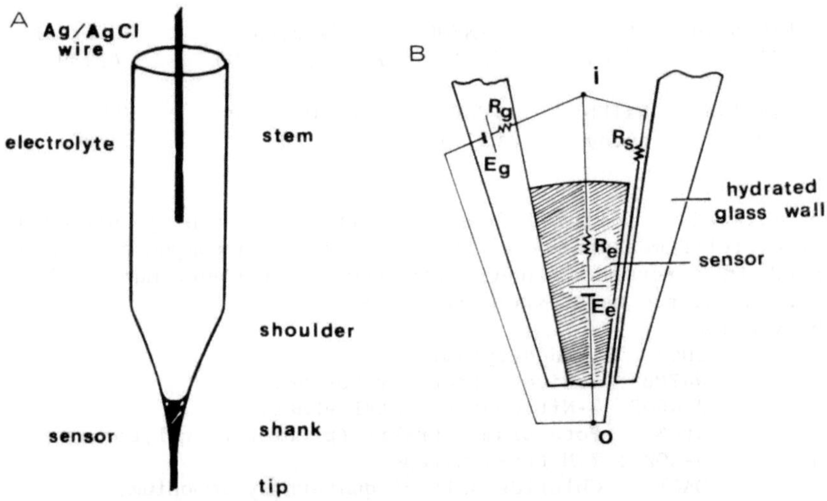

Fig.1. A. Components of a LISM. B. Schematic representation of the tip of a LISM showing the possible conductive pathways. E_e and R_e represent the potential difference and resistance across the sensor, R_s is a simple conductive shunt between the sensor and the inner surface of the glass micropipette, and E_g and R_g denote a conductive pathway through the hydrated glass surface (or microcracks produced by pulling), all with possible ionic selectivity. i and o are the filling solution and the outer (test) solution respectively (Armstrong and Garcia-Diaz, 1980).

Single barrelled microelectrodes

It is recommended that glass capillaries be cleaned before pulling (with strong acid or boiling in soap or alcohol, followed by repeated rinsing in distilled water and drying) to remove residues and other impurities. For the silanization there is no apparent advantage of cleaning the glass, since most of the inner glass surface at the taper is newly exposed after the pulling (Munoz et al., 1983). Single barrelled micropipettes for the fabrication of LISMs are pulled from conventional glass capillaries (Thomas, 1978; Purves, 1981; Halliwell and Whitaker, 1987). Borosilicate glass capillary tubing (OD 1.2 or 1.5 mm) with or without inner fibers is fine. Tip size and taper are critical parameters. Too large a tip diameter results in membrane damage upon cell impalement; with too small a tip, the high resistance of the ISM renders it useless. As with silanization (see below), the proper values are to be found by trial and error.

Silanization

Since the glass surface of a micropipette is hydrophilic, water will rapidly displace any water-immiscible liquid, such as the ion-sensitive cocktail, from the tip of the micropipette. To ensure proper adhesion between the droplet of ion-sensitive cocktail and the inner glass surface of the micropipette it is necessary to make the latter hydrophobic. This is done by exposing it to silanes. Silanization is one of the most critical steps in the preparation of LISMs. A number of factors, including the relative humidity of the surrounding atmosphere, are important, particularly with fine-tipped (<1μm) microelectrodes. Thus, silanization usually tends to be an empirical process, with the optimal conditions being determined by trial and error. With poor silanization the aqueous filling or test solution displaces the droplet of ion-sensitive cocktail from the tip. In less extreme cases insufficient silanization can contribute to two unwanted phenomena: surface conductance through an aqueous layer between the ion-sensitive solution and the inner wall of the micropipette and conductance across the hydrated glass wall of the micropipette itself (Fig.1B). Both of these phenomena can contribute to deviations of the slope of the electrode potential from its theoretical value and to lower selectivity than the nominal or bulk selectivity of the ion-sensitive cocktail (Armstrong and Garcia-Diaz, 1980; Tsien and Rink, 1981).

There are many different silanization methods (Thomas, 1978; Ammann, 1986). Basically the process involves exposing the inner glass surface of the micropipettes to a silicon compound at high temperature. The relative efficacy of different silanizing agents has been determined by Munoz et al., (1983). We will describe here two methods that we have found simple and quite efficient. One is a dip or liquid method, the other a vapor method.

i) Dip method

After pulling, the micropipette tips are immersed in a fresh solution (2.5-4.0 % V/V) of a silicon polymer (DC 1107, Dow Corning) in acetone for approximately 7 sec. The micropipettes are then placed horizontally on a hot plate and cured for at least 20 min at 250°C. They are then stored in dry air in a jar. Similar methods but involving different silicon compounds and solvents are described in the literature (Thomas, 1978).

ii) Vapor method

Once drawn the micropipettes are inserted, tips upwards, in holes drilled in a Teflon stopper (Fig.2A). Approximately 0.1 ml of Bis-(dimethylamino)-dimethyl-silane (Fluka) is placed in a beaker, which is then closed with the stopper holding the micropipettes. This assembly is positioned inside a dessicator connected to a vacuum line and the silane is allowed to react with the glass of the micropipettes for 2-3 min. The micropipettes are then cured on a hot plate (or in an oven) at 250°C for 1 hr. Here again there are several variations of the method (Thomas, 1978; Ammann, 1986).

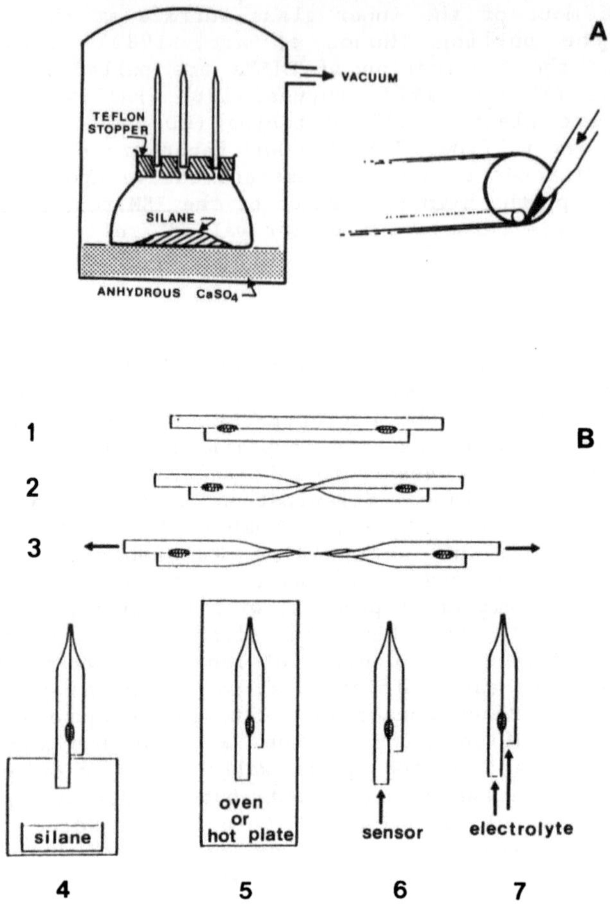

Fig.2. A. Silanization set-up and filling of micropipettes with inner fibers. B. Steps in the fabrication of double-barrelled LISMs (Zeuthen, 1980).

Filling pipettes

Filling a micropipette which contains an inner fiber is simple. A drop of the ion-sensitive cocktail is placed, with the help of a fine glass capillary, at the corner between the wall of the micropipette and the inner fiber (Fig.2A). Usually after a few hours (or overnight) the solution moves to the tip, forming a column 0.5-2 mm long which has to be inspected under the microscope to ensure that no air bubbles are present. With micropipettes without inner fibers a small volume of the ion-sensitive cocktail is injected in the micropipette shank, as close to the tip as possible, using a very fine plastic tubing or hand-drawn glass capillary. To help the solution fill the shank a finely drawn glass capillary is poked through the cocktail and moved back and forth, under low power microscopic observation. If the tip is not plugged the air bubbles trapped near the tip will be driven out by the cocktail.

The last step is to fill the micropipettes with an appropriate electrolyte solution (such as 0.5 M KCl or NaCl). This is usually done under microscopic observation by using a finely drawn glass capillary or plastic tubing connected to a syringe. The tip of this tubing must be small enough to reach the meniscus of the cocktail inside the micropipette shank, so that no air bubbles are left between the cocktail and electrolyte. It is also possible to fill the micropipettes by sucking the electrolyte and then the cocktail through the tip (Tsien and Rink, 1980; Levy and Fein, 1985).

Double-barrelled microelectrodes

These can be made with two parallel barrels glued together or with a single glass tubing with a longitudinal partition (the so-called theta tubing). The parallel barrel arrangement can be fabricated in the lab by gluing together two glass capillaries with fast setting epoxy. This permits using barrels of different diameters, in order to obtain bigger tips for the ion-sensitive barrel. The procedure for the fabrication of the micropipettes is outlined in Fig. 2B (Zeuthen, 1980). After gluing the barrels they are heated at the middle and twisted 360° (there are pullers specially fabricated for this operation, e.g. Narishige PD-5). This procedure is required to prevent the separation of the tips which otherwise occurs after pulling. The longer barrel is inserted, tip upwards, in holes drilled in a Teflon stopper, similarly to the procedure described before for single barrelled micropipettes. The glass is exposed for 1 to 3 min to silane vapor (e.g.: dimethyl-dichloro-silane or bis-dimethylamino -dimethyl-silane) and then baked at 250 ^{o}C for 1 hr. The long barrel is then filled with the ion sensitive cocktail, as previously described for the single barrelled micropipettes. Finally, the reference barrel is filled in a similar way with electrolyte solution. It is recommended to use a solution not containing the ion for which the cocktail is sensitive in order to prevent interference from electrolyte leakage. Thus when using Cl-selective microelectrodes, the reference barrel is usually filled with 4M K acetate. This may require correcting the measured values of membrane potential for a junction potential. Theta glass tubing can be silanized by the same procedure after cutting a piece of a barrel at one end. Thick septum theta glass is preferable (Brown and Flaming, 1977). Another method to silanize double-barrelled micropipettes, using silane vapor applied to the back of the micropipette tip from a fine capillary, was designed by Coles and Tsacopoulos, 1977 (see also Thomas, 1978 or Ammann, 1986).

Double- and single-barrelled microelectrodes can have a very large resistance and behave erratically. Breaking or bevelling the tip improves stability and overall performance of the ISM. When advancing against tissue the tip will occasionally break just enough to permit recordings with no excessive membrane damage. But controlled breaking (Tripathi et al., 1985) or bevelling (Ammann, 1986) is preferable.

Calibration

To determine the activity of an ion i in a sample of unknown composition, the potential difference V_i between an electrode selective to i and a reference non-selective electrode is measured. The activity of i in the sample, a_i, is then read from a calibration curve of V_i vs. log a_i obtained with test solutions of known composition. Ideally, the reference electrode should be insensitive to the composition of the solutions. Calomel electrodes connected to the solutions via a flowing KCl bridge or a wide (>1mm ID) agar-3M KCl bridge are used to minimize junction potentials at the tip of the bridge.

At calibration the characteristics of the electrode are tested: sensitivity S or slope (in mV/decade change in activity), selectivity k_{ij} against potential interfering ions, response time, limit of detection and electrical resistance. In the initial stage of using ISMs it is recommended that a few of each batch of microelectrodes be tested for their performance before trying to use them in a cell. However the actual calibration for the calculation of intracellular activities should be done after cell recording. The reason for this is that often the microelectrode tip breaks when is advanced against the tissue altering the microelectrode characteristics. Most calibration procedures are variations of two different methods: the separate solution method (SSM) and the fixed interference method (FIM). However, the most appropriate for intracellular measurement is an empirical method derived from the FIM that we will discuss last.

Separate solutions and fixed interference methods

In the SSM the electrode potential is measured in pure solutions of the test ion i and in pure solutions of the interfering ion j, covering the range of interest (Fig.3A). Then, if V_i' and V_j' are the electrode potentials measured at the same activity of i and j, respectively, from Eq.2:

$$k_{ij} = 10^{(V_j' - V_i')/S} \qquad (4)$$

In practice, however, it is often found that the electrode response deviates from the Nicolsky equation and the slope S is not the same when calibrating in solutions of the main ion i or the interfering ion j. Thus, Eq.4 gives a crude estimate of k_{ij} but cannot be used reliably to calculate

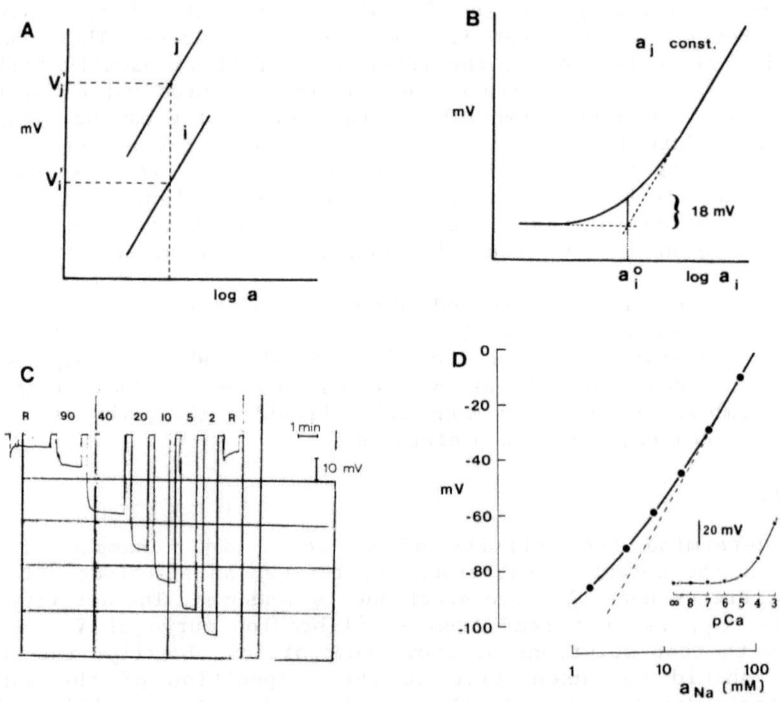

Fig.3. A. Calibration curves in separate solutions of the test ion i and the interfering ion j. B. Fixed interference method for calibration. C. Calibration of Na-selective microelectrodes in Ringer (110 mM Na, 1 mM Ca) and in solutions with [Na]+[K]=125 mM and pCa=7.0 (EGTA 10 mM, $CaCl_2$5 mM, pH=7.29). [Na] values in mM are shown above traces; R=Ringer. D. Plot of voltage vs. Na activity for the calibration shown in C. The value in Ringer is taken as zero. The dotted line is the Nernstian response (59 mV/decade). Insets: Calibration of Na microelectrode in solutions with varying pCa (at constant [Na]=10 mM and [K]=115 mM).

intracellular ionic activities. Since the solutions employed with this method are of varying ionic strength, activity coefficients must be calculated using the Debye-Hückel formalism or taken from tables (Conway, 1952; Parsons, 1959).

In the FIM the electrode is calibrated in solutions of different activities of i containing a fixed activity of the interfering ion, a_j (Moody and Thomas, 1971). Fig.3B shows a calibration curve obtained in this way. As a_i decreases the electrode response deviates from the Nernstian straight line as a consequence of the interference by j, approaching at low a_i an asymptotic value that represents complete interference by j. If a_i^o is the activity of i corresponding to the intersection of this asymptote with the extrapolated Nernstian response, then from Eq.2: $k_{ij}=a_i^o/a_j$. In practice one does not need to determine the complete interference curve. It can be shown using Eq.2 (Armstrong and Garcia-Diaz, 1980) that a_i^o is that value of a_i for which the difference in potential between the FIM calibration curve and the extrapolated Nernstian line is S·log2 (theoretically 18 mV at room temperature for $z_i=z_j=1$). The value a_i^o is also the limit of detection of the microelectrode (Ammann, 1986), which clearly depends on the interfering ions present. Values of k_{ij} obtained with the FIM are usually larger than those measured with the SSM. Actually k_{ij} is concentration dependent (Bolton and Vaughan-Jones, 1977; Edelman et al., 1978; Armstrong and Garcia-Diaz, 1980) so that the use of a single value of k_{ij} to estimate the interfering effect of j on V_i over a wide range of activities may lead to significant errors in the determination of a_i.

Empirical method

Because of the indeterminacy of k_{ij} and the limited applicability of the Nicolsky equation several investigators have designed empirical methods that do not rely on a single estimate of k_{ij}. The electrodes are calibrated in solutions of varying activity of i and with the activities of interfering ions similar to their intracellular activities. The intracellular activity of the test ion is determined by direct interpolation of the electrode potential (minus the membrane potential) from the calibration curve. With this method the effects of interfering ions present in the extracellular medium (e.g. Na for the K electrode, Ca for the Na electrode) are irrelevant. It requires, however, an estimate of the intracellular activities of interfering ions. As an example Fig.3 C and D shows the calibration of a Na-selective microelectrode used in frog skin (Garcia-Diaz et al., 1986). Panels C and D show the calibration in solutions of variable [Na] with the free Ca concentration kept constant at pCa=7.0 (pCa=-log[Ca]). Na was substituted for K ([Na]+[K]= 125mM) so as to keep the ionic strength and activity coefficient (0.77) near constant. In the range of [Na] from 2 to 10 mM (the intracellular range) the K activity was 95-88 mM, which is close to the intracellular level of K in these cells. Thus the interference by intracellular K is mimicked in the calibration curve. Another interfering ion in these electrodes is Ca, but only in the mM range. The inset in panel D shows the calibration of the same Na electrode in solutions with varying pCa (made according to the method of Tsien and Rink, 1980) at constant [Na]=10 mM and [K]=115 mM. For pCa≤5 the electrode shows a strong dependence on pCa while in the range pCa≥6 it hardly responds to Ca, indicating that physiological changes in intracellular Ca will not affect the measurement of a_{Na}^c. The interference of extracellular Ca on the electrode potential does not affect the determination of a_{Na}^c, because this is done by direct interpolation from the calibration curve at pCa=7.0.

The same method is applied for calibration of other ISMs. For example, K-selective microelectrodes are calibrated in solutions of constant [Na]+[K] (Garcia-Diaz et al., 1985). In the range of intracellular

K activities (≈100mM) interference by intracellular Na is negligible and the electrode potential follows a Nernstian straight line response. The intracellular K activity is calculated by interpolation from the calibration curve. Note again that this method avoids the use of the Nicolsky equation (and k_{KNa}, which is not even calculated) to correct for the effect of extracellular Na. Similarly, Cl-selective microelectrodes are best calibrated in solutions containing a fixed amount (about 10 mM) of HCO_3. These electrodes are also prone to interference by unknown intracellular anions (Bolton and Vaughan-Jones, 1977).

A set of calibrating solutions for Ca-selective microelectrodes was designed by Tsien and Rink (1980). The free Ca concentrations were buffered with a mixture of 10mM buffer and 5mM Ca to maximize buffering power. The desired free Ca concentrations were obtained by choosing a Ca buffer with the appropriate pKa and by adjusting the solution pH. The variation in pH of the calibrating solutions is unimportant because of the high selectivity of the microelectrodes against H. All the solutions contain about 125mM K to mimic the intracellular content and to keep the ionic strength constant.

Calibration curves are plotted, and results given, either in activity or free ion concentration. In the first case an activity coefficient is calculated for the calibrating solutions. No assumption is made about the intracellular activity coefficient. In the second case, the value calculated from the ISM reading and the calibration curve is taken as the intracellular free ion concentration, if one assumes an intracellular activity coefficient equal to that of the calibrating solutions. Both methods are valid when using calibrating solutions of constant ionic strength; the results should differ by a factor equal to the activity coefficient of the calibrating solutions. Regarding the nomenclature, it is advisable to follow The International Union of Pure and Applied Chemistry (IUPAC) recommendations (see Guibault et al., 1976; IUPAC, 1979).

MEASUREMENT OF INTRACELLULAR ION ACTIVITIES

Experimental set-up

A vibration-resistant set-up is required for microelectrode work (see Purves, 1981). The simplest arrangement is to bolt the manipulator and perfusion chamber to a metal baseplate and to isolate the system from the table underneath with inflated bicycle tires or solid rubber balls. The table itself must be sturdy, e.g. a balance table. Commercial air tables are available, but are expensive. The set-up must be electrically isolated with a metal enclosure (Faraday cage) to minimize noise pick-up. The cage must be grounded and care taken to avoid ground loops (Purves, 1981). Perfusion bottles and waste reservoirs should be inside the cage. That part of the equipment which may be accidentally or intentionally touched during recording (Faraday cage, perfusion bottles and valves, binocular) should be supported separately from the baseplate and table. For work in small cells (≤20μm diameter) a remote control (pneumatic, hydraulic or electrical) manipulator greatly improves the quality and number of successful recordings. Stepping motor or piezoelectric devices are particularly good for penetrating cells in a single step without dimpling. Microelectrode holders are made of Lucite or, even better, Teflon. Holders and the outside glass of the micropipettes should be kept clean.

The calibration of ISMs can be done in the same experimental set-up. Because it is not desirable to expose the tissue to the calibrating solutions, a device adjacent to the perfusion chamber must be used, for example a simple Lucite block with a groove. The test solutions can be quickly changed with a multi-way valve. The microelectrode tip is positioned in the groove and a saturated KCl bridge located downstream connects the solutions to the reference calomel or Ag/AgCl electrode.

Electronics and instrumentation

Basic electronic requirements

Because of the high resistance of ISMs (ca. $10^{11}\Omega$) it is necessary to use a preamplifier (input stage or headstage) with high input impedance and low bias current. Typically these should be $>10^{14}\Omega$ and ≤ 50 fA, respectively. The amplifier input impedance and the microelectrode resistance form a voltage divider. In order to measure most of the microelectrode voltage, the amplifier input resistance should be >100 times the electrode resistance. All amplifiers draw some current from the source (in this case the microelectrode). This input bias current produces a voltage drop across the microelectrode resistance equal to $R_{ME} \cdot i_{bias}$. Any change in R_{ME}, e.g. by exposing the microelectrode to solutions of different composition, will result in voltage changes. Therefore a low i_{bias} is required to minimize these voltage changes. There are FET-input amplifiers, e.g. Analog Devices AD515L, suitable for this purpose. The preamplifier circuit should be enclosed in a metal box (headstage or probe)

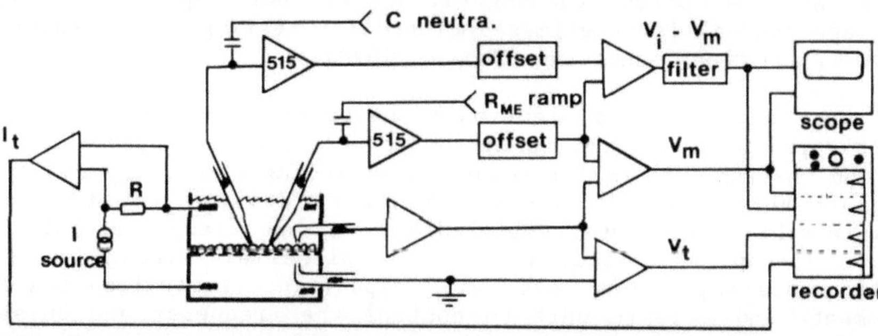

Fig.4. Diagram of an experimental arrangement for electrophysiological recordings in epithelia. A capacitance neutralization circuit is included in the ISM preamplifier. The conventional microelectrode amplifier has a current ramp injection (R_{ME} ramp) for measuring microelectrode resistances. When the microelectrode tips are in the external solution the signals from both microelectrodes can be zeroed with the offset null circuits. Differential amplifiers and filter are also shown.

located as close as possible to the microelectrode. Other features worth including in the electronic set-up (see Fig.4) are a capacitance neutralization circuit to speed up the response time of ISMs, a microelectrode resistance measuring circuit, offset null circuits and low-pass filters (Thomas, 1982; Garcia-Diaz et al., 1984). To measure the resistance of conventional microelectrodes either a ramp or a square pulse current injection can be used. However, there is not a convenient method to measure the high resistance of ISMs during an experiment. It can be done during calibration by shunting its input with a high value resistor and measuring the voltage attenuation (Thomas, 1978). The books by Purves (1981) and Thomas (1982) as well as the chapter by Ogden (1987) contain several circuits of interest for anyone using microelectrodes. A very useful book for practical circuit design and information on components is Horowitz and Hill (1989).

Set up for recording in epithelia

Epithelia are usually studied with separate solutions perfusing the apical and basolateral surfaces. Transepithelial electrodes are placed in each solution to measure voltage (V_t) and to pass current (I_t). Current or voltage pulses are applied to measure the transepithelial conductance ($g_t=\Delta I_t/\Delta V_t$) and, from the microelectrode potential, ΔV_m, the voltage divider ratio, $F_a=\Delta V_m/\Delta V_t$. In tissues with low extracellular conductance F_a is a good estimate of the fractional resistance of the cell membranes. That is, if V_m is measured with reference to the apical solution, $F_a=R_a/(R_a+R_b)$, where R_a and R_b are the lumped equivalent resistances of the apical and basolateral membranes respectively. Measurement of F_a is important, not only for the estimation of membrane conductances, but also to assess the proper intracellular location of the microelectrode, as will be discussed in the next section. A diagram of a simple circuit (modified from Giraldez, 1984) is shown in Fig.4. The potential measuring electrodes could be either calomel or Ag/AgCl wires connected to the solutions via KCl-agar bridges. The current passing electrodes (Ag/AgCl rings or pellets) are connected to a floating current source which permits the application of intermittent pulses, measured as a voltage drop across the resistor R (about 100 Ω). Note that the arrangement shown in Fig.4 is an extension of those used in non-epithelial systems (Thomas, 1978; Vaughan-Jones and Aickin, 1987). Since in the latter case there is only one extracellular solution the V_t amplifier and the current passing circuit are not required. A calibrator (DC voltage source) is sometimes included for testing (i.e. measuring ISM resistance) and troubleshooting (Thomas, 1978).

A more sophisticated electronic device described by Garcia-Diaz et al. (1984) provides alternative microelectrode and transepithelial current pulses and on-line outputs for recording of voltages, g_t, R_{ME} and F_a, based in sample-and-hold circuitry. Another device (available from Biomedizinische Instrumente) employed for voltage clamped preparations is made up of separate modules: a transepithelial voltage clamp, two microelectrodes amplifiers, each with capacitance neutralization and R_{ME} measurement, and a logic unit to control the frequency and duration of pulses. This instrument provides also outputs for the on-line recording of all the parameters of interest.

Criteria for the acceptance of recordings

Measuring membrane potentials

Microelectrode recording is difficult and subject to artifacts that introduce serious errors in the measurement of membrane potentials and intracellular ion activities. This is particularly true when working with small cells. Several criteria have been developed in order to recognize the presence of artifacts. The basic criteria are a sharp deflection of the microelectrode voltage on penetration, stability of the recording and return to the original baseline (within a few mV) on withdrawal. This normally ensures against drift of the microelectrode voltage and obvious membrane damage, as in those cases where the membrane potential depolarizes progressively upon penetration (Armstrong and Garcia-Diaz, 1981). An additional concern is the change in junction potential when the tip is inserted in the intracellular space (Vaughan-Jones and Aickin, 1987; Halliwell and Whitaker, 1987; Purves, 1981). Normally microelectrodes are filled with 3M KCl in order to minimize junction potentials. Leakage of KCl from the microelectrode tip can then be a problem, particularly in small isolated cells, so that a lower concentration (0.5 or 1M) is often used. Alternative electrolyte solutions are also used when leakage can interfere with the measurement of the activity of a certain ion, particularly when using double-barrelled microelectrodes. For example, K_2SO_4 or K acetate are used for the reference microelectrode or barrel when measuring

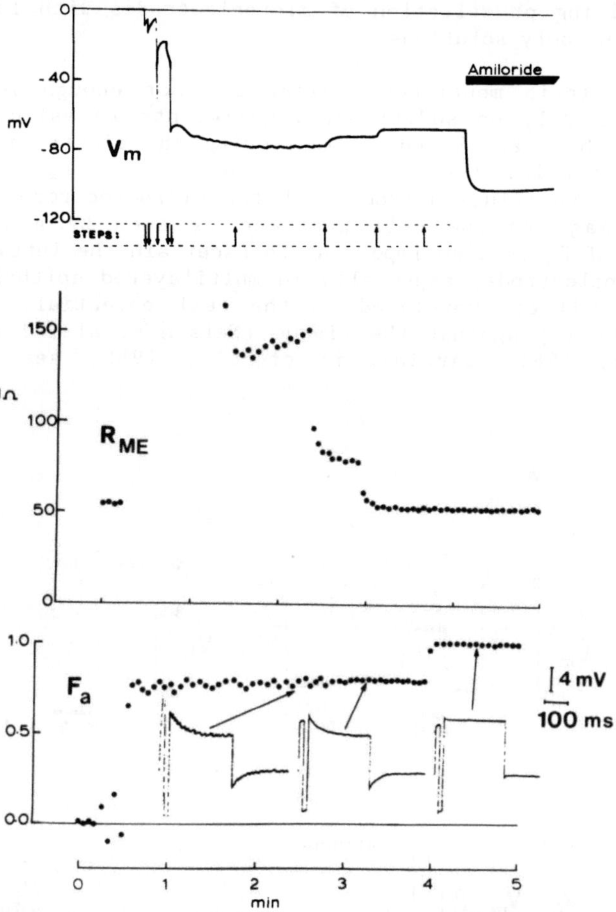

Fig.5. Artifactual offset potentials recorded in frog skin epithelium with a 55 MΩ microelectrode filled with 1.5M KCl. Downward arrows (shown below the V_m panel) indicate advancing steps of 3μm each. Upward arrows denote withdrawing steps of 0.5μm each. During the advance through the cornified layer of the skin R_{ME} went off scale and F_a, although extremely noisy, did not indicate the crossing of a resistance barrier. At the last advance F_a increased to 0.7 and V_m jumped to a negative value, indicating cell penetration. However V_m was noisy and R_{ME} still off scale. Successive withdrawals of the microelectrode reduced R_{ME} and the noise in V_m. Note that at the second and third backward steps V_m depolarized. These changes of V_m are attributed to decreases in artifactual offset potentials arising from partial occlusion of the microelectrode tip. Insets in F_a graph show oscilloscope traces of the V_m responses to microelectrode current injection (to measure R_{ME}) and to ΔV_t pulses. Amiloride (20 μM) was added to the apical solution during the period marked by the bar, producing the expected hyperpolarization and increase of F_a to near 1.0 (Garcia-Diaz et al., 1985).

intracellular Cl. In this case there may be large changes in junction potentials when the tip is advanced from the extracellular solution into the intracellular medium. The values of V_m obtained must then be corrected by measuring the change in potential between the extracellular solution and a solution mimicking the intracellular ion content, using a floating KCl bridge or similar device as reference. Tip potentials, particularly large in high resistance microelectrodes, present an additional problem (Purves,

1981). Bevelling or selection of microelectrodes with low tip potentials seem to be the only solutions.

The criteria mentioned earlier are not enough in most cases and investigators rely on additional requirements to validate intracellular recordings. These are based on monitoring the electrode resistance, R_{ME}, and, in epithelia, the voltage divider ratio, F_a. With the voltage reference in the solution from which the microelectrode is advanced (i.e. apical), damage to the cell membrane is seen by a low value of F_a. Measurement of F_a is also important to ascertain the intracellular location of the microelectrode, especially in multilayered epithelia, since abrupt changes in voltage, unrelated to the cell potential, often occur when advancing the tip against the tissue (Nelson et al., 1978; Armstrong and Garcia-Diaz, 1981; Garcia-Diaz et al., 1985; see also Fig.5).The

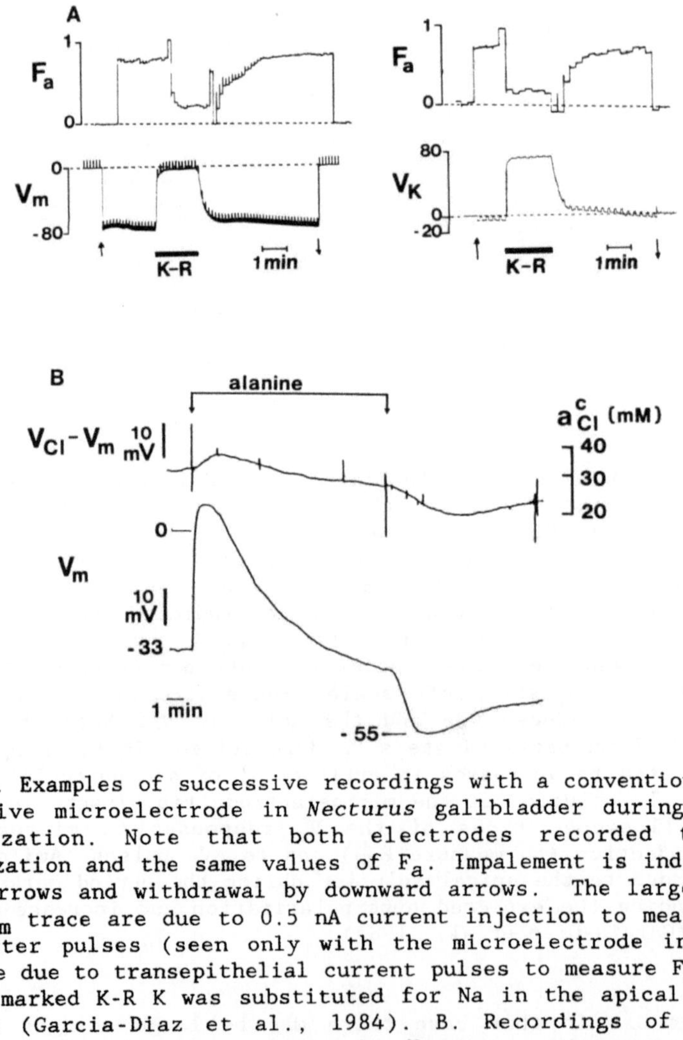

Fig.6. A. Examples of successive recordings with a conventional and a K-selective microelectrode in *Necturus* gallbladder during rapid K depolarization. Note that both electrodes recorded the same depolarization and the same values of F_a. Impalement is indicated by upward arrows and withdrawal by downward arrows. The larger pulses in the V_m trace are due to 0.5 nA current injection to measure R_{ME}. The shorter pulses (seen only with the microelectrode inside the cell) are due to transepithelial current pulses to measure F_a. During the bar marked K-R K was substituted for Na in the apical Ringer's solution (Garcia-Diaz et al., 1984). B. Recordings of membrane potential and cell Cl activity in *Necturus* enterocytes with a double-barrelled microelectrode. The addition of alanine produced a biphasic response of V_m: fast depolarization followed by slower repolarization. Note that the differential trace V_{Cl}-V_m remains constant during the initial depolarization. Later V_{Cl}-V_m changes reflecting the alterations in cell Cl (Giraldez and Sepulveda, 1986).

artifactual nature of these voltage jumps is indicated by a value of F_a near zero. Only after crossing a resistance barrier (e.g., apical membrane in Fig.5) does F_a reach a significant value. Membrane crossing cannot be recognized by monitoring only R_{ME} since in highly coupled cell systems, such as epithelia, the input resistance of the cell membrane will increase the apparent R_{ME} by less than 1-2 MΩ. On the other hand, in non-coupled cell systems the input resistance of the cell membrane will increase R_{ME} on impalement. This can be misleading, however, since often the microelectrode tip is partially occluded by cellular elements after penetration causing an increase in R_{ME} and, usually, an artifactual hyperpolarization of the microelectrode voltage and excessive noise of the oscilloscope trace of V_m. On occasion, as in the example shown in Fig.5, it is possible to reduce R_{ME} by slightly withdrawing the microelectrode or even by overcompensating briefly the negative capacitance. Slight withdrawal can be done more easily by employing a sensitive remote control micromanipulator. When successful, the decrease in R_{ME} is associated with a reduction in noise and a drop of the microelectrode potential, sometimes by as much as 20 mV, to stable values which presumably reflect the cell potential more accurately. That this depolarization is not due to unsealing of the microelectrode is indicated by constancy of F_a (Fig.5).

Also useful in ascertaining the intracellular location of the microelectrode is observation on the oscilloscope of the voltage deflections in V_m induced by transepithelial pulses. Inside the cell ΔV_m should follow a characteristic time course, reflecting the capacitative decay associated with the series arrangement of apical and basolateral membranes (Garcia-Diaz and Essig, 1985). This is shown by the insets below the F_a trace in Fig.5 for recordings in frog skin. The time course is clearly affected by manoeuvres altering the membrane conductances (e.g. addition of drugs or ionic substitutions).

Other criteria rely on some conductance property of the cell membrane. For instance in frog skin, the inhibition of apical Na entry by amiloride or Na substitution hyperpolarizes V_m to values around -100 to -120 mV and increases F_a to near 1.0 (see Figs.5 and 7). In *Necturus* gallbladder, which exhibits a large apical K conductance, K for Na substitution should cause a significant depolarization and drop in F_a (see Figs.6A and 9). These manoeuvres sometimes cause unsealing of the microelectrode, perhaps by altering cellular volume.

Measuring ionic activities

In general most of the above criteria, with the exception of the continuous measurement of R_{ME} which is not feasible with ISMs, are applied to recordings made with ion-selective microelectrodes. To measure F_a with the slow responding ISMs, it is often necessary to increase the pulse duration. Additional requirements are: i) constancy of the calibration curve before and after impalement (although, as explained before, breaking of the tip on penetration sometimes requires the use of only the calibration obtained after the withdrawal of the ISM) and ii) equal response of the conventional and ISM to changes in membrane potential induced by transepithelial pulses or fast solution changes, such as depolarization by K in gallbladder (Fig.6A), by sugar or aminoacid exposure in small intestine (Fig.6B) or hyperpolarization by amiloride or Na substitution in frog skin (Fig.7).

MEASURING FLUXES WITH ISMs

As with other methods for measuring ionic concentrations, it is possible to calculate net transmembrane fluxes from the change in the intracellular concentration of a given ion measured with ISMs. The amount of i in the cell, m_i^c(moles) is:

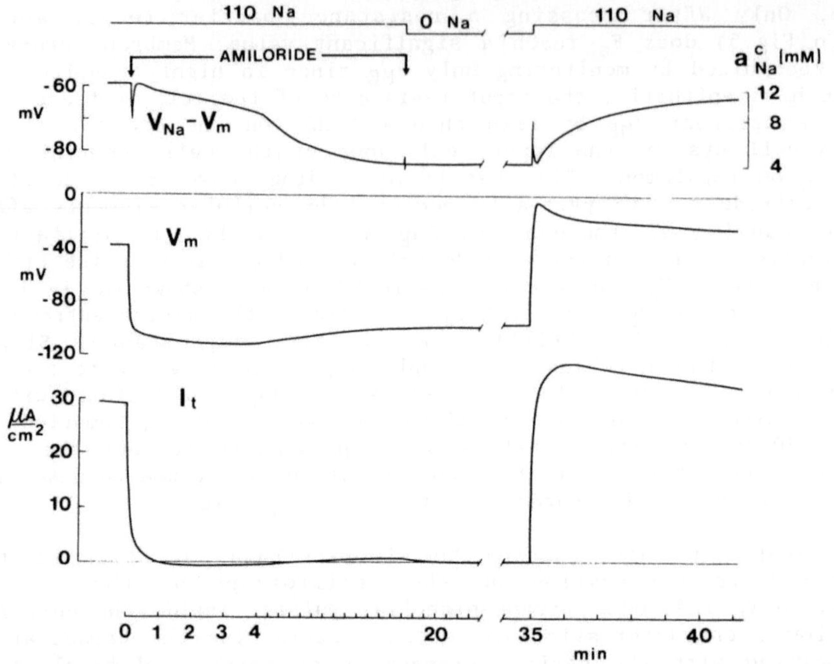

Fig.7. Example of the simultaneous recording of intracellular Na activity and membrane potential in frog skin using two separate microelectrodes. I_t is the transepithelial current. The experiment was done with the transepithelial voltage clamped to 0. Cell Na activity decreases following inhibition of apical entry and recovers after restoration of apical Na. Note the changes in time scale (Garcia-Diaz et al., 1986).

$$m_i^c = c_i^c \ v^c \qquad (5)$$

where c_i^c (mole cm^{-3}) is the concentration of i and v^c the cell volume. Differentiating Eq.5 we obtain an expression for the rate of change of the amount of i in the cytoplasm:

$$dm_i^c/dt = (dc_i^c/dt) \ v^c + (dv^c/dt) \ c_i^c \qquad (6)$$

Dividing Eq.6 by the membrane surface area and using Eq.1 we obtain:

$$J_i = (1/\gamma_i^c) \ [(da_i^c/dt) \ h + (dh/dt) \ a_i^c] \qquad (7)$$

where J_i is the net flux of ion i (rate of change of the amount of i in the cell per unit of membrane surface area), h is the volume to surface ratio of the cell (cm) and a_i^c and γ_i^c the activity and activity coefficient of the ion i in the cytosol. Several points about the use of Eq.7 are worth discussing. First, this equation assumes that changes in m_i^c are solely due to transmembrane fluxes of i and that no buffering, sequestration or consumption/production of i occurs. What we are actually measuring with ISMs is the appearance and disappearance of i in the cytosol. Thus, any exchange between intracellular compartments or chemical reactions in which ion i participates will look like net flux across the plasma membrane.

Chemical buffering or restricted intracellular compartments containing the ion at levels different from that of the cytosol affect mostly pH and Ca measurements. In the former case the buffer capacity can be calculated from experiments (Roos and Boron, 1981), but for Ca the calculation is not possible due to the multiple nature of the Ca buffering systems. Another source of uncertainty is the value of the intracellular activity coefficient. Since the relevant variable in the calculation of fluxes is concentration (Eq.6), but the ISM measures activity, we need to estimate γ_i^c and assume that it remains constant.

Finally, the use of Eq.7 requires a knowledge of the cell volume and of its change during the experiment. A value for h can be obtained by microscopic measurements of cell dimensions. Alternatively, using assumptions or actual knowledge about cell geometry, one can calculate cell volume and surface area. What is actually done is to model the cells as known geometric figures and calculate volume to surface ratios (Thomas, 1972; Vaughan-Jones, 1979; Giraldez, 1984). The value of h for a sphere is r/3 and for a cylinder is r/2. For an epithelial sheet it is simply its height. A more critical problem, however, is to evaluate how cell volume changes affect measurements throughout the experiment. As changes in cell volume depend on the fraction of the total salt content that is lost or gained by the cell, if the cell is large and the fluxes are small, then the rate of change in cell volume is small and, as a first approximation, one can ignore the second term on the right hand side of Eqs.6 and 7. This can be safely done, for instance, with Cl depletion experiments in cardiac Purkinje muscle fibers where cells are large and changes in a_i^c are due to equimolar exchange (Vaughan-Jones, 1979). However for small cells a measure of dh/dt is preferable. If this is not possible, one at least must have an idea of maximal limits for the possible error introduced by changes in volume and to assess the direction of these changes. A way of estimating dv^c/dt from measurements of ionic activities has been obtained from mass balance equations (Giraldez, 1984; Alvarez-Leefmans et al., 1988):

$$(1/v^c)\, dv^c/dt = -\,(\Sigma z_i da_i^c/dt)\; /\; (\Sigma z_i a_i^c) \qquad (8)$$

where a_i^c denotes the intracellular activities of those ions able to permeate the cell membrane. This equation has been used to estimate the increase in cell volume after ouabain poisoning of epithelial cells and the results compare well with optical measurements, as shown in Fig.8A. In some cases the solution of Eq.8 is simplified. For example, in dorsal root ganglion neurons, Cl movements upon extracellular Cl removal are electroneutral and coupled to Na and K (Alvarez-Leefmans et al., 1988). For isotonicity to hold, the sum $a_{Na}^c + a_K^c$ should remain constant and, therefore, $da_{Na}^c/dt + da_K^c/dt = 0$. Hence Eq.8 reduces to:

$$(1/v^c)\, dv^c/dt = (da_{Cl}^c/dt)\; /\; (a_{Na}^c + a_K^c - a_{Cl}^c) \qquad (9)$$

The solution of this equation for v^c as a function of time can be obtained knowing the function that relates a_{Cl}^c with t. Fig.8B shows the calculated change in cell volume obtained by fitting to a single exponential the decrease in a_{Cl}^c after removal of external Cl.

An interesting way of addressing the problem of volume changes is the direct approach followed by Reuss (1985). The technique is based on the extreme sensitivity that K microelectrodes made with the Corning exchanger have for quaternary ammonium ions. The cells are loaded with the impermeant ion tetramethylammonium (TMA) by exposing them to the ionophore nystatin. Once nystatin is removed, changes in cell volume are monitored by measuring the intracellular activity of TMA with the K electrode (see article by Reuss in this book).

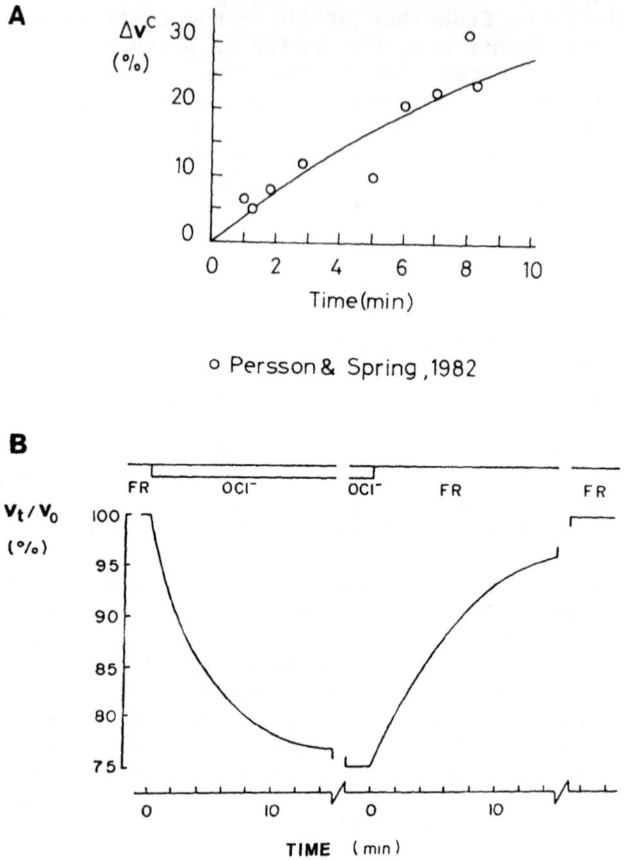

Fig.8. A. Percentage increase in the volume of *Necturus* gallbladder cells after exposure to serosal ouabain. The line represents the increase in volume calculated using Eq.8. The circles are values measured with an optical method (Persson and Spring, 1982). B. Relative changes in cell volume of frog dorsal root ganglion neurons following depletion and restoration of external Cl. The volume changes were calculated according to Eq.9, fitting to single exponentials the measured changes in a_{Cl}^c (Alvarez-Leefmans et al., 1988). FR: frog Ringer's solution.

Describing the transients

In a transient experiment one analyzes the reaction of the cell when it is perturbed from a given steady-state. The power of transient analysis experiments resides in the fact that the initial rates of change in concentration or electrical potential, calculated by extrapolation to t=0, characterize the original steady-state. For instance, we can think of a parallel array of transport systems which in the steady-state keep the intracellular concentration of the ion constant by the balance of all fluxes:

$$dc_i^c/dt = (1/h)\ (J_i^1 + J_i^2 + \ldots + J_i^n) = 0 \qquad (10)$$

Imagine that at t=0 we are able to specifically block one transport system, say J_i^1, and look at dc_i^c/dt. If we are able to measure dc_i^c/dt in an

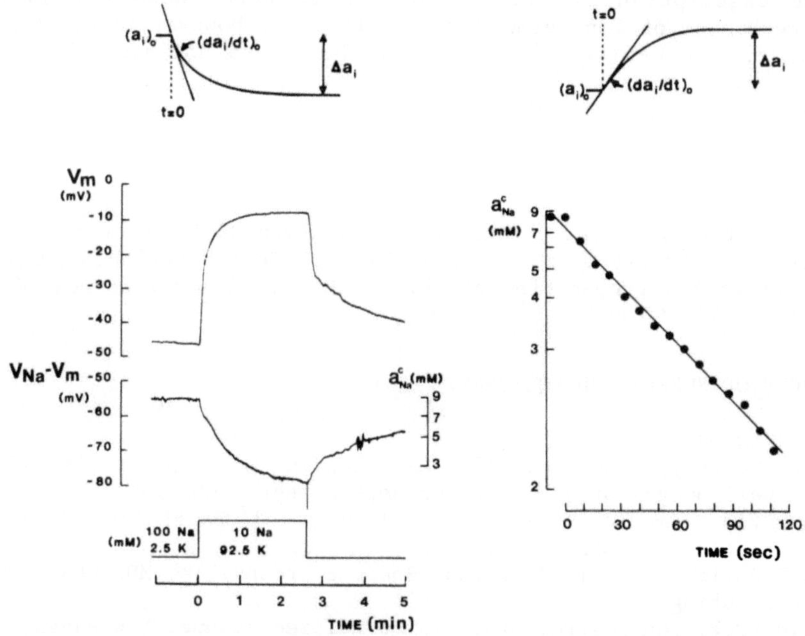

Fig.9. Top: Initial rates of change and total changes of cell activity following depletion (left) and restoration (right). Bottom, left: Cell membrane potential and $a_{Na}{}^{c}$ changes in *Necturus* gallbladder following substitution of apical solution Na for K. Right: Semi-log plot of the change in activity vs. time for the experiment shown at left.

interval which is short enough so as to ensure that little else has changed in the cell, we can then take the remaining J_i's as constants and hence $dc_i{}^{c}/dt$ becomes equal to $J_i{}^{1}/h$. Thus we can look at a single transport system, before others change in a significant way. The critical problem now is to have enough time resolution and a good empirical mathematical description of the transient so as to allow the measurement of the desired initial changes (at t=0). The following example deals with the net Na efflux from a cell after substituting extracellular Na by K while monitoring the fall in $a_{Na}{}^{c}$ with a Na-selective microelectrode, as shown in Fig.9. We want now to calculate the initial rate of change in $a_{Na}{}^{c}$, $(da_{Na}{}^{c}/dt)_o$, which is proportional to the initial efflux of Na and which can be used in the equations discussed above. The conventional way of obtaining initial rates is by plotting the instantaneous value of $a_{Na}{}^{c}$ against time and fitting a continuous function through these data, as shown in the lower right panel of Fig.9. In general, intracellular space filling and emptying can be reasonably well described by a single exponential of the form (Giraldez, 1984; Garcia-Diaz et al., 1986):

$$(a_i{}^{c})_t = [(a_i{}^{c})_o-(a_i{}^{c})_\infty]e^{-kt}+(a_i{}^{c})_\infty \qquad (11)$$

i.e.:

$$(a_i{}^{c})_t - (a_i{}^{c})_\infty = \Delta a_i{}^{c}\ e^{-kt} \qquad (12)$$

where the subscripts o and ∞ refer to the initial and final steady-state values of $a_i{}^{c}$, $\Delta a_i{}^{c}$ is their difference and k is the rate constant (k=1/tau) which characterizes the process. The semi-log plot of the change in $a_i{}^{c}$ against time will give a straight line with a slope k. The initial rate of change in activity can then be calculated as $(da_i{}^{c}/dt)_o = -k\ \Delta a_i{}^{c}$.

The above description of the changes in activity does not imply any specific mechanism of ion permeation. It is true, however, that if the net uptake or loss of Na (or other ion considered) is dominated by a single mechanism that happens to follow first order kinetics, then the transient has to be a single exponential.

ACKNOWLEDGEMENTS

Thanks to Gregor Jones, Simon Levy and Alvin Essig for comments on the manuscript. J.F. Garcia-Diaz supported by U.S. Public Health Service grant DK 39214 during the preparation of this work. F. Giraldez acknowledges the support of DGICYT, PB 86/0326 (Spain).

SOME SOURCES OF SUPPLY AND EQUIPMENT

Glass capillaries

Glass Company of America, 1835 Mays Landing Road, Hillville, NJ 08332, USA. (Thin wall omega-dot #1211; thick septum theta tubing).

Clark Electromedical Instruments, P.O. Box 8- Pangbourne, Reading RG8 7HU, England. (Theta tubing TGC 200-15).

R & D Optical Co., 15931 Batson Road, Spercenville, MD 20868, USA. (Thick septum theta tubing).

World Precision Instruments, 375 Quinnipiac Avenue, New Haven, CT 06513, USA / 291 Cricklewood Lane, Childs Hill, London NW2 2JL, England / Rohrbacher Strasse 27, D6900, Heidelberg, FRG. (A variety of glass capillaries).

Frederick Haer & Co., Brunswick, ME 04011, USA. (Borosilicate and aluminosilicate glasses in a variety of conformations).

Ion-selective cocktails

Fluka Chemical Corp., 980 South Second Street, Ronkonkoma, NY 11779, USA. (Neutral carrier based sensors for H, Na, Ca, Li and Mg; see Table I for catalogue numbers).

CIBA-Corning, 63 North Street, Medfield, MA 02052, USA. (K and Cl liquid ion-exchangers; see Table I for catalogue numbers).

Silanes and other chemicals

Fluka Chemical Corp.: Address above. (A full line of chemical reagents for the fabrication of LISM is marketed under the name Selectophore; ask for Handbook).

Dow Corning Corp., Midland, MI 48640, USA. (Silicone polymer 1107).

Pierce Chem. Co., Box 117, Rockford, IL 61105, USA / Pierce Europe b.v., P.O. Box 1512, 3260 BA oud Beijerland, Holland. (Chloro- and amino-silanes).

Pullers, bevellers and electronic equipment

World Precision Instruments: Addresses above. (A variety of electrophysiological equipment: puller, beveller, electrode amplifiers, manipulators, etc.).

Narishige, One Plaza Road, Greenvale, NY 11548, USA. (Manipulators, pullers, bevellers, etc.).

Biomedizinische Instrumente, Sudetenstrasse 6, D8034, Germering/Munich, FRG. (Manipulators, stepping motor and piezoelectric manipulators, epithelial voltage clamp and microelectrode amplifiers).

Burleigh Instruments, Inc., Burleigh Park, Fishers, NY 14453, USA. (PZ-550 Inchworm piezoelectric manipulator).

Stoelting, 1350 South Kostner Avenue, Chicago, IL 60623, USA. (Manipulators, etc.).

Miscellaneous.

E.W. Wright, 760 Durham Road, Guilford, CT 06437, USA. (Microelectrode holders).

Analog Devices, Two Technology Way, P.O.Box 280, Norwood, MA 02062, USA / Central Avenue, East Molessey, Surrey, KT8 OSM, England. (High input impedance amplifiers).

Burr-Brown, International Airport Industrial Park, P.O. Box 11400, Tucson, AZ 85734, USA / 1 Millfield House, Woodshots Meadow, Watford, Hertfordshire, WD1 8YX, England. (Operational amplifiers).

REFERENCES

Alvarez-Leefmans, F.J., Gamiño, S.M., Giraldez, F. and Nogueron, I., 1988. Intracellular chloride regulation in amphibian dorsal root ganglion neurones studied with ion-selective micro-electrodes. J. Physiol.(London) 406:225-246.

Ammann, D., 1986. Ion-selective microelectrodes. Principles, design and application. Springer-Verlag, Berlin.

Armstrong, W.McD. and Garcia-Diaz, J.F., 1980. Ion-selective microelectrodes: theory and technique. Federation Proc.:39:2851-2859.

Armstrong, W.McD. and Garcia-Diaz, J.F., 1981. Criteria for the use of microelectrodes to measure membrane potentials in epithelial cells. In Epithelial Ion and Water Transport (A.D.C. Macknight and J.P. Leader, eds.), pp.43-53. Raven Press, New York.

Bockris, J.O'M. and Reddy, A.K.N., 1973. Modern Electrochemistry. Vol. 1. Plenum/Rosetta Edition. Plenum Press. New York.

Bolton,T.B. and Vaughan-Jones, R.D., 1977. Continuous direct measurement of intracellular chloride and pH in frog skeletal muscle. J. Physiol.(London) 270:801-833.

Brown, K.T. and Flaming, D.G., 1977. New microelectrodes techniques for intracellular work in small cells. Neuroscience 2:813-827.

Coles, J.A. and Tsacopoulos, M., 1977. A method of making fine double-barrelled potassium-sensitive micro-electrodes for intracellular recording. J. Physiol.(London) 270:12P-14P.

Coles, J.A. and Tsacopoulos, M., 1979. Potassium activity in photoreceptors, glial cells and extracellular space in the drone retina: changes during photostimulation. J. Physiol.(London) 290: 525-549.

Conway, B.E., 1952. Electrochemical data. Elsevier, Amsterdam.

Edelman, A., Curci, S., Samarzija, I. and Frömter, E., 1978. Determination of intracellular K activity in rat kidney proximal tubular cells. Pflügers Arch. 378:37-45.

Eisenman, G., 1967. Glass Electrodes For Hydrogen and Other Cations. Principles and Practice. M. Dekker, Inc.. New York.

Eisenman, G., 1968. Similarities and differences between liquid and solid ion exchangers and their usefulness as ion specific electrodes. Anal. Chem. 40:310-320.

Eisner, D.A., Lederer, W.J. and Vaughan-Jones, R.D., 1981. The dependence of Na pumping and tension on intracellular sodium activity in voltage-clamped sheep Purkinje fibres. J. Physiol.(London) 317:163-187.

Garcia-Diaz, J.F., Baxendale, L.M., Klemperer, G. and Essig, A., 1985. Cell K activity in frog skin in the presence and absence of cell current. J. Membrane Biol. 85:143-158.

Garcia-Diaz, J.F. and Essig, A., 1985. Capacitative transients in voltage-clamped epithelia. Biophys. J. 48:519-523.

Garcia-Diaz, J.F., Klemperer, G., Baxendale, L.M. and Essig, A., 1986. Cell sodium activity and sodium pump function in frog skin. J. Membrane Biol. 92:37-46.

Garcia-Diaz, J.F., Stump, S. and Armstrong, W.McD., 1984. Electronic device for microelectrode recordings in epithelial cells. Am. J. Physiol. 246:C339-C346.

Giraldez, F., 1984. Active sodium transport and fluid secretion in the gall-bladder epithelium of *Necturus*. J. Physiol.(London) 348:431-455.

Giraldez, F. and Sepulveda, F.V., 1986. Chloride activity in *Necturus* enterocytes. J. Physiol.(London) 380:26P.

Gorman, A.L.F., Levy, S., Nasi, E. and Tillotson, D., 1984. Intracellular calcium measured with calcium-sensitive micro-electrodes and arsenazo III in voltage-clamped *Aplysia* neurones. J. Physiol.(London) 353:127-142.

Guibault, G.C., Durst, R.A., Frant, M.S., Freiser, H., Hansen, E.H., Light, T.S., Pungor, E., Rechnitz, G., Rice, N.M., Rohm, T.J., Simon, W. and Thomas, J.D.R., 1976. Recommendations for nomenclature of ion-selective electrodes. Pure Appl. Chem. 48:127-132.

Halliwell, J.V. and Whitaker ,M.J., 1987. Using microelectrodes. In Microelectrodes Techniques: The Plymouth Workshop Handbook (N.B. Standen, P.T.A. Gray and M.J. Whitaker, eds.), pp.1-12. The Company of Biologists Ltd., Cambridge.

Horowitz, P. and Hill, W., 1989. The Art of Electronics. Cambridge University Press, Cambridge.

Hu, Z., Buhrer, T., Muller, M., Rusterholtz, B., Rouilly, M. and Simon, W., 1989. Intracellular magnesium ion selective microelectrode based on a neutral carrier. Anal. Chem. 61:574-576.

IUPAC, 1979. Commission on analytical nomenclature (prepared for publication by G.C. Guibault). Recommendations for publishing manuscripts on ion-selective electrodes. Ion-selective Electrode Rev. 1:139.

Koryta, J. and Stulik, K., 1983. Ion-Selective Electrodes. Second Edition. Cambridge University Press. Cambridge.

Lanter, F., Steiner, R.A., Ammann, D. and Simon, W. 1982. Critical evaluation of the applicability of neutral carrier based Ca^{2+} selective microelectrodes. Anal. Chim. Acta 135:51-59.

Lev, A.A. and Armstrong, W.McD., 1975. Ionic activities in cells. Curr. Top. Membranes & Transp. 6:59-123.

Levy, S. and Fein, A., 1985. Relationship between light sensitivity and intracellular free Ca concentration in *Limulus* ventral photoreceptors: a quantitative study using Ca-selective microelectrodes. J. Gen. Physiol. 85:805-841.

Morris, M.E. and Krnjevic, K. (editors), 1987. Ion-Selective Microelectrodes and Excitable Tissues. Can. J. Physiol. Pharmacol. 65:867-1100.

Moody, G.J. and Thomas, J.D.R., 1971. Selective ion-sensitive electrodes. Merrow Pub.Co.Ltd., Watford, England.

Munoz, J.L., Deyhimi, F. and Coles, J.A., 1983. Silanization of glass in the making of ion-sensitive microelectrodes. J. Neurosci. Meth. 8:231-247.

Nelson, D.J., Ehrenfel, J. and Lindemann, B., 1978. Volume changes and potential artifacts of epithelial cells of frog skin following impalement with microelectrodes filled with 3M KCl. J. Membrane Biol. 40 (Special Issue):91-119.

O'Doherty, J., Garcia-Diaz, J.F. and Armstrong, W.McD., 1979. Sodium-selective liquid ion-exchanger microelectrodes for intracellular measurements. Science 203:1349-1351.

Ogden,D.C., 1987. Microelectrode electronics. In Microelectrodes Techniques: The Plymouth Workshop Handbook (N.B. Standen, P.T.A. Gray and M.J. Whitaker, eds.), pp.199-227. The Company of Biologists Ltd., Cambridge.

Parsons, R., 1959. Handbook of Electrochemical Constants. Butterworths Scientific Publications. London.

Persson, B.-E. and Spring, K.R., 1982. Gallbladder epithelial cell hydraulic water permeability and volume regulation. J.Gen.Physiol. 79:481-505.

Purves, R.D., 1981. Microelectrode methods for intracellular recording and ionophoresis. Academic Press, London.

Reuss,L., 1985. Changes in cell volume measured with an electrophysiologic technique. Proc. Natl. Acad. Sci. 82:6014-6018.

Robinson, R.A. and Stokes, R.H., 1959. Electrolyte Solutions. Second Edition (revised). Butterworths Scientific Publications. London.

Roos, A. and Boron, W.F., 1981. Intracellular pH. Physiol. Rev. 61:296-434.

Schefer, U., Ammann, D., Pretsch, E., Oesch, U. and Simon, W., 1986. Neutral carrier based Ca-selective electrode with detection limit in the sub-nanomolar range. Anal.Chem. 58:2282-2285.

Thomas, M.V., 1982. Techniques in calcium research. Academic Press, London.

Thomas, R.C., 1972. Intracellular sodium activity and the sodium pump in snail neurones. J. Physiol.(London) 220:55-71.

Thomas, R.C., 1978. Ion-sensitive intracellular microelectrodes. How to make and use them. Academic Press, London.

Tripathi, S., Morgunov, N. and Boulpaep, E.L., 1985. Submicron tip breakage and silanization control improve ion-selective microelectrodes. Am. J. Physiol. 249:C514-C521.

Tsien, R.Y., 1980. Liquid sensors for ion-selective microelectrodes. TINS 3.:219-221.

Tsien, R.Y., 1983. Intracellular measurements of ion activities. Ann. Rev. Biophys. Bioeng. 12:91-116.

Tsien, R.Y. and Rink, T.J., 1980. Neutral carrier ion-selective microelectrodes for measurement of intracellular free calcium. Biochim. Biophys. Acta 599:623-638.

Tsien, R.Y. and Rink, T.J., 1981. Ca-selective electrodes: A novel PVC-gelled neutral carrier mixture compared with other currently available sensors. J.Neurosci.Meth. 4:73-86.

Vaughan-Jones, R.D., 1979. Non-passive chloride distribution in mammalian heart muscle: microelectrode measurement of the intracellular chloride activity. J. Physiol.(London) 295:83-109.

Vaughan-Jones, R.D. and Aickin, C.C., 1987. Ion-selective microelectrodes. In Microelectrode Techniques: The Plymouth Workshop Handbook. (N.B. Standen, P.T.A. Gray and M.J. Whitaker, eds.), pp.137-167. The Company of Biologists Ltd., Cambridge.

Walker, J.L.,Jr., 1971. Ion specific liquid ion exchanger microelectrodes. Anal.Chem. 43:89A-93A.

Zeuthen, T., 1980. How to make and use double-barrelled ion-selective microelectrodes. Curr. Top. Membranes & Transp. 13:31-47.

12

VOLTAGE-DEPENDENT IONIC CHANNELS: "WHOLE-CELL" RECORDING BY PATCH-CLAMP TECHNIQUES

José López-Barneo

Departamento de Fisiología y Biofísica
Facultad de Medicina, Universidad de Sevilla
Avda. Sánchez Pizjuán, 4
41009 Sevilla, Spain

INTRODUCTION

Ionic channels are integral membrane proteins that mediate the passive flux of ions across the cell membrane. This family of proteins is broadly distributed, from procaryotes to mammals, and participates in a number of cell functions including electrical signalling, secretion, motility, and growth and proliferation. Most known ionic channels are gatable, i.e. they can change their three-dimensional structure and adopt an open or a closed state. Ions only diffuse through open channels and thus channel activation results in the generation of transmembrane ionic currents. What induces ionic channels to change their conformation is either the binding of a ligand (as for example in the acetylcholine receptor) or a change in the membrane potential. Channels activated by the second mechanism are the "voltage-dependent" channels which have charged domains in their molecule that sense the variations in the transmembrane electric field.

Voltage-dependent sodium and potassium channels were first studied by Hodgkin and Huxley (1952) in their classical work on the squid giant axon. They used the "axial wire voltage-clamp" technique to record the ionic currents generated in response to a step change in membrane potential. For many years quantitative voltage-clamp studies were mostly limited to large invertebrate preparations and therefore properties of ionic channels of other cells were not well known. Since the discovery of the "patch-clamp" technique (Hamill et al., 1981) it has become possible to perform accurate current measurements in cells as small as 5-10 μm in diameter. This has considerably increased the interest in cell electrophysiology and attracted many investigators to a field where, in recent years, the recording of ion channel activity can be used in conjunction with biochemical, pharmacological, and molecular biology techniques.

Cell Membrane Transport, Edited by D. L. Yudilevich *et al.*
Plenum Press, New York, 1991

This chapter focuses on the study of voltage-gated channels as evidenced by whole-cell recording of membrane ionic currents. The first section describes elementary technical aspects that I believe are of importance for a proper understanding of the patch-clamp technique. Afterwards, some of the protocols followed in the separation and characterization of the different ionic currents are presented with experimental data obtained in our laboratory from several types of mammalian cells. In the last section the fact that voltage-dependent ionic channels can be modulated is briefly discussed and illustrated with an example.

GENERAL TECHNICAL DESCRIPTION

Preparation

The "whole-cell" variant of the patch-clamp technique has been successfully used for recording of ionic currents in enzymatically dispersed cells from a large variety of tissues and in clonal cell lines. Cells must have a membrane clean enough to allow the sealing of the recording micropipette and the isolation of a membrane patch. The whole-cell recording mode is established after disruption of the membrane patch, which is usually achieved by applying suction to the pipette interior (Hamill et al., 1981). In most cases, sealing of the recording pipette to the membrane requires that the isolated cells should be firmly attached to the substrate. When Petri dishes are used for plating most cells fasten themselves to the bottom in a few hours. However if the cells are plated on slivers of glass coverslips (which is very convenient because they can be transferred one at a time to the recording chamber during the experiment) attachment is less firm, though it can be improved by treating the pieces of glass with poly-L-lysine.

Accurate whole-cell current recording requires, as in other voltage-clamp techniques, a fast and uniform control of the membrane potential. This is easily accomplished in spherical cells of 8-12 μm in diameter using pipettes of a relatively large size. In these cells charging of membrane capacitance is fast, and, in addition, complete dialysis of the aqueous phase of the cytosol by the solution contained in the micropipette is achieved in less than 20 s (Fenwick et al., 1982; Matteson and Armstrong, 1984). These conditions favour a low and uniform access resistance (resistance between the pipette tip and the inner surface of the membrane) and a good space clamp. Stable recordings can be obtained in cells with a diameter as small as 4-5 μm, although they are less convenient because of the difficulties in approaching them under the microscope. In cells larger than 20-30 μm membrane capacitance is large and therefore the clamp is slow. This inconvenience can be, however, partially overcome by using low resistance electrodes and special charging procedures (Armstrong and Chow, 1987; Swandulla and Armstrong, 1988).

Experimental Set Up

Most equipment and accessories needed for patch clamping are now customized and commercially available. It

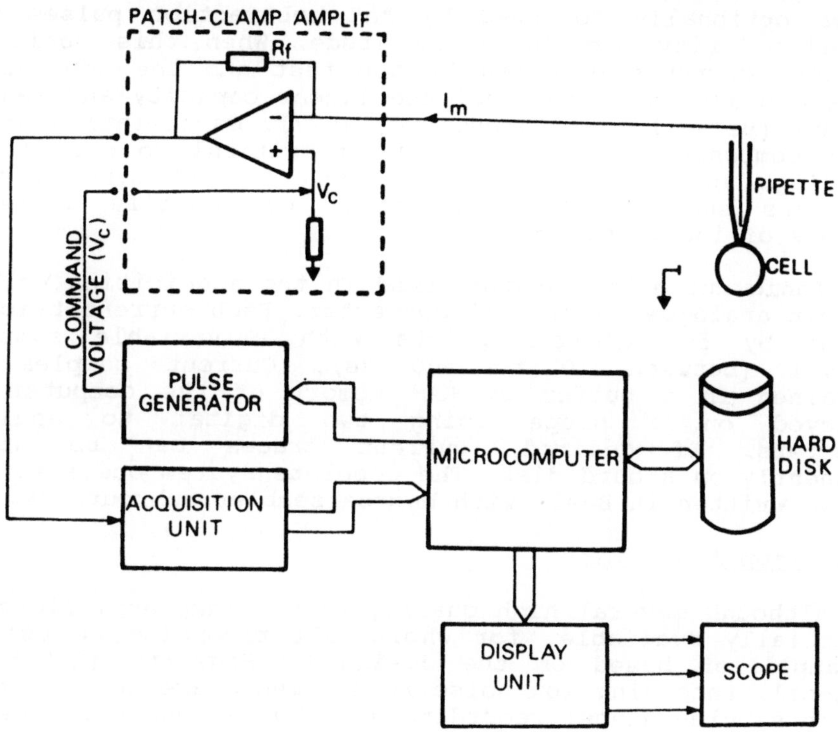

Fig. 1. Block diagram of the system used for acquisition and display of transmembrane ionic currents. I_m = membrane current; R_f = feedback resistance of the current to voltage converter; V_c = command voltage. (Modified from Ureña et al., 1989a).

is, however, less usual to find in the market a complete system adequated for quantitative studies of whole-cell currents. This experimentation requires the application to the cell membrane of complicated voltage pulse protocols, fast digitization of current signals generated by the cell and the conditioning of the signals before they are displayed on scope or stored for subsequent analysis. Here, I will briefly discuss the major features of a system built in our laboratory (Ureña et al., 1989a) which was used for the acquisition and analysis of most of the current traces presented in this report.

Fig. 1 shows the complete block diagram of the system used for activation, acquisition and display of membrane current signals. It is based on an IBM PC-AT computer and built on two IBM Prototype Cards. Command voltage pulses are obtained from a programmable pulse generator and applied to the non-inverting input of the patch-clamp amplifier. The pulser has an 8-bit digital to analogue converter controlled by the computer through a programmable peripheral interface (Peatman, 1977; Sargent and Shoemaker, 1984). The converter operates in bipolar mode and can generate a variable number of consecutive steps of variable amplitude (± 200 mV in steps of 2 mV) and duration (from 0.1 to 1000 ms). The test pulses can be optionally followed by "n" subtraction pulses with opposed polarity and "n_{th}" amplitude. When this option is selected, currents elicited by the test and the subtraction pulses are averaged to cancel the linear capacity and leakage currents (Bezanilla and Armstrong, 1977). Cancellation of the linear components of the current is critical to obtain good recordings and for an accurate resolution of tail currents which, as shown later, are useful to identify different subtypes of ionic channels.

Ionic currents are digitized in the acquisition unit by a 12 bit analogue to digital converter. Each current trace is defined by 500 digital points with a variable sampling interval (between 10 to 500 μs). Current samples are maintained in a buffer of RAM memory of the computer and displayed on a scope using two digital to analogue converters. If desired, current traces can be stored permanently on a hard disk. The complete system operates by a program written in Basic with Macroassembler subroutines.

<u>Patch-Clamp Amplifier</u>

Although several high quality patch-clamp amplifiers are commercially available, for whole-cell recording we use our own amplifier based on the design by Sigworth (1983). In whole-cell recording low noise requirements are less critical than in single-channel recording and thus emphasis can be put on clamp speed and frequency response.

The diagram of the headstage of the patch-clamp amplifier is shown in Fig. 2. It is based on a low-noise operational amplifier (Burr Brown, OPA 111) wired as a current to voltage (I/V) converter. Current generated in the cell (I_m) is collected by the micropipette which is electrically connected to the negative input of the amplifier. The amplifier produces a voltage at the output (V_o) that applied through the feedback loop nulls any voltage change at the summing point (s). Thus, V_o is related to I_m by:

$$I=-V_o/R_f$$

where R_f is the feedback resistance of the I/V converter.

Membrane voltage is measured between the external electrode (connected to ground) and the voltage at the summing point. This last voltage value is, in ideal conditions (with low resistance electrodes and low access resistance), approximately the same as the voltage at the inner surface of the membrane. Membrane potential can then be controlled by applying command pulses to the positive input of the amplifier which, in turn, will automatically produce the same voltage change at the negative input.

Frequency response is mainly limited by the value of the distributed capacitance (C_f) of the feedback resistance and by the time required to charge membrane capacitance. Therefore, if the current size is not too small, it is convenient to use relatively low R_f values (between 10 and 100 Mohm) and small cells. The clamp can be additionally speeded up by ballistic charging, directly applying a

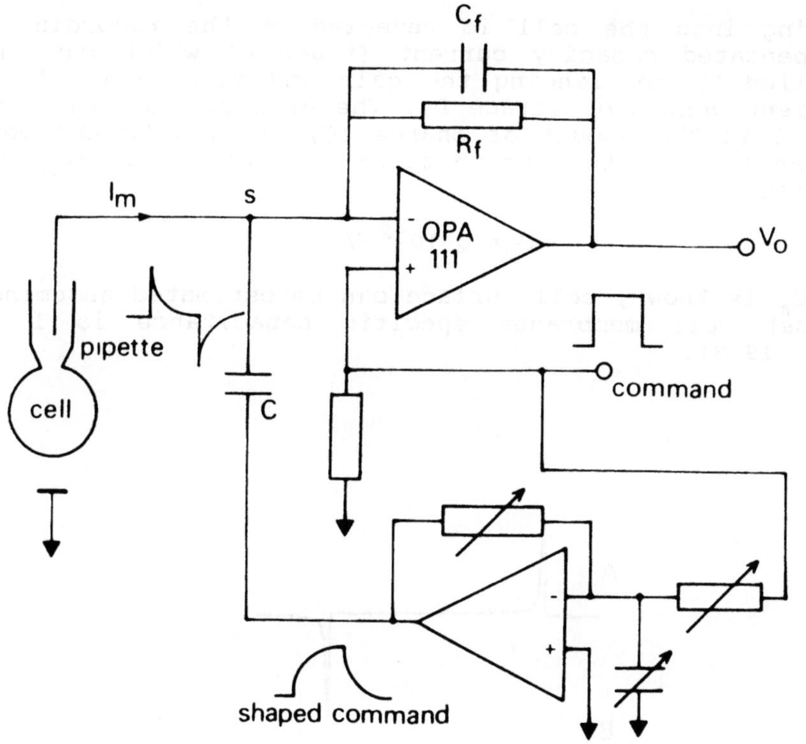

Fig. 2. Schematic diagram of the headstage of the patch-clamp amplifier and the transient generator. I_m = membrane current; V_o = output voltage; R_f = feedback resistance; C_f = distributed capacitance of the feedback resistance; S = summing junction of the operational amplifier; C = 0.1 pF capacitor (see text for further details).

differentiated shaped command pulse obtained from a transient generator (see Fig. 2) to the membrane. Supercharging, or "fast charging" of membrane capacitance can be used in large cells to improve clamp speed. This last procedure basically consists of adding to the command pulse an initial "spike" of much larger amplitude lasting a few μs (for details see Armstrong and Chow, 1987).

Ballistic charging of the membrane with a transient signal of variable time constant also helps to cancel the transient capacitiy currents and to record ionic currents in isolation. Fig. 3 shows current traces elicited by a 10 mV square command pulse of 2 ms while establishing the whole-cell configuration. After forming a tight seal between the pipette and the membrane (trace A) large current transients are generated at the onset and at the end of the pulse due to charging of electrode capacitance (that in this experiment was relatively large because the level of the solution in the chamber was high). Trace B illustrates that these transients can be perfectly compensated by applying a transient signal of opposed polarity to the negative input of the amplifier.

Breaking into the cell is revealed by the recording of an uncompensated capacity current (trace C) which can also be cancelled by rebalancing the gain and time constant of the transient generator (trace D). The area of the transients in trace C is the amount of charge (Q) applied to the membrane and can be used to measure membrane capacitance (C_m) by the relation:

$$C_m = Q/10^{-2}\ V$$

Once C_m is known, cell surface can be estimated assuming that in most cell membranes specific capacitance is 1 uF/cm^2 (Cole, 1968).

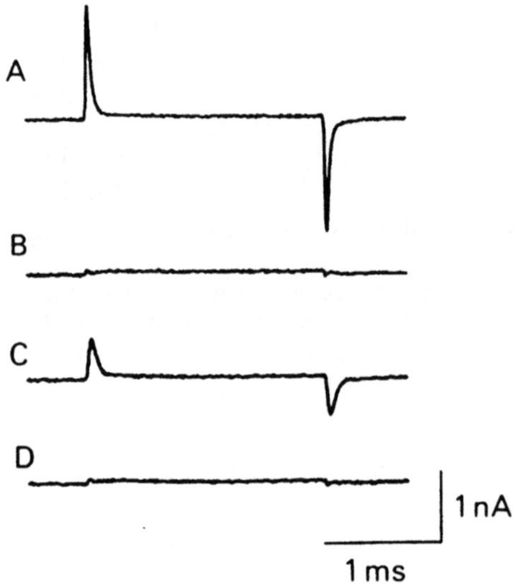

Fig. 3. Transient capacity currents elicited by a 10 mV voltage pulse while establishing the whole-cell recording configuration. (Explanation in the text).

When the amplitude of the current recorded is, for example, 1 nA and the series resistance (resistance between the inner surface of the membrane and the input of the operational amplifier) is in the range of a few Mohm, membrane voltage can be several mV apart from the value at the summing point, which is the actual voltage kept under experimental control. The errors due to the series resistance can be minimized by recording currents of small amplitude and by the use of electrodes of low resistance (which is responsible for more than half of the total series resistance). Most patch-clamp amplifiers are equipped with electronic compensation of series resistance. This consists of adding to the square command pulse a voltage signal that is a fraction of the output current (see also Sigworth, 1983).

Electrodes

Electrodes used for patch-clamping are fabricated from glass capillaries by a double pull and fire polished with a microforge mounted on an inverted microscope. We have used both soft hematocrit capillaries and Kimax 51 borosilicate glass. For whole-cell recording we have not found measurable differences between the two types of glass although it has been reported that hard glass may have electrical properties that favour low noise recordings (Sakmann and Neher, 1983). Some types of soft glass can release trace quantities of heavy metals which produce modification of the ionic currents (Cota and Armstrong, 1988; Furman and Tanaka, 1988). These effects are avoided if a significant concentration of a chelator, such as EGTA or BAPTA, is added to the internal solution.

As previously discussed, electrodes prepared for whole-cell recording should have a low resistance. For cells between 10-14 μm we have used electrodes that, when filled either with KCl or CsCl, have resistances between 1 and 2.5 Mohm.

SEPARATION AND CHARACTERIZATION OF IONIC CURRENTS

Current Components and Strategies for their Separation

Voltage-dependent ionic currents are usually recorded by stepping the membrane potential to a depolarized value. In excitable cells membrane depolarization activates several types of ionic channels and thus total currents are a mixture of individual components carried by the different ionic species. Fig. 4 is an example of typical whole-cell currents recorded from a mammalian central neuron. The holding potential was -80 mV and membrane voltage was stepped to +20 mV. Current generated during the pulse (indicated between the arrows) has three major components: a) almost immediatelly after the change in membrane potential there is a fast transient inward current; b) in less than 1 ms the current goes from inward to outward and reaches an almost steady value by the end of the pulse; and c) at the instant of repolarization current suddenly changes in direction and an inward tail current is observed.

Individual current components can be recorded in isolation by the use of agents that specifically block a type of ion channel or by substitution of permeant ions by impermeant ones. This procedure is illustrated in Fig. 5 with recordings from type I cells of the carotid body which are electrically excitable (López-Barneo et al., 1988). The composition of the solutions (external//internal) used in each case are indicated near each trace. In A, using the standard external and internal solutions, the three current components are clearly present. Replacement of internal K^+ by Cs^+ (which is a blocker of potassium channels, Adelman and

Senft, 1966; Bezanilla and Armstrong, 1972) produces the disappearance of the outward component, indicating that is was due to activation of potassium channels. An inward current with a fast and slow component and the tail (B) are still present. The fast component of the current is abolished by external application of tetrodotoxin (TTX), a specific blocker of sodium channels (Narahashi et al., 1964), while the slow inward current and the tail, representing the activity of calcium channels, remain (C).

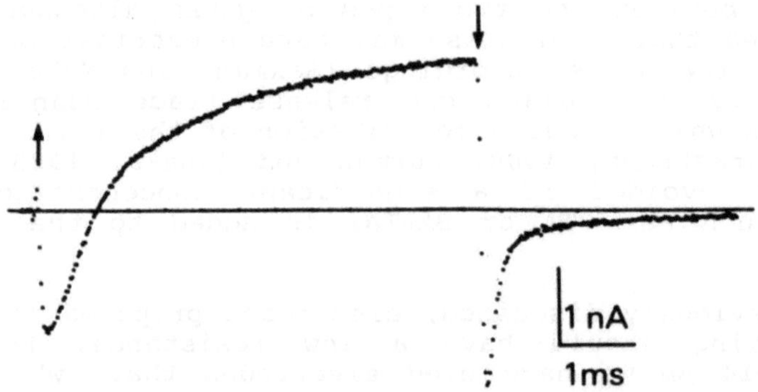

Fig. 4. Whole-cell currents recorded in a mammalian septal neuron. Holding potential was -80 mV and pulse voltage +20 mV. Solutions (mM) 130 Na, 2.7 K, 2.5 Ca, 1 Mg // 130 K, 2 Mg, 10 EGTA (external // internal).

Thus, these experimental protocols reveal that total current is due to the flow of ions through sodium, calcium and potassium channels. These three types of ionic currents can be recorded in most excitable membranes (Hille, 1984), but in some preparations, as for example in adrenocortical cells (Tabares et al., 1989), there are no sodium channels. In contrast, in other membranes, as in both myelinated and unmyelinated nerve fibers, calcium channels are absent or their density is very low. As will be shown later in some cells there are several subtypes of voltage-dependent ionic channels which can be separated by their different kinetic and pharmacological properties.

<u>Sodium Currents</u>

Sodium currents can be recorded in isolation using internal solutions with K^+ replaced by Cs^+ and external solutions where Ca^{2+} is replaced by Mg^{2+}, or a mixture of Mg^{2+} and Co^{2+}. These two divalent cations are well known blockers of calcium channels (Kohlhardt et al., 1972). Fig. 6 shows an isolated sodium current elicited in a dispersed septal neuron by depolarization from -80 to 0 mV. External Na^+ ions were partially replaced by the large impermeant cation $Tris^+$ to reduce current amplitude and to minimize errors due to uncompensated series resistance. The sodium current has a characteristic fast activation time course and

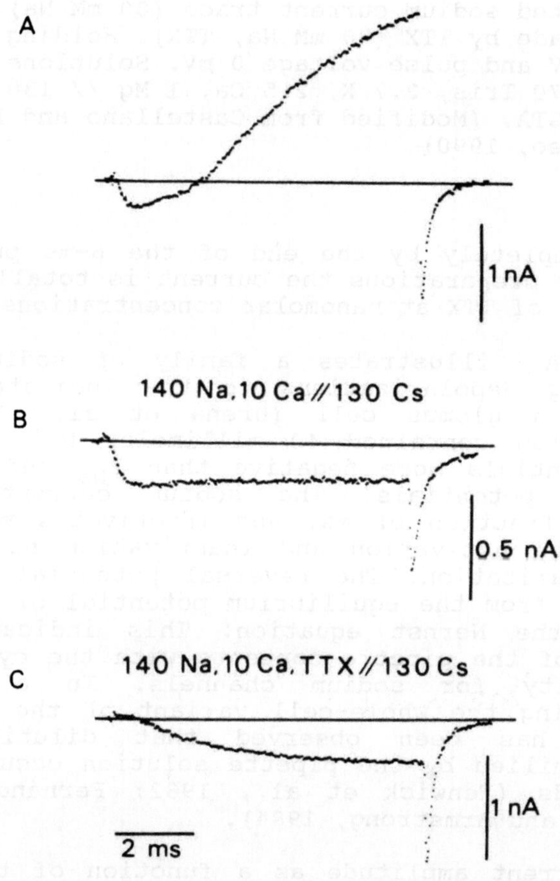

Fig. 5. Isolation of the different major components in whole-cell currents. The composition of the solutions (in mM) are indicated on top of each current trace. (From Ureña et al., 1989b).

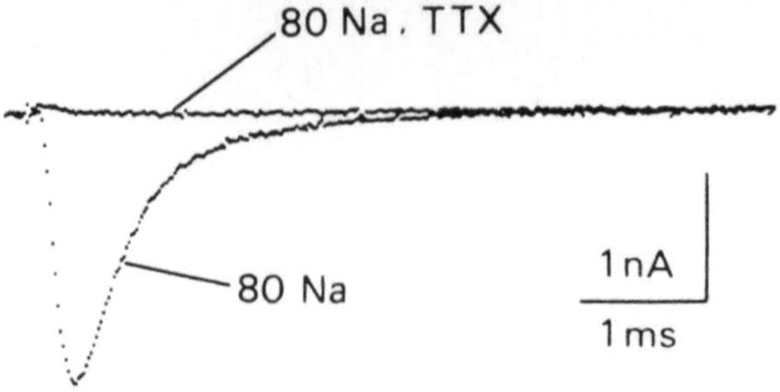

Fig. 6. Isolated sodium current trace (80 mM Na) and blockade by TTX (80 mM Na, TTX). Holding potential -80 mV and pulse voltage 0 mV. Solutions (mM) 80 Na, 70 Tris, 2.7 K, 2.5 Ca, 1 Mg // 130 Cs, 2 Mg, 10 EGTA. (Modified from Castellano and López-Barneo, 1990).

inactivates completely by the end of the 6-ms pulse. As it occurs in other preparations the current is totally abolished by the presence of TTX at nanomolar concentrations.

Fig. 7A illustrates a family of sodium currents recorded during depolarizations to the indicated membrane potentials in a glomus cell (Ureña et al., 1989b). The internal solution contained 40 millimolar Na^+. Current is inward at potentials more negative than E_{Na} and outward at more positive potentials. The sodium current turns on rapidly, in a fraction of ms, and inactivates more slowly. Kinetically both activation and inactivation speed up with membrane depolarization. The reversal potential is +32 mV, only 1 mV away from the equilibrium potential of Na^+ ions as predicted by the Nernst equation. This indicates a good equilibration of the pipette contents with the cytosol and a high selectivity for sodium channels. In a number of experiments using the whole-cell variant of the patch-clamp technique it has been observed that dilution of the intracellular milieu by the pipette solution occurs in a few tens of seconds (Fenwick et al., 1982; Fernández et al., 1984; Matteson and Armstrong, 1984).

Peak current amplitude as a function of the membrane potential is plotted in Fig. 7B. The activation threshold is around -40 mV and maximal current amplitude is obtained by depolarizations to 0 or +10 mV. Sodium currents in most cells studied are qualitatively similar but there are some differences among the various preparations in the voltage at which the maximal inward current amplitude is obtained and in the voltage-dependence of inactivation. These differences may have physiological interest but their molecular bases are unknown (Bezanilla and Armstrong, 1977; Matteson and Armstrong, 1984; Moolenaar and Spector, 1978; Sah et al., 1988; Ureña et al., 1989b).

<u>Calcium Currents</u>

As previously shown in Fig. 5, calcium currents are studied in isolation with internal solutions containing Cs^+ and/or N-methylglucamine (NMG) to block the potassium

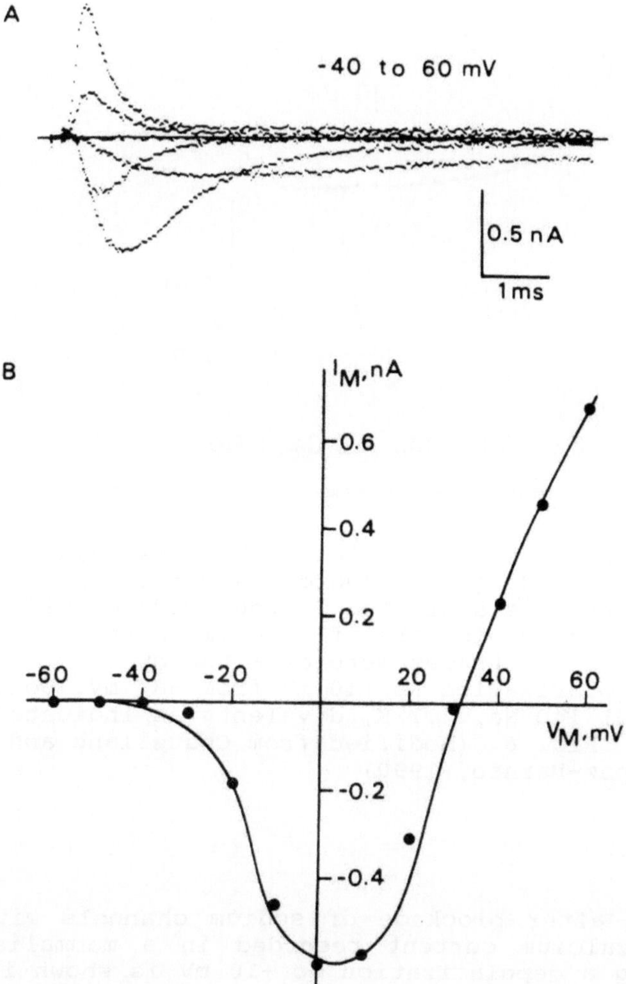

Fig. 7. Current-voltage relation for the sodium channels. A. Superimposed sodium currents recorded at various membrane potentials. B. Peak sodium current amplitude (ordinate) as a function of the step membrane potential (abscissa). Solutions (mM) 140 Na, 0.5 Ca, 9 Mg, 2.7 K // 40 Na, 90 Cs, 2 Mg, 10 EGTA. (From Ureña et al., 1989b).

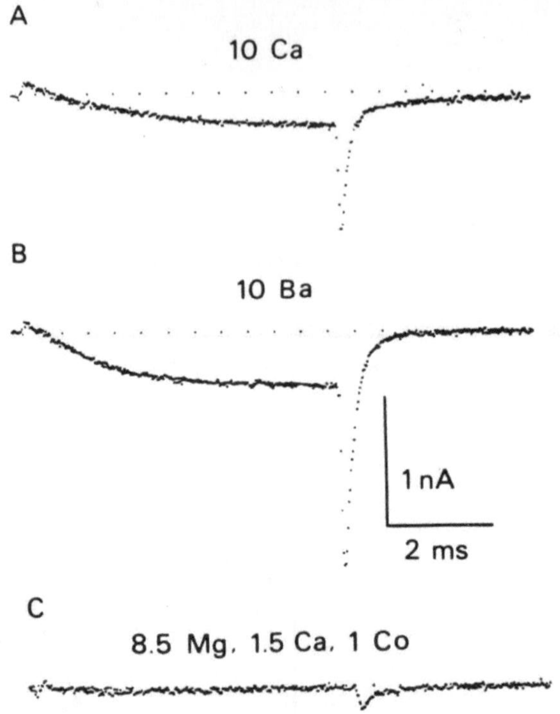

Fig. 8. Current through calcium channels using Ca (A) and Ba (B) as charge carriers. The current is abolished in the presence of inorganic blockers of calcium channels (C). The three traces were obtained by depolarization to +10 mV from -80 mV. Solutions (mM) 140 Na, 2.7 K, divalents as indicated // as in Fig. 6. (Modified from Castellano and López-Barneo, 1990)

channels and after blockade of sodium channels with TTX. An example of calcium current recorded in a mammalian central neuron during a depolarization to +10 mV is shown in Fig. 8A. The current activates slowly, reaches an almost steady value by the end of the 6-ms pulse, and at the instant of repolarization a large tail current due to the sudden change in the electrochemical gradient of Ca^{2+} ions is generated. The time course of the tail reflects the closure of calcium channels that were open during the pulse. Besides its characteristic time course the calcium current can also be identified from the fact that in most preparations it is larger when Ba^{2+} ions are the charge carrier instead of Ca^{2+} (Fig. 8B; Hagiwara et al., 1974). In addition, the current disappears when external Ca^{2+} is replaced by Co^{2+} and/or Mg^{2+} (Fig. 8C), which are inorganic blockers of calcium channels.

Fig. 9A illustrates calcium currents recorded at various membrane potentials in an adrenocortical cell, which as indicated before, lacks sodium channels and the inward current is only mediated by calcium channels (Tabares et al.,

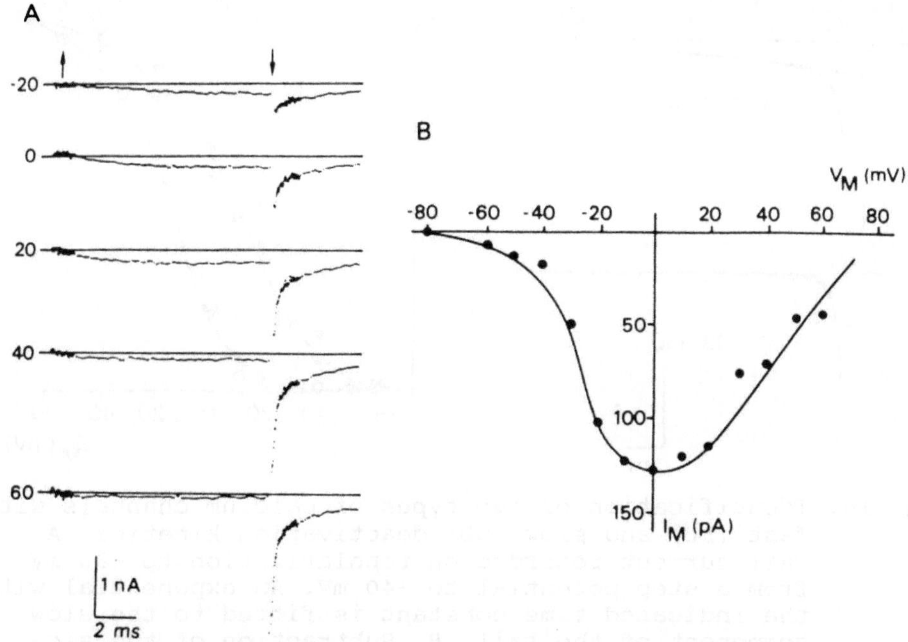

Fig. 9. Current-voltage relation for the calcium channels. A. Calcium currents recorded at the membrane potentials indicated on the left of each trace. B. Inward current measured at the end of the 10 ms pulses (ordinate) as a function of the membrane potential (abscissa). Solutions (mM) 140 Na, 10 Ca // 70 Cs, 70 NMG, 2 Mg, 2 Mg-ATP, 10 EGTA. (From Tabares et al., 1989).

1989). Current elicited during the pulse (between the arrows) activates more rapidly at more depolarized levels. Tails generated at the end of the pulse also increase in amplitude with membrane depolarization as the driving force for Ca^{2+} ions increases. Fig. 9B is the current-voltage relation obtained by measuring current amplitude at the end of the 10 ms pulse.

In a number of preparations studied so far it is well established that the total calcium current is due to the activity of at least two types of calcium channels that differ in their kinetic and pharmacological properties (Carbone and Lux, 1984; Nowycky et al., 1985; Matteson and Armstrong, 1986). In Fig. 9A it is clearly seen that after small depolarizations the decay of the tails follows a slow, single exponential time course but after steps to more positive potentials a fast component appears. Each component of the tail activates at a different threshold and reflects the closing of fast (FD) and slow (SD) deactivating calcium channels (Matteson and Armstrong, 1986). The two components can be isolated by following the method illustrated in Fig. 10. Part A is a tail current with a biphasic decay recorded on returning to the holding potential of -80 mV after a 10 ms depolarization to +40 mV. An exponential with a time

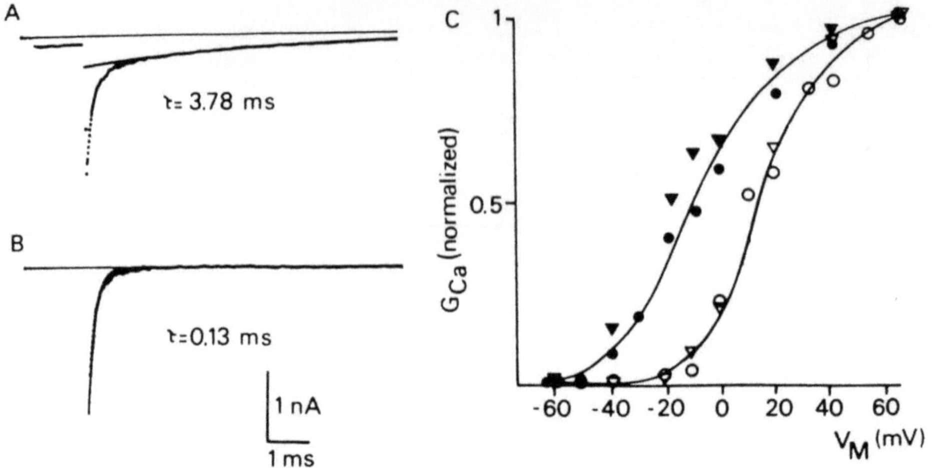

Fig. 10. Identification of two types of calcium channels with fast (FD) and slow (SD) deactivation kinetics. A. Tail current recorded on repolarization to -80 mV from a step potential to +40 mV. An exponential with the indicated time constant is fitted to the slow component of the tail. B. Subtraction of the slow component from the total current and isolation of a fast component. C. Conductance-voltage relations of SD (filled symbols) and FD (open symbols) channels with data from two cells. Solutions as in Fig. 9. (From Tabares et al., 1989).

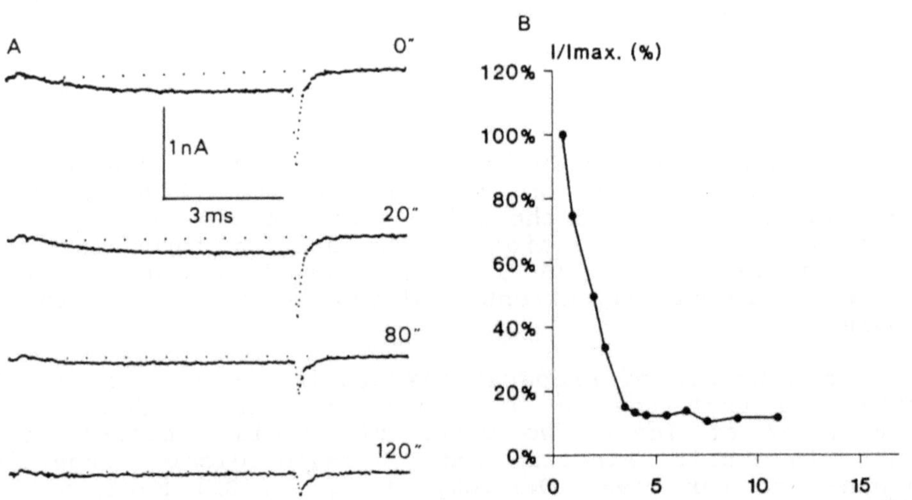

Fig. 11. Wash-out of calcium channels. A. Calcium currents recorded by depolarization to +20 mV at different times after establishment of the whole-cell recording configuration. B. Normalized current amplitude at the end of the pulse (ordinate) is plotted as a function of time (abscissa). Solutions as in Fig. 8 but with 10 mM external Ca and 150 nM Ca in the internal solution.

constant of 3.78 ms is fitted to the slow component and extrapolated back to the instant of repolarization. A fast component, obtained by subtracting the slow exponential from the tail current, can also be fitted by a single exponential with a time constant of 0.13 ms (B). In C, normalized SD and FD conductances (which are proportional to the slow and fast components of the tails) are plotted as a function of the step membrane potential. This plot clearly illustrates that the activation threshold of SD channels is about 30 mV lower than for FD channels and that half maximal activation is between -15 and -10 mV for SD channels and between +10 and +15 mV for FD channels.

Besides their differences in deactivation time course, SD and FD channels, probably equivalent to T and L (Nowycky et. al, 1985) or low and high voltage activated channels (Carbone and Lux, 1984), also differ in their activation and inactivation time courses and in their pharmacological sensitivity.

In most cells studied with the whole-cell recording technique calcium currents decrease progressively in amplitude after the initiation of the dialysis of the cytosol. This phenomenon is referred to as "wash-out" of calcium channels and is due to the decrease in the number of functional FD channels. SD channels are apparently unaffected by intracellular dialysis (Cota, 1986; Fedulova et al., 1985; Forscher and Oxford; 1985). Wash-out of calcium channels is illustrated in Fig. 11 with recordings obtained from a glomus cell, which have mainly the FD channel type (Ureña et al., 1989b; Rodríguez-Benot et al., 1989). Part A shows calcium currents recorded during depolarization to +20 mV immediately after breaking into the cell (trace at 0") and 20, 80 and 120 seconds later. Normalized current amplitude as a function of time is represented in plot B. The figure shows that in the conditions of this experiment (with 150 nanomolar Ca^{2+} in the pipette solution) current almost totally disappears 3 to 4 min after the initiation of the cell dialysis.

Intracellular dialysis in certain experimental conditions also produces modifications in the voltage-dependent parameters of sodium, calcium and potassium channels (Cota, 1986; Fernández et al., 1984: Tabares et al., 1989). The molecular basis of these phenomena are not well understood yet although wash-out of the L-type calcium channel seems to be related to dephosphorylation of the channels by calcium-activated phosphatases. Buffering of internal Ca^{2+} and/or addition of exogenous nucleotides drastically retard the disappearance of calcium currents. In glomus cells, for example, calcium currents can be recorded without alteration for 30-40 min with internal solutions containing 10 millimolar EGTA, 3 millimolar ATP and 10 ng/ml of the catalytic subunit of cAMP-dependent protein kinase (Rodríguez-Benot et al., 1989).

Potassium Currents

Potassium currents are recorded in isolation with high K^+ internal solutions and after blockade of sodium and calcium channels. In both vertebrate and invertebrate

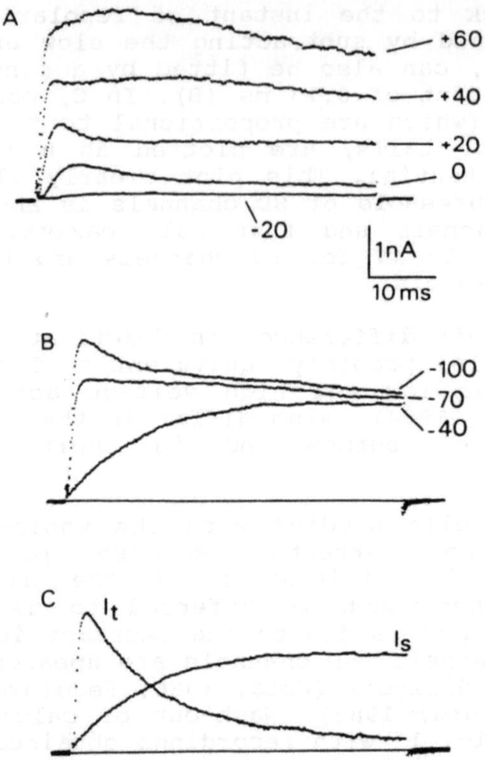

Fig. 12. Separation of two components in the outward potassium current. A. Currents recorded during voltage steps to the indicated membrane potentials. B & C. Procedure followed in the separation of the fast (I_t) and slow (I_s) currents. Solutions (mM) 140 Na, 2.5 Ca, 2.7 K, TTX // 140 K, 2 Mg, 10 EGTA. (Modified from Castellano et al., 1989).

preparations there are a large variety of different voltage-dependent potassium channels that differ in their kinetic and pharmacological properties (Hille, 1984; Rudy, 1988). In whole-cell recording the different sets of potassium channels contribute to the total outward current and therefore special procedures must be followed to identify the individual components.

An example of an outward potassium current which is produced the activity of more than one potassium channel type is illustrated in Fig. 12A with data obtained from dispersed pinealocytes (Castellano et al., 1989). Membrane potential was changed to the values indicated in the figure from a holding potential of -70 mV. It can be clearly observed that

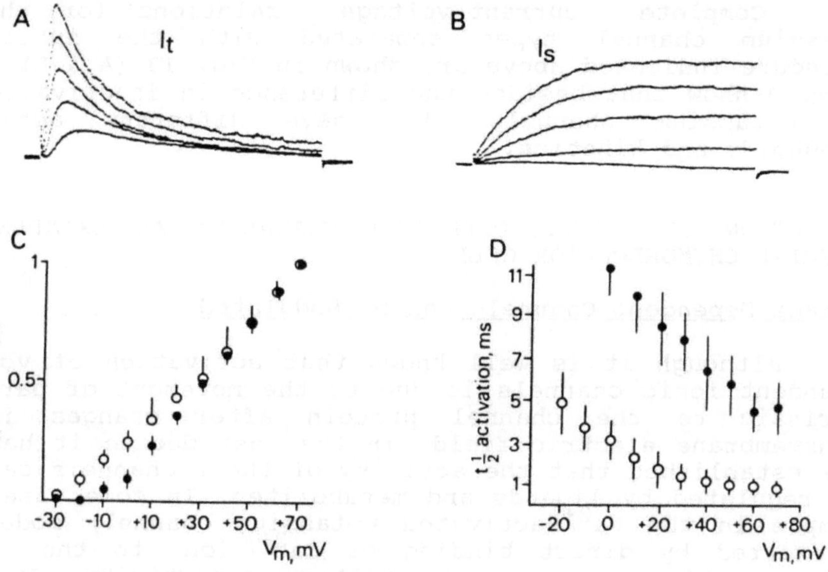

Fig. 13. Current-voltage relations for the two components of outward potassium current. A & B. I_t and I_s current traces at various membrane potentials. C & D. Peak current amplitude (C) and time to reach half activation (D) as a function of membrane potential. Open symbols (I_t) and filled symbols (I_s). Solutions as in Fig. 12. (From Castellano et al., 1989).

the current has a complex time course. With small depolarizations it reaches a peak and subsequently partially inactivates but with more positive pulses there is, after the initial rising phase, a second rise followed by a partial and slow inactivation. These recordings suggest that a transient current is activated at a threshold lower than the voltage change required to elicit a second and more sustained current.

Separation of the two current components is illustrated in Fig. 12B. The three currents were generated by depolarization to 50 mV delivered from a holding potential of -70 mV (middle trace) or after prepulses that changed the holding potential to -100 (upper trace) or -40 (lower trace) mV for 140 ms. The current traces have the same amplitude at the end of the 50 ms pulse but a fast activating component, clearly present when activation is from a holding potential of -100 mV, disappears when membrane potential is held at -40 mV. These two components, representing a transient (I_t) and a slow (I_s) activating potassium current, can be easily separated as shown in Fig. 12C. The slow component is the current recorded from a holding potential of -40 mV (voltage at which the transient current is inactivated) and the fast component is obtained by subtracting the slow current from the total current generated from a holding potential of -100 mV (voltage at which inactivation of I_t is totally removed).

Complete current-voltage relations for the two potassium channel types separated with the subtraction procedure indicated above are shown in Fig. 13 (A & B). Plots C and D show that besides the difference in inactivation the two potassium channels also have different activation thresholds and kinetics.

MODULATION OF VOLTAGE-DEPENDENT CHANNELS: AN EXAMPLE FROM ARTERIAL CHEMORECEPTOR CELLS

Voltage-Dependent Channels can be Modulated

Although it is well known that activation of voltage-dependent ionic channels is due to the movement of particles intrinsic to the channel protein after changes in the transmembrane electric field, in the last decade it has been well established that the activity of these channels can also be regulated by ligands and metabolites. In some cases (for example in the Ca^{2+}-activated potassium channel) modulation is exerted by direct binding of Ca^{2+} ions to the channel protein, which enhances its voltage sensitivity. In other channels (for example potassium and calcium channels regulated by neurotransmitters and neuropeptides) ligands interact with specific membrane receptors and the action on the channels is exerted through mediators. The mediator can be a subunit of a G protein or cyclic nucleotides, which produce phosphorylation of the channels by activation of protein-kinases.

In the whole-cell recording mode chemical regulation of the voltage-dependent currents can be studied if the modulator does not require the synthesis of soluble molecules, since it must be kept in mind that in this technique the cytosol is rapidly dialyzed by the pipette solution. In the next paragraphs, to illustrate an example of modulation of a voltage-dependent current, I will briefly describe the protocol followed to study the interaction of O_2 with potassium channels in the chemoreceptor cells of the carotid boby.

In Arterial Chemoreceptor Cells Environmental O_2 Selectively Reduces a Voltage-Dependent Potassium Current

It is known that type I cells of the carotid body release neurotransmitters in response to a decrease in partial oxygen pressure (pO_2). These transmitters activate the afferent fibers of the sinus nerve, which convey the information to the CNS and induce a hyperventilatory response (see for example Eyzaguirre and Zapata, 1968). We have recently shown that type I cells have sodium, calcium and potassium channels and that pO_2 selectively regulates a potassium current (López-Barneo et al., 1988).

The effect of lowering O_2 tension from 150 (control) to 10 (low pO_2) mmHg on the whole-cell currents of type I cells is shown in Fig. 14. Part A shows sodium and calcium currents recorded in a cell after blockade of potassium channels with internal Cs^+. Sodium currents recorded after blockade of calcium channels with 0.5 mM Cd^{2+} are shown in B. In both cases currents were elicited by 7-ms depolarizations to 0 mV

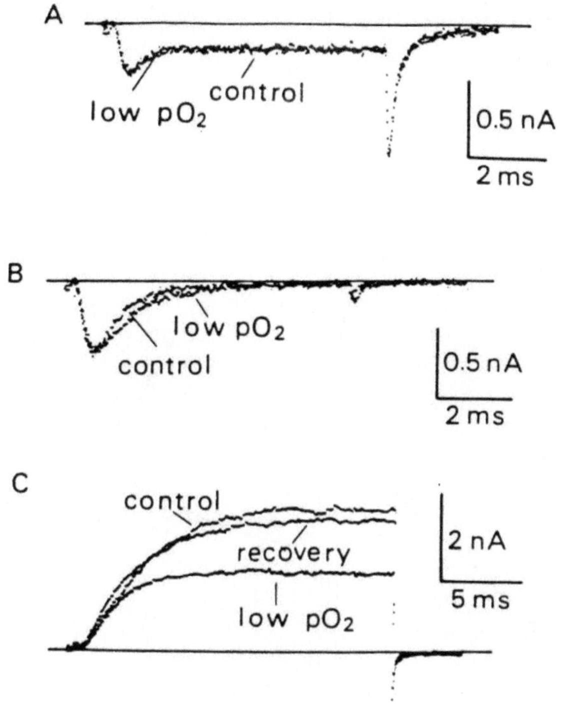

Fig. 14. Effect of lowering environmental oxygen tension on the whole-cell currents of arterial chemoreceptor cells.
A. Superimposed recordings of sodium and calcium currents. B. Recordings of sodium currents in isolation.
C. Reversible reduction of the potassium current by hypoxia. (Modified from López-Barneo et al., 1988).

in the control solution and 3 minutes after switching to low pO_2. In these two figures control and low pO_2 recordings are almost perfectly superimposed, which indicates that neither sodium nor calcium channels are affected by hypoxia. A decrease of pO_2 had, however, a marked effect on potassium currents (Fig. 14C). Exposure to low pO_2 produced a reduction in the amplitude of the outward current recorded by a depolarization to 40 mV which reverted almost completely after we reintroduced a solution with normal O_2 content.

These results, observed in every type I cell tested (n > 50), show that the potassium current in glomus cells is modulated by oxygen. This effect is seen in well dialyzed cells where the internal solution had 10 millimolar EGTA to buffer ionic calcium and no added ATP. This suggests that the effect of oxygen may not require a soluble intracellular mediator and could be explained by a direct interaction with the channels or by acting through a G protein (López-López et al., 1989).

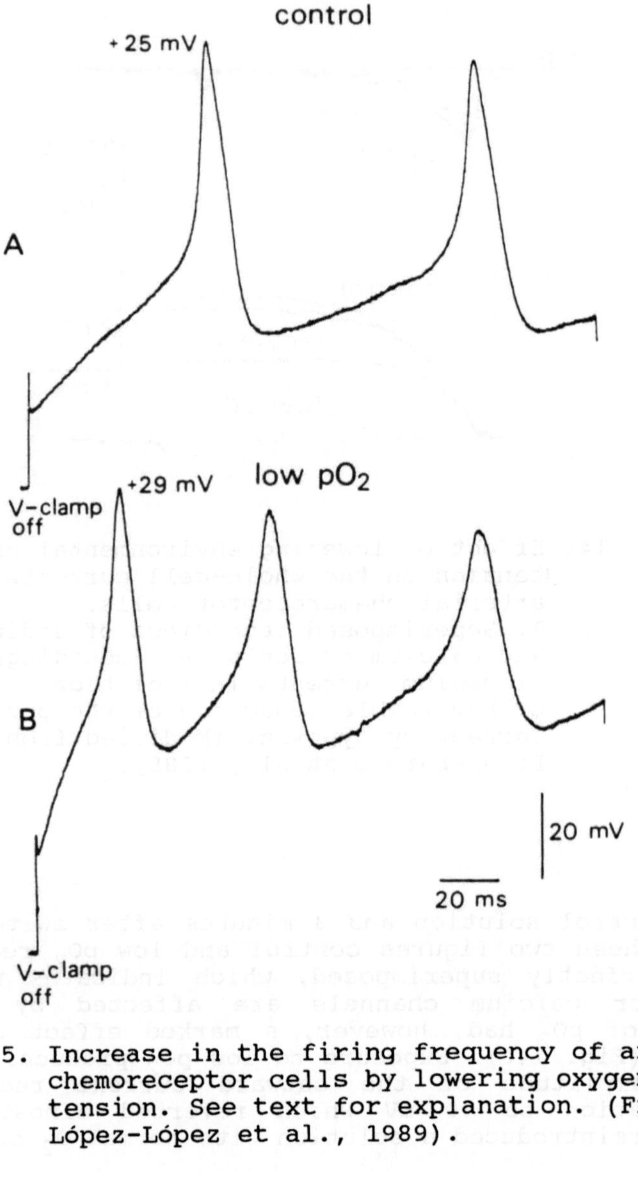

Fig. 15. Increase in the firing frequency of arterial chemoreceptor cells by lowering oxygen tension. See text for explanation. (From López-López et al., 1989).

Inhibition of the Potassium Current Produces an Increase in the Firing Frequency of the Cells

The decrease in the potassium current by low pO_2 in type I cells has a marked effect on their electrical behavior. This fact is illustrated in Fig. 15 where in a cell previously subjected to voltage-clamp, the membrane voltage in current-clamp mode was also recorded. In the control experimental situation (A) switching from voltage- to current-clamp produces a slow depolarization that after reaching a threshold elicits the generation of action potentials repetitively. After exposure to a pO_2 of 10 mmHg (B) the same experimental protocol gives rise to a faster pacemaker depolarization which results in a higher firing frequency. The hypoxic stimulus also increases the value of the overshoot potentials. Thus, low pO_2, by reducing a voltage-dependent potassium current, produces in chemoreceptor cells an increase in the action potential amplitude and in the slope of pacemaker potentials that determines the firing frequency of the cells. These two effects increase Ca^{2+} entry into the cytosol and enhance transmitter release.

ACKNOWLEDGEMENTS

The author wishes to thank his colleagues for their collaboration and comments on the manuscript. This research was supported by the Dirección General de Investigación Científica y Técnica (Grant PB86-0250).

REFERENCES

Adelman, W.J., and Senft, J.P., 1966, Voltage clamp studies of the effect of internal caesium ions on sodium and potassium currents in the squid giant axon, J. Gen. Physiol., 50:279-293.

Armstrong, C.M., and Chow, R.H., 1987, Supercharging: a method for improving patch-clamp performance, Biophys. J., 52:133-136.

Bezanilla, F., and Armstrong, C.M., 1972, Negative conductance caused by entry of sodium and caesium into the potassium channels of squid axons, J. Gen. Physiol., 60:588-608.

Bezanilla, F., and Armstrong, C.M., 1977, Inactivation of the sodium channels. I. Sodium current experiments, J. Gen. Physiol., 70:549-566.

Carbone, E., and Lux, H.D., 1984, A low voltage-activated calcium conductance in embryonic chick sensory neurons, Biophys. J., 46:413-418.

Castellano, A., López-Barneo, J., and Armstrong, C.M., 1989, Potassium currents in dissociated cells of the rat pineal gland, Pflügers Arch., 413:644-650.

Castellano, A., and López-Barneo, J., 1990, Sodium and calcium currents in dispersed mammalian septal neurons, J. Gen. Physiol., in press.

Cole, K.S., 1968, "Membranes, Ions, and Impulses," University of California Press, Berkeley.

Cota, G., 1986, Calcium channel currents in Pars Intermedia cells of the rat pituitary gland, J. Gen. Physiol., 88:83-105.
Cota, G., and Armstrong, C.M., 1988, Potassium channel "inactivation" induced by soft-glass patch pipettes, Biophys. J., 53:107-110.
Eyzaguirre, C., and Zapata, P., 1968, A discussion of possible transmitter or generator substances in carotid body chemoreceptors, in: "Arterial Chemoreceptors," R.W. Torrance, ed., Blackwell Scientific Publications Inc., Oxford.
Fedulova, S.A., Kostyuk, P.G., and Veselovsky, N.S., 1985, Two types of calcium channels in the somatic membrane of new-born rat dorsal root ganglion neurons, J. Physiol., 359:431-446.
Fenwick. E.M., Marty, A., and Neher, E., 1982, Sodium and calcium channels in bovine chromaffin cells, J.Physiol., 331:599-635.
Fernández, J., Fox, A.P., and Krasne, S., 1984, Membrane patches and whole-cell membranes: a comparison of electrical properties in rat clonal pituitary (GH_3) cells, J. Physiol., 356:565-585.
Forscher, P., and Oxford, G.S., 1985, Modulation of calcium channels by norepinephrine in internally dialyzed avian sensory neurons, J. Gen. Physiol., 85:743-763.
Furman, R.E., and Tanaka, J. C., 1988, Patch electrode glass composition affects ion channel currents, Biophys. J., 53:287-292.
Hagiwara, S., Fukuda, J., and Eaton, D.C., 1974, Membrane currents carried by Ca, Sr, and Ba in barnacle muscle fiber during voltage clamp, J. Gen. Physiol., 63: 564-578.
Hamill, O.P., Marty, A., Neher, E., Sakmann, B., and Sigworth, F., 1981, Improved patch-clamp techniques for high-resolution current recording from cells and cell-free membrane patches. Pflügers Arch., 391: 85-100.
Hille, B., 1984, "Ionic Channels of Excitable Membranes," Sinauer Associates Inc., Sunderland.
Hodgkin, A.L., and Huxley, A.F., 1952, A quantitative description of membrane current and its application to conduction and excitation in nerve, J. Physiol., 117:500-544.
Kohlhardt, M., Bauer, M., Krause, H., and Fleckenstein, A., 1972, Differentiation of the transmembrane Na and Ca channels in mammalian cardiac fibres by the use of specific inhibitors, Pflügers Arch., 335:309-322.
López-Barneo, J., López-López, J.R., J. Ureña, and C. González, 1988, Chemotransduction in the carotid body: K current modulated by pO_2 in type I chemoreceptor cells, Science, 241:580-582.
López-López, J.R., González, C., Ureña, J. and López-Barneo, J., 1989, Low pO_2 selectively inhibits K channel activity in chemoreceptor cells of the mammalian carotid body, J. Gen. Physiol., 93:1001-1015.
Matteson, D.R., and Armstrong, C.M., 1984, Na and Ca channels in a transformed line of anterior pituitary cells, J. Gen. Physiol., 83:371-394.

Matteson, D.R., and Armstrong, C.M., 1986, Properties of two types of calcium channels in clonal pituitary cells, J. Gen. Physiol., 87:161-182.

Moolenaar, W.H., and Spector, I., 1978, Ionic currents in cultured mouse neuroblastoma cells under voltage clamp, J. Physiol., 278:265-286.

Narahashi, T., Moore, J.W., ans Scott, W.R., 1964, Tetrodotoxin blockade of sodium conductance on lobster giant axons, J. Gen. Physiol., 47:965-974.

Nowycky, M.C., Fox, A.P. and Tsien, R.W., 1985, Three types of neuronal calcium channels with different calcium agonist sensitivity, Nature, 316:440-443.

Peatman, J.B., 1977, "Microcomputer-based design," McGraw Hill, New York.

Rodríguez-Benot, A., Bolaños, P., and López-Barneo, J., 1989, Corriente de calcio en células glómicas y su dependencia de metabolitos intracelulares, Actas I Congreso Iberoamericano de Biofísica, p. 156.

Rudy, B., 1988, Diversity and ubiquity of K channels, Neuroscience, 25:729-749.

Sah, P., Gibb, A.J., and P. W. Gage, 1988, The sodium current underlying action potentials in guinea pig hippocampal neurons, J. Gen. Physiol., 91:373-393.

Sakmann, B., and Neher, E., 1983, "Single-Channel Recording," Plenum Press, New York.

Sargent, M., and Shoemaker, R.L., 1984, "The IBM personal computer from the inside-out," Addison Wesley, New York.

Sigworth, F.J., 1983, Electronic design of the patch-clamp, in: "Single-Channel Recording," B. Sakmann and E. Neher, eds., Plenum Press, New York.

Swandulla, D. and Armstrong, C.M., 1988, Fast-deactivating calcium channels in chick sensory neurons, J. Gen. Physiol., 92:197-218.

Tabares, L., Ureña, J., and López-Barneo, J., 1989, Properties of calcium and potassium currents of clonal adrenocortical cells, J. Gen. Physiol., 93: 495-519.

Ureña, J., Mateos, J.C., and López-Barneo, J., 1989a, Low-cost system for automated acquisition, display and analysis of transmembrane ionic currents, Med.& Biol. Eng. & Comput., 27: 94-98.

Ureña, J., López-López, J.R., González, C., and López-Barneo, J., 1989b, Ionic currents in dispersed chemoreceptor cells of the mammalian carotid body, J. Gen. Physiol., 93:979-999.

13

ELECTROPHYSIOLOGICAL METHODS IN THE STUDY OF WATER TRANSPORT ACROSS CELL MEMBRANES

Luis Reuss and Calvin U. Cotton

Department of Physiology and Biophysics
University of Texas Medical Branch
Galveston, Texas

INTRODUCTION

The study of water transport across the plasma membrane of animal cells has clearly lagged behind work on transport of ions and nutrients, although cell volume and hence water transport are essential for normal cell function. New physiological, biophysical and molecular approaches have been applied to the problem of water transport across cell membranes by several groups of investigators in recent years, resulting in significant progress in understanding the pathways and mechanisms of water transport and the nature and mechanisms of cell volume regulation. The purpose of this article is to review the use of electrophysiological approaches in the assessment of water transport across cell membranes.

The major problems facing a cell physiologist interested in cell membrane water transport can be summarized as follows:

1) How do cells maintain and regulate their volume? Two sub-topics can be identified: First, all cells undergo colloid-osmotic swelling because of the higher concentration of impermeant ions in the intracellular compartment; ion pumps and ion leak pathways must extrude ions for cell volume to be maintained. Second, many cells exhibit volume regulation upon primary acute changes in cell volume (ΔV_{cell}) induced osmotically; major issues are the mechanisms of regulatory volume increase (RVI) and regulatory volume decrease (RVD).

2) Do changes in V_{cell} contribute to pathophysiological conditions and how? A number of pathological conditions can potentially involve cell volume changes as a major contributing factor to abnormal function, including sickle cell anemia, hypoxia, trauma, and toxic damage.

3) What are the mechanisms of water transport in specialized cells such as those of transporting epithelia and osmoreceptors? How is water transport regulated in these systems?

These questions have been addressed experimentally at different levels. In some instances, good phenomenological descriptions have been obtained, e.g., on the transport processes involved in acute cell volume regulation upon osmotic swelling or shrinkage. However, the intracellular signals involved in adaptive processes and in regulation remain largely unknown.

Cell Membrane Transport, Edited by D. L. Yudilevich *et al.*
Plenum Press, New York, 1991

Complete answers to these questions will require detailed understanding of both pathways and mechanisms of water transport.

BASIC MECHANISMS OF WATER TRANSPORT ACROSS CELL MEMBRANES

The physiologist is interested in two major mechanisms by which water molecules traverse lipid membranes: diffusion and osmosis. We will review the physical bases of these processes, in elementary terms, restricting the discussion to a lipid membrane separating two well-mixed solutions.

Diffusion is consequence of random motion of molecules in space. A net flux of a diffusing uncharged molecule under isosmotic and isothermal conditions will occur if there are differences in the chemical potential of the solute between the two fluid compartments bathing the lipid membrane (the chemical potential is $\mu_s = RT \ln C_s$, where R and T are the gas constant and the absolute temperature, respectively, and C_s is the solute concentration). Consider the simple case of a membrane separating a solution (solute concentration = C_s) from pure water. If the membrane is permeable to the solute, a net flux will occur from solution side to water side, because the probability of solute movement in that direction is finite, whereas the probability of solute movement in opposite direction is zero. The flux is directly proportional to the concentration difference and to the permeability of the membrane, i.e., the property of the membrane of allowing solute permeation. The case of water diffusion is better understood if tracer water is present on only one side of the membrane. The appropriate relationship is

$$J_w^* = P_{dw} \cdot \Delta C_w^* \qquad (1)$$

where J_w^* is the flux of tracer water, P_{dw} is the diffusive water permeability coefficient and ΔC_w^* is the difference in concentration of tracer water between the two fluid compartments.

If tracer water is present on both sides, the unidirectional fluxes are $J_w^*(1 \rightarrow 2) = P_{dw} \cdot C_w^*(1)$ and $J_w^*(2 \rightarrow 1) = P_{dw} \cdot C_w^*(2)$. $J_w^*(\text{net})$ is the difference between the two unidirectional fluxes.

In Eq. (1) P_{dw} is the diffusive water permeability coefficient, which has two meanings: an operational one ($P_{dw} = J_w^*/\Delta C_w^*$), i.e., P_{dw} is the coefficient relating flux to concentration difference, and a physicochemical one ($P_{dw} = D_w \beta_w / \delta$), which relates the permeability coefficient to the properties of the membrane and the diffusing substance. The three parameters on the right hand side of the equation describing P_{dw} are the diffusion coefficient of water in the membrane (D_w, $cm^2 \cdot s^{-1}$), the lipid/water partition coefficient (β_w, adimensional, which denotes the ease with which water dissolves in the membrane), and the membrane thickness (δ, cm). The units of P are $cm \cdot s^{-1}$.

Most membranes of animal cells have very high diffusive water permeability, several orders of magnitude greater than the permeability to even the most permeant ions. The main reason is the difference in partition coefficients. Water dissolves significantly in lipid, in contrast with ions, which are highly hydrophilic and permeate mostly via water filled pathways in the membrane (pores or channels) or other specific transport sites (carriers or pumps). Because of the high P_{dw} and since under most conditions transmembrane differences in water concentration are small, net water fluxes are also small, but unidirectional diffusional water fluxes are large. However, in the presence of a transmembrane water chemical potential difference, and if the membrane has a $P_{dw} > 0$, a diffusive water flux will occur.

<u>Osmosis</u>, in contrast with diffusion, is not the result of independent motion of particles, but of flow of molecules which interact with each other. In a solution, such interaction is expressed as both water-water and water-solute friction. If osmotic water flow exists across a membrane, the membrane must contain a domain that allows for such an interaction between solvent molecules or solvent and solute molecules. This is in contrast with diffusion, in which the permeation of particles occurs by independent motion. Hence, osmotic water flow across membranes occurs via pores in which friction takes place among water molecules (viscous water flow) and among water and solute molecules (solvent drag).

The existence of frictional interaction influences the relationship between the water flux and the geometry of the pathway. Consider the case of a porous membrane (cylindrical pores of density = n per cm^2, length = l, radius = r) exposed to a unit concentration difference of impermeant solute (=ΔC) between the two sides. In the case of diffusion, for 1 cm^2 of membrane

$$J_w = P_{dw} \cdot \Delta C_w \tag{2}$$

and since
$$P_{dw} = \frac{n \cdot D_w \cdot \Pi r^2}{l} \tag{3}$$

$$J_w = \frac{n \cdot \Pi r^2}{l} D_w \Delta C_w \tag{4}$$

whereas for osmosis in the same pore, the equation is

$$J_w = \frac{8 \Pi r^4}{l} \Delta C \frac{RT}{V_w} \tag{5}$$

and the permeability term is given by

$$P_f = P_{os} = L_p \frac{RT}{V_w} \tag{6}$$

where the Greek character Π denotes the number 3.1415... (elsewhere we use π for osmotic pressure), L_p is the hydraulic permeability coefficient, and V_w is the partial molar volume of water (for further details, see Finkelstein, 1987, or Reuss and Cotton, 1987).

Note that the diffusive water flow is proportional to r^2 whereas the osmotic water flow is proportional to r^4. Hence, the coefficient relating driving force (ΔC) and J_w cannot be the same for osmotic and diffusive flows. This difference has been used as a test for the presence of pores in artificial and biological membranes.

<u>Reflection Coefficients and Effective Osmotic Pressure</u>

At equal concentrations, diverse solutes can have different effects on the osmotic flow across a membrane. Impermeant solutes exert a maximum effect ($J_v = L_p \Delta\pi$, where $\Delta\pi$ is the difference in osmotic pressure, approximately given by $\Delta\pi = RT \Delta C_s$). Permeant solutes have varying effects, depending on their permeability coefficients, relative to the water permeability coefficient. Staverman's reflection coefficient of the solute (σ_s) is the parameter used to assess this effect. The effective osmotic pressure is not $\Delta\pi$, but $\sigma_s\Delta\pi$. When the solute is impermeant, $\sigma_s = 1$, when the solute is permeant, $\sigma_s < 1$, and can be zero when the solute permeability is very high, so that in the presence of a ΔC the volume flows of water in one direction and of solute in the opposite direction cancel each other.

Cell swelling can be elicited by reducing the absolute external osmolality (e.g., by reducing the $[NaCl]_o$), but also by decreasing the reflection coefficient of the membrane for the solute, at constant external total osmolality. Cells can swell if external NaCl is substituted with urea, keeping the external osmolality constant (because $\sigma_{NaCl} > \sigma_{urea}$), or if the membrane permeability to external solutes is increased, among other ways by using ionophores. For instance, cells will swell if σ_{NaCl} is reduced while the σ_s of intracellular K salts remains unchanged.

THE STATE OF WATER INSIDE THE CELL

In this discussion we consider the cell interior as a rather simple solution (whose concentration of colloidal particles is relatively high), in an isotropic fluid compartment devoid of matrix. Clearly this view is an oversimplification. The cell interior, i.e., nucleus, cytoplasm and organelles, is most likely a gel with a network-like matrix containing fixed negative charges. The physics of this system is complicated and beyond the scope of this article.

Briefly, if the network is cross-linked, its equilibrium volume depends on the external osmolality, the net charge of the chains, the concentration and distribution of diffusible ions (counter-ion concentration within the gel is higher than co-ion concentration), and the hydrostatic pressure within the gel. In turn, some of these parameters depend on other factors. For instance changes in pH and divalent ion concentration can alter the net charge of the chains and hence the physical properties of the gel. An interesting corollary of this view is that the osmotic behavior of cells is not completely dictated by the properties of the cell membrane, i.e., the intracellular compartment may have a paramount role, which so far has received little attention (see Lechene, 1985).

VOLUME MARKERS AND THE MEASUREMENT OF THEIR CONCENTRATION WITH ION-SENSITIVE MICROELECTRODES

Volume markers are substances that distribute homogeneously in a fluid compartment, are chemically stable in the compartment (i.e., in the case of cells they are not synthesized, bound or metabolized), and are not transported (by active or passive mechanisms) by the membrane limiting the compartment. Volume markers have been widely used to measure fluid compartments in whole animals (Edelman and Liebman, 1959). Tracer water can be used to determine total body water content. Radioactive sulfate and cell membrane-impermeant, non-metabolized sugars (sucrose, mannitol) can be employed to determine the extracellular fluid compartment, since they do not penetrate cells. Serum albumin tagged with radioactive iodine can be used to determine total plasma volume.

The general principle in these studies is that under appropriate conditions the total amount (Q_m) of marker in the compartment is constant and known. Therefore, if the concentration (C_m) is measured, the volume of the compartment (V_c) can be determined from

$$V_c = Q_m/C_m \tag{7}$$

Ion-sensitive microelectrodes based on organic ligands (ion exchangers or neutral carriers) have been used to measure intracellular chemical activities (and calculate intracellular concentrations) of Na^+, K^+, H^+, Cl^-, divalent cations, and other ions (for review, see Ammann, 1986). In addition, in recent years fluorescent ligands with high ionic selectivity have been developed and used widely to measure a variety of intracellular ionic activities (for reviews, see Tsien, 1983, 1989).

Microelectrode measurements of intracellular ion activities require cell impalement with a double-barrel microelectrode or with two single-barrel microelectrodes, one ion-sensitive and the other one to measure the membrane voltage. Frequently, the ion-sensitive ligand responds to the ion of interest and also to interfering ions. If the selectivity coefficient of the electrode is independent of concentrations, the chemical activity of the ion of interest can be calculated using the Nicolsky-Eisenman equation:

$$E = E_0 + S \log(a_i + k_{ij}\, a_j) \tag{8}$$

where E is the measured voltage difference between the ion-selective microelectrode and the reference microelectrode (used to measure cell membrane voltage) E_0 is a characteristic constant of the microelectrode, S is another constant (the microelectrode "slope"), a_i is the activity of the ion of interest, k_{ij} is the selectivity coefficient (j/i), and a_j is the intracellular activity of the interfering ion.

Provided that the cell under study can be impaled with microelectrodes, the use of an intracellular ion as a volume marker provides a method to assess changes in cell water volume with excellent sensitivity and time resolution. In principle, this can be done with native ions or with extrinsic ions.

Native ions have been used to assess cell volume changes in epithelial cells (Zeuthen, 1982, Tripathi and Boulpaep, 1988). The advantage of this approach is the simplicity of the experimental procedure, which does not require to load the cell with the probe. The major disadvantage is that most native ions that can be measured with microelectrodes are transported by the cell membrane and therefore the necessary assumption of constancy of the amount of marker in the intracellular compartment does not hold. Nevertheless, if the ion fluxes during the measurement are small relative to the water fluxes, the error is small and the method can yield a reasonable estimate of the change in cell volume. In many cell types experimentally-induced changes in cell volume are followed by cell volume regulation, which depends on activation or inactivation of ion transport processes. This results in changes in the intracellular contents of ions such as K^+, Na^+ and Cl^-, thus invalidating their use as markers for cell water volume.

Extrinsic ions have the advantage that they can be chosen to be impermeant, ideal volume markers, but this technique requires loading of the cells, demonstration that upon completion of the loading procedure the cell membrane integrity has been restored, and demonstration that the probe ion is inert, i.e., does not interfere with the process under study. In the next section, we will describe the use of tetramethylammonium as a marker for cell water volume in epithelial cells.

USE OF TETRAMETHYLAMMONIUM TO ASSESS CHANGES IN CELL VOLUME

In studies of epithelial water transport we have used tetramethylammonium (TMA^+) as a marker for cell water volume; K^+-selective microelectrodes are used to monitor intracellular TMA^+ activity during experimental perturbations. TMA^+ was chosen because the native TMA^+ permeability of Necturus gallbladder is low and because K^+-selective electrodes have a high selectivity coefficient for quarternary ammonium compounds over K^+ (i.e. 10^2-10^3) (Neher and Lux, 1973, Reuss, 1985).

Methods

The gallbladder of Necturus maculosus is removed, opened, mounted horizontally, and bathed on both sides with a NaCl Ringer solution (usually, in mM: NaCl 90, $NaHCO_3$ 10, KCl 2.5, $CaCl_2$ 1.8, $MgCl_2$ 1, NaH_2PO_4 0.5, gassed with 1% CO_2/99% air, pH ca. 7.60). The transepithelial voltage (V_{ms}) is measured with macroelectrodes in the bathing solutions, and the cell membrane voltages (apical = V_{mc}, basolateral = V_{cs}) are measured referring the intracellular voltage to the respective adjacent solution (Reuss and Finn, 1975a,b). In addition, the differential ion-selective electrode voltage (V_{TMA}-V_{cs}) is amplified, displayed on an oscilloscope, digitized, and stored in a microcomputer for subsequent analysis.

Double-barrel microelectrodes (one barrel responsive to TMA^+, the other one to measure voltage) are prepared from fused borosilicate glass (each barrel 1.0 mm O.D.; 0.5 mm I.D.; Hilgenberg, Malsfeld, Federal Republic of Germany). The fused doublet is twisted 360° upon softening by heat, allowed to cool, and then pulled (PD-5 microelectrode puller, Narishige, Tokyo, Japan).

The voltage-measuring barrel is partially filled with deionized water and the other barrel is partially filled with hexamethyldisilazane (Sigma Chemical Co., St. Louis, MO). The pipettes are placed tip-up on a hot plate under a stream of hot air until first the silane and then the water evaporate. The pipettes are then heated for an additional 1-2 hours. The voltage-sensing barrel is filled with a 1 M Na formate/0.01 M KCl solution. A drop (~1 μl) of K^+-selective cocktail (potassium tetrakis [p-chlorophenylborate] 5 mg in 0.1 ml of 3-nitro-O-xylene) is introduced into the hydrophobic barrel with a thin needle allowing the tip to fill by capillarity. This barrel is then back-filled with NaCl Ringer solution and chloridized silver wires soldered to gold pins are sealed into each barrel with dental wax.

The TMA^+-selective microelectrodes are calibrated with solutions simulating intracellular composition, i.e., nominally Ca^{2+}-free, containing 120 mM KCl plus 0.1-20 mM TMACl. Slopes range from 54 to 62 mV/log [TMA^+] (Fig. 1) and the selectivity ratio for (TMA^+/K^+) is 10^2 to 10^3. In most instances, double-barrel microelectrodes do not sense low [TMA^+] (< 20 mM) unless the tips are broken by impaling the tissue several times. Useful electrodes had resistances, measured in NaCl Ringer solution, of 80-150 MΩ (reference barrel) and 5-10 GΩ (TMA^+-selective barrel). Measurements can also be made with two single-barrel microelectrodes (reference and ion-selective). However, due to the small signal (e.g. 20% change in volume results in ~5 mV electrode signal), the difference voltage must be stable and the noise level low.

The normally impermeant cation TMA^+ is loaded into the cytosol by exposing the apical surface of the epithelium to the pore-forming polyene antibiotic nystatin (40-60 U/ml; Sigma, St. Louis, MO) in the presence of a mucosal solution in which NaCl has been replaced with TMA_2SO_4, adjusting the osmolality with sucrose. The loading procedure can be monitored by recording the cell membrane depolarization and the fall in apparent ratio of cell membrane resistances (apical:basolateral

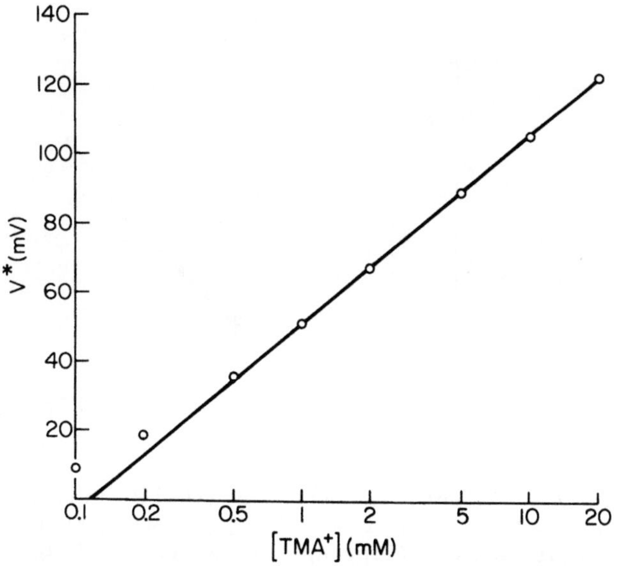

Figure 1. Calibration of a TMA^+-sensitive microelectrode. V* is the voltage recorded by the microelectrode in a rapid-flow chamber with respect to a calomel half-cell connected to the solution by a saturated-KCl, flowing junction. The rate of exchange of the solution was high, to avoid changes in $[K^+]$. The calibration solutions consisted of 120 mM KCl plus the TMACl concentration given in the abscissa. In the range 1-20 mM, the slope was 54 mV/log$[TMA^+]$. Reprinted from Reuss, 1985.

= R_a/R_b, see Reuss and Finn, 1975a,b, Cotton et al., 1989) or by monitoring directly intracellular $[TMA^+]$ with the ion-selective microelectrode. After 5-10 min, nystatin is removed and mucosal superfusion with TMA_2SO_4 Ringer solution is continued until membrane voltages and resistance ratio (R_a/R_b) recover (usually 60-90 min) (Fig. 2). At the end of this procedure, the intracellular $[TMA^+]$ is usually 5-15 mM, a range of concentrations that can be accurately sensed with insignificant contamination of the signal by other intracellular cations.

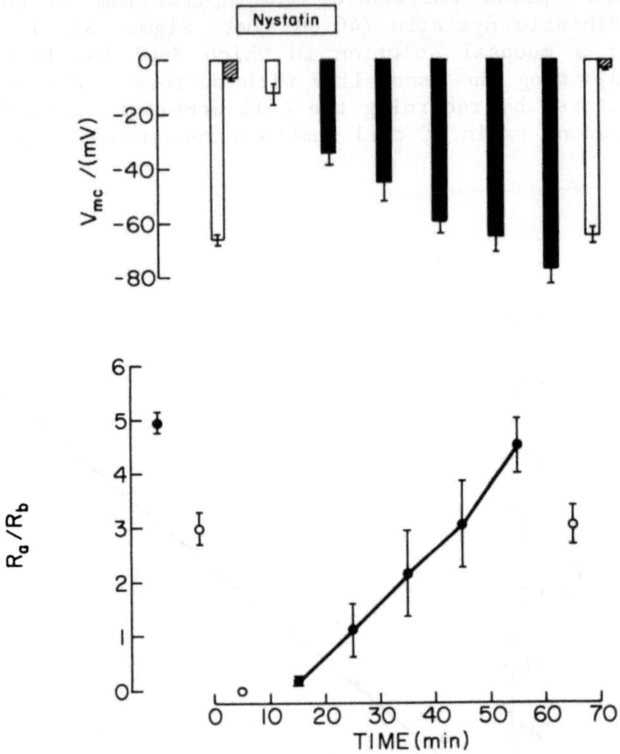

Figure 2. Effects of nystatin (60-180 units/ml) on electrical properties of *Necturus* gallbladder epithelial cells. (Top) Apical membrane voltage (V_{mc}) (cell-mucosal bathing medium) with NaCl-Ringer solution on both sides (open bars), during an increase in mucosal solution [K^+] to 82.5 mM (hatched bars), and during mucosal exposure to TMA sulfate-Ringer's solution (solid bars). (Bottom) Apparent ratio of cell membrane resistances (R_a/R_b), calculated from the voltage deflections elicited across each cell membrane by transepithelial current pulses of ca. 50 $\mu A \cdot cm^{-2}$, after correction for series resistances. Open circles, values during bilateral exposure to NaCl-Ringer's solution. Filled circles, values during mucosal exposure of TMA sulfate. Nystatin caused a large cell membrane depolarization, a mucosa-negative change in transepithelial voltage, and a fall in R_a/R_b to a value not different from zero. Replacing mucosal NaCl with TMA sulfate, resulted in slow recovery of V_{mc} and R_a/R_b toward control values. Forty minutes after removal of the antibiotic, the tissues were exposed again to NaCl-Ringer solution on both sides. V_{mc}, R_a/R_b, and the change in apical membrane voltage produced a high mucosal [K^+] (difference between open and hatched bar) were not different from the analogous values measured before nystatin treatment (pair analysis). Means ± SEM are shown; n = 6 experiments. Reprinted from Reuss, 1985.

Calculations of Intracellular [TMA$^+$] and Changes in Cell Water Volume

Intracellular [TMA$^+$] is calculated from the differential ion-selective voltage ($V^*_{TMA} = V_{TMA} - V_{cs}$) for each data point in the record according to Equation 9:

$$C(t) = C(0)\ 10^{(\Delta V^*_{TMA}/S)} \tag{9}$$

where $C(t)$ = concentration at TMA$^+$ at time = t(mM), $C(0)$ = concentration of TMA$^+$ at time = 0 (beginning trace), $\Delta V^*_{TMA} = V^*_{TMA}(t) - V^*_{TMA}(0)$, and S = slope of the electrode (mV/log[TMA$^+$]).

Changes in cell water volume are calculated from Equation 10:

$$\Delta V(t) = \left[V(t) - V(0)\right] = V(0)\left[\frac{[TMA^+](0)}{[TMA^+](t)} - 1\right] \tag{10}$$

where $\Delta V(t)$ = cell water volume change (nl/cm^2), $V(t)$ = cell water volume at time = t (nl/cm^2), $V(0)$ = initial cell water volume, [TMA$^+$](0) = intracellular [TMA$^+$] at t = 0 (mM), [TMA$^+$](t) = intracellular [TMA$^+$] at time = t (mM). The absolute value for initial cell water (on average 2870 nl/cm^2) must be assessed from independent measurements, such as morphometry in combination with wet/dry weight determination.

Typical examples of records of intracellular [TMA$^+$] changes upon changes in external solution osmolality are shown in Fig. 3.

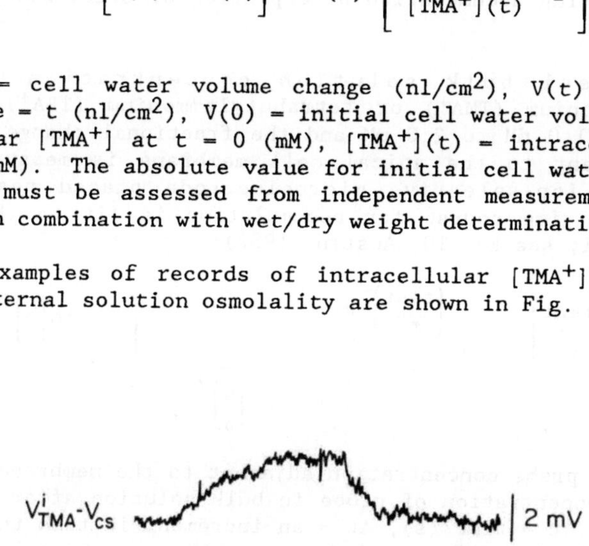

Figure 3. Changes in intracellular differential ion-sensitive voltage ($V^i_{TMA} - V_{cs}$) in response to anisosmotic mucosal bathing solutions. Upper trace: the mucosal bathing solution osmolality was transiently increased from 202 to 234 mosmol/kg. Lower trace: the same cell was exposed transiently to a hyposmotic mucosal solution (202 mosmol/kg to 162 mosmol/kg). The voltage traces indicate cell shrinkage and swelling, respectively.

Unstirred fluid layers (USL) can pose significant problems in studies of epithelial salt and water transport (Barry and Diamond, 1984). In this section we describe the use of extracellular ion-selective microelectrodes to assess the time course of the accumulation or depletion of a solute at the membrane surface, to estimate effective USL thickness, and to measure indirectly the osmolality of the fluid bathing the cell surface. The time course for change in solute concentration at the membrane surface depends on convective mixing in the bulk solution (time constant = τ) and diffusion across the USL. The latter is determined by the USL thickness (δ) and the diffusion coefficient of the solute (D_s). A complete description of the situation requires assessment of τ and δ and knowledge of D_s. Experimental determinations used in the past to assess δ (see House, 1974) frequently neglected the distinction between bulk mixing and diffusion. Since the time course for solute accumulation or depletion was entirely attributed to the USL, the results obtained with one solute result in significant error in predicting accumulation or depletion of solutes with different D_s.

The mucosal bulk solution concentration of either tetramethylammonium (TMA^+) or tetrabutylammonium (TBA^+) is rapidly increased from 1.0 mM to 2.0 mM and the fractional change in chemical activity adjacent to the apical cell membrane is measured with an extracellular ion-selective microelectrode placed near the cell membrane. The time course for accumulation is fit to the following equation (Eq. 11; see Eq. 10, Austin, 1932):

$$\frac{C_{(t+\Delta t)}}{C_{(\infty)}} = 1 - e^{-\left(\frac{t+\Delta t}{\tau}\right)} - \frac{e^{-\left(\frac{t+\Delta t}{\tau}\right)}}{\tau} \sum_{m=1}^{m=\infty} A \, \frac{1}{D\left(\frac{Q}{\delta}\right)^2 - \frac{1}{\tau}} \left[1 - e^{-\left(D\left(\frac{Q}{\delta}\right)^2 - \frac{1}{\tau}\right)(t+\Delta t)}\right] \quad (11)$$

where $C_{(t+\Delta t)}$ = probe concentration adjacent to the membrane at time = t (mM), $C_{(\infty)}$ = concentration of probe in bulk solution after steady-state is reached (mM), t = time (s), Δt = an increment in time that allows us to experimentally identify t = 0 (s), τ = time constant for the change in probe concentration at the interface between bulk solution and USL (s), m = 1,2,3..., $A = 4(-1)^{1+m}/\Pi(2m-1)$, $Q = (2m-1)\Pi/2$, D = diffusion coefficient for TMACl (1.39×10^{-5} cm^2/s) or TBACl (0.76×10^{-5} cm^2/s), δ = unstirred layer thickness (cm).

Methods

The preparation is mounted as described above. The solution in the upper (mucosal) compartment has the shape of a segment of a sphere with the inflow pipette positioned tangentially to the solution's edge in a plane parallel to the apical surface. The suction pipette is placed at the apex of the dome of solution. With this arrangement and a flow rate of 20-30 ml/min, bulk solution mixing is rapid and homogeneous.

The extracellular ion-selective microelectrode is connected to a high input-impedance electrometer and is referenced to the Ag-AgCl pellet in the serosal compartment. The tip of the microelectrode is positioned within ~1-3 μm of the apical cell membrane. Transepithelial voltage (V_{ms}) and differential ion-selective electrode voltage (V_i-V_{ms}) are amplified, displayed on an oscilloscope, digitized and stored in a microcomputer for subsequent analysis.

Single-barrel ion-selective microelectrodes are constructed from borosilicate glass with internal fiber (1 mm OD, 0.5 mm ID, Glass Company of America, Bargaintown, NJ). The pipettes are pulled with a horizontal, two-stage puller (model PD-5, Narishige, Tokyo, Japan), and the tips are broken to a diameter of 2-4 μm by advancing the pipette perpendicularly onto a polished stainless steel surface, under microscopic observation. The pipettes are then placed tip-up on a perforated aluminum block, covered with a glass jar, and baked in an oven at 200°C for at least 2 hr. A small volume (~0.1 ml) of hexamethyldisilazane (Sigma, St. Louis, MO) is introduced into the jar; the silane vapors react with the glass, rendering the surface of the pipettes hydrophobic. The pipettes are baked for an additional 2 hr period before filling. At the end of this period, they are removed from the oven and allowed to cool, and a small aliquot (~1 μl) of ion-sensitive cocktail (potassium tetrakis (p-chlorophenyl borate), 5 mg in 0.1 ml of 3-nitro-O-xylene, plus 10% wt/vol polyvinyl chloride and 20% wt/vol tetrahydrofuran) is introduced into the back of the pipette with a thin needle. The tip fills by capillarity. The pipettes are placed in a jar with desiccant for 36-48 hours to allow the polyvinyl chloride (PVC) to polymerize. Incorporation of PVC into the microelectrode cocktail is necessary to polymerize the resin and prevent its loss from the large tip of the microelectrode. The electrode is back-filled with NaCl Ringer solution and a chloridized silver wire soldered to a gold pin is sealed in place with dental wax. Resistances range from 1 to 5 GΩ when the electrode is immersed in NaCl Ringer solution.

Electrodes are calibrated with solutions that contain NaCl Ringer solution plus 0.5 to 20 mM TMACl or TBACl. Slopes are 59 ± 3 mV/log [TMA^+] or [TBA^+] and the selectivity ratios are: $TMA^+/K^+ = 10^2\text{-}10^3$, $TMA^+/Na^+ = 10^3\text{-}10^4$, TBA^+/Na^+ or $TBA^+/K^+ > 10^6$.

For measurements of USL apparent thickness, the mucosal perfusate is rapidly changed with a pneumatic driven valve (Rheodyne Inc., Cotati, CA) from NaCl Ringer containing 1 mM TMACl or TBACl to NaCl Ringer containing 2 mM TMACl or TBACl and the differential voltage (V^*_{ion}) output is recorded. The experiment is repeated with the same flow rates and pipette geometry with solutions containing the other probe (TMACl or TBACl).

The voltage traces (V^*_{TMA} and V^*_{TBA}) are converted to TMA^+ or TBA^+ concentrations according to:

$$C(t) = C(0)\ 10^{(\Delta V^*_{ion}/S)} \tag{12}$$

where C(t) = concentration of TMA^+ or TBA^+ at time = t (mM), C(0) = concentration of TMA^+ or TBA^+ at time = 0 (mM), $\Delta V^*_{ion} = V^*_{ion}(t) - V^*_{ion}(0)$ (mV), and S = electrode slope (mV/log concentration). The fractional concentration of TMA^+ or TBA^+ is calculated for each time point according to:

$$FC(t) = \frac{C(t) - C(0)}{C(\infty) - C(0)} \tag{13}$$

where C(t) = concentration of TMA^+ or TBA^+ at time = t (mM), C(0) = concentration of TMA^+ or TBA^+ at time = 0 (= 1 mM), and $C(\infty)$ = concentration of TMA^+ or TBA^+ at the steady state after solution change (= 2 mM). The data, expressed as fractional concentrations, are transferred to a computer for curve fitting to Equation 11. The best-fit values for the free parameters Δt, τ, and δ are determined.

If a single probe is used (either TMA^+ or TBA^+) multiple solutions can be obtained from the fitting routine. However, if the time courses for both probes are obtained under identical experimental conditions, a unique solution is obtained (Fig. 4). Once τ and δ have been determined, the parameters governing the time courses for convective mixing in the bulk solution and diffusion in the USL are known. The time course for accumulation of any molecular species whose diffusion coefficient is known can be calculated from these parameters.

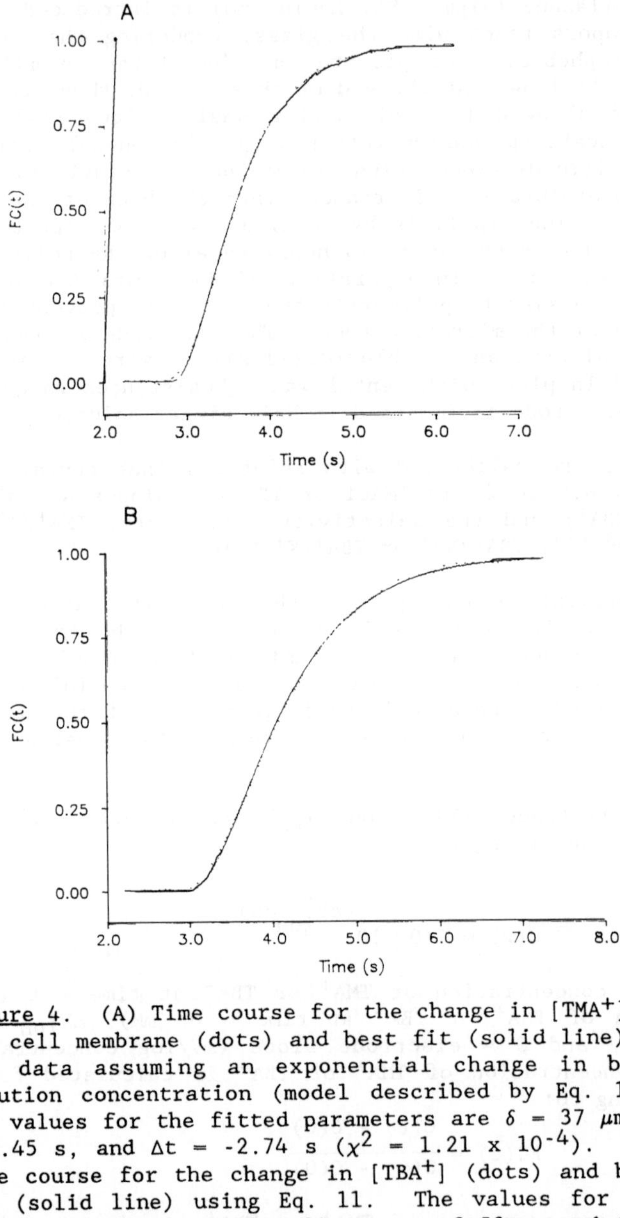

Figure 4. (A) Time course for the change in [TMA^+] at the cell membrane (dots) and best fit (solid line) to the data assuming an exponential change in bulk solution concentration (model described by Eq. 11). The values for the fitted parameters are δ = 37 μm, τ = 0.45 s, and Δt = -2.74 s (χ^2 = 1.21 x 10^{-4}). (B) Time course for the change in [TBA^+] (dots) and best fit (solid line) using Eq. 11. The values for the fitted parameters are δ = 34 μm, τ = 0.52 s, and Δt = 2.89 s (χ^2 = 1.43 x 10^{-4}). The values of δ and τ in A and B can be compared since the traces were obtained sequentially under identical superfusion conditions. From Cotton and Reuss (1989), by copyright permission of Rockefeller University Press.

DETERMINATION OF CELL MEMBRANE OSMOTIC WATER PERMEABILITY

In the preceding two sections we described intracellular and extracellular ion-selective microelectrode techniques to measure changes in cell water volume and unstirred layer thickness, respectively. In this section we illustrate the combined application of extracellular and intracellular techniques to measure cell membrane osmotic water permeability in Necturus gallbladder epithelium. To this end, the extracellular ion-selective microelectrode is used to indirectly monitor changes in osmolality at the apical cell membrane upon exposure to mucosal anisosmotic bathing solutions. Simultaneously, changes in cell water volume are assessed by measuring changes in intracellular TMA^+ activity with a double-barrel ion-selective microelectrode.

Since the diffusion coefficients for TBACl (0.76×10^{-5} cm^2/s) and sucrose (0.52×10^{-5} cm^2/s) are similar, TBA^+ is used as a tracer to monitor [sucrose]. For instance, if the bulk solution [sucrose] and [TBA^+] are simultaneously increased, the time courses for arrival of

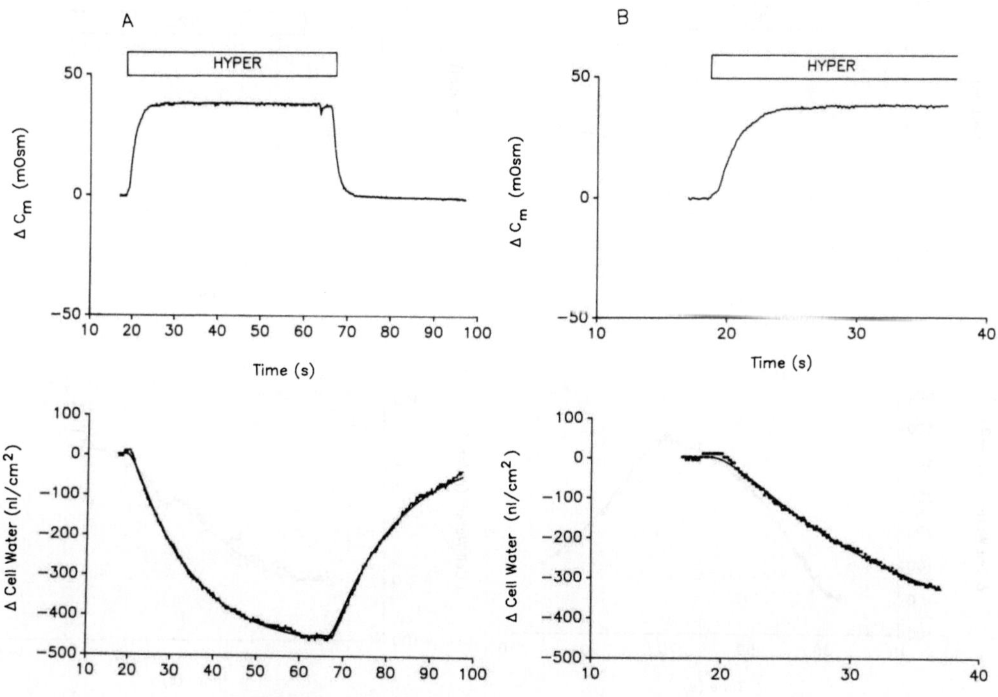

Figure 5. Change in mucosal solution osmolality, cell water volume, and the fitted change in cell water volume in response to exposure to a hyperosmotic mucosal solution. (A) The mucosal solution osmolality was increased from 206 to 244 mosmol/kg water. The initial cell water volume was assumed to be 2,870 nl/cm^2. The data for the complete experiment (~70 s) are fit to the model described in Methods. The estimates of apical and basal L_p's are 0.87×10^{-3} and 0.09×10^{-3} $cm \cdot s^{-1} \cdot (osmol/kg)^{-1}$, respectively. (B) Same experiment as above except that the fit spans only 20 s. The estimates of apical and basal L_p's are 0.90×10^{-3} and 0.10×10^{-3} $cm \cdot s^{-1} \cdot (osmol/kg)^{-1}$ respectively. From Cotton et. al. (1989), by copyright permission of the Rockefeller University Press.

each solute at the apical cell membrane are similar. Therefore, by measuring TBA^+ activity adjacent to the cell membrane we can calculate the change in [sucrose] and thus the osmolality as a function of time after the solution change. In principle, a variety of osmotic solute/probe pairs could be used if their diffusion coefficients are similar and the extracellular ion-selective microelectrode has sufficient selectivity and adequate response time for the probe molecule.

The two measurements yield continuous records of osmolality at the apical cell membrane and changes in cell water volume (Figs. 5 and 6). Using a simple compartmental model of the epithelium, the change in driving force ($\Delta\pi$) can be related to the change in cell volume to calculate the hydraulic permeability coefficients (L_p) of the apical and basolateral cell membranes (Cotton et al., 1989). In brief, the L_p of the apical membrane determines the initial rate of cell swelling or shrinkage, whereas the L_p's of both membranes determine the remainder of the volume change and the apparent steady-state change before compensatory mechanisms. Our results in <u>Necturus</u> gallbladder epithelium

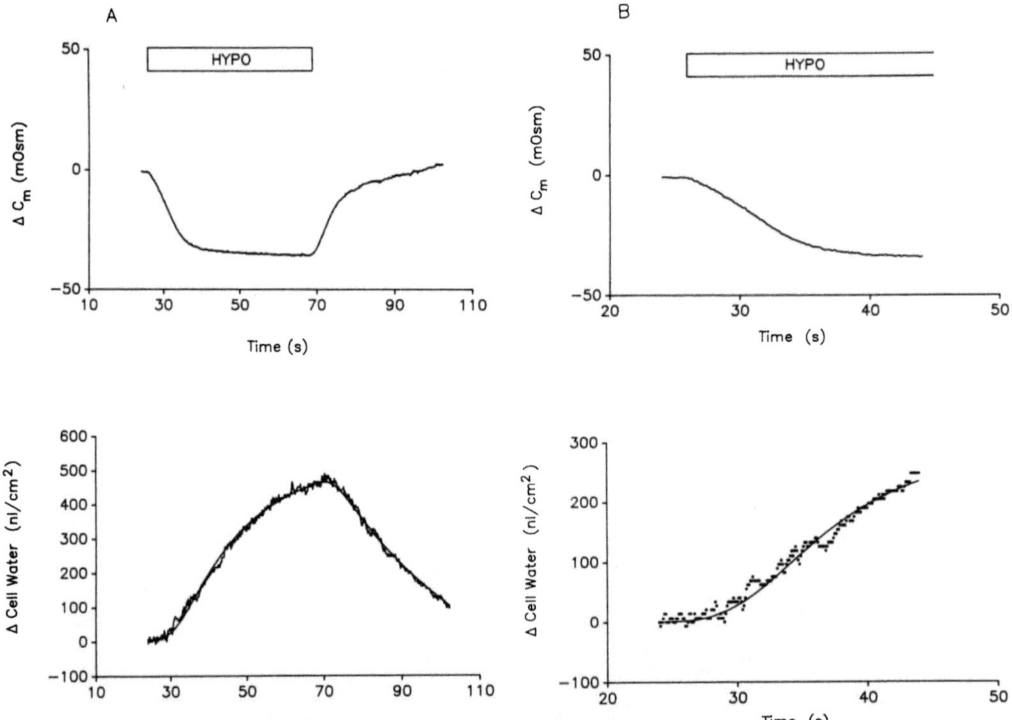

Figure 6. An example of the change in mucosal solution osmolality, cell water volume, and the fitted change in volume in response to exposure to a hyposmotic mucosal solution. (A) The mucosal solution osmolality was decreased from 206 to 163 mosmol/kg water. The initial cell water volume was assumed to be 2,870 nl/cm^2. The data for the complete experiment (~70 s) are fit to the model described in Methods. The estimates of apical and basal L_p's are 0.72×10^{-3} and 0.08×10^{-3} $cm \cdot s^{-1} \cdot (osmol/kg)^{-1}$, respectively. (B) Same experiment as above except that only 20 s of the data are fit to the model. The estimates of apical and basal L_p's are 0.69×10^{-3} and 0.17×10^{-3} $cm \cdot s^{-1} \cdot (osmol/kg)^{-1}$ respectively. From Cotton et al. (1989), by copyright permission of the Rockefeller University Press.

indicate that the cell membranes have L_p's of the order of 10^{-3} $cm \cdot s^{-1} \cdot (osmol/kg)^{-1}$. Calculations based on the spontaneous rate of fluid absorption indicate that the osmotic driving force required to account for water absorption is of the order of only 1-2 mosmol/kg across each membrane.

APPLICATION TO OTHER SYSTEMS

Since its introduction, the intracellular TMA^+ technique has been applied to Necturus gallbladder epithelial cells (Reuss, 1985, Cotton et al., 1989) and to frog spinal cord motoneurons (Serve et al., 1988, Ballanyi and Grafe, 1988). Recent experiments have suggested the applicability of this method to assess cell volume changes in squid giant axons, barnacle muscle fibers (Russell, personal communication), frog dorsal root ganglion cells and snail neurons (Alvarez-Leefmans and Reuss, unpublished observations). Application of this method to cells amenable to impalements with microelectrodes permits measurements of cell volume changes with excellent sensitivity and time resolution. In addition, combining cell volume measurements with intracellular-microelectrode determinations of intracellular ionic activities will permit assessment of net ion fluxes with a time resolution of a few seconds, in a non-destructive fashion.

In cells difficult to impale with microelectrodes because of size, input resistance or other factors, optical techniques based on the same principle used in our experiments can be in principle developed.

ACKNOWLEDGMENTS

This work was supported by National Institute of Diabetes and Digestive and Kidney Diseases Grant DK38588.

REFERENCES

Ammann, D., 1986, Ion-selective microelectrodes, Springer-Verlag, Berlin.

Austin, J.B., 1932, Temperature distribution in solid bodies during heating or cooling. A correction. Physics 3:179-184.

Ballanyi, K., and Grafe, P., 1988, Cell volume regulation in the nervous system, Renal Physiol. Biochem., 1:114-141.

Barry, P.H., and Diamond, J.M., 1984, Effects of unstirred layers on membrane phenomena, Physiol. Rev., 64:763-873.

Cotton, C.U., and Reuss, L., 1989, Measurement of the effective thickness of the mucosal unstirred layer in Necturus gallbladder epithelium, J. Gen. Physiol., 93:631-647.

Cotton, C.U., Weinstein, A.M., and Reuss, L., 1989, Osmotic water permeability of Necturus gallbladder, J. Gen. Physiol., 93:649-679.

Edelman, I.S., and J. Leibman, 1959, Anatomy of body water and electrolytes, Am. J. Med., 27:256-277.

Finkelstein, A., 1987, Water movement through lipid bilayers, pores and plasma membranes, John Wiley & Sons, New York.

House, C.R., 1974, Water transport in cells and tissues, Edward Arnold Ltd., London.

Lechene, C., 1985, Cellular volume and cytoplasmic gel, Biol. Cell, 55:177-180.

Neher, E., and Lux, H.D., 1973, Rapid changes of potassium concentration at the outer surface of exposed single neurons during membrane current flow, J. Gen. Physiol., 61:385-399.

Reuss, L., 1985, Changes in cell volume measured with an electrophysiologic technique, Proc. Natl. Acad. Sci. USA, 82:6014-6018.

Reuss, L., and Cotton, C.U., 1988, Isosmotic fluid transport across epithelia, Contemp. Nephrol., 4:1-37.

Reuss, L., and Finn, A.L., 1975a, Electrical properties of the cellular transepithelial pathway in Necturus gallbladder. I. Circuit analysis and steady-state effects of mucosal solution ionic substitutions, J. Membr. Biol., 25:115-139.

Reuss, L., and Finn, A.L., 1975b, Electrical properties of the cellular transepithelial pathway in Necturus gallbladder. II. Ionic permeability of the apical cell membrane, J. Membr. Biol., 25:141-161.

Serve, G., Endres, W., and Grafe, P., 1988, Continuous electrophysiological measurements of changes in cell volume of motoneurons in the isolated frog spinal cord, Pflügers Arch. Eur. J. Physiol., 411:410-415.

Tripathi, S., and Boulpaep, E.L., 1988, Cell membrane water permeabilities and streaming currents in Ambystoma proximal tuble, Am. J. Physiol., 24:F188-F203.

Tsien, R.Y., 1983, Intracellular measurements of ion activities, Ann. Rev. Biophys. Bioeng., 12:91-116.

Tsien, R.Y., 1989, Fluorescent probes of cell signaling, Ann. Rev. Neurosci., 12:227-253.

Zeuthen, T., 1982, Relations between intracellular ion activities and extracellular osmolarity in Necturus gallbladder epithelium, J. Membr. Biol., 66:109-121.

Part 4: SPECTROSCOPIC METHODS

Chapter 14

A practical introduction to the use of intracellular fluorescent indicators
Miguel Valdeolmillos and David Eisner

Chapter 15

Biomedical NMR spectroscopy and transport
Susan Wray

Chapter 16

NMR studies of solute transport
Glenn King and C. A. Richard Boyd

Chapter 17

Optical and NMR spectroscopy in the study of secretion of exocrine pancreas
Tomio Kanno

14

A PRACTICAL INTRODUCTION TO THE USE OF INTRACELLULAR FLUORESCENT INDICATORS

M. Valdeolmillos* and D.A. Eisner**

*Departamento de Fisiologia, Facultad Medicina
Universidad de Alicante. Spain
**Department of Physiology, University College London
Gower St. London WC1E 6BT. Great Britain.

1. INTRODUCTION

The concentrations of intracellular ions such as Ca^{2+}, Na^{+} and H^{+} are important in the regulation of cell function. Much effort has therefore been expended in measuring them. However, until the last few years, most direct methods for measurement of these concentrations were only suitable for large cells. The situation has changed completely with the introduction of fluorescent indicators. Here we describe the properties of some of these indicators.

This chapter is designed to be a very basic and simple introduction to the use of fluorescent indicators and is specifically aimed at the novice who is considering entering this field. A large fraction of the space is devoted to considering the various parts of the apparatus needed to make reliable fluorescence measurements.

We will concentrate on the use of fluorescent indicators, without mentioning other methods of measurement such as metallochromic and bioluminescent indicators in the case of Ca and ion- sensitive microelectrodes which are available for most ions. This does not imply that such probes are not important, simply that fluorescent probes are much more widely used, a trend that continues to grow. We will not discuss intracellular imaging techniques as they are the subject of another chapter in this book.

The first part of the chapter will describe the characteristics of some Ca and pH indicators, with special emphasis on practical points. The second part will deal with the apparatus necessary to make fluorescence measurements from single, isolated cells.

During the last years many interesting reviews on fluorescent indicators have appeared, among them the interested reader is invited to consult the following general reviews: Borle & Snowdowne, 1986; Cobbold & Rink, 1987; Tsien et al, 1985; Tsien, 1988; duBell et al, 1988). For an introduction to the general principles of fluorescence the book of Lakowicz (1983) is an excellent source.

Cell Membrane Transport, Edited by D. L. Yudilevich *et al.*
Plenum Press, New York, 1991

When a fluorophore (fluorescent substance) is excited by light, its energy level is raised. Fluorescence occurs when an excited electron returns to the ground energy level. Under normal conditions, the photons emitted during the fluorescence process are less energetic (and therefore have a longer wavelength) than the absorbed photons. This wavelength shift is know as the Stokes shift. In general, the fluorescence lifetime (the period of time that a fluorophore remains in the excited state) is of the order of nanoseconds. Another important concept in fluorescence is the quantum yield of the fluorophore which is the ratio of the number of photons emitted to the number absorbed.

For a fluorescent compound to be a useful indicator, the fluorescence intensity must change as a function of the concentration of the measured ion. A useful property is that, in addition to a change in the magnitude of the fluorescence, either the excitation or emission spectrum should change in shape thus allowing dual wavelength measurements.

2. CALCIUM-SENSITIVE FLUORESCENT INDICATORS

Quin-2

Until recently the most popular calcium indicator was Quin-2 (Tsien et al, 1982). This is a tetracarboxylic derivative of the Ca-chelating agent EGTA. Quin-2 binds calcium with a stoichiometry 1:1. When it is excited by light at 340 nm it emits light at 500 nm: the intensity of this fluorescence is a saturating function of calcium as shown in Fig. 1. The dissociation constant of Quin-2 for Ca is 115 nM. This makes it a suitable indicator of even small changes in the resting $[Ca^{2+}]_i$ (50-200 nM) but less suitable for larger Ca transients reached, for example, during muscle contraction.

The most serious problem with Quin-2 is that the amount of indicator that has to be present in the cell in order to obtain an adequate signal to noise

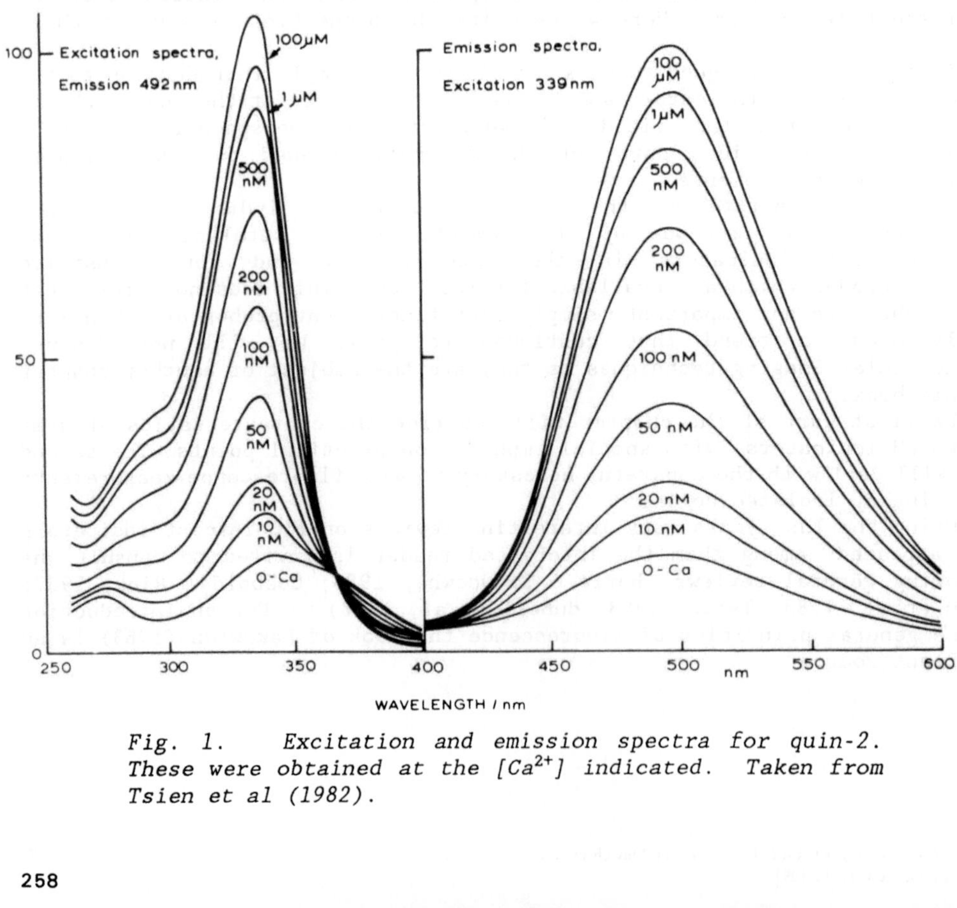

Fig. 1. Excitation and emission spectra for quin-2. These were obtained at the $[Ca^{2+}]$ indicated. Taken from Tsien et al (1982).

ratio is very high, of the order of 0.5-3 mM. This concentration of Quin-2 has a significant Ca-buffering effect which will slow down and attenuate rapid changes of $[Ca^{2+}]_i$. This additional buffering should, however, have little effect on the steady state level of $[Ca^{2+}]_i$. Although the selectivity of Quin-2 for Ca over Mg is very high (pK_{Ca}- pK_{Mg} = 8.7), the fact that the intracellular Mg^{2+} concentration is four orders of magnitude greater than that of Ca^{2+} means that Quin-2 fluorescence due to Mg is not negligible. The affinities of Quin-2 for other metals such as Fe, Mn and Zn are even greater than for Ca. This interference can be a limiting factor in certain cells. For instance, in Ehrlich and Yoshida carcinoma cells it is necessary to chelate heavy metals in order to obtain reliable Ca-dependent fluorescent signals (Arslan et al., 1984).

The great advantage of Quin-2 is the ease with which it can be incorporated into the cells. Tetracarboxylate indicators are membrane impermeant molecules. However the carboxylate groups can be masked with acetoxymethyl (AM) groups, resulting in a liposoluble molecule which is able to diffuse into the cell. Once inside the cell, the AM groups are hydrolyzed by esterases present in the cytoplasm. Quin-2 is then regenerated and trapped inside the cell where it seems to be uniformly distributed in the cytoplasm.

One of the side effects of the AM group cleavage is the release of formaldehyde, a highly toxic product for the cell. The extended use of Quin-2 has led to the discovery of many side effects on cell function. To name only a few, it is known that Quin-2 loading decreases catecholamine (Knight & Kesteven, 1983) and prolactin (Albert & Tashjian, 1984) secretion as well as decreasing or slowing down platelet aggregation (Johnson et al., 1985). In lymphocytes, platelets and erythrocytes it decreases ATP levels and induces NADH accumulation in the latter cells (Garcia-Sancho, 1984).

In order to calibrate the signal in terms of intracellular Ca, it is necessary to know the amount of indicator present. At the end of the experiment, Quin-2 is released from the cells using a detergent and then the fluorescence is measured at 0 Ca (in the presence of EGTA) and saturating Ca. The calculated calcium concentration will depend on the concentration of intracellular Mg, the presence of interfering metals and the photobleaching of the indicator during the experiment (Tsien et al., 1982). This form of calibration is only feasible in a cuvette, where the dye released by the detergent is not lost, but it would not be possible in a microscope where the cells are superfused continuously.

Fura-2 and Indo-1

Fura-2 and Indo-1 are also both derivatives of EGTA (Grynkiewicz et al, 1985). Fura-2 is 30 times brighter than Quin-2. This is due to a better extinction coefficient and quantum efficiency. An important consequence of this improvement is that, in order to obtain the same signal as with Quin-2, much less fluorescent probe has to be inside the cells, avoiding to a great extent Ca-buffering problems.

A second advantage with respect to Quin-2 is that when fura binds to calcium there is a large shift of the excitation spectrum (Fig. 2A). In the case of Indo-1 the shift occurs in the emission spectra (Fig 2B). The use of double excitation (Fura-2) or double emission (Indo-1) spectroscopy increases the precision of the measurement, reduces the effects of instrumentation drifts or movement artifacts and, more importantly, permits the quantification of the signal without knowing the concentration of intracellular indicator. The fluorescent signal is also independent of cell thickness.

The K_d of Fura-2 and Indo-1 for Ca are respectively 224 and 250 nM making these indicators more suitable than Quin-2 for the measurement of larger intracellular Ca responses. The discrimination for Ca over Mg is approximately 4 times better. The affinity of Fura-2 for Zn is still quite high, of the order of 2 nM, but this represents an improvement of 40 times compared with Quin-2.

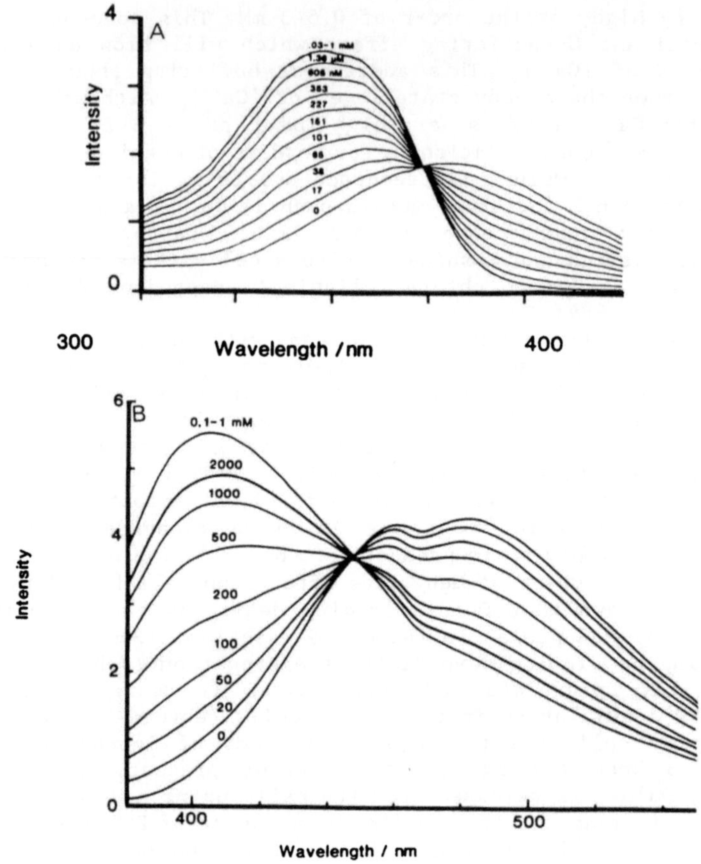

Fig.2. Spectral properties of fura-2 and indo-1. <u>*A*</u> *Excitation spectrum of fura-2. This was obtained at the [Ca^{2+}] indicated (nM).* <u>*B.*</u> *Emission spectrum of indo-1. From Grynkiewicz et al (1985).*

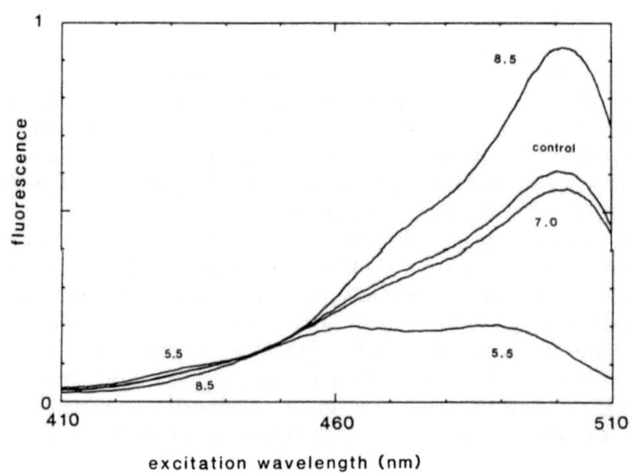

Fig. 3. <u>*In vivo*</u> *excitation spectrum for BCECF from a single isolated rat ventricular myocyte. The trace labelled "control" was obtained under control conditions. The others were obtained after adding nigericin at the external pH indicated in the presence of 140 mM* $[K^+]_o$*. Taken from Eisner et al., 1989b.*

Both Fura-2 and Indo-1 can be incorporated into the cells as AM derivatives. This loading can be improved by adding the detergent Pluronic F127 to the incubation medium (Owen, 1988). There are reports indicating compartmentalization of Fura-2 in organelles, for instance mitochondria, in a way that remains Ca-sensitive (Gunter et al., 1988; Davis et al., 1987). Other problems include the finding of partial hydrolysis products which are highly fluorescent but Ca insensitive (Powis, 1988). The amount of such products seems to be very dependent on the cell type under study (Oakes et al., 1988). A further problem is related to the photobleaching that occurs during the experiment. This should not be a problem by itself when the ratio method is being used. It seems, however, that as a consequence of the photobleaching, a fluorescent product is generated that shows different sensitivity to Ca compared to the free indicator (Becker & Fay, 1987). As far as the possible binding of these indicators to intracellular membranes or molecules is concerned the reports in the literature are very contradictory, ranging from results that indicate a high degree of intracellular binding accompanied by spectral changes (Popov et al., 1988; Konishi et al., 1988) to lack of evidence of such a binding (Grynkiewicz et al., 1985; Timmerman & Ashley, 1986). In cardiac myocytes loaded with Fura-AM, we have found that 70% of the fluorescence can be released with a time constant of 7.5s by permeabilization of the membrane with saponin. Eventually all the fluorescence is lost, indicating therefore a low degree of compartmentalization. Other reports have indicated that the degree of compartmentalization is critically dependent upon the incubation time in the presence of the AM form (Eng et al., 1989).

In several studies, intracellular Ca has been measured in cells which are simultaneously patch-clamped. In this protocol, Fura-2 free acid is one of the constituents of the patch pipette solution. This approach seems to largely avoid the problems of compartmentalization -as well as partial hydrolysis. One study (Almers & Neher, 1985) has compared AM and free acid loaded cells and found important differences.

Other fluorescent Ca indicators

A new set of Ca-indicators has recently become available (Tsien & Minta, 1987; Minta et al., 1989 and see also Molecular Probes catalogue). Their principal advantage is that they are excited at longer wavelengths. This allows glass rather than quartz optics to be used and, also, decreases contamination from NADH fluorescence. For instance, the so-called Fluo-3 is excited at 490 nm with peak emission at 525 nm. Another advantage is that they have larger K_d's for Ca: the dissociation constant for Ca at 22°C is 0.45 μM. Although the increase in fluorescence produced by Ca binding is of the order of 40 times, this does not shift the spectra, precluding the use of the ratio method.

3. ABSOLUTE CALIBRATION OF INTRACELLULAR INDICATORS

If fluorescence is measured at two wavelengths (1 and 2) then for any $[Ca^{2+}]_i$, the ratio of these signals can be calculated:

$$R = F_1/F_2$$

$[Ca^{2+}]_i$ can be calculated from the following equation (Grynkiewicz et al., 1985):

$$[Ca^{2+}]_i = K_d*\beta\{(R-R_{min})/(R_{max}-R)\}$$

where R_{min} is the value of R in the absence of Ca and R_{max} the value at saturating $[Ca^{2+}]_i$. K_d is the dissociation constant of the indicator for Ca. The multiplying factor (β) takes account of the fact that the apparent K_d of the ratio is different from that for either single wavelength. β is given by $F_2(0Ca)/F_2(saturating\ Ca)$. β is 1 if F_2 is measured at an isosbestic wavelength.

In order to use the above equation one must have values for R_{max}, R_{min} and K_d. There are, in principle, two ways to obtain these. One involves measuring fluorescence at two wavelengths in solutions containing different Ca concentrations at a constant fluorophore concentration. Such solutions should be designed to mimic the cytoplasm. The ratios of the fluorescence at the two wavelengths during the experiment are then compared with those obtained using the "in vitro" calibration curve. This calibration depends upon the fact that the fluorescent indicator behaves in the same way inside the cell as in the artificial calibrating solutions, an assumption with much contradictory evidence (Popov et al., 1988).

The second method involves calibrating the cell signal "in situ". The method consists of permeabilizing the cell to calcium using a Ca- ionophore and then adding saturating concentrations of calcium and EGTA in order to obtain the maximum (R_{max}) and minimum (R_{min}) of the fluorescence ratio. The most difficult problem is to ensure that the Ca permeability of the surface membrane is sufficient to overcome the regulatory actions of surface membrane and organelle Ca pumping. The available ionophores are A23187 (in particular the non-fluorescent brominated derivative: Deber et al., 1985) and ionomycin. In our hands ionomycin gives more reliable results. In order to facilitate the action of the ionophores, some researchers recommend depleting the cell of energy using metabolic poisons before the calibration is done (Li et al., 1987). Despite the use of many precautions one finds that the maximum fluorescence ratio obtained with an ionophore in the presence of saturating Ca can sometimes be smaller than the ratio obtained during the experiment, for instance, when the cell is depolarized in the presence of high extracellular potassium, concluding therefore that the ionophore was not able to equilibrate cytoplasm with extracellular Ca concentration.

Whichever method is used to calibrate, one must allow for the effects of autofluorescence if this makes a significant contribution to the total fluorescence readings. One way to estimate the magnitude of this autofluorescence is to quench the signal due to the indicator in the presence of Mn (Hesketh et al., 1983; Macintyre et al., 1988). Mn has another use for cuvette studies inasmuch as it quenches any indicator which will have leaked out of the cell. The resulting fluorescence should be mostly due to autofluorescence, providing that the background signal is comparatively small. It is not known to what extent partial hydrolysis products of the indicators are quenched by Mn. Among the natural fluorophores most likely to be the source of autofluorescence, the more important are: (i) NADH which is highly fluorescent, it is excited at 340nm and emits at 450nm (31) (ii) Proteins, which are excited near 280nm and fluoresce at 320-350nm. 90% of the protein fluorescence is due to the amino acid tryptophan. (iii) Other fluorescent molecules are riboflavin, FAD and FMN with excitation at 450nm and emission at 515nm. The problem of autofluorescence is exacerbated if it changes during an experiment. For example changes in the redox state of the cell will change the [NAD]/[NADH] ratio and therefore the autofluorescence (Chance et al., 1962).

A final complication is provided by the fact that various drugs and other compounds are fluorescent. Examples are: quinine and quinidine; tri-fluoroperazine and chlorpromazine. One must also be sure that compounds do not interfere with the fluorescent indicator. An example of this is the quench of indo-1 by caffeine (Donoso et al., 1989b).

Despite the uncertainties with regard to the calibration, the data obtained with fluorescent measurements at two wavelengths are perfectly reliable for qualitative or semi-quantitative purposes.

4. FLUORESCENT pH INDICATORS

Fluorescein derivatives

There are various indicators available for the measurement of intracellular pH. To date, most work has used carboxy-fluorescein (Thomas et al., 1979) and its derivatives. The fluorescence of this compound is affected by pH. One disadvantage is that the indicator leaks out of the cell. This problem was reduced with another derivative bis(carboxyethyl)carboxyfluorescein (BCECF) (Rink et al., 1982). Other derivatives have also been used (Chaillet & Boron, 1985). The pK of BCECF is 7, therefore ideal for the measurement of intracellular pH in the physiological range. BCECF is loaded by incubation with the AM derivative. In theory pH indicators should suffer from many of the problems mentioned above for Ca indicators. There are, however, two important differences that make the use of pH indicators much simpler. Firstly the cellular pH buffer capacity is of the order of 15-25 mEq. and therefore adding say 3 mM of a pH buffer will not appreciably change the cell buffer capacity. The second advantage concerns calibration of the signal. This can be performed by means of the K^+/H^+ exchanger Nigericin, which seems to work consistently in a wide variety of cells. BCECF is used with two excitation wavelengths: 490-500 nm where fluorescence is increased by increasing pH and a lower wavelength (440- 450) where changes of pH either have no effect or the opposite effect on fluorescence. Emission is measured at 530 nm. Figure 3 compares the emission spectra of BCECF in solution and inside the cell, both having similar peaks at pH 7.

Other fluorescent pH indicators

The major disadvantage of BCECF is that changes of pH_i affect the excitation rather than the emission spectrum. This therefore necessitates changing between two exciting wavelengths rather than simply splitting the emitted light. 3,6-dihydroxyphthalonitrile (Kurtz & Balaban, 1985) is a pH indicator which changes its emission spectrum with pH. It has two pK_as (pH 5 and pH 7.9). More recently other pH-sensitive indicators have become available (see Molecular Probes catalogue). Two families of compounds are based on seminaphthofluorescein (SNAFL) and seminapthorhodafluor (SNARF). These have the advantage that both the excitation and emission spectra shift with pH. A minor disadvantage of these compounds is that they have pK_a's greater than 7.5.

Calibration

Several methods are available for calibration. The in vitro method suffers from the same problems as those discussed above for Ca indicators. In contrast the in situ calibration is easy to perform using nigericin. In order to use this, external K is increased to equal the internal concentration and then pH_o altered. This sets pH_i equal to pH_o. One disadvantage of this technique is the need to elevate external K. This depolarizes the cell membrane and, also, generally involves the removal of external Na. Both of these effects tend to increase $[Ca^{2+}]_i$. In order to avoid this, we perform the calibration in a Ca-free solution.

Recently new methods have been introduced for the calibration of pH signals (Szatkowski & Thomas, 1986) some of them specifically suitable for fluorescent indicators. We have described a method that consists of adding mixtures of weak acids and bases in order to find one that produces no change in intracellular pH. This is a "null method" that can be used at any point during the experiment as it is fully reversible. The values of intracellular pH obtained with this method are in very good agreement with those obtained with nigericin in many different cells (Eisner et al., 1989a).

5. OTHER FLUORESCENT INDICATORS

Tsien has recently produced a series of dyes which measure intracellular Na, Mg and K (Minta & Tsien, 1989). The Na indicator (SBFI) has been used to measure $[Na^+]_i$ in a variety of cells (e.g. Moore et al, 1988). We have used it to measure resting $[Na^+]_i$ in isolated cardiac myocytes (Donoso et al., 1989a). A variety of Na ionophores such as nystatin and gramicidin is available for in situ calibration. Dyes which are quenched by halide ions such as 6-methoxy-N-(sulfopropyl)quinolinium (SPQ:Illsley & Verkman, 1987) can record $[Cl^-]_i$. The lack of a membrane permeant form of the indicator and the fact that the spectrum does not shift with Cl-binding makes these indicators less useful than some of the others discussed.

6. APPARATUS

Epifluorescence set-up

The apparatus is built around a fluorescence microscope (Fig.4). This must be designed to transmit all required wavelengths. In particular the use of indicators which require excitation in the ultra-violet (as short as 340 nm

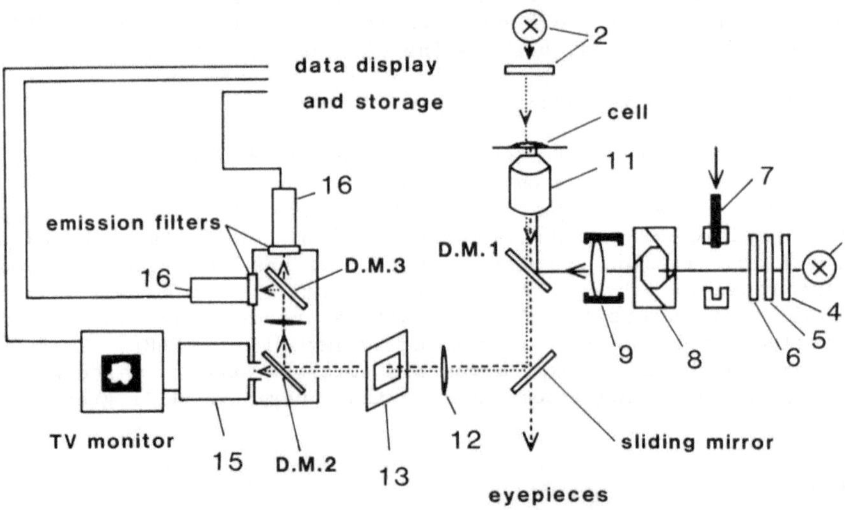

Fig. 4. Schematic diagram of apparatus based around an inverting micrsocope. The numbers refer to the sections in the text and represent the following components: (2) Microscope illuminator and long pass (red) filter. (3) Fluorescence excitation light (Xe or Hg). (4) Heat absorbing filters. (5) Neutral density filters. (6) Excitation interference filter. (7) Electronic shutter. (8) Excitation field diaphragm. (9) Epifluorescence adaptor and focussing lens. (11) Microscope objective. (12) Output lens. (13) Output diaphragm. (14) Interface to photomultipliers and video camera. (15) Video camera. (16) Photomultipliers. (DM) Dichroic mirrors.

for fura-2 or indo-1) means that these wavelengths must be transmitted by the microscope. While some attenuation can be corrected for by increasing the intensity of illumination, the microscope optics must still transmit a significant fraction of the light.

The basic principle of epifluorescence is illustrated in Fig.5. The short wavelength light from the light source is reflected through 90° by the dichroic mirror and then focussed onto the specimen by the objective. The emitted longer wavelength light from the specimen now passes straight back through the dichroic and is then sent to the eyepieces or to any suitable camera port. Until recently the main use of fluorescence microscopes was for fluorescent histology in which pictures were taken under fluorescent light often to look for the distribution of fluorescently labelled compounds. The need to either excite at two wavelengths or record emission at two wavelengths often means that the microscope must be modified. Various items are commercially available. However in many cases it is better and, invariably less expensive to build components oneself. In the following sections we will describe the various components of a suitable set-up built around a standard fluorescence microscope.

Microscope illuminator (2) This light should not interfere with the measurements of fluorescence. This can be achieved by putting a long-pass filter in front of the illuminator. The cut-off wavelength should be considerably longer than that of the emitted fluorescent light. It should ideally also be longer than the wavelength which is reflected by the dichroic to the PMT. The problem is made easier by the fact that the wavelength optimum of video cameras is much longer than that of PMTs.

Fluorescence light source (3). The choice is basically between a xenon or a mercury source. If high intensity UV illumination is required the mercury may be the best. However most of its UV spectrum is in the form of a few discrete peaks. If one wants to use wavelengths other than these peaks, special blocking filters with high rejection must be used to remove the peaks. In contrast a xenon light has a more continuous spectrum. In general, a xenon lamp will be the UV source of election if different indicators are going to be used. As far as the excitation intensity is concerned, we have found that in general it is necessary to decrease excitation light in order to avoid photobleaching problems (see below). Both xenon and mercury lamps are high pressure lamps and have to be handled with extreme care.

Microscope manufacturers generally make light sources which fix directly onto the back of the microscope. Once the arc is aligned (using adjustment screws provided) its alignment is fixed. One can also buy a third party light source although one must then fix it to the microscope. Even if one buys a light source designed to fit directly onto the microscope, one will often want to extend the mount so that neutral density and interference filters can be added. In particular if dual wavelength excitation is required, modification will certainly be required.

Heat-absorbing filters (4) Interference filters are expensive. They work partly by reflecting but also by absorbing unwanted light. It is this absorption that heats them up and can often lead to damage. This problem can be reduced by adding a suitable glass filter in the light path.
A large variety of such filters is available and generally cost less than 10% of an interference filter of the same size. The exact one used depends on the application. These filters block light of longer than a certain wavelength. They also cut out shorter wavelengths and must therefore be selected so that this is not a problem.

Neutral density filters (5) The intensity of the exciting light must be kept to the minimum compatible with the measurements required. This is for two reasons. (i) Intense light (particularly UV) damages cells. (ii)

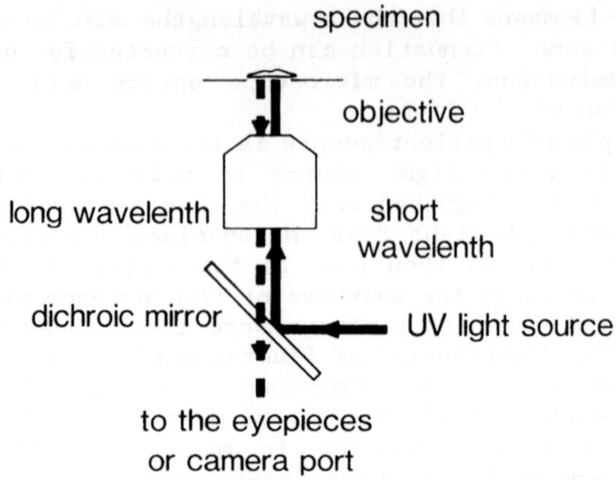

Fig. 5. The use of a dichroic mirror in epifluorescent microscopy. The diagram shows the short wavelength exciting light (solid) reflected by the dichroic to the objective. The longer wavelength emitted light is transmitted by the dichroic.

Fluorescent indicators are bleached by bright light. These problems can be reduced by attenuating the light with neutral density filters. The designation "neutral" means that their transmission is not a function of wavelength and therefore that they do not affect the spectral properties of the system. These are general referred to as "N.D. x" (x = - $\log_{10}$ transmission). Therefore an N.D. 1 filter transmits 0.1 of the incident light. It should be noted that glass N.D. filters are only of neutral density over the range of light transmitted completely by glass and that, at wavelengths less than about 400 nm, quartz N.D. filters must be used. We find it convenient that the N.D. filters be easily removable. This allows one to therefore add the largest N.D. filter consistent with measurement.

<u>Excitation interference filter (6)</u> This is designed to transmit the desired range of wavelengths to the specimen. We will consider the exact requirements for this filter later. If one wishes to use an indicator such as BCECF or fura- 2 which requires dual excitation then this must be allowed for here. There are various methods. The simplest is to have two filters on a slider which can be moved manually. Alternatively the filters can be moved by a solenoid or a motor. These methods generally take of the order of 100 msec to move the filter. More sophisticated versions are available commercially. Thus one can purchase filter wheels which rotate at several hundered Hz. One potential problem with such wheels is that each filter is only in place for a fraction of the total time thus dereasing the average light intensity. Other systems split the light into two beams and chop between the two with an oscillating mirror.

<u>Electronic shutter (7)</u> It is often useful to be able to stop the exciting light reaching the preparation. One does not always want to make continuous measurements and cutting off the light stops bleaching in between measurement periods. Most microscopes are supplied with a metal shutter on a slider which can be placed in the path of the exciting light. This is often inconvenient to use and we have found it better to add an electronic shútter.

<u>Excitation field diaphragm (8)</u> This can be useful to restrict the exciting light to the region of the cell of interest. There are two reasons for this. (i) One does not want to bleach or damage other cells. (ii) The greater the area of the exciting light, the larger the background signal. The diaphragms

on most microscopes are circular. While this is useful for spherical cells, it is less satisfactory for rectangular cells such as cardiac myocytes. One option would be to add a rectangular diaphragm here. Instead we generally prefer to ignore problem (i) and circumvent (ii) by restricting the area from which light is recorded (see below).

Epifluorescence adaptor (9) This, in combination with the microscope objective forms an image of the field diaphragm on the specimen. While one can do this oneself with a suitable lens, it is often simplest to use the manufacturers. With some microscopes (e.g. Nikon Diaphot) the epifluorescence lens is held by the same piece that has the light source holder.

Dichroic Mirror (10) This must have a region of half transmission between the wavelength of the exciting light and that of the emitted light. There are several standard dichroic mirrors available for most microscopes generally designed for the fluorescent probes used for histology such as rhodamine and fluorescein, as well as for Ca and pH indicators. Dichroics can be made to order by many companies for little more than that charged for off the shelf components.

Microscope objective (11) It is important to bear in mind the importance of numerical aperture (N.A.) in the choice of objective. For a given magnification the intensity of illumination via epifluorescence is proportional to $(N.A.)^2$. For a given fluorescent excitation the intensity of the collected fraction of emitted light is also proportional to $(N.A.)^2$. Therefore the total emitted light signal will be proportional to $(N.A.)^4$ whereas the bleaching and other light induced damage will only go as $(N.A.)^2$. Therefore for a given level of illumination and therefore of bleaching the fluorescent signal will be 4.7 x greater for a typical 1.3 N.A. immersion objective compared to a dry 0.6 N.A. objective.

The size of the specimen will influence the choice of magnification of the objective. On many systems the camera or video port to which (see later) the photomultiplier is often connected contains only the central region of the image. Therefore if the cell of interest occupies the whole of the diameter of the visual field it may be larger than the field of the video camera. At first sight, the easiest way to deal with this problem may appear to be to use a lower power objective. Unfortunately the N.A. of 10 and 20x objectives is generally lower than that of 40 x objectives. It therefore makes more sense to use a 40x objective and decrease the magnification at some later stage.

A complication comes if one wishes to make phase contrast measurements with high N.A. objectives. The 40x immersion generally requires a different phase ring and larger condenser than that generally available on inverted microscopes. A suitable condenser must have a higher N.A. and, if a reasonable working distance is required above the preparation, a large diameter condenser lens must be used.

Output lens (12) For the system described, the light which leaves the microscope must, as well as falling on the photomultipliers, form an image on the face of a video- camera. Some form of lens assembly is required to make sure that the image is the right size for the video camera and the PMT. We use a projection lens provided by Nikon. If different size cells are used it would be useful to be able to change the degree of magnification. This could of course be done by changing the intermediate lenses between the microscope and subsequent apparatus. One particularly convenient way of doing this would be do use two zoom camera lenses coupled together. Changing the relative focal lengths will then change the final effective magnification.

Output Diaphragm (13) This restricts the area from which fluorescence is measured and therefore eliminates interference from other cells and decreases the background. For spherical cells a simple and inexpensive circular diaphragm would do. However for other cells we find a rectangular diaphragm to be useful. Obviously the rays leaving the microscope must form an image in the plane of the diaphragm. This must then be refocused (with another lens) onto the video camera or subsequent stage.

Interface to video-camera and PMT (14) In our application the video camera receives long wavelength light and the shorter wavelength light goes to the PMTs. This is achieved with a (long pass) dichroic. We use a 610 nm dichroic so that the long wavelength light in front of the normal microscope illuminator, in fact red light (see above) is transmitted to the camera whereas the shorter wavelength emitted fluorescence light is reflected. It should be noted that no dichroics are perfect. In particular they do not reflect at all wavelengths below the nominal value. In our application with indo-1 it is necessary to reflect light at wavelengths as short as 400 nm to the PMT. The 610 nm dichroic has to be specially treated to allow this. If purchasing any dichroic it is therefore always advisable to check the exact characteristics. This problem could of course be overcome by having a dichroic with a shorter cut-off. It would then, however, not be suitable for an indicator such as BCECF (emitted light at 530 nm). A different sort of dichroic will be needed for some of the recently introduced long-wavelength indicators.

The video camera must be mounted such that the light forms an image on its active face when the microscope is focused such that the image is in focus for an observer looking through the eyepieces.

The shorter wavelength light to the photomultipliers may be split again by another dichroic if dual emission is required. For example, for use with indo-1, we use a dichroic at 450 nm. The light then goes to the two photomultipliers with filters in front of them. It is useful if the box holding these dichroic mirrors allows for easy adjustment of the position of the dichroics so that the same portion of the visual field is seen by each photomultiplier. It is also useful if the filters in front of the PMTs can be changed easily. The system must also be light tight.

The optical path to the PMTS may be such that the divergent rays that leave the microscope form a large (out of focus) image. This may mean that the image of the whole visual field is too large to fit on the face of the PMT. This may be an advantage since one does not always wish to record from the whole of the field. If however the recorded region is too small it can be enlarged by adding a convex lens in the light path after the dichroic mirror and before the PMTs to decrease the image size.

Video Camera (15) In the present application the video camera is only used to form a standard transmitted light image and not to image the fluorescent light. It does not therefore need the sensitivity demanded by the latter application. Nevertheless it is useful to be able to keep the light level as low as possible and a reasonably sensitive camera is therefore useful. A camera with a nuvicon tube is appropriate.

Photomultipliers (16) Emitted light is detected by photomultiplier tubes (PMT). Basically a PMT is a source of current which is proportional to the light intensity. It consists of a photocathode and a series of dynodes responsible for signal amplification. Incident photons eject photoelectrons that collide with the first dynode due to the imposed negative potential. The collision of one photoelectron with the first dynode will cause 5 to 20 additional electrons, This process is repeated along the dynode chain until a pulse of current arrives at the anode. The total amplification of the signal depends upon the potential difference between anode and cathode. The time that a photon of light takes from the cathode to the current pulse at the anode is of the order of 20 nsec. PMT's do not respond to the same light wavelength with the same efficiency, therefore PMT characteristics should be checked for each application. An important requirement for reliable PMT function is to use a stable high voltage power supply.

<u>Choice of filters and dichroic mirrors</u> Interference filters generally have both the centre wavelength and the half band width specified. For dichroics the wavelength at which 50% of the light is transmitted is specified. The steepness of the dichroic is also important. Several criteria govern the choice of appropriate dichroics and filters.

(i) The most obvious criterion is set by the spectral properties of the indicator used. One must select wavelengths which excite and are emitted by the indicator.
(ii) The dichroic must have a centre wavelength between the wavelengths used for excitation and emission.
(iii) There should be no cross talk between the excitation and emission filters. Specifically the longest wavelength transmitted by the excitation filter should be shorter than the shortest wavelength transmitted by the emission filter. To assess this one may need more information than just the half pass band width of the filter. This condition is easy to satisfy with an indicator such as fura-2 where one excites at 340 and 380 nm and measures at 500 nm but much harder for BCECF where one excites at, say, 430 and 500 nm and measures at 530 nm. In the latter case one requires filters with fairly sharp spectra. The degree of cross talk can generally be estimated by measuring the level of fluorescence with no fluorescent object in the set-up.
(iv) In order to maximize the amount of light collected (and therefore the signal-to-noise ratio) the emission filter should record as much light as possible. For an indicator such as fura-2 or BCECF where one measures at only one emission wavelength one should try to record as much of the emission spectrum as possible. Obviously there is no point in trying to record all since this would also increase the background. However it makes sense to use for fura-2 a 500 nm emission filter with a half band width of 40 nm. For BCECF a much narrower emission filter must be used to avoid overlap with the excitation filter. For an indicator with two emission wavelengths (indo-1) one has two contrasting requirements. (a) The narrower the width of each emission filter the greater the degree of spectral selection and therefore the greater the fractional change, in this case, $[Ca^{2+}]_i$. (b) The wider the width the more light is measured and therefore the better signal to noise.

<u>Signal-to-noise</u> There are three possible sources of noise in light measurements. (i) Variations in the dark current or background of the PMT. For studies using fluorescent indicators the light levels are generally much higher than the background of the photomultiplier. For example with an indo-2 loaded cardiac cell we obtain photomultiplier currents of 200 nA without significant bleaching. Even the comparatively inexpensive PMTs used have dark currents of 0.1 nA and this is therefore not usually a problem. (ii) Variations in the background light due to the normal microscope illuminator or ambient lights. The effect of ambient lights can be minimized by enclosing the microscope in a reasonably light proof enclosure. There is no need to overdo this. Generally photographic blackout type curtains will be perfectly adequate. The effects of the microscope illuminator are greatly decreased by the long pass filter (see above). (iii) Noise due to the quantal nature of light. Specifically if there are n photons in any counting period the noise as reflected by the standard deviation (s.d.) is $\sqrt{n}$. The signal to noise is therefore $\sqrt{n}$ and therefore the larger the light signal the better the signal to noise. It is this problem which is the greatest in practice.

Electronics Signals generated by PMT's can be measured in two different ways. In the "photon counting" mode each current pulse generated at the anode is counted. This method is very sensitive and is used when light levels are very low, however the amplifier and discriminator required are very expensive. The "analog" method consists of averaging current pulses at the anode and then measuring it .This method involves a current to voltage converter with a finite time constant (say 0.1 -1 msec). This gives a voltage which is a function of light intensity. Since the intensity of the light and the time resolution required can vary greatly between experiments it is useful to have variable filtering and gains. For the light levels measured in an standard fluorescent experiment the analog mode is satisfactory.

For experiments involving two photomultipliers one needs two such current to voltage converters. It is then useful to have a way to display the ratio of the two signals in real time. There are two ways of doing this. (i) using a digital to analog converter and computer. (ii) Using an analog circuit. We find the latter particularly convenient. Ready made divide amplifiers are available. Full circuit diagrams of appropriate electronics are freely available from the authors.

ACNOWLEDGEMENTS

Work from the authors' laboratories was supported by grants from the British Heart Foundation and the Wellcome Trust.

REFERENCES

Almers, W. and Neher, E. (1985). The calcium signal from Fura-2 loaded mast cells depends strongly on the method of dye loading. FEBS Letters 192, 13-18

Arslan, P.,De Virgilio, R. Y., Tsien, R. Y., and Pozzan, T.(1984). Cytosolic Ca^{2+} homeostasis in Ehrlich and Yoshida carcinomas. A new membrane-permeant chelator of heavy metals reveals that these ascite tumor cells have normal cytosolic free Ca^{2+}. J. Biol. Chem. 260,2719-2727.

Albert, P. R., and Tashjian, A. H. (1984).Thyrotropin-releasing hormone-induced spike and plateau in cytosolic free Ca^{2+} concentration in pituitary cells. Relation to prolactin release. J. Biol. Chem. 259, 5827-5832.

Borle, A. B., and Snowdowne, K. W. (1986). Methods for the measurement of intracellular ionized calcium in mammalian cells: comparison of four classes of calcium indicators. In Calcium and cell function ed Cheung, W.Y. vol 7, pp 159-200. Orlando: Academic Press.

Becker, P. L., and Fay.,F. (1987). Photobleaching of fura-2 and its effects on determination of calcium concentrations. Am. J. Physiol. 253, C613-C618.

du Bell, W. H., Philips, Ch. P., and Houser, S. R. (1988). A technique for measuring cytosolic free Ca^{2+} with Indo-1 in feline myocytes. in "Biology of isolated adult cardiac myocytes" eds Clark, W.A., Decker, R.S. & Borg, T.K. Elsevier Science Publishing Co.

Chaillet, J.R. & Boron, W.F. (1985). Intracellular calibration of a pH-sensitive dye in isolated, perfused salamander proximal tubules. J. gen. Physiol. 86, 765-794.

Chance, B., Cohen, P., Jobsis, F. & Schoener, B. (1962). Intracellular oxidation-reduction states in vivo. Science (Wash D.C.) 137, 499-508.

Cobbold, P. H. and Rink, T. J. (1987). Fluorescence and bioluminiscence measurements of cytoplasmic free calcium. Biochem. J. 248, 313-328.

Davis, M. H., Altschuld, R. A., Jung, D. W., and Brierley, G. P., (1987). Estimation of intramitochondrial pCa nd pH by fura-2 and 2,7 biscarboxyethyl-5(6)-carboxyfluorescein (BCECF) fluorescence. Biochem. Biophys. Res. Comm. 149, 40-45.

Deber, C. M., Tom-Kun, J., Mack, E., and Grinstein, S. (1985). Bromo A23187: a fluorescent calciun ionophore for the use with fluorescent probes. Anal. Biochem. 146, 349-352.

De feo, T. T., Briggs, G. M. and Morgan, K. G. (1987). Ca^{2+} signals obtained with multiple indicators in mammalian vascular muscle cells. Am. J.Physiol. 253, H1456-H1461.

Donoso, P. Eisner, D.A. & O'Neill, S.C. (1989a). Measurement of intracellular [Na^+] in isolated rat cardiac myocytes using the fluorescent indicator SBFI. J. Physiol. 418, 48P.

Donoso, P., Eisner, D.A., O'Neill, S.C. & Valdeolmillos, M. (1989b). the fluorescent probe Indo-1 can indicate $[Ca^{2+}]_i$ and caffeine concentration simultaneously. J. Physiol. 416, 42P.

Eisner, D. A., Kenning, N. A., O'Neill, S. C., Pocok, G., Richards, C. D., and Valdeolmillos, M., (1989a). A novel method for absolute calibration of intracellular pH indicators. Pflugers Arch. 413, 145-157.

Eisner, D. A., Nichols, C. G., O'Neill, S. C., Smith, G. L., and Valdeolmillos, M., (1989b). The effects of metabolic inhibition on intracellular calcium and pH in isolated rat ventricular myocytes. J. Physiol. 411, 393-418.

Eng, J., Lynch, R. M., and Balaban, R. S. (1989). Nicotinamide adenine dinucleotide fluorescence spectroscopy and imaging of isolated cardiac myocytes. Biophys. J. 55, 621-630.

Garcia-Sancho, J. (1984). Inhibition of glycolysis in the human erytrocyte by formaldehyde and Ca-chelator esters, J. Physiol. 357, 99.

Grynkiewicz, G., Ponie, M., and Tsien, R. Y. (1985). A new generation of Ca^{2+} indicators with greatly improved fluorescence properties. J. Biol. Chem. 260, 3440-3450.

Gunter, T. H., Restrepo, D., and Gunter, K. K. (1988). Conversion of esterified fura-2 and indo-1 to Ca^{2+} sensitive forms by mitochondria. Am. J. Physiol. 255, C304-C310.

Hesketh, T.R., Smith, G.A., Moorc, J.P., Taylor, M.V. & Metcalfe, J.C. (1983). Free cytoplasmic calcium concentration and the mitogenic stimulation of lymphocytes. J. biol. Chem. 258, 4876-4882.

Illsley, N.P. & Verkman, A.S. (1987). Membrane chloride transport measured using a chloride-sensitive fluorescent probe. Biochemistry 26, 1215-1219.

Johnson, P. C., Ware, J. A., Cliveden, P. B., Smith, H., Dvorak, A. H., and Salzman, E. W. (1985). Measurement of ionized calcium in blood platelets with the photoprotein aequorin. Comparison with quin-2. J. Biol. Chem. 260, 2069-2076.

Knight, D. E., and Kesteven, N. T. (1983). Evoked transient, intracellular free Ca^{2+} changes and secretion in isolated bovine adrenal medullary cells. Proc. R. Soc. London Ser. B 218, 177-199.

Konishi, M., Olson, A., Hollingworth, S., and Baylor, S. M. (1988). Myoplasmic binding of fura-2 investigated by steady-state fluorescence and absorbance measurements. Biophys. J. 54, 1089-1104.

Kurtz, I. & Balaban, R.S. (1985). Fluorescence emission spectroscopy of 1,4-dihydroxyphthalonitrile. A method for determining intracellular pH ni cultured cells. Biophys. J. 48, 499-508.

Lakowicz, J. R. (1983). "Principles of fluorescence spectroscopy" Plenum Press N.Y.

Li, Q., Altschuld, R.A. & Stokes, B.T. (1987). Quantitation of intracellular free calcium in single adult cardiomyocytes by Fura-2 fluorescence microscopy: calibration of fura-2 ratios. Biochem. Biophys. res. Comm. 147, 120-126.

Macintyre, E.A., Tatham, P.E.R., Abdul-Gaffar, R. & Linch, D.C. (1988). The effects of pertussis toxin on human T lymphocytes. Immunology 64, 427-432.

Minta, A., Kao, J.P. & Tsien, R.Y. (1989). Fluorescent indicators for cytosolic calcium based on rhodamine and fluorescein chromophores. .J biol. Chem. 264, 8171-8178.

Minta, A. & Tsien, R.Y. (1989). Fluorescent indicators for cytosolic sodium J. biol. Chem. 264, 19449-19457

Moore, E.D.W., Tsien, R.Y., Minta, A. & Fay, F.S. (1988). Measurement of intracellular sodium with SBFP. A newly developed sodium sensitive fluorescent dye. FASEB J. 2, A754.

Oakes, S. G., Martin, W. J., Lisek, . A., and Powis, G. (1988). Incomplete hydrolysis of the calcium indicator precursor Fura-2 pentaacetoxymethyl ester (Fura-2 AM) by cells. Anal. Biochem. 169, 159-166.

Owen, Ch. S. (1988). Quantitation of lymphocyte intracellular free calcium signals using indo-1. Cell Calcium 9, 141-149.

Popov, E. G., Gavrilov, I. Yu., Pozin, E. Ya., and Gabbasov, Z. A. (1988). Multiwavelength method for measuring concentration of free cytosolic calcium using the fluorescent probe indo-1. Arch. Biochem. Biophys. 261, 91-96.

Rink, T. J., and Pozzan, T. (1985). Using Quin-2 in cell suspensions. Cell Calcium 6, 133-144.

Rink, T. J., Tsien, R. Y., and Pozzan, T. (1982). Cytoplasmic pH and Mg^{2+} ni lymphocytes. J. Cell. Biol. 92 , 189-196.

Szatkowski, M. S., and Thomas, R. C. (1986). New method for calculating pH_i from accurately measured changes in pH_i induced by a weak acid and base. Pflugers Arch. 407, 59-63.

Thomas, J.A., Buchsbaum, R.N., Zimniak, A. & Racker, E. (1979). Intracellular pH measurements in Ehrlich ascites tumour cells utilizing spectroscopic probes generated in situ. Biochemistry 18, 2210-2218.

Timmerman, M. P., and Ashley, C. C. (1986). Fura-2 difussion and its use sa an indicator of transient free calcium changes in single striated muscle cells. FEBS Lett. 209, 1-8.

Tsien, R. Y. (1988). Fluorescence measurement and photochemical manipulation of cytosolic free calcium. Trends Neuros. 11, 419-424.

Tsien, R.Y. & Minta, A. (1987). J. cell. Biol. 105, 89a

Tsien, R. Y., Pozzan, T., and Rink, T. J. (1982). Calcium homeostasis in intact lymphocytes: cytoplsmic free calcium monitored with a new intracellularly trapped fluorescent indicator. J. Cell Biol. 94,325-334.

Tsien, R. Y., Rink, T. J., and Poenie, M. (1985). Measurement of cytosolyc free calcium Ca^{2+} in individual small cells using fluorescence microscopy with dual excitation wavelengths. Cell Calcium 6, 145-157.

15

BIOMEDICAL NUCLEAR MAGNETIC RESONANCE SPECTROSCOPY AND TRANSPORT

Susan Wray

Physiology Department
The University of Liverpool
P.O. Box 147
Liverpool L69 3BX

INTRODUCTION

Nuclear Magnetic Resonance (NMR) spectroscopy, can be used for transport studies either directly, by elucidating metabolic pathways, transport into and out of organs and organelles, or indirectly via determination of ATP concentration, intracellular pH, viscosity, flux rates and thermodynamic information. Previously the tool of the chemist, NMR has now found many applications in biology and medicine. This advance has been made possible largely due to technological advances, allowing wider bores and stronger magnetic fields to be achieved. In this chapter I will firstly outline briefly the theory behind NMR so that its advantages and limitations can be appreciated. I will then detail, with examples, the information available from NMR Spectroscopy. This will be followed by discussion of the more biologically useful NMR nuclei. My emphasis throughout will be on *in vivo* applications and how NMR may be useful for transport studies. This chapter is not a comprehensive review of NMR spectroscopy nor is it written for the expert. I have erred towards being didactic and have not given references to back up every statement made, for which I apologise to all concerned. I hope it shines some light on the technique and that it may even persuade the reader to have a go!

BASIC THEORY OF NMR SPECTROSCOPY

NMR spectroscopic data can be obtained only from those atoms whose nuclei possess the quantum property known as *spin*, (I). Examples of such nuclei are: 1-H, 13-C, 14-N, 15-N, 19-F, 23-Na, 31-P and 35-Cl. I is integral for nuclei with even mass number, zero for nuclei with even numbers of both protons and neutrons and half - integral for nuclei with odd mass number. When $I = {}^1/_2$ high resolution spectra can usually be obtained e.g. 1-H, 13-C and 31-P, (see Table 1). A magnetic component is associated with this spin. If such a nucleus is placed in a magnetic field, it will precess around the field. The nuclei do not all align with the applied field, but can have one of two alignments, (for a spin, of 1/2). These two alignments differ in their energy level, (ΔE), and there is a population difference between the two alignments; more nuclei are at the lower energy level. The populations of nuclei in the various

Cell Membrane Transport, Edited by D. L. Yudilevich *et al.*
Plenum Press, New York, 1991

alignments are given by the Boltzmann distribution. The lower energy level is populated with nuclei aligned with the field ($I = +1/_2$).

The ΔE between the two energy states is proportional to the magnitude of the applied magnetic field, βo. Since by Planck's law energy is proportional to frequency (V) then

$$Vo \propto \beta o \qquad .....1$$

where Vo is the resonance exciting frequency of an applied magnetic field (β_1). The proportionality constant is the magnetogyric ratio γ, which is directly related to the nuclear magnetic moment. Thus

$$V = \gamma o \beta o \qquad .....2$$

This equation is called the Lamor Equation. Values of γ can be found in Table 1.

If a brief pulse of electromagnetic energy (β_1) is applied to the nuclei, some of them will absorb it and be transposed to the higher energy level. When this pulse is finished, the excited nuclei will return to the lower energy level, ("relax"), emitting a weak radiofrequency (RF) signal. This is the NMR signal which is detected. The signal (also known as a "free induction decay", FID), is then Fourier transformed in the frequency domain. This resonance frequency is quite different for different nuclei. In addition, the resonance frequency is also slightly different for the *same* nucleus in different compounds. This is because the magnetic flux density experienced is altered slightly by the different environments, e.g. circulating electrons, local diamagnetic shielding and paramagnetic centres. This shielding (σ), in turn alters ("shifts") the resonance frequency, thus the Lamor equation (equation 2) becomes:

$$V = \gamma o \ \beta o \ (1 - \sigma) \qquad .....3$$

By detecting these different frequencies, known as chemical shifts, we can identify the different compounds of the same nucleus. Chemical shifts from a chosen reference frequency are usually expressed in units of parts per million (p.p.m.). The NMR frequencies are in the radio frequency range i.e. MHz.

It was stated above that when the RF pulse is over the nuclei return to their equilibrium position. The processes that govern this return are called relaxation mechanisms and are of considerable importance in NMR. The main "route" for this relaxation is via interactions between the nuclear spins and the rest of the molecule, (which is called a lattice) and is known as spin - lattice relaxation or longitudinal relaxation and is characterized by a time constant T_1. There will also be interactions between neighbouring nuclear spins, referred to as spin-spin or transverse relaxation, which again has a characteristic time constant, T_2. These relaxation measurements are of practical concern because they govern the selection of the RF pulse interval, as discussed earlier. T_1 values are needed to allow reaction rate constants to be calculated in saturation transfer experiments (see later), and provide information about rotational motion and microviscosity. T_2 measurements can provide information about diffusion rates and molecular structure.

The magnetic field is constantly present in modern NMR spectrometers. The field strength is large, (e.g. 1 - 10 teslá (T); 1T = 10 K-Gauss).

This is up to 100,000 times greater than the earth's magnetic field. This means that there is a "no go" area for metal objects around the spectrometer! Bore size varies from 10 to 100 cm and the magnet bore can be orientated vertically or horizontally. The spectrometer is connected to a computer which can perform line broadening to reduce noise, the Fourier transform, signal enhancement, integration, curve fitting, as well as controlling the acquisition parameters during data collection.

LIMITATIONS

(i) Low Signal Intensity

As the nucleus interacts only weakly with the applied field, NMR frequencies are low (MHz) and signal intensities are also low. To maximise intensity, the applied magnetic field is made as high as is technically, logistically or ethically possible, (around 1 - 10 T). Such magnets require superconducting coils and are expensive. Even with such large magnetic fields, the low signal intensities set detection limits. The answer to the commonly asked question; "what concentration can be detected?", depends upon several factors including field strength, data collection time, NMR visibility (see below) and sample size. A rough guide however is 0.5 mM for 31-P and 0.1 mM for 1-H.

(ii) NMR Visibility

Substances have to be sufficiently mobile, (in general, free in solution), to give discernable NMR signals. With large molecules only that portion which is mobile enough will be NMR visible and give a signal. Thus some, (e.g. phosphorus in bone), are too immobilized to yield measurable signals, whereas others (eg. the phosphorus of PCr) are in relatively small compounds in an aqueous environment and are therefore able to translate or rotate rapidly and give a narrow signal. However, the above should not be taken to imply that all nuclei in large molecules are "NMR invisible". It may be that the nucleus is in a small part of a molecule that is able to rapidly rotate relative to the rest of the large molecule. An example of this is glycogen - the large storage molecule for glucose. Glycogen contains carbon which is NMR visible because of rapid intra-molecular motion - see section on 13-C. With nuclei possessing spins $> 1/_2$ there may also be an NMR invisible component to some signals. The reasons for this are complicated and not fully understood, except to say that they arise from the electrical quadrupolar nature of such nuclei and the NMR acquisition parameters usually used.

(iii) Useful Nuclei

Some of the nuclei which possess spin are not naturally abundant, e.g. 13-C and 15-N (see Table 1). This may mean, that unless they can be introduced into the biological system of interest, they are generally unsuitable for NMR experiments unless signal collection can be performed over many hours. For example 13-C is only 1.1% naturally abundant. However, some substances are present in such high concentrations that the 13-C NMR signals can be detected without too much difficulty e.g. glycogen in skeletal muscle and liver. Likewise, other nuclei may be 100% naturally abundant, but not present in most biological systems, e.g. 19-F. Again to yield NMR signals, 19-F has to be introduced into the system of interest. This can be done readily with isolated organ preparations and cell suspensions. In intact animals 19-F-containing inhalation anaesthetics e.g. halothane have been used to monitor cerebral haemodynamics and the metabolism of various 19-F-containing anti-tumour drugs in body tissues has also been followed, see section on 19-F NMR.

As was noted earlier, to give NMR signals, the nucleus must possess spin. The spin of some nuclei is complicated e.g. 39-K is 3/2, 17-0 is 5/2. This can result in signals so broad (>100ppm) as to be useless. This is the case for many of the resonances of 14-N, the naturally abundant form of nitrogen. It has a spin of 1 i.e. quadrupolar moment, and has thus found relatively little use in biological NMR. However, some metabolites e.g. urea and betaine, by virtue of their symmetry around the N atom, do give narrow 14-N signals and will be discussed in the section on nitrogen NMR.

In addition to all these factors there is also a difference in inherent sensitivity of NMR nuclei. This sensitivity is expressed relative to 1-H. Table 1 below gives the sensitivity of the commonly used nuclei, which are discussed in this chapter. Sensitivity is expressed at constant field for equal number of nuclei. The table also gives the % natural abundance for the nuclei. The final column shows the product of sensitivity and natural abundance, termed relative sensitivity.

TABLE 1. Useful NMR parameters.

Nucleus	Spin	Magnetogyric Ratio MHz/T	Sensitivity	Natural Abundance %	Relative Sensitivity
1-H	1/2	42.57	1.00	99.98	100
13-C	1/2	10.71	1.59×10^{-2}	1.11	1.76×10^{-2}
14-N	1	3.08	1.01×10^{-3}	99.63	1.01×10^{-1}
15-N	1/2	4.31	1.04×10^{-3}	0.37	3.85×10^{-4}
19-F	1/2	40.06	0.83	100	83
23-Na	3/2	11.26	9.25×10^{-2}	100	9.25
31-P	1/2	17.23	6.63×10^{-2}	100	6.63

It can be seen from Table 1 that relative sensitivity is often the most useful parameter. For example the sensitivity of 14- and 15-N are virtually identical but as 15-N is so very sparse (0.37%) its relative sensitivity and hence usefulness may be much less than 14-N. The relative sensitivity however, does not take any account of how quickly nuclei loose their excitation i.e. relaxation mechanisms as described in the section on NMR theory. In general, if they are rapid, then RF pulses can be applied more frequently and hence signal built up more rapidly than if they are slow. So if one was to consider for example 31-P and 14-N, the table above shows us that 31-P is an order of magnitude more sensitive than 14-N. However, for many compounds 14-N has short T_1 values which permits rapid pulsing, say once every 0.2s compared to 31-P where pulsing is commonly once every 2s. Thus the lower sensitivity of 14-N is offset and the two nuclei become more comparable.

In general NMR is not a particularly rapid technique because (i) time must be allowed for sufficient nuclei to lose their excitation and be capable of absorbing energy again and (ii) sufficient R.F. pulses must be applied to build up a good enough signal to noise ratio given the low inherent sensitivity of the technique. Thus the time scale is usually minutes rather than seconds. However if an event is repetitive e.g. a beating heart or stimulated nerve then NMR signal collection can be gated to part of the event e.g. the first or last seconds and the signal built up by repeating the event several hundred times. In this way the time *resolution* can be very good, even if the *duration* of the data collection is long.

From all the above it can be seen that the limitations of the NMR technique depend very much upon the specific question and experiment. What may be hopeless for one study may be the method of choice for another!

ADVANTAGES

The reason why NMR spectroscopy has blossomed in the biomedical field is attributable to the fact that it is a *non-invasive* technique. This means that it can be used, to study, not only isolated cells and tissues, but also intact animals and humans. This allows many measurements to be made on the same preparation, along with other monitoring. For example, the amount of work done by the heart can be measured at the same time as several metabolites, e.g. ATP or phosphocreatine (PCr) are being detected; then the effects of ischaemia and recovery can be examined. Being non-destructive and, as far as it is possible to say, harmless, NMR spectroscopy and imaging can be used for human studies. This has led to it providing unique biochemical and physiological information on man. This is particularly true for studies on the brain, which is difficult to probe with any conventional biochemical techniques, due to its protective coating of skull, the blood-brain barrier and the lack of dispensable areas for biopsies! A good illustration of these points is provided by a study on human neonates (Reynolds et al., 1988, Delpy et al., 1987, Tofts & Wray, 1985). NMR has been used to study cerebral hypoxic-damage around the time of birth. This is a major cause of handicap in survivors of neonatal intensive care units. Firstly, the normal levels of *metabolites* in the *living* human brain were established. Then changes in these metabolites with *development* were noted, so that ultimately the cerebral consequences of hypoxia in the babies could be determined. By being able to make measurements *sequentially* over several days the time scale of these metabolite changes has been established. The ratio of PCr/Pi in the brain of these babies (see the next section) is now being used to provide *prognostic* information. In the future, NMR will be used to monitor the effects of *therapeutic interventions* such as calcium channel blockers, NMDA receptor antagonists and free-radical scavengers. This will follow from NMR studies on animals to establish the *mechanism* of neuronal damage in hypoxia.

The following sections will outline the type of information available from NMR spectra. The discussion will be focused around 31-P NMR since this is the most widely used NMR nucleus in biomedical spectroscopy. Following this the particular usefulness of other nuclei will be described.

WHAT INFORMATION CAN BE OBTAINED BY NMR?

(i) <u>Metabolites</u>

Figure 1 shows a 31-P NMR spectrum from rat uterus. I will use this as the basis for describing how metabolic information is obtained.

The X-axis is frequency in ppm. The Y-axis is signal intensity. The figure shows peaks from several phosphorus metabolites in the smooth muscle which constitutes over 90% of the cellular element of the uterus. There are 3 peaks labelled NP, these are nucleotide triphosphates, one from each of the phosphorus atoms in the molecule. These peaks are essentially ATP although there may be a contribution of about 10% from other nucleotides. This is because the resonance positions of CTP, UTP and GTP cannot be resolved from ATP. The extreme right-hand peak is used as the measure of ATP because it alone does not contain a co-resonance from any ADP which may also be present in tissue. Clear peaks from PCr and inorganic phosphate (P_i) can be seen. There is also a peak from phosphomonoesters (PME). These include phosphoethanolamine and phosphocholine. As with nucleotide triphosphates, these two substances resonate at only very slightly different frequencies and cannot be separated at the field strength used in this experiment (4.7 T). Note also that we do not see metabolite peaks from membrane phospholipids as they are too immobile, nor from cyclic phosphates since they are not

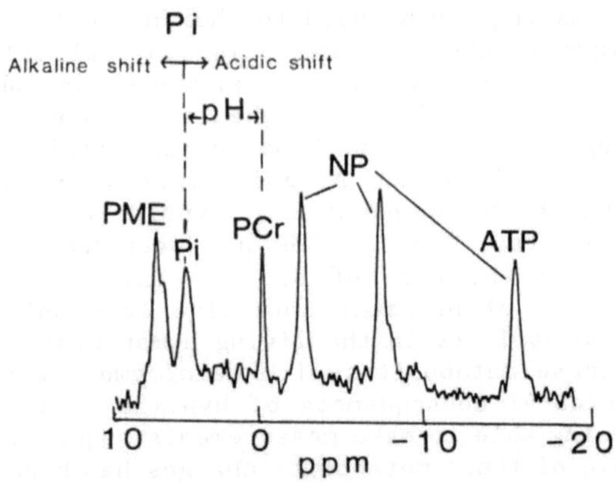

Fig. 1. ^{31}P NMR spectrum from rat uterus. The y axis is frequency in parts per million (ppm) and the x-axis is signal intensity. Peaks can be seen from the major, mobile phosphorus containing metabolites in the uterus. NP; nucleotide phosphates, of which ATP is the major contributor, PCr; phosphocreatine, Pi; inorganic phosphate. PME; phosphomonoesters. The Pi peak shifts with intracellular pH. Field strength 4.7 T, 300 scans, repetition time 0.25 s. From Wray, 1988a.

present in a large enough concentration. Other naturally occurring phosphorus metabolites can be detected in different tissues, e.g. phosphodiesters like glycerylphosphoryl choline in brain, diphosphoglycerates in red blood cells and sugar phosphates (glycolytic intermediates) in metabolically inhibited preparations.

All the above metabolites are detected *directly*. The concentrations of other metabolites can be estimated indirectly. The best example is provided by ADP. ADP is normally present in concentrations too low to be detected by NMR, i.e. μM rather than mM. However, if we have obtained the [ATP] and [PCr] and pH (see below) by NMR, and know the [Cr], then ADP can be calculated under conditions in which the creatine kinase reaction is at or near equilibrium.

$$\text{i.e.} \quad [ADP] = \frac{[ATP][Cr]}{K\,[PCr][H^+]} \quad ....4$$

where K is the equilibrium constant for the creatine kinase reaction. (This calculation depends upon all the substrates being in the same compartment and free to interact.)

Knowledge of [ADP] allows one to calculate the free energy of hydrolysis from ATP, which is the driving force for many reactions (see the section on thermodynamics). It is also worth noting that when [ADP] was calculated by NMR, the values obtained (20 μM) were an order of magnitude less than those obtained previously using other biochemical techniques. This is because NMR *free* ADP, whereas conventional biochemical assays had measured *total* ADP. This latter figure includes large amounts of ADP bound to proteins which are "NMR invisible". Since it is only the free ADP which takes place in cellular reactions and is able to influence the activity of rate limiting enzymes, the free ADP, determined by NMR is a more useful measurement. Having obtained [ADP] this can in turn be used to calculate AMP, which is in equilibrium with ADP through the myokinase reaction.

(ii) <u>Quantification of metabolites from NMR spectra</u>

In most studies undertaken and especially those concerned with transport, there is a need to know the concentration of substances and if those concentrations change. As NMR spectroscopy has progressed from being a "novel" to a "routine" biomedical technique, so too has concentration data on metabolites progressed from being qualitative to quantitative (see Tofts & Wray, 1988). Several techniques have been described for obtaining absolute quantification of metabolites, directly with NMR.

The basis for calculating metabolite concentrations is that the metabolite peak areas are directly related to concentration. Thus the first hurdle to be overcome in obtaining accurate concentration measurements is the accurate measurement of peak areas. This may sound simple, i.e. integration, triangulation, cut and weigh or curve fitting. However, because *in vivo* spectra usually have noisy baselines, broad peaks which overlap, which may in addition be superimposed upon a very broad hump (due to bone or lipids) and whose lineshape is unknown, the matter of peak area measurement is not so straightforward and there is still debate about which method is the best.

Having obtained area measurements the effect of signal saturation has next to be taken into account. In most NMR experiments there is an optimal pulse repetition rate where maximal signal is obtained in shortest unit time. This time will be less than that for *all* nuclei to have returned to their unexcited state i.e. some nuclei will still be 'saturated' and not respond to the next radio frequency pulse. The proportion of saturated nuclei will differ from one metabolite to another, as their chemical environment is different and thus their relaxation mechanisms will differ (see section on Basic Theory of NMR). Thus the peak areas of different metabolites will underestimate concentration by different amounts, which must be corrected for. This is done by using signal saturation correction factors obtained in separate experiments, where the interval between pulses is sufficiently long e.g. 20 s for all nuclei in all metabolites to have returned to their unexcited state. It is because this interval is so long, that one normally pulses every 1 s say, and then corrects for saturation.

The simplest way of obtaining quantitative data from the NMR experiment is to express the metabolite areas as ratios e.g. PCr/Pi or as fractions of the total spectral integral. Often a particular metabolite may be taken as a reference if it can be considered to stay constant; ATP may fulfil this role. If [ATP] is known then the concentration of the other metabolites can be calculated. When this kind of method is not practical, then in experiments on isolated tissues or cells, reference compounds can be put in bathing solutions or glass capillaries.

More recently methods have been described which utilize the size of the 1-H tissue water concentration signal to calibrate signals from the nucleus of interest (Wray & Tofts, 1986). There are technical problems in using a different nucleus in the standard, from that of the metabolite of interest, but these can be overcome. One has to match the spatial sensitivity of the two nuclei in the tissue by matching the nutation (flip) angle for the two nuclei (This is the degree to which the magnetization is "flipped" from the βo field). With increasing availability of combined imaging - spectroscopy machines, methods are being developed to obtained metabolic information from spatially localised signals. Such techniques may allow the question of tissue heterogeneity to be addressed and will add considerably to the impact of NMR spectroscopy.

(iii) Intracellular pH

Intracellular pH (pH_i) measurements can be made by NMR spectroscopy and this has turned out to be one of its major uses. The basis for the pH measurement is to have a compound whose resonance frequency depends on hydrogen ion concentration. The pK of the compound should be in the physiological range. For 31-P NMR spectroscopy Pi provides such an indicator, (see Fig 1). The NMR signal from Pi is a single peak whose frequency is determined by the relative amounts of basic and acidic Pi, i.e.

$$H_2PO_4^{-1} \equiv H^+ + HPO_4^{2-}$$

Once the Pi frequency has been calibrated in solutions at different pH's, then its position in subsequent experiments can be used to determine pH. This is done by using a variation of the Henderson - Hasslebach equation:

$$pH = pK + \log \frac{\delta - \delta_i}{\delta_2 - \delta} \quad ......5$$

Where δ is the Pi chemical shift observed in the spectrum and δ_1 and δ_2 are the Pi chemical shifts in the acidic and basic forms respectively.

Care must be taken with this calibration to ensure suitable ionic strength, magnesium concentration and temperature are used as all these may also affect the Pi resonance position. Pi is the most commonly used resonance for pH_i measurements, but in theory any peak which is sensitive to pH could be used. In 1-H NMR the histidine resonances of carnosine have been used. The section on 19-F NMR describes the use of specially designed indicators for pH measurements.

As pH_i has been shown to be a key player in several important cellular events, from fertilization to failure of muscle contraction, there is considerable interest in its measurement. Reviewing the literature on pH_i measurements reveals that NMR values are in agreement with those obtained by other methods i.e. microelectrodes and fluorescent indicators. Initial criticisms regarding Pi perhaps reporting *extracellular* or organelle phosphate were shown not to be correct by performing "double label" type NMR experiments, (see Gadian et al., 1982 for full discussion). The pH_i was calculated from both P_i and accumulated deoxy-glucose 6 phosphate in isolated hearts. The latter substance is considered to be entirely cytoplasmic. There was no significant differences between the pH_i value calculated by both substances. The advantages of obtaining pH_i by NMR spectroscopy are (i) other relevant information may be obtained e.g. [ATP]. (ii) It can be used on cells which are inaccessible to microelectrodes due either to size or connective tissue or contractible behaviour. (iii) It can be combined with other measurements on tissues e.g. force production, lactic acid efflux, which are often not possible when pH_i is measured in single cells. One particular area where NMR pH_i measurements have been most useful is smooth muscle, where the cells are small and the tissue active. For example from NMR studies in the uterus pH_i changes with pregnancy and parturition have been reported, pH regulation studied and the effect of metabolic inhibition on pH_i, force and lactate production established (Dawson & Wray, 1985, Wray, 1988b, Wray, 1990). No previous pH_i measurements existed for the uterus.

(iv) <u>Intracellular Magnesium</u>

The same principles can be applied to measuring free intracellular Mg ($Mg^{2+}{}_i$) as were described for pH_i In this case it is the resonance position of ATP which is sensitive to the amount of Mg bound to it. As with pH, the calibration curve, is all important and ionic strength, temperature, other cations and pH will all affect it. The ATP peaks are often broad in *in vivo* studies which can make precise measurements impossible, especially as the expected frequency shift with Mg is small. The measurements which have been made e.g. in striated muscle reveal that ATP is predominantly (≃95%) complexed to Mg (Dawson & Wilkie, 1984). This determination is important, since it is as Mg-ATP that ATP undergoes its energy-providing reactions. Free intracellular magnesium can be obtained from the measurement of the ratio of Mg-bound to free ATP using the binding constant for the reaction. The values obtained vary depending upon the equilibrium constant used. For example in cardiac muscle, taking ATP as being 90% bound to Mg, different constants will give a free intracellular $[Mg^{2+}]$ of between 0.5 and 1.5 mM. These figures appear higher than those obtained using microelectrodes. It is unclear at the moment why this should be so or which method is more accurate.

Having pointed out the limitations of the NMR estimates, it is also fair to say that in many instances no other techniques or measurements are available. Given the importance of free Mg to many cellular enzymes e.g. adenylate kinase, and its competition with Ca for binding sites, it is clear that the NMR measurements are an important addition to our knowledge. The same principles outlined here can be adapted to measure $[Ca]_i$ via 19-F NMR. This work is described in the section on 19-F NMR.

(v) Reaction Rates and Exchange Mechanisms

Kinetic studies of chemical exchanges can be made by NMR using magnetization transfer experiments. By these methods the unidirectional rates of enzyme catalysed reactions in living systems can be measured, (see Alger & Shulman, 1984a).

The fundamental principle is to *selectively* introduce a non-equilibrium nuclear spin magnetization into a nucleus in a compound which exchanges with nuclei in another unperturbed compound. If this exchange is sufficiently rapid (compared to the T_1 relaxation time), then the flow of the non-equilibrium spin magnetization between the two compounds can be observed. From this the rate constant for the chemical reaction between the two compounds can be calculated. The simplest and most used magnetization transfer technique is the "saturation transfer" experiment. I will therefore, use this to illustrate the practical usefulness of these techniques, and the creatine kinase (CK), reaction in muscle as an example of its application:-

$$ADP^{-} + PCr^{2-} + H^{+} = ATP^{2-} + Cr \qquad ....6$$

There is much interest in this enzyme system because it has been generally *assumed* that this reaction is always close to equilibrium and that the forward and backward exchange fluxes are considerably faster than the rate at which ATP is hydrolysed - hence accounting for the observation that there is very little net ATP breakdown during contraction. However, there has been no direct test of the assumption because by conventional means it is difficult to measure rates *in vivo* that are at equilibrium or in the steady state. But these conditions are ideal for saturation transfer NMR. Since calculations of cytoplasmic phosphorylation potential and intracellular pH have been based upon the assumption that the CK reaction is at or near equilibrium there was a clear case for using saturation transfer NMR to measure the CK reaction.

The NMR experiment performed first was on muscle (Gadian et al., 1981). The terminal (γ) phosphate of ATP was selectively "saturated". This was done by using a continuous RF irradiation at the frequency where the γ-ATP resonates. The effect of this is to remove the normal distribution difference between low and high energy spins. The γ-ATP nuclei are saturated. Thus when a (normal) RF pulse is applied these saturated nuclei of γ-ATP cannot respond. The NMR spectrum obtained therefore lacks the γ-ATP peak, as shown in Fig. 2. Now because there is chemical reaction between PCr and ATP, the *saturation* introduced into

ATP will be *transferred* to PCr. This will cause the PCr signal to decrease (for the same reason the γ-ATP peak decreased) and eventually reach a new steady state level. The first order rate constant for the PCr to ATP direction is thus obtained, e.g. 1.7 x 10^{-3} M.s.$^{-1}$ (Gadian et al., 1981). The PCr signal can also be selectively saturated and the reduction in the γ-ATP signal observed to obtain the rate constant for the ATP to PCr direction. Most of the studies so far have concluded that the CK reaction is in equilibrium at least in resting muscle (and brain). (In contracting muscle the PCr to ATP rate has been measured to be several times faster than the reverse rate. However, a discussion of the possible causes for this observation is beyond the remit of this chapter). The correctness of the equilibrium model means that free ADP concentrations can also correctly be calculated.

Limitations to magnetization transfer studies arise if enzyme - bound intermediates have fast relaxation mechanisms. The saturation transfer experiment is only sensitive to rate constants which are similar to the relaxation (T_1) rate constants, (around 0.1 to 3 seconds for most

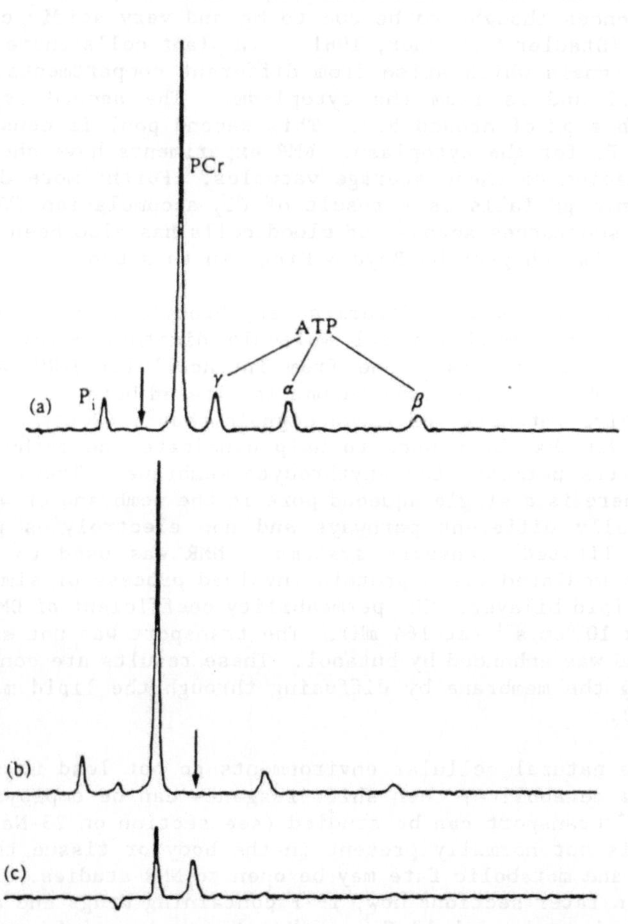

Fig. 2. ^{31}P NMR spectra of frog gastrocnemii obtained with selective irradiation applied at the frequencies indicated by arrows. The irradiation applied in (a) acts as the control for (b), in which selective irradiation was applied to the γ-phosphate of ATP. The difference between (a) and (b), given in (c), gives the extent of saturation transfer from the γ-phosphate of ATP to PCr. Field strength 4.7 T, 24-64 scans, repetition time 12 s. (Adapted from Gadian et al., 1981).

biological molecules). It should be noted that rate constants in this range are difficult to measure by any method other than NMR i.e. too fast for isotope flux. Another limitation is that the reactions must be at or near equilibrium to get accurate NMR measurements. Rate-controlling enzymes in biochemical paths are often not near equilibrium. The reactions studied must also be exhibiting pseudo first order kinetics.

(vi) Transport of Metabolites Across Membranes

When the internal and external NMR signals from metabolites can be distinguished, then transport of the metabolite across cellular membranes can be studied. The internal and external signals of the same metabolite may have different chemical shifts, and hence be distinguishable if they are in sufficiently different environments, as discussed earlier or if an agent can be added to "shift" one of them (see later). For example, it has been shown that ATP in synaptic vesicles is distinguishable from ATP in the rest of the tissue (electric organ from Torpedo) due to chemical shift differences thought to be due to Mg and very acidic conditions in the vesicles (Stadler & Fuldner, 1981). In plant cells there are at least 2 clear Pi signals which arise from different compartments. One is at around pH 7.3 and is from the cytoplasm. The second is from Pi in vacuoles with a pH of around 5.7. This second pool is considered to be a reserve of Pi for the cytoplasm. NMR experiments have shown transport of Pi from medium to these storage vacuoles. Furthermore during anoxia the cytoplasmic pH falls as a result of CO_2 accumulation (Vogel, 1987). Transport of substances across red blood cells has also been studied e.g. alanine, see also chapter by Boyd & King, in this book.

NMR has been used to determine erythrocyte volume (see Berling & Hall, 1989). The small neutral molecule dimethyl methyl phosphonate (DMMP) gives 31-P signals; one from intracellular DMMP and one from extracellular DMMP. As red cell volume is altered both the chemical shift and the relative intensity of these signals change (Potts et al., 1989). In addition DMMP has been used to help elucidate the pathways by which non-electrolytes permeate the erythrocyte membrane. There is debate as to whether there is a single aqueous pore in the membrane or whether there are functionally different pathways and non electrolytes permeate via specific facilitated transport systems. NMR was used to see if DMMP transport was mediated via a protein involved process or simply diffuses through the lipid bilayer. The permeability coefficient of DMMP was found to be ~ 9.5×10^{-5}cm s^{-1} (at 164 mM). The transport was not saturable (up to 600 mm) and was enhanced by butanol. These results are consistent with DMMP crossing the membrane by diffusing through the lipid matrix (Potts et al., 1989).

When the natural cellular environments do not lead to separate NMR signals for a metabolite, then shift reagents can be employed. In this way Na^+ and K^+ transport can be studied (see section on 23-Na NMR). When a substance is *not* normally present in the body or tissue then its rate of transport and metabolic fate may be open to NMR studies. For example, I will show in later sections how, 19-F containing drugs and anaesthetics can be so monitored, and 13-C enrichment can be used as a metabolic marker.

(vii) Thermodynamic Information

From the preceding sections it can be seen that NMR spectroscopy measures, either directly or indirectly several parameters needed for a thermodynamic approach to transport problems. Classically the free energy change for ATP hydrolysis is calculated from the equation:

$$\Delta G_{ATP} = \Delta G^{o}_{ATP} + RT \ln \frac{[ATP]}{[ADP] \times [Pi]} \qquad ....7$$

where ΔG_{ATP} is the standard free energy of hydrolysis of ATP, and a value appropriate for temperature, pH and Mg should be used. The concentrations of ATP, Pi and ADP can be determined by NMR, hence ΔG_{ATP} can be calculated. This value (in kJ mol^{-1}) shows the maximum work theoretically obtainable per mol of ATP hydrolysed. This figure can then be related to energy requiring processes under consideration e.g. transport of substances against concentration gradients. Let us take as an example of this sarcoplasmic reticulum (S.R.) pumping of calcium i.e. from myoplasm to S.R. The S.R. pump transfers 2 calcium per ATP split. Given myoplasmic [Ca] as 0.2 μM and S.R. [Ca] as 1 mM the free energy changed required of ATP hydrolysis, is 43 KJ mol^{-1}. Now it is known that during hypoxia cardiac failure occurs. Is this because there is a decline in the ΔG_{ATP}? NMR studies (Allen et al., 1985) have shown that while [ATP] may fall only slightly during this period [ADP] and [Pi] rise and, from equation 7 that their rise will lower ΔG_{ATP}. Calculations suggest that this will be from 60 to about 47 kJ mol^{-1}, (depending on pH_i, ADP calculation etc). From these calculations the authors concluded that S.R. calcium transport should not be compromised in hypoxia and is not therefore a likely cause of cardiac failure in hypoxia. A similar NMR study of skeletal muscle showed that only a small fall in ΔG_{ATP} occurs during fatigue and was unlikely to be the cause of fatigue (Dawson et al., 1978). It is suggested that this type of approach, where NMR measured parameters can be taken at several different experimental points and then related to work e.g. transport, may shed light in other areas of transport studies.

1-H NMR

Hydrogen is ubiquitous in biological systems. Furthermore 1-H is a naturally abundant and sensitive NMR nucleus. These features have made it ideal for NMR *imaging* of the body. However biomedical NMR spectroscopic studies have been hampered because the *water* signal (55 M) is so enormous compared to that of all other substances, that their detection ordinarily becomes impossible. Drying samples and then using organic solvent or replacing H_2O with D_2O is one approach to the water problem - but is of rather limited usefulness. Recently several water-suppression techniques have been described. These work by either selectively pre-saturating the water resonance as described already and/or minimizing excitation of the H_2O signal by using a pulse sequence which provides almost zero excitation at the water frequency (a"1331" sequence, see e.g. Gadian, 1982). These pulse sequences largely "remove" the water signal allowing other 1-H metabolites to be observed. As there are so many 1-H metabolites, signal overlap and identification may still cause problems. Several spectral editing techniques have been described which can help to overcome this problem. For example, there will be some resonances which have magnetic interactions ("coupling") with nearby nuclei. Spectra can be accumulated under conditions that remove some particular interaction between nuclei (decoupled). If unmodified spectra are subtracted from the decoupled ones, the difference spectra will show only nuclei in which the decoupled interaction is present (see e.g. Williams et al., 1985).

The potential of 1-H spectroscopy is enormous although it has not yet caught up with 31-P spectroscopy in biomedicine. Studies have been

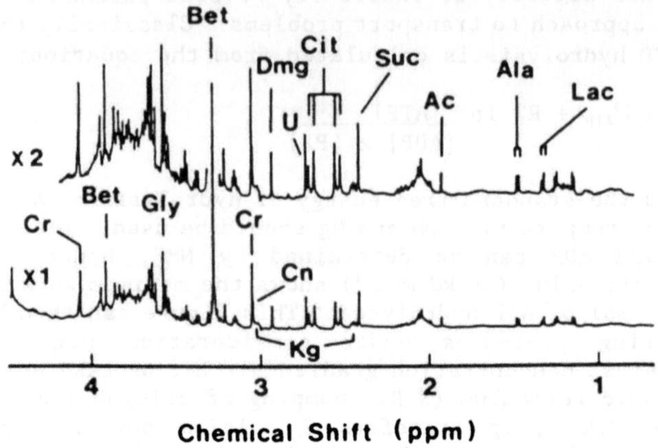

Fig. 3. The 1-H NMR spectrum, in the aliphatic region, from urine of a normal but preterm infant. The multiple peaks obtained should be compared to those in the 31-P spectrum of Fig. 1. Note also the chemical shift range of just 3 ppm. Field strength 11.7 T, 1 scan (gated water suppression). From Brown et al., 1989.

made of many body fluids in health and disease,(e.g. Brown et al., 1989). Figure 3 shows part of the NMR spectrum of urine from a premature baby. *In vivo* studies of various organs have also been made. In brain for example accumulation of lactate during hypoxic episodes has been demonstrated, *in vivo* (e.g. Behar et al., 1984). Changes in 1-H metabolites with benign and malignant cerebral tumours are being described. The application of 1-H NMR to difussional studies of water and solvent transport will be dealt with in a separate chapter by Boyd and King in this book.

13-C NMR

As mentioned earlier, 13-C is *not* the naturally abundant isotope of carbon and its sensitivity is less than that of 31-P. In addition coupling with nearby protons reduces the 13-C signals by splitting resonances into multiple smaller ones. This makes 13-C NMR spectroscopy difficult. To be successful the 13-C containing compounds must be present at high concentration. If this is not the case then the only chance is to introduce ("enrich") 13-C compounds.

Because carbon is present in many compounds and there is the potential to investigate the transport of carbon atoms through metabolic paths, effort has been expended and considerable progress made in 13-C spectroscopy. It should be noted that the chemical shift range of 13-C is large, around 200 ppm. This means that, in contrast to 1-H (range 10 ppm) the many 13-C resonances are likely to be well resolved in the spectra. Furthermore the *absence* of 13-C metabolites in spectra, unless enrichment occurs, can be advantageous in that simpler spectra are obtained and changes easily monitored. Finally, using variations of the spectral editing techniques described for 1-H (see above), 13-C metabolites can be seen via their coupling to 1-H. The decoupling and spectral subtraction techniques described for 1-H were homo-nuclear i.e. 1-H to 1-H interactions. The same principles can be used for two different nuclei (heteronuclear editing), exploiting the coupling between 13-C and 1-H. 13-C nuclei are irradiated, and the 1-H spectra observed. Thus one gains the great sensitivity of the 1-H nucleus and the selectivity of the 13-C label.

Natural abundance (i.e. not enriched), 13-C studies have been made mainly on either glycogen or fats, because of their high concentration (50 mM) (Alger & Shulman, 1984b). As explained earlier glycogen is 100% NMR visible. This was quite a startling discovery given its M.W. of ~ 10^7, and has been explained by glycogen molecules having rapid internal segmental motion (Tr $\simeq 10^{-8}$s) (Alger & Shulman, 1984b). Many *in vivo* studies have now been made on glycogen dynamics in man and animals under a variety of physiological and pathological conditions e.g. glucagon, phosphorylase kinase - deficiency. Similarly the effect of dietary fat intake on fatty acid metabolism and membrane composition is an example of natural abundance 13-C spectroscopy of fats.

While there was agreement that the decoupling process described above, would undoubtably improve 13-C spectra in man, there has been debate as to whether too much power would be deposited in the tissue being studied. However working at low field strength (e.g. 1.5 T) and optimizing RF coil configurations it appears possible to perform safe

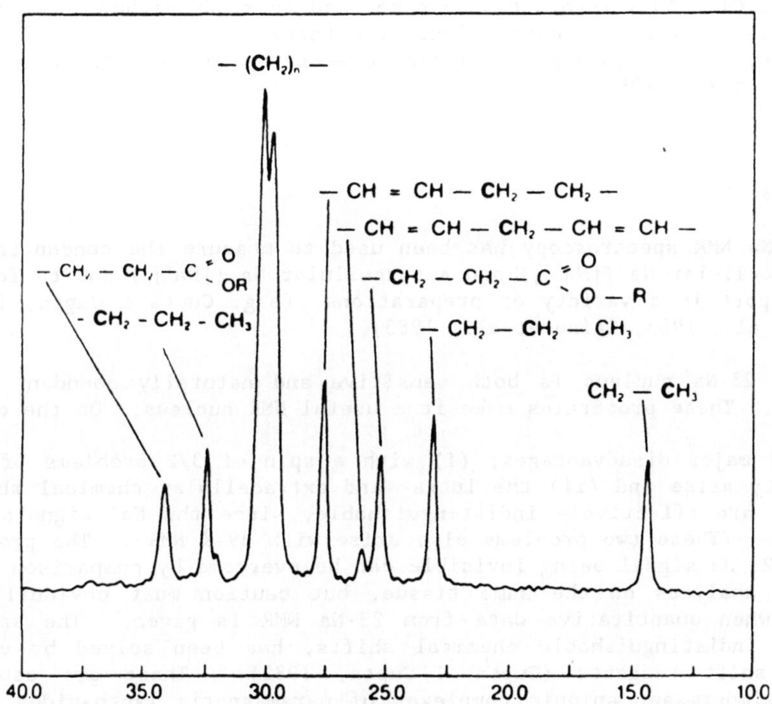

Fig. 4. The methyl and methylene part of a proton decoupled 13-C NMR spectrum from human calf subcutaneous adipose tissue. Field strength 1.5 T, 180 scans, Repetition time 3 s. From Heerschap et al., 1989.

broadband proton decoupling in humans. This has recently been done for muscle, liver and subcutaneous fat and an example is shown in Fig. 4. With this technique signals from histidine, amino acids, ethanolanine, choline and creatine became visible, along with the expected glycogen and triacylglycerol carbon resonances (Heerschap et al., 1989).

In contrast to these *in vivo* studies, natural abundance 13-C NMR spectra from excised tissues have shown a great many 13-C resonances e.g. 55 in rat brain, including many amino acids. However the spectra take at least 8 hr to accumulate making it impractical for many studies, (Barany et al., 1985).

The use of 13-C enriched compounds is at present limited to animals because of the high cost of 13-C enriched compounds. One estimate is that it would cost $10,000 to study man by 13-C enriched NMR spectroscopy (Shulman, 1987). In animals 13-C labelled compounds can be used to examine fluxes and products of various metabolic pathways. The rate of incorporation or degradation of label is observed along with its position in the carbon skeleton. An example of 13-C enrichment studies is in the study of metabolism of brain tumours, *in vivo*. Tumour metabolism is thought to be more anaerobic than that of normal tissue. An infusion of 13-C glucose labelled at C1 into rats showed the production of C3 labelled lactic acid, C1 labelled glycogen and C2, C3 and C4 glutamate/glutamine. Glutaminolysis was found to contribute 9% to the lactate produced by the tumour cells (Ross et al., 1988). A large number of fluxes in metabolic paths are currently being studied by NMR. Another example of the use of 13-C spectroscopy has been to study lipase activity. NMR is able to distinguish whether fatty acids are bound to a glycerol backbone or are free fatty acids due to their different chemical shifts of C1. This information can be used to study either the rate of hydrolysis of triacylglycerols into free fatty acids or triacylglycerol synthesis, e.g. triacylglycerol lipase activity in the plasma of rats (Diamand et al., 1989).

23-Na NMR

23-Na NMR spectroscopy has been used to measure the concentration of intracellular Na ($[Na^+]_i$) and extracellular Na ($[Na^+]_o$) and to follow Na transport in a variety of preparations, (e.g. Gupta & Gupta, 1982, Civan et al., 1983, Ogino et al., 1983).

The 23-Na nucleus is both sensitive and naturally abundant (see Table 1). These properties make it a useful NMR nucleus. On the other hand
it has 2 major disadvantages; (i) with a spin of 3/2 problems of NMR visibility arise and (ii) the intra- and extracellular chemical shifts of 23-Na are effectively indistinguishable, since the Na^+ signals co-resonate. (These two problems also arise with 39-K NMR). The problem of some 23-Na signal being invisible can be overcome by comparison with chemical analyses on the same tissue, but caution must obviously be applied when quantitative data from 23-Na NMR is given. The second problem, indistinguishable chemical shifts, has been solved by using aqueous shift reagents (Gupta & Gupta, 1982). These are usually membrane-impermeant anionic complexes of paramagnetic lanthanides e.g. dysprosium tripolyphosphate (DyTP). The shift reagent, by binding Na close to the paramagnetic moiety, alters the environment of the cation sufficiently that its resonance frequency is shifted by several ($\simeq$ 10) ppm from the unadulterated cation signal. Thus the intra and extracellular 23-Na signals can be separated, as can be seen in Fig. 5. Obviously such reagents must remain either entirely inside or outside the

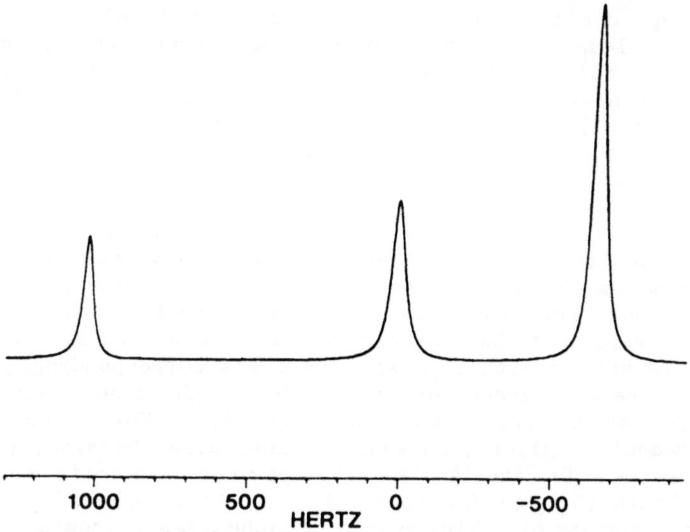

Fig. 5. 23-Na NMR spectrum of Na loaded red blood cell suspension. The peak at 0 Hz is from intracellular Na. The peak at 700 Hz is from extracellular Na. (Reference compound at 1050 Hz.) Field strength 9.4 T, 500 scans, repetition time, 0.2 s. From Knubovets et al., 1989.

cell and be stable if quantification is to be undertaken. In whole organ studies there may be a question mark over whether the shift reagent has penetrated all the extracellular space, especially when this is somewhat tortuous e.g. the heart. Furthermore, the reagents must not be toxic to cells. Whereas most studies on DyTP in muscle and brain have suggested that toxicity is not a problem, there is evidence from *in vivo* studies that there is stability problems when tissues have pyrophosphatase activity and there is evidence of toxicity in the rat kidney, (Endre, et al, 1989). An additional problem with 23-Na NMR is that the extracellular Na signal is so much larger than the intracellular, so that the latter may be difficult to detect, under physiological conditions.

23 Na NMR spectroscopy has been used to study ion transport in several preparations. Using red blood cells as our example, several workers have now used 23-Na to measure transmembrane sodium fluxes. These studies have all used shift reagents and human erythrocytes but have described quite different intracellular sodium NMR "visibility", from 75% to 100%. This problem remains unresolved. Red blood cells have characteristically low Na^+ fluxes. Nevertheless ouabain-sensitive and insensitive Na^+ fluxes have been measured with NMR, and a maximal rate of Na^+ efflux of around 10 $mmole^{-1}$ h^{-1} l^{-1} cells measured (Knubovets et al, 1989).

14-N and 15-N NMR

As nitrogen is widely distributed in biological systems and both 14-N and 15-N possess spin there has been considerable interest in the possibilities of Nitrogen NMR spectroscopy in biomedicine and plant physiology. However, there are considerable difficulties and it is fair to say that, so far, it has found little application.

It can be seen from Table 1 that these nuclei are not very

Rapid pulsing, due to relative short T_1's can help offset this to some extent and, at least for 14-N which is naturally abundant, signals can be obtained in a relatively short period (minutes). But for 14-N the quadrupolar spin means that most metabolites give signals that are too broad (100 p.p.m.) to be useful. The extremely low abundance of 15-N coupled with low sensitivity make signal detection in living systems, without enrichment, too much of an uphill struggle.

Having said what the drawbacks are, I would now like to briefly point out what can be done. With 14-N some compounds give surprisingly narrow lines. This is thought to be because these compounds are roughly symmetrical about the 14-N atom, (in terms of electronic environment). This in turn means that the electric quadrupole constant is smaller, hence relaxation is not so efficient and lines are correspondingly narrower. Thus for a select group of compounds, high resolution 14-N NMR spectroscopy can be performed, (see Fig. 6). This group includes, nitrates, ammonium, glycerylphospho-choline, urea, betaine, choline and some amino acids. To date 14-N spectroscopy has been performed on liver, muscle and brain (Balaban & Knepper, 1983). An example of its use was a study of the response of living muscle (*Xenopus*, gastrocnemius) to osmotic stress. It is known that *Xenopus* responds to the increased osmolarity by concentrating urea in its tissues. It is also known that urea can denature proteins. A 14-N NMR study showed that in vivo the effects of urea are counterbalanced by production of glycerylphosphocholine and betaine. Both substances increased in the gastrocnemii with increasing plasma urea concentration. $NaNO_3$, which does not penetrate muscle fibres and is ordinarily absent from animal cells was used as a reference for chemical shifts and concentration (Wray & Wilkie, 1987).

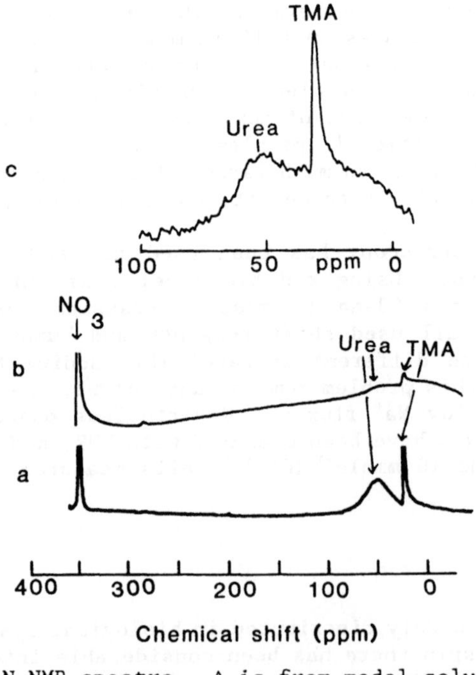

Fig. 6. 14-N NMR spectra. A is from model solution, B & C are from *Xenopus* gastrocnemius. C is an expanded version of B. TMA; trimethylamine. The nitrate signal is truncated. Field strength 4.7 T, 400 scans repetition time 1 s. (Wilkie & Wray, 1987).

Biochemical 15-N NMR studies have been made for over a decade for example to study proteins and nucleic acids, and usually involve 15-N enrichment (Mason, 1981). As nitrogen lacks a reasonably long-lived radioisotope, 15-N is a potentially important tracer of nitrogenous metabolism. The first use of 15-N NMR spectroscopy in whole organ metabolism has been reported; in rat liver, *in vivo* using 15-N glycine enrichment (Grunder et al., 1989). The fungus *neurospora crassa* provides a good example of how 15-N can be used to study transport. When fed on medium containing 0.2% 15-NH_4Cl many well resolved amino acids were seen in the resulting 15-N spectrum. The effect of different nitrogen sources in the growth medium, on the rate of assimilation of 15-N into glutamic acid, alanine and glutamine was followed. Competition for glutamic acid was found to limit alanine synthesis

when glutamine synthesis was allowed to occur (Kanamori & Roberts, 1983). Thus 15-N NMR could be useful for studying nitrogen regulation and transport *in vivo*.

19-F NMR

After 1-H, 19-F is the most sensitive NMR nucleus. This along with a spin of 1/2 and 100% natural abundance make it an attractive NMR nucleus. As pointed out earlier, there are no endogenous 19-F signals as biological systems do not contain fluorine. This limitation can be turned to an advantage by introducing 19-F into a biological system and measuring transport and metabolism without complications from existing signals. In this way 19-F NMR has found several useful applications which include pharmacokinetics, pH_i and $[Ca^{2+}]_i$ measurements.

Several drugs and anaesthetic agents contain fluorine. Therefore if 19-F NMR signals can be detected, their transport, distribution and metabolism can be monitored. The commonly used anaesthetic gases halothane and isoflurane have multiple fluorine nuclei. The appearance of the anaesthetic in the brain and correlation with depth of anaesthesia *in vivo* has been monitored (e.g. Evers et al., 1987). In a recent study, cerebral blood flow (cbf), was calculated by following the wash out of halothane from the brain, upon cessation of its administration. Changes in cbf with various drugs and physiological stimuli e.g. hypercapnia were measured and were in good agreement with published values using more established methods. Care has to be taken to minimise any 19-F signal from cranium and scalp by choosing a pulse angle which picks up very little signal from the surface (Rudin & Sauter, 1989).

So far two types of drugs have been studied with 19-F; a variety of anti-cancer drugs and aldose reductase inhibitors used in diabetes. One of the first *in vivo* 19-F NMR studies was on the clinically important antitumour drug, fluorouracil (Stevens et al, 1984). Investigators were able to establish how well the drug was taken up by an implanted tumour in rats and how it was handled by the liver. Interestingly when 2 analogues of fluorouracil, floxuridine and doxifluridin which are also potent anticancer drugs, were studied, fluoro nucleotides were *not* seen in the tumour spectra. This suggests that these drugs may be exerting their effect via other mechanisms e.g. by production of fluoro-deoxyuridine monophosphate which inhibit thymidine synthetase. There have also been 19-F studies of polyamine-related, fluorine containing compounds which are potential tumour markers e.g. difluoromethylornithine which is an enzyme activated, irreversible inhibitor of ornithine decarboxylase (Hull et al, 1988). In diabetes, the aldose reductase sorbitol pathway has been implicated in the pathogenesis of various diabetic complications and hence there is much interest in clinical agents capable of inhibiting aldose reductase, e.g. sorbinol and cyclandelate. For the reason already outlined, NMR can add significantly to our knowledge of *in vivo* pharmacokinetics, and has been used recently to look at aldose reductase inhibitors. The study found significant discrepancies between the *in vitro* and *in vivo* potencies (Kwee & Nakada, 1989).

Measurement of pH_i with 31-P NMR spectroscopy has already been described. However there are tissues where Pi is present at extremely low levels making it unsuitable as a pH_i indicator. This was one of the main reasons behind the development of 19-F pH indicators. The 19-F nucleus was chosen for the reason already mentioned i.e. high sensitivity, $I = 1/2$, and 100% natural abundance. The principle of the pH_i determination is exactly the same as for Pi i.e. the chemical shift of the 19-F signal must be strongly dependent on whether a nearby acidic group is protonated. The large chemical shift range of 19-F, (see Fig.7), is another advantage to pH determination with this nucleus. Unlike Pi, the 19-F indicators have to be introduced into the system and proved to be non-toxic, not interfere with normal cellular functioning, and have pK's in the physiological range. In the last few years much effort and research has gone into the development and improvement of such indicators e.g. fluoroanilines (Deutsch & Taylor, 1987). These indicators have been used in a variety of tissues and cells including, lymphocytes, hepatocytes and frog skin.

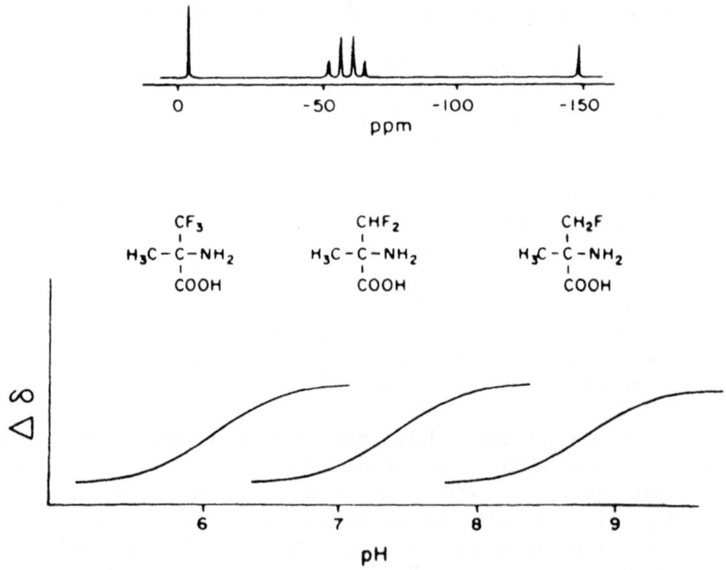

Fig. 7. 19-F NMR resonance frequencies and pH dependence of the chemical shifts (centre peak spacing for the difluoromethyl alanine) of trifluoro-α-methylalanine, difluoro-α-methylalanine and monofluoro-α-methylalanine. The structure of each molecule is shown with its corresponding resonance frequency above. The pH titration curves are shown below. From Deutsch & Taylor, 1987.

The enormous importance of calcium in controlling many cellular activities has meant that many techniques have been developed for its measurement, (see chapter by Valdeolmillos & Eisner, this book). Using NMR spectroscopy, calcium (and other divalent cations) can be measured by combining it intracellularly with a 19-F containing compound, fluoro-BAPTA (FBAPTA, Smith et al, 1983). BAPTA is a fluorescent Ca indicator and it and FBAPTA are analogues of EGTA. The basis of the technique is that there is a large chemical shift when FBAPTA binds calcium. Thus two signals are obtained; one from FBAPTA and one from Ca-FBAPTA. From these signals the free and complexed FBAPTA area can be calculated. The free [Ca] can then be determined via the binding constant for the reaction. The effects of Mg^{2+} below 10mM and other contaminating divalents e.g. Zn, Fe do not appear to upset the measurement of [Ca] (Smith et al., 1983). As with fluorescent Ca indicators, FBAPTA can be introduced into the cell as an ester with the ester groups being hydrolysed inside the cell to release the impermeant indicator. The disadvantage of this method of measuring $[Ca^{2+}]_i$ is that mM concentrations of the chelator must be used, thus significantly increasing

intracellular Ca buffering. It may also alter cell function. (These considerations may apply but to a lesser degree, to fluorescent measurements). The advantage over fluorescent methods is that inaccessible tissues or opaque cells may be studied. In addition the 19-F NMR measurements provide a *direct* identification of the bound ions and does not suffer from autofluorescence. The approach is being extended to monitor other ions via 19-F signals e.g. magnesium (Rotevatn, et al., 1989).

CONCLUSIONS

This chapter has outlined the basis of NMR spectroscopy and detailed the type of information obtained from spectroscopy. It has also shown the usefulness and limitations of the NMR nuclei which are being used for biological studies. *Inter alia* I have attempted to emphasise how this technique is applicable to transport studies. In the future these applications should expand allowing new avenues of research to open up.

REFERENCES

Alger, J.R. & Shulman, R.G., 1984a. NMR studies of enzymatic rates *in vitro* and *in vivo* by magnetization transfer. Q. Rev. Biophys. 17, 83-124.

Alger, J.R. & Shulman, R.G., 1984b. Metabolic applications of high-resolution 13C nuclear magnetic resonance spectroscopy. Brit. Med. Bull. 40, 160-164.

Allen, D.G., Morris, P.G., Orchard, C.H. & Pirolo, J.S., 1985. A nuclear magnetic resonance study of metabolism in the ferret heart during hypoxia and inhibition of glycolysis. J. Physiol. 361, 185-204.

Balaban, R.S. & Knepper, M.A., 1983. Nitrogen-14 nuclear magnetic resonance spectroscopy of mammalian tissues. Am. J. Physiol. 245, C439-444.

Barany, M., Ans, C. & Chang, Y.C., 1985. Natural abundance ^{13}C NMR of brain. Mag. Res. Med. 2, 289-295.

Behar, K.L., Rothman, D.L., Shulman, R.G., Petroff, O.A.C. & Prichard, J.W., 1984. Detection of cerebral lactate *in vivo* during hypoxemia by 1H NMR at relatively low field strengths (1.9T). Proc. Natn. Acad. Sci. USA. 81, 2517-2519.

Berling, C. & Hall, C.D., 1989. An NMR assessment of the rheological properties of blood and its constituents:A review. NMR Biomed. 1, 1-6.

Brown, J.C.C., Mills, G.A., Sadler, P.J & Walker, V., 1989. ^{1}H NMR studies of urine from premature and sick babies. Mag. Res. Med. 11, 193-201.

Civan, M.M., Degani, H., Margalit, Y. & Shporer, M. 1983. Observations of 23-Na in frog skin by NMR. Am.J. Physiol. 245,C213-C219.

Dawson, M.J., Gadian, D.G. & Wilkie, D.R., 1978. Muscular fatigue investigated by phosphorus nuclear magnetic resonance. Nature 274, 861-866.

Dawson, M.J. & Wilkie, D. R., 1984. Muscle and brain metabolism studied by 31-P nuclear magnetic resonance In:Recent Adv. Physiol. 10, 247-276.

Dawson, M.J. & Wray, S., 1985. The effects of pregnancy and parturition on phosphorous metabolites in rat uterus studied by ^{31}P nuclear magnetic resonance. J. Physiol. 368, 19-31.

Delpy, D.T., Cope, M.C., Cady, E. Hamilton, P. Reynolds, E.O.R., Wray, S. & Wyatt, J., 1987. Cerebral monitoring in neonates by NMR and NIR spectroscopy. Scand J. Clin. Chem. 47, Suppl. 188, 9-17.

Deutsch, C. J. & Taylor, J.S., 1987. Intracellular pH as measured by ^{19}F NMR. Annals. New York Acad. Sci. 508, 33-47.

Diamand, R. J., Bradbury, E. M. & Cox, K.L., 1989. Determination of triaclyglycerol lipase activity using carbon-13-labelled triacylglycerols and nuclear magnetic resonance spectroscopy: Evidence that hepatic lipase hydrolyses medium-chain triacylglycerols. Mag. Res. Med. 9, 273-277.

Endre, Z.H., Allis, J.L. & Radda, G.K., 1989. Toxicity of dysprosium shift reagents in the isolated perfused rat kidney. Mag. Res. Med. 11, 267-274.

Evers, A.S., Berkowitz, B.A. & d'Avignon, D.A., 1987. Correlation between the anaesthetic effect of halothane and saturable binding in brain. Nature 328, 157-170.

Gadian, D.G. 1982. Nuclear Magnetic Resonance and its Applications to living Systems. Clarendon Press, Oxford.

Gadian, D.G., Radda, G.K., Brown, T.R., Chance, E.M., Dawson, M.J. & Wilkie, D.R., 1981. The activity of creatine kinase in frog skeletal muscle studied by saturation-transfer nuclear magnetic resonance. Biochem. J. 194, 215-228.

Gadian, D.G., Radda, G.K., Dawson, M.J. & Wilkie, D.R. 1982. pH measurements of cardiac and skeletal muscle using 31-P NMR. p 61-77 In: Intracellular pH:Its Measurement, Regulation and Utilization in Cellular Function, edited by R. Nuccitelli and D.W. Deamer. Liss, New York.

Grunder, W., Krumbiegel, P., Buchali, K. & Blesin, H.J., 1989. Nitrogen -15 NMR studies of rat liver *in vitro* and *in vivo*. Phys. Med. Biol. 34, 457-463.

Gupta, R.K. & Gupta, P. 1982. Direct observation of resolved resonances from intra- and extracellular sodium-23 ions in NMR studies of intact cells and tissues using dysprosium (111) tripolyphosphate as paramagnetic shift reagent. J. Magn. Res. 47, 344-350.

Heerschap, A., Luyten, P.R., Vander Heyden, J.I., Oosterwaac, L.J.M.P. & Denhollander, J.A., 1989. Broadband proton decoupled natural abundance ^{13}C NMR spectroscopy of humans at 1.5T. NMR Biomed. 2, 124-132.

Hull, W.E., Kunz, W., Port, R.E. & Seiler, N., 1988. Chain-fluorinated polyamines as tumour markers. III. Determination of geminal difluoropolyamines and their precursor 2-2-difluoroputrescine in normal tissues and experimental tumours by in vitro and in vivo ^{19}F NMR spectroscopy. NMR Biomed. 1, 11-19.

Kanamori, K. & Roberts, J.D., 1983. ^{15}N NMR studies of biological systems. Acc. Chem. Res. 16, 35-41.

Knubovets, T.L., Revazov, A.V., Sibeldina, L.A. & Eichhoff, V., 1989. ^{23}Na NMR measurement of the maximal rate of active sodium efflux from human red blood cells. Mag. Res. Med. 9, 261-272.

Kwee, I.L. & Nakada, T., 1989. *In vivo* pharmacokinetics of aldose reductase inhibitors: 3-fluoro-3 deoxy-D-glucose NMR studies in rat brain. NMR Biomed. 2, 44-46.

Mason, J. 1981. Nitrogen nuclear magnetic resonance spectroscopy in inorganic, organometallic, and bioinorganic chemistry. Chem. Rev. 8, 205-227.

Ogino, T., den Hollander, J.A., Shulmann, R.G. 1983 ^{39}K, ^{23}Na, ^{31}P NMR studies of ion transport in Saccharomyces cerevisiae. Proc. Natn. Acad.Sci. USA, 80, 5185-5189.

Ogino, T., Shulman, G.I., Avison, M.S., Gullans, S.R., Den Hollander, J.A. & Shulman, R.G., 1988. Proc. Natl. Adac. Sci. USA. 82, 1099 -

Potts, J.R., Kirk, K. & Kuchel, P.W., 1989. Characterization of the transport of the nonelectrolyte dimethyl methylphosphonate across the red cell membrane. NMR Biomed. 1, 198-204.

Reynolds, E.O.R., Wyatt, J.S., Azzopardi, D., Delpy, D.T., Cady, E.B., Cope, M. & Wray, S., 1988. New non-invasive methods for assessing brain oxygenation and haemodynamics. Brit Med. Bull. 44, 1052-1075.

Ross, B.D., Higgins, R.J., Boggan, J.E., Willis, J.A., Knittel, B. & Unger, S.W., 1988. Carbohydrate metabolism of the rat C6 glioma. An *in vivo* ^{13}C and *in vitro* ^{1}H magnetic resonance spectroscopy study. NMR Biomed. 1, 20-26

Rotevatn, S., Murphy, E., Levy, L.A., Raju, B., Liberman, M. & London, R.E. 1989. Cytosolic free magnesium concentration in cultured chick heart cells. Am. J. Physiol. 257, C141-C146.

Rudin, M. & Sauter, A., 1989. Non-invasive determination of cerebral blood flow changes by 19-F NMR spectroscopy. NMR Biomed. 2, 98-103.

Shulman, R.G., 1987. Contributions of ^{13}C and ^{1}H NMR to Physiological control. Annals New York Acad. Sci. 508, 10-15.

Smith, G.A., Hesketh, R.T., Metcalfe, J.C., Feeney, J. & Morris, P.G., 1983. Intracellular calcium measurements by ^{19}F NMR of fluorine-labelled chelators. Proc. Natn. Acad. Sci. 80, 7178-7182.

Stadler, H. & Fuldner, H.H., 1981. 31-P NMR analysis of ATP in synaptic vesicles and its relationship to "in vivo" conditions. Biomed. Res. 2, 673-676.

Stevens, A.N., Morris, P.G., Iles, R.A., Sheldon, P.W. & Griffiths, J.R., 1984. 5- fluorouracil metabolism monitored by ^{19}F NMR. Br. J. Cancer. 50, 113-117.

Tofts, P.J. & Wray, S. 1985. Changes in brain phosphorus metabolites during the postnatal development of the rat. J. Physiol. 359, 417-429.

Tofts, P.S. & Wray, S., 1988. A critical assessment of methods of measuring metabolite concentrations by NMR spectroscopy. Mag. Res. Med. 6, 84-86

Vogel, H.S., 1987. NMR studies of phosphite uptake and storage in plant cells and algae in "Physiological NMR Spectroscopy" ed. S.M. Cohen Annals New York Acad. Sci. 508, 164-175.

Wilkie, D.R. & Wray, S. 1987. 14-nitrogen nuclear magnetic (NMR) spectroscopy of isolated living Xenopus muscle. J. Physiol. 390, 227P.

Williams, S.R., Gadian, D.G., Proctor, E., Sprague, D.B. & Talbot, D.F. 1985. Proton NMR studies of muscle metabolites *in vivo*. J. Mag. Res. 63, 406-411.

Wray, S., 1988a. Regulation of intracellular pH in rat uterine smooth muscle, studied by 31-phosphorus NMR spectroscopy. Biochem. Biophys. Acta. 972, 299-301.

Wray, S., 1988b. Smooth muscle intracellular pH: Measurement, regulation and function. Am. J. Physiol. 254, C213-C225.

Wray, S., 1990. The effects of metabolic inhibition on uterine metabolism and intracellular pH in the rat. J. Physiol. 423, 411-423.

Wray, S. & Tofts, P. 1986. Direct *in vivo* measurement of metabolite concentrations using ^{31}P nuclear magnetic resonance spectroscopy. Biochem. Biophys. Acta. 886, 399-405.

Wray, S. & Wilkie, D.R. 1987. Some novel uses for 14-nitrogen NMR spectroscopy. Soc. Mag. Res. Biol. Med. 6th Annual Meeting, 2, 597.

16

PROTON NMR STUDIES OF TRANSMEMBRANE SOLUTE TRANSPORT

G. F. King[1] and C.A.R. Boyd[2]

1) Dept of Biochemistry, Univ. of Sydney, Sydney, NSW 2006, Australia

2) Dept of Human Anatomy, University of Oxford, South Parks Road Oxford OX1 3QX, England

INTRODUCTION

Although the ability of nuclear magnetic resonance (NMR) spectroscopy to monitor chemical reactions non-invasively in intact cells and tissues has been well documented (Kuchel, 1981; Gadian, 1982; Brindle and Campbell, 1987) surprisingly few attempts have been made to review the wealth of NMR techniques available for studying transmembrane solute transport. It is the aim of this Chapter to summarise these techniques and to provide practical examples of each to guide the non-expert NMR user in the choice of appropriate NMR experiments for the transport system to be investigated. Discussion will be restricted to ^{1}H NMR studies, since ^{1}H (proton) nuclei are the most NMR-responsive and are present in most biologically important molecules; those readers interested in studies employing other nuclei, such as the alkali metal ions, should consult an excellent recent review (Kirk, 1990).

GENERAL CONCEPTS IN ^{1}H NMR STUDIES OF BIOLOGICAL SYSTEMS

The theory underlying the application of NMR to biological systems has been extensively reviewed (Kuchel, 1981; Brindle and Campbell, 1987) and will not be reiterated in detail. NMR spectroscopy involves excitation of nuclei to a high energy state by absorption of radio-frequency (rf) radiation. The frequency of rf radiation required to excite a particular nucleus is known as its resonance frequency, which determines the chemical shift position of the signal from that nucleus in the frequency-domain spectrum. The longitudinal relaxation rate, $1/T_1$, characterises the rate of return of the excited nuclei to thermal equilibrium. The transverse relaxation rate, $1/T_2$, characterises the rate of loss of 'phase coherence' of the population of excited nuclei and thus determines the duration for which the NMR signal can be observed. There are two major problems which need to be addressed in ^{1}H NMR studies of biological systems:

High concentration of water protons

Proton NMR spectra of cell suspensions obtained in non-deuterated solvents are dominated by a signal (resonance) from water protons. The concentration of water protons is about 110 M whereas most metabolites are present at millimolar concentrations or less; the ratio of the ^1H_2O resonance to that of most metabolites in thus some 10^5:1. This represents a significant problem for the spectrometer's analogue-to-digital converter and, to avoid poor digitisation of weak metabolite resonances, the ^1H_2O signal needs to be significantly reduced. A common method of achieving this is to wash cells in ^2H_2O (heavy water). However, this may introduce kinetic isotope effects and in any transport mechanism where proton transfer is likely to be rate-limiting, this procedure should be avoided.

A more useful (and cheaper) approach which is available on modern spectrometers is to suppress the water resonance by selective irradiation immediately prior to the observation pulse; this enables adequate suppression to be achieved without the complications inherent in those methods designed for selective non-excitation of water (Kuchel, 1989). However it should be realised when employing selective irradiation that resonances of protons in rapid exchange with water will also be suppressed.

Spectral overlap

The ^1H NMR spectrum of a suspension of cells obtained using a single rf collection pulse consists of an amorphous envelope of overlapping resonances; a classic example is the spectrum obtained in this way from human erythrocytes (Fig. 1A).

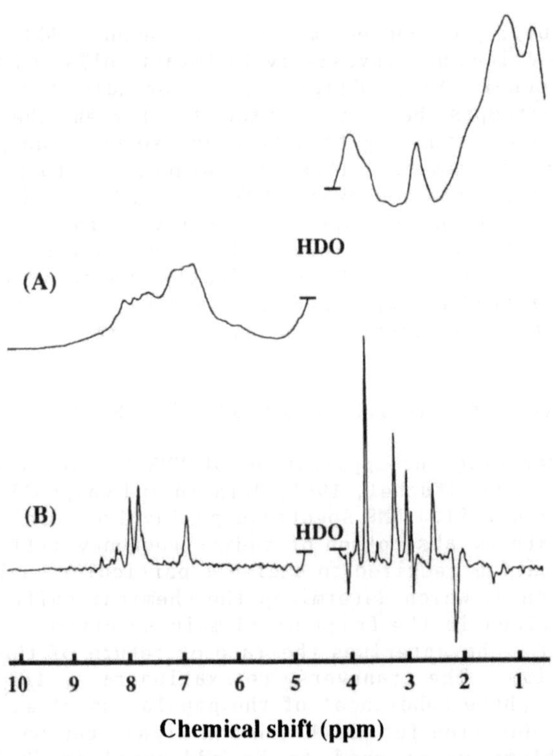

Fig. 1. 270 MHz ^1H NMR spectra of glucose-depleted human erythrocytes obtained at 37 ^{0} C using (A) a single rf pulse: (b) the spin-echo pulse sequence with 2τ = 120 ms. The sharp peaks in (B) correspond to resonances of small metabolites such as glutathione, creatine, ergothioneine, and lactate. HDO denotes the residual water signal obtained after washing the cells in deuterated water. The chemical shift, measured in parts per million (ppm) is relative to 2,2-dimethyl-2-silapentane 5-sulphonate (DSS). Adapted from Figs. 1 and 2 of Brown, et al., (1977)

The envelope of resonances is dominated by signals from proteins and membrane phospholipids which obscure metabolite resonances. There are two methods which allow selective editing of the spectrum so that only resonances from small metabolites are observed: (i) The spin-echo method (Brown et al., 1977) in which the spectrum is collected using the spin-echo pulse sequence (Rabenstein, 1978). This pulse sequence excites the nuclei and then allows a short time (2τ) for transverse relaxation (T_2) prior to spectral acquisition. Large, immobile species such as proteins and membrane lipids undergo rapid T_2 relaxation relative to small metabolites because of their slow solution tumbling (correlation) times. Hence, judicious choice of the period allowed for relaxation will lead to discrimination against resonances from large molecules. This is exemplified in Fig. 1B, which shows a proton spin-echo NMR spectrum of human erythrocytes obtained using 2τ = 120 ms. The spin-echo technique has been used to acquire ^{1}H NMR spectra from a variety of cell systems, including erythrocytes (Brown et al., 1977), enterocytes (Boyd et al., 1988), hepatocytes (Nicholson et al., 1985), and brain homogenates (Middlehurst et al., 1984). (ii) The protein-saturation-pulse method (Rabenstein, 1984) in which protein resonances are selectively irradiated just prior to spectral acquisition; this reduces the intensity of most of the protein envelope by transfer of saturation through cross-relaxation. This technique has proved particularly useful for acquisition of spectra from erythrocytes (Rabenstein, 1984; Guy et al., 1986) since the ^{1}H NMR spectrum of these cells is dominated by resonances from 5mM intracellular haemoglobin (Hb); consequently irradiation of a single Hb resonance eliminates the greater part of the protein envelope. However, this technique does not reduce the intensity of resonances from membrane phospholipids and is unlikely to be successful in cells where there is no single dominant protein. It has not found widespread application in cellular systems other than erythrocytes.

COMPARTMENTAL DIFFERENTIATION IN NMR STUDIES OF MEMBRANE TRANSPORT

All experimental methods for measuring a membrane transport process require that there be some means of differentiating between the populations of the transported (permeant) species on either side of the membrane. As summarised in Fig. 2, most, but not all, NMR-based methods rely on transmembrane differences in either the chemical shift (resonance frequency), nuclear relaxation rate, or diffusion properties of the transported species as the basis for distinguishing between its internal and external populations. This section, which is a necessary prelude to consideration of the various NMR methods available for monitoring membrane transport processes, is devoted to factors which facilitate these differences. Only a brief summary of those factors shown to have practical relevance to proton NMR studies is given below; for a more general listing see the review by Kirk (1990).

Factors facilitating transmembrane differences in chemical shift

Preamble NMR transport studies in which compartmental discrimination is based on transmembrane chemical shift differences provide simultaneous measurement of the internal and external concentrations of the permeant. This provides two distinct advantages over NMR studies employing other methods of compartmental discrimination. Perhaps the most obvious is that the transport process can be monitored in two ways using either the internal or external resonances of the permeant, thus providing an internal check on the measured transport rate. The second advantage is not widely appreciated and relates to quantitation of the permeant concentration in each compartment. There are numerous experimental methods for achieving this (Tofts and Wray, 1988), but most authors have relied on some form of intensity calibration obtained by adding known concentrations of permeant to the cell or vesicle system (Beilharz et al., 1984); however, as discussed below, this method is often based on unfounded assumptions.

Fortunately, the ability to observe separate resonances from each compartment simultaneously allows derivation of a simple expression based on the principle of conservation of mass from which the external and internal permeant concentrations can be calculated (King et al., 1986; Vandenberg et al., 1986). If we define P_{total} as the total moles of permeant added to the system and P_{ext} and P_{int} as the moles of permeant in the external or internal compartments, respectively, at any time t after addition, then:

$$P_{total} = P_{ext} + P_{int} \tag{1}$$

The NMR signal intensity (I) is proportional to the amount of permeant giving rise to the signal. If we define the coefficient of proportionality (or NMR extinction coefficient) as ζ (mol/intensity units) then from Eqn (1) we have:

$$P_{total} = \zeta_{ext}I_{ext} + \zeta_{int}I_{int} \tag{2}$$

Rearranging Eqn (2) gives:

$$I_{int} = -(\zeta_{ext}/\zeta_{int})I_{ext} + P_{total}/\zeta_{int} \tag{3}$$

Thus, simple regression of a straight line onto all pairs of I_{int} versus I_{ext} data will yield the 'best fit' NMR extinction coefficients from the intercept (P_{total}/ζ_{int}) and slope $-(\zeta_{ext}/\zeta_{int})$. If V_{ext} and V_{int} are the volumes of the external and internal compartments, respectively, then the respective permeant concentrations in these compartments equals ($\zeta_{ext}I_{ext}/V_{ext}$) and ($\zeta_{int}I_{int}/V_{int}$).

Use of 'best-fit' NMR extinction coefficients has several advantages over the intensity calibration method. First, the intensity calibration method often assumes that ζ_{ext} and ζ_{int} are equal, which is often not true, especially in spin-echo experiments (see section on paramagnetic relaxation reagents). On the other hand, the extinction coefficient method does not assume any particular relationship between ζ_{ext} and ζ_{int}. Second, in many instances, the intensity calibration method assumes that there is a unique set of extinction coefficients for the resonances of a permeant. In practice, these may vary from sample to sample due to variations, for example, in paramagnetic impurities and magnetic field gradients. This problem is obviated in the extinction coefficient method because a new set of extinction coefficients is calculated for each experiment. Inaccuracies should only arise using this method if the extinction coefficients change during the course of the transport experiment.

<u>pH</u> A ^{1}H nucleus proximal to a titratable group will give rise to a resonance whose chemical shift varies markedly when the pH is varied within a pH unit of the pK_a. This occurs because the protonated and non-protonated species exchange rapidly on the NMR timescale such that a single resultant chemical shift for the nucleus is obtained which is the population-weighted average of the signals from the two species. Altering the pH in the region of the pK_a will significantly alter the relative amounts of the protonated and non-protonated forms which in turn alters the weighted average chemical shift of the nucleus; this forms the basis by which the ^{1}H NMR chemical shifts of the ring protons of imidazole (pK_a = 7.0) have been used to monitor the intracellular pH of erythrocytes (Rabenstein and Isab, 1982).

Populations of a permeant in the internal and external compartments may thus have different chemical shifts if a transmembrane pH gradient exists and if the pH in at least one compartment is near the pK_a of a titratable group on the permeant. Application of this technique to cell suspensions would generally require that the permeant contains a titratable group with a pK_a in the range 6-8 since, although most cell-types maintain a significant transmembrane pH gradient, both the intra- and extracellular pH are close to neutral; consequently, very few ^{1}H NMR studies on cell suspensions have used pH as the basis for distinguishing between the internal and external populations of the permeant. Conversely artificial phospholipid vesicles are more tolerant to extremes of pH and the intra- and extravesicular pH, as well as the transmembrane pH gradient, can be imposed in such a way as to satisfy the necessary conditions for discriminating between the intra- and extravesicular populations of a permeant. For example, in a suspension of small unilamellar vesicles (SUVs) prepared with an intravesicular pH of 7.0, Cramer and Prestegard (1977) were able to discriminate between the intra- and extravesicular ^{1}H NMR resonance of fumarate (pK_a = 5.9) by adjusting the extravesicular pH to 4.7 or 5.5 (i.e., in the region of the pK_a of fumarate).

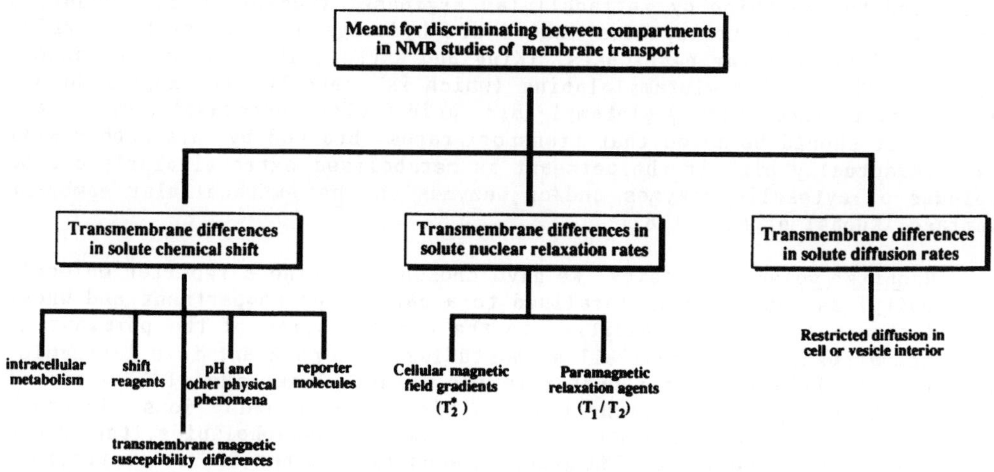

Fig. 2. A summary of the major experimental methods (shown in boxes) for discriminating between compartments in proton NMR studies of membrane transport. Factors facilitating these means of compartmental discrimination are also listed.

Paramagnetic shift reagents The NMR resonance of a nucleus can be significantly altered by close spatial proximity to a variety of paramagnetic ions, especially the lanthanides. The strict requirement for close spatial proximity has resulted in most success being obtained in NMR spectra of alkali metal ions (e.g., Na^{+}) when the alkali metal ion forms a complex with an anionic chelate of the paramagnetic ion; a good example is the large upfield (low frequency) shift of the ^{23}Na resonance resulting from Na^{+} binding to the $Dy(PPP_i)_2^{7-}$ complex (PPP_i=tripolyphosphate) (Gupta and Gupta, 1982; King et al., 1988).

Unfortunately, there have been few reports of significant shifts of ^{1}H NMR resonances induces by paramagnetic ions; the major reason is the difficulty of facilitating specific interactions between proton nuclei and paramagnetic ions. Alger and Prestegard (1979) showed that Pr^{3+} caused a significant downfield (high frequency) shift of the methyl resonance of acetate; consequently, the addition of $Pr(NO_3)^3$ to the extravesicular space of a suspension of large unilamellar vesicles (LUVs) enabled the intra- and extravesicular acetate resonances to be separated. More recently, it has been reported that the 1,8-diamino-3,6,10,13,16,19-hexazobicyclo[6,6,6] icosane cobalt chloride complex specifically interacts with and shifts the ^1H and ^{13}C NMR resonances of α-hydroxycarboxylates such as lactate (Hockings et al., 1989). It remains to be seen how useful this complex will be for ^1H NMR transport studies.

Intracellular metabolism Permeants which are substrates for intracellular biochemical pathways may undergo enzymic modification upon entering the cell. Such covalent modification is likely to alter the chemical shift of ^1H NMR resonances arising from nuclei close to the modification site, enabling spectral discrimination between the unaltered extracellular permeant and the covalently modified intracellular species. This procedure has been used to monitor the uptake by human erythrocytes of arginine (which is converted to ornithine by intracellular arginase) (Kuchel et al., 1984), a variety of di- and tripeptides (which are hydrolysed to constituent amino acids by intracellular peptidases) (King and Kuchel, 1984, 1985; Vandenberg et al., 1985), and γ-glutamylalanine (which is converted to oxoproline and alanine by intracellular γ-glutamyl-amino acid cyclotransferase) (York et al. 1984). It should be noted that transport rates obtained by this method will be artefactually high if the permeant is metabolised extracellularly due to release of cytosolic enzymes and/or enzymes on the extracellular membrane surface (Kuchel et al., 1987).

Reporter molecules/moeity We have chosen to define a reporter molecule (or moeity) as one which is localised to a particular compartment and whose ^1H NMR chemical shift is sensitive to the concentration of the permeant in that compartment. In almost all such studies, the permeant is a lanthanide ion (usually Pr^{3+}) and the reporter is the trimethylammonium (TMA) group of membrane-bound phosphatidylcholine. Since lanthanide ions interact specifically with the phosphate groups of membrane phospholipids, the ^1H NMR chemical shift of the nearby TMA group is sensitive to the Pr^{3+} concentration. Thus, so long as the $[Pr^{3+}]$ differs in the internal and external compartments, the chemical shifts of the TMA resonances from the inner and outer lipid monolayers will be different; calibration of the internal $[Pr^{3+}]$ is usually based on the chemical shift difference between these resonances (Hunt, 1980; Hunt and Jones, 1982; Hunt et al., 1978, 1984; Jones et al., 1985).

EDTA has also been used as a ^1H NMR reporter molecule (Yoon and Sharp, 1985). The two methylene resonances of EDTA, which is membrane-impermeant, shift 0.5 ppm upfield (i.e., to lower frequency) as a result of Ca^{2+} binding (Nicholson et al. 1983). Yoon and Sharp (1985) resealed chromaffin granule ghosts in the presence of EDTA to provide a means of discriminating between intra- and extravesicular Ca^{2+}; in this system external Ca^{2+} is NMR-invisible while the internal Ca^{2+} concentration can be gauged from the intensity of resonances from the intravesicular Ca^{2+}-EDTA complex.

When the transmembrane movement of protons is being studied any species in the internal and/or external compartment which contains a titratable group with a pK_a near the pH of the compartment in which it is located will act as the reporter. Thus Yoon and Sharp (1985) were able to monitor the ATPase-mediated flux of protons into EDTA-loaded chromaffin granule ghosts (initial intravesicular pH = 7.0) by monitoring the ^1H NMR chemical shifts of the methylene protons of internal (free) EDTA (pK_a = 6.6). Clearly transmembrane proton fluxes are readily monitored by any technique capable of monitoring compartmental pH changes; however, NMR has the particular advantage of allowing fluxes of protons and other permeants to be examined simultaneously. For example in the study of Yoon and Sharp (1985) the marked difference in the chemical shifts of the methylene protons of free EDTA and the Ca^{2+}-EDTA complex was used simultaneously to monitor Ca^{2+} entry (as described above); this enabled the authors to show that Ca^{2+} uptake in chromaffin granule ghosts is not linked to proton extrusion. Simultaneous ^1H NMR measurement of proton and permeant uptake should be particularly useful for measuring the stoichiometry of proton cotransport systems; we have had some success in preliminary ^1H NMR experiments designed to measure the cotransport of protons and peptides in suspensions of isolated enterocytes (Quirk et al., 1990).

Magnetic susceptibility It is doubtful whether compartmental chemical shift discrimination based on transmembrane magnetic susceptibility differences are of general application in ^1H NMR studies of membrane transport; they are only mentioned here for the sake of completeness and as a prelude to a subsequent discussion of their relationship to magnetic field gradients across cell and vesicle membranes (see section on paramagnetic relaxation reagents).

The chemical shift of a nucleus is proportional to the net magnetic field which it experiences, which in turn depends (amongst other factors) upon the bulk magnetic suceptibility (χ) of the medium in which the nucleus resides. Thus, if the internal and external compartments have different magnetic susceptibilities, the internal and external chemical shifts for all nuclei will be different. The extent of the compartmental chemical shift difference depends intimately on the geometry of the cells or vesicles and the extent of the transmembrane difference in magnetic susceptibility ($\Delta\chi$). For most cells, $\Delta\chi$ is minimal and the intra- and extracellular chemical shifts are not sufficiently different for practical use as a means of compartmental discrimination. However, because of the high concentration of intracellular haemoglobin, $\Delta\chi$ may be quite significant in the case of erythrocytes; this is especially true if the haemoglobin is in the slightly paramagnetic deoxy form (Fabry and San George, 1983).

Theoretical calculations in which oblate spheroids with χ equivalent to deoxygenated red cells were suspended in a solution with χ equal to that of physiological saline indicated that the intra- and extracellular chemical shifts of a solute could differ by as much as 0.064 ppm depending on the orientation of the spheroid with respect to the field (Kuchel and Bulliman, 1990). Interestingly, this chemical shift difference is of the same order of magnitude as that observed for small phosphoryl compounds in suspensions of oxygenated human erythrocytes (Labotka and Omachi, 1987; Kirk and Kuchel, 1988). To date, no studies have specifically reported compartmental ^1H chemical shift differences resulting from transmembrane differences in magnetic susceptibility; however, $\Delta\chi$ is presumably at least partly responsible for the separate intra- and extracellular resonances obtained for dimethylsulphoxide and α-aminoisobutyrate in suspensions of anoxic chicken erythrocytes (Brown et al., 1983).

Factors contributing to transmembrane differences in nuclear relaxation rates

Paramagnetic relaxation reagents (T1 and T2 differences) The dominant relaxation mechanism for ^1H nuclei is generally the interaction between fluctuating dipole moments of adjacent nuclei. However, the rate of both T_1 and T_2 relaxation can be enhanced by addition of paramagnetic molecules which contain unpaired electron spin. The magnetic dipole moment of an unpaired electron spin is 2000 times more intense than typical nuclear magnetic dipole moments; consequently, proximity of a nucleus to an unpaired electron spin provides a very effective means of dipole-dipole relaxation (Freeman, 1988).

Compartmental relaxation rates are usually differentiated by adding an impermeant paramagnetic species to one of the compartments. For example, dextran-magnetite (Ashley and Goldstein, 1980) and Mn^{2+}, either free (Conlon and Outred, 1972) or bound to albumin (Fabry and Eisenstadt, 1975), have been found to reduce the extracellular T_2 of water protons when added to the external compartment of cell suspensions. In experiments with vesicles, the intracellular T_2 can be decreased by resealing in the presence of paramagnetic reagents such as the $Gd(EDTA)_2^{5-}$ complex (Bar-On and Degani, 1985); very recently Steward et al (1990) have used this method to measure the diffusive water permeability of a complex epithelium, the rabbit salivary gland. In some studies the aim of adding the paramagnetic species to a particular compartment is to decrease T_2 to such an extent that the linewidth ($=1/\pi T_2$) is so broad that NMR resonances from solutes in that compartment have negligible amplitude (Degani et al., 1981).

Magnetic field gradients (T_2* differences) The amplitude of an NMR signal obtained using the spin-echo pulse sequence (see section 2.2) is reduced if the solute diffuses through magnetic field gradients during the time required to produce the echo (Brindle et al., 1979). For an uncoupled nucleus diffusing through a linear magnetic field gradient, the spin-echo amplitude at time 2τ after the initial rf pulse, denoted by $I(2\tau)$, is given by (Brindle et al., 1979; Brown, 1983; Kirk, 1990):

$$I(2\tau)=I(0)\exp[(-2\tau/T_2) - (2D\gamma^2G^2\tau^3/3)] = I(0)\exp(-2\tau/T_2^*) \quad (4)$$

where (I)0 is the spin-echo amplitude when $\tau = 0$, γ is the magnetogyric ratio, D is the diffusion coefficient, G is the magnitude of the magnetic field gradient, and T_2 is the intrinsic transverse relaxation time of the nucleus. In the simplified expression, T_2^* is simply defined as the time constant characterising the decay of signal amplitude during the spin-echo experiment, which includes contributions from both the intrinsic T_2 and from spin diffusion through the magnetic field gradients.

Non-uniform magnetic fields commonly occur in suspensions of cells and vesicles as the result of differences in the bulk magnetic susceptibility of the inner and outer compartments (see Section 3.1.5). Studies on diamagnetic glass beads (Glasel and Lee, 1974) and model calculations using oblate and prolate spheroids (Kuchel and Bulliman, 1990) indicate that the magnetic field is always uniform in the inner compartment and inhomogeneous in the external compartment; to illustrate this Fig. 3 shows the lines of magnetic force for two different orientations (with respect to the external magnetic field) of a prolate spheroid with a magnetic susceptibility significantly different to the external compartment. Thus, T_2^* will always be shorter for nuclear species in the external compartment (because G is larger) and consequently the amplitude of an extracellular resonance will always be less than that of the corresponding intracellular resonance at any given value of τ in the spin-echo experiment. In fact, experiments with chicken erythrocytes (Brown, 1983) demonstrated that T_2^* for all molecules is determined primarily by their compartmental location so that the spatial location of a molecule can be assigned simply on the basis of its measured T_2^*.

The compartmental difference in T_2^* can be accentuated by the addition of impermeant paramagnetic species to the extracellular medium to increase external magnetic susceptibility and hence the magnitude of the magnetic field gradient. The most commonly used reagents are ferrioxamine (Fe^{3+} ligated to the iron-chelator desferrioxamine) (Jones and Kuchel, 1980; King et al., 1983) and the Dy^{3+}-diethyltriaminepenta-acetate (Dy-DTPA) complex (Brown et al., 1983; Brindle et al., 1979).

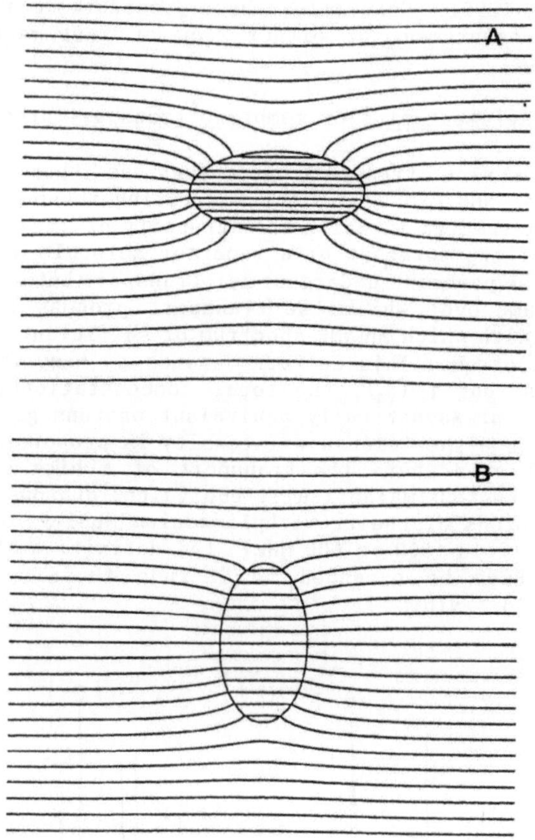

Fig. 3. Calculated magnetic lines of force for a prolate spheroid in a uniform magnetic field with its main axis oriented (A) parallel and (B) perpendicular to the field. For these calculations (Kuchel and Bulliman, 1990), the magnetic permeability of the spheroid interior was set to 100 times that of the exterior. The equally spaced lines of force inside the spheroid indicate that the interior field is uniform; this is not true of the external space, especially at the spheroid interface.

Factors contributing to transmembrane differences in diffusion properties

Whereas the diffusion of a molecule in the external space of a suspension of cells or vesicles is relatively unrestricted, diffusion within the internal compartment may be restricted. Consequently for an intracellular molecule if diffusion is studied over a longer period than the time taken for the molecule to traverse the internal compartment, the apparent translational diffusion coefficient will be less than that obtained for unrestricted diffusion.

PROTON NMR METHODS FOR MONITORING MEMBRANE TRANSPORT

The versatility of NMR spectroscopy for membrane transport studies lies in its ability to monitor solute permeabilities spanning a wide range of timescales. Specifically, membrane transport processes occurring over timescales ranging from milliseconds to hours can be monitored by appropriate

NMR procedures (Kirk, 1990). Generally, these procedures can be divided into those which enable measurement of either slow or fast membrane transport processes.

Methods enabling measurement of slow membrane transport processes

We shall define slow transport processes as those occurring over timescales of > 1 min. NMR procedures for monitoring these processes rely on acquiring a series of spectra in which the intensity or chemical shift of one or more resonances of the permeant or a reporter molecule varies in a time-dependent manner as the permeant undergoes net transport between compartments. The minimum time-frame over which the transport process can be monitored depends on the speed with which an NMR spectrum of sufficient quality (signal-to-noise) can be collected. This in turn depends on many factors including the magnetic field strength (B_0), the total concentration of the permeant ([P]), and the number of magnetically equivalent protons giving rise to the NMR resonance (N), since the resonance intensity is proportional to B_0, [P], and N. In an ideal case where the transport of a high concentration of permeant containing numerous magnetically equivalent protons (e.g., the nine equivalent methyl protons of the trimethylammonium moeity of choline) could be monitored at high field (500 or 600 MHz), the definition of slow transport processes could perhaps be expanded to include those occurring over a timescale of 30 sec to 1 min.

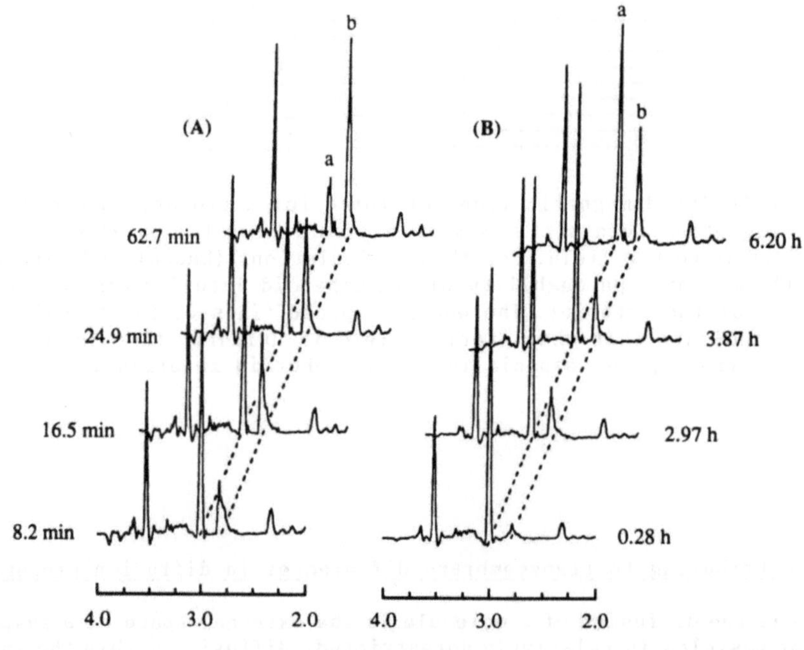

Fig. 4. Series of 400 MHz ^1H spin-echo NMR spectra (2τ = 120 ms) obtained from (A) a lysate of human erythrocytes to which arginine (~9 mM) was added at t = 0 min; (B) intact human erythrocytes to which arginine (~11.5 mmol/l extracellular fluid) was added at t = 0 min. The spectra were obtained at 37°C with 1 s presaturation of the residual water resonance. The time adjacent to each spectrum represents the mid-point of the spectral acquisition period; note the more protracted time scale for the intact cell experiment. Arginase activity is readily monitored Chemical shifts are relative to DSS. Adapted from Fig.1 of Kuchel et al. (1984)

Methods based on transmembrane differences in chemical shift The slow transport of a solute exhibiting separate internal and external NMR resonances can be readily monitored from the time- dependent changes in the resonance intensity of either or both of these resonances. For example this technique has been used to monitor the uptake of arginine by human erythrocytes (Kuchel et al., 1984). Fig. 4A shows a series of spin-echo ^1H NMR spectra obtained when arginine was added to a lysate of human erythrocytes; the arginase-catalysed conversion of arginine to ornithine is readily monitored from the decreasing intensity of the arginine δ-CH_2 resonance and the concomitant increase in intensity of the ornithine δ-CH_2 resonance. Fig. 4B shows part of the spectral timecourse obtained when arginine was added to a suspension of intact human erythrocytes. Conversion of arginine to ornithine is again apparent, but the timecourse is clearly more protracted.

The rate of arginine-to-ornithine conversion in the cell suspension was interpreted as resulting from rate-limiting transport of arginine into the cell followed by its intracellular hydrolysis (Kuchel et al., 1984); this interpretation is valid so long as the rate of the intracellular hydrolysis is substantially faster than the transport rate (King and Kuchel, 1984). Since this is so intracellular arginine hydrolysis may be considered to be effectively instantaneous, so that the increase in ornithine resonance intensity reflects arginine entry. Concentrations were calibrated using the 'extinction coefficient method' (see section "Preamble") and, based on the time-dependence of the extracellular arginine concentration, the authors calculated an arginine transport rate of 0.5 mmol/h per litre of cells, which compares well with a previous estimate (0.4 mmol/h per litre of cells) obtained using radiotracer methods (Young and Ellory, 1977).
from the declining intensity of the arginine δ-CH_2 resonance (a) and the concomitant increase in intensity of the ornithine δ-CH_2 resonance (b).

Other slow transport processes monitored using NMR timecourse methods based on compartmental chemical shift differences (where the nucleus monitored is on the transported species) include the uptake of dipeptides (King and Kuchel, 1984, 1985), tripeptides (Vandenberg et al., 1985), and γ-Glu-Ala (York et al., 1984) by human erythrocytes, and the diffusion of maleate and fumarate into vesicles (Cramer and Prestegard, 1977; Prestegard et al., 1979).

Slow transport processes have also been monitored using NMR spectroscopy by utilising compartmental differences in the chemical shifts of reporter nuclei (i.e. nuclei not covalently attached to the permeant). Transmembrane Pr^{3+} fluxes have been determined utilising the trimethylammonium (TMA) group of membrane-bound phosphatidylcholine as a reporter moeity (Hunt, 1980; Hunt and Jones, 1982; Hunt et al., 1978; Jones et al., 1985). As discussed briefly in section on report molecules/moeity, addition of paramagnetic Pr^{3+} to a suspension of phosphatidylcholine (PC) vesicles induces a downfield shift of the methyl proton resonance resulting from the TMA moeity of phosphatidylcholine in the outer lipid monolayer (TMA_{ext}); this resonance is consequently shifted well away from the TMA resonance resulting from the inner monolayer (TMA_{int}) (Fig. 5A,B).

In the absence of mediators of Pr^{3+} transport, the chemical shift separation (Δδ) between these resonances remains constant for many hours (Hunt et al., 1978; Jones et al., 1985), indicating that PC vesicles are impermeant to Pr^{3+}. However, Δδ decreases in a time-dependent manner when the (slow) transport of Pr^{3+} into vesicles is mediated by a variety of agents such as indole-3-acetic acid (Fig. 5C); this is due to a gradual downfield shift of TMA_{int} as the intravesicular $[Pr^{3+}]$ increases. Due to the high V_{ext}/V_{int} ratio in these experiments, $[Pr^{3+}]_{ext}$ is essentially constant throughout the timecourse. Thus, a calibration curve relating Δδ to $[Pr^{3+}]_{int}$ can be obtained by preparing a series of vesicle suspensions in which $[Pr^{3+}]_{ext}$ is constant and the concentration of entrapped intravesicular Pr^{3+} is varied (Hunt, 1980; Hunt and Jones, 1982; Hunt et al., 1978, 1984).

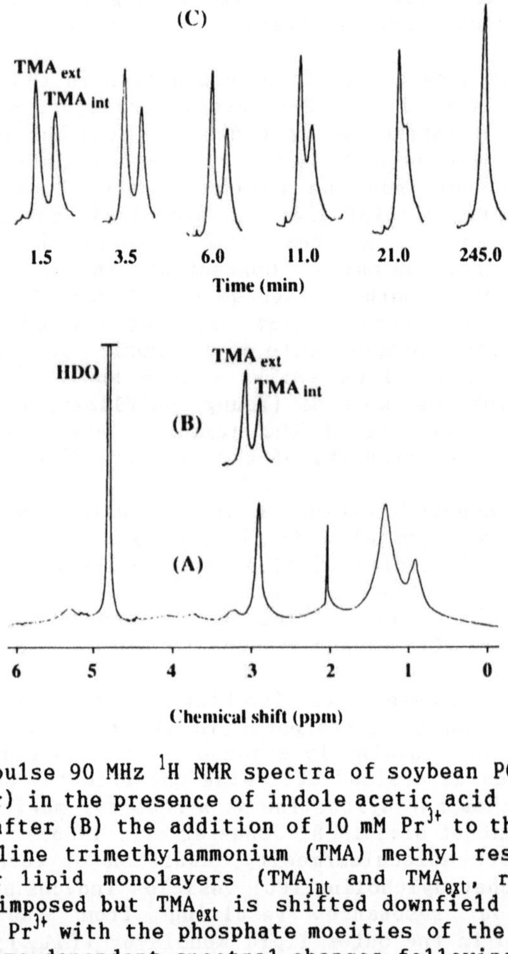

Fig. 5. Single pulse 90 MHz ^1H NMR spectra of soybean PC vesicles (28 nm average diameter) in the presence of indole acetic acid (0.5 molar ratio) before (A) and after (B) the addition of 10 mM Pr^{3+} to the extravesicular milieu. The choline trimethylammonium (TMA) methyl resonances from the inner and outer lipid monolayers (TMA_{int} and TMA_{ext}, respectively) are initially superimposed but TMA_{ext} is shifted downfield upon interaction of paramagnetic Pr^{3+} with the phosphate moeities of the outer monolayer. (C) shows the time-dependent spectral changes following addition of Pr^{3+} at t = 0 min: TMA_{int} gradually shifts downfield, without a change in intensity, until it finally merges with TMA_{ext}. Adapted from Figs 2 and 3 of Jones et al. (1985)

Thus a distinction should be drawn between this methodology, which relies on time-dependent changes in the chemical shift of reporter nuclei, and experiments involving direct observation of internal and external permeant resonances in which the transport rate is obtained from changes in resonance intensity. The timecourse depicted in Fig. 5C, where the internal TMA resonances shifts smoothly downfield towards the external TMA resonance without a change in intensity or linewidth, is characteristic of a "slow-leakage" transport mechanism in which, for all vesicles, one or at most a few ions at a time are transported (Ting et al., 1981; Westerhoff, 1982). However, the appearance of the NMR spectral timecourse can vary markedly depending on the exact mode of ion transport; Ting et al. (1981) summarises and provides practical examples of the modes of transport which can be distinguished using this NMR technique and develops a theoretical framework with which NMR spectral timecourses can be interpreted in terms of fundamental parameters of the transport system.

Methods based on transmembrane differences in T_2^* Transmembrane solute fluxes can be monitored from changes in ^1H NMR spin-echo intensity as the solute is transported between compartments, provided that the magnetic susceptibilities of the internal and external compartments differ significantly enough to cause a compartmental variation in T_2^* (see section 3.2.2). As discussed in Section 3.1.6, transmembrane differences in magnetic susceptibility rarely enable compartmental chemical shift discrimination in ^1H NMR spectra and consequently the internal and external resonances of the solute are almost always superimposed in these studies.

Brindle and coworkers pioneered this NMR technique with their study of D- and L-alanine uptake by human erythrocytes (Brindle et al., 1979). As shown in Fig. 6A, addition of L-alanine to a suspension of human erythrocytes led to a time-dependent increase in spin-echo intensity of the alanine β-CH_3 resonance as the amino acid was transported into the cell; this occurs because of the longer T_2^* inside the cells relative to the extracellular compartment (see Section 3.2.2). The change in spin-echo intensity was accentuated when the external magnetic susceptibility was increased by addition of Dy-DTPA.

A procedure for extracting the permeability coefficient for the membrane transport process from the NMR data can be derived as follows: For 'zero-trans' influx, the total amount (mol) of permeant in the internal compartment (P_{int}) at any time t (in sec) is given by the equation (Kirk, 1990):

$$P_{int}(t) = \{V_{int} (P_{int} + P_{ext})/(V_{int} + V_{ext})\} \{1-\exp[-PA((1/V_{ext}) + (1/V_{int}))t]\} \quad (5)$$

where P_{ext} is the amount of permeant in the external compartment (mol), P is the permeability coefficient (cm/s), A is the total membrane surface area (cm^2) across which the flux occurs, and V_{int} and V_{ext} are the volumes (cm^3) of the internal and external compartments, respectively. If we define the NMR extinction coefficients as in Section 3.1.1, then the change in signal intensity when a mole of permeant moves from the external to internal compartment is simply $\zeta_{int} - \zeta_{ext}$. Hence, if we define $P_{int}(t)$ as the moles of permeant in the internal compartment at any time t, and I(0) and I(t) as the signal intensity at zero time and time t, respectively, then:

$$P_{int}(t) \ (\zeta_{int} - \zeta_{ext}) = I(t) - I(0) \quad (6)$$

If $P_{int,eq}$ and I(eq) are the moles of permeant in the internal compartment and the signal intensity at chemical equilibrium, respectively, then:

$$P_{int,eq}(\ _{int} - \ _{ext}) = I(eq) - I(0) \quad (7)$$

Furthermore, if we define:

$$\Delta(t) = P_{int}(t)/P_{int,eq} = [I(t) - I(0)]/[I(eq) - I(0)] \quad (8)$$

then substituting into Eqn (6), rearranging, and defining t½ as the time at which Δ (t) = 0.5, yields the following equation for the permeability coefficient:

$$P=(\ln 2/t½)\ [(A/V_{ext}) + (A/V_{int})]^{-1} \quad (9)$$

Thus, assuming A, V_{int} and V_{ext} are known, P is simply obtained by noting the value of t½ from a plot of Δ(t) against time. This is shown in Fig. 6B for the L-alanine influx experiment described above. This method has been used to measure D- and L-alanine (Brindle et al., 1979), lactate (Brindle et al., 1979), and choline (Jones and Kuchel, 1980) transport by human erythrocytes, the uptake of DMSO, 3-O-methylglucose, thymidine, and α-aminoisobutyrate by chicken erythrocytes (Brown et al., 1983), and the diffusional permeability of glycerol (Brown et al., 1982). A modification of the method has been used to monitor Gly-Gly uptake by human erythrocytes (King et al., 1983).

<u>Methods based on transmembrane differences in T_2</u> These NMR methods are usually based on complete supression of the intensity of permeant resonances in either the internal or external compartment using impermeant paramagnetic reagents. In an unusual variation of this method, Degani and coworkers (Degani, 1978; Degani et al., 1981) measured the transport of the paramagnetic reagent itself, Mn^{2+}, by utilising its effect on the T_2 of TMA methyl proton resonances arising from the inner and outer monolayer of vesicle membranes (TMA_{int} and TMA_{ext}, respectively).

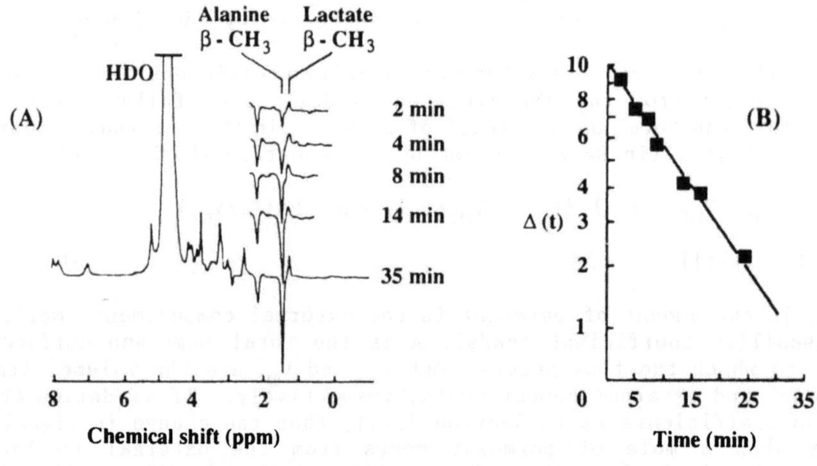

Fig. 6. (A) A series of 270 MHz ^1H spin-echo NMR spectra (2τ = 120 ms) obtained at 37°C from a suspension of ^2H_2O-washed human erythrocytes (haematocrit = 80%) to which L-alanine (~27 mmol/l extracellular fluid) was added at t = 0 min. The β-methyl resonances of L-alanine and D-lactate are marked, as is the residual water resonance (HDO). The time adjacent to each spectrum represents the mid-point of the spectral acquisition period. Note the time-dependent increase in the intensity of the L-alanine β-methyl resonance. A logarithmic plot of Δt-versus-time obtained for the alanine β-methyl resonance from this experiment is shown in (B): t½ estimated from this plot is ~11 min. Adapted from Figs. 3 and 4 of Brindle et al. (1979).

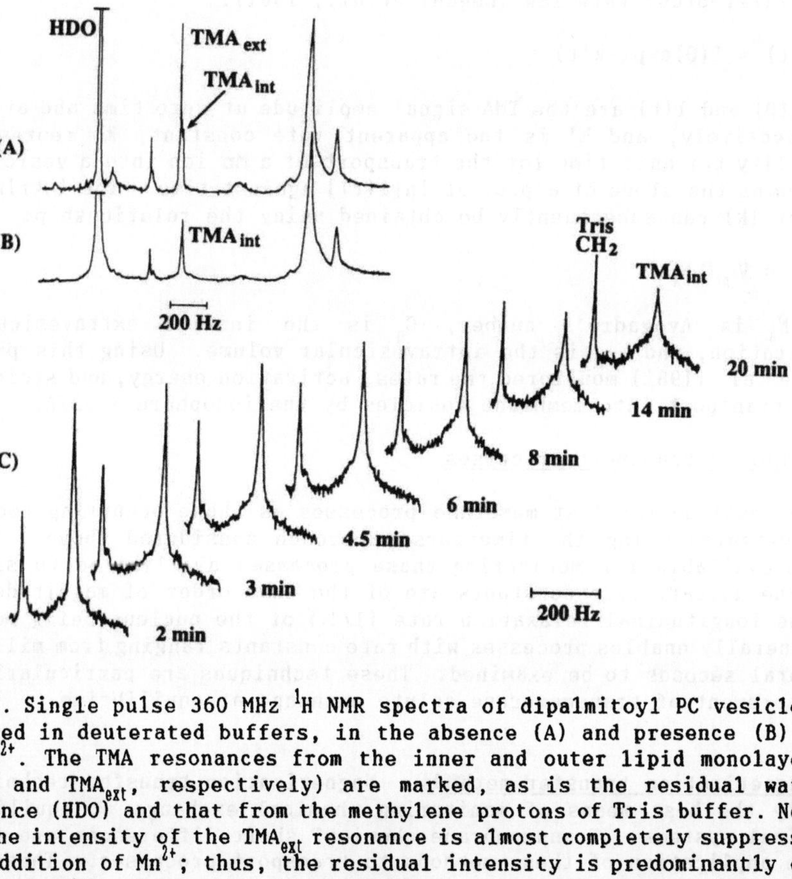

Fig. 7. Single pulse 360 MHz ^1H NMR spectra of dipalmitoyl PC vesicles, prepared in deuterated buffers, in the absence (A) and presence (B) of 1mM Mn^{2+}. The TMA resonances from the inner and outer lipid monolayers (TMA_{int} and TMA_{ext}, respectively) are marked, as is the residual water resonance (HDO) and that from the methylene protons of Tris buffer. Note that the intensity of the TMA_{ext} resonance is almost completely suppressed upon addition of Mn^{2+}; thus, the residual intensity is predominantly due to TMA_{int}. (C) shows the time-dependent spectral changes following addition of Mn^{2+} at t = 0 min: the intensity of TMA_{int} gradually diminishes without a concomitant change in linewidth or chemical shift. Adapted from Figs. 1 and 2 of Degani et al. (1981).

In this experiment there is no shift reagent in either compartment, so the TMA_{int} and TMA_{ext} resonances are almost coincident. However, addition of millimolar concentrations of Mn^{2+} to the vesicle suspension causes a dramatic reduction in the T_2 of TMA_{ext} with the result that this resonance is broadened to the extent that the signal amplitude is negligible. The remaining amplitude is thus due to TMA_{int} (Fig. 7A,B). Because the internal vesicle volume is sufficiently small the influx of one Mn^{2+} is sufficient to suppress the intensity of TMA_{int}. Thus, as Mn^{2+} permeates there is a gradual decline in the TMA resonance amplitude, without a concomitant increase in linewidth (Fig. 7C).

The time dependence of the TMA signal amplitude is described by the simple first-order rate law (Degani et al., 1981):

$$I(t) = I(0)\exp(-k't) \tag{10}$$

where I(0) and I(t) are the TMA signal amplitude at zero time and at any time t, respectively, and k' is the apparent rate constant; k' represents the probability per unit time for the transport of a Mn ion into a vesicle and is obtained as the slope of a plot of ln[I(t)] against time. The intrinsic rate constant (k) can subsequently be obtained using the relationship:

$$k' = V_{int} C_0 N_A k \tag{11}$$

where N_A is Avogadro's number, C_0 is the initial extravesicular Mn^{2+} concentration, and V_{int} is the intravesicular volume. Using this procedure, Degani et al. (1981) monitored the rates, activation energy, and stoichiometry of Mn^{2+} transport into membrane vesicles by the ionophore X-537A.

Fast membrane transport processes

We shall define fast membrane processes as those occurring too rapidly to be measured using the timecourse approach considered above. Most NMR methods available for monitoring these processes are limited to situations where the unitary rate constants are of the same order of magnitude or less than the longitudinal relaxation rate ($1/T_1$) of the nucleus being monitored; this generally enables processes with rate constants ranging from milliseconds to several seconds to be examined. These techniques are particularly suited to measurement of transmembrane solute exchange at equilibrium.

Magnetization transfer methods   Magnetization transfer techniques are probably the best means of monitoring chemical exchange at equilibrium in biological systems. Transmembrane chemical shift differences are mandatory for the application of these methods to transport processes. These involve either saturating or inverting a resonance from permeant residing in one compartment and then monitoring the transfer of saturation or inversion to the resonance of the same nucleus from permeant in another compartment.

Alger and Prestegard (1979) used the technique of inversion transfer to monitor the equilibrium exchange of acetic acid in a suspension of vesicles to which $Pr(NO_3)_3$ had been added to generate a transmembrane chemical shift of the ^1H methyl resonance (see section on paramagnetic shift resonances). In their experiments, the intravesicular acetic acid resonance was inverted (i.e., the resonance appears 180° out-of-phase in the NMR spectrum) by a long, 'soft' rf pulse (Alger and Prestegard, 1977), then a delay was allowed for exchange of 'inversion-labelled' intravesicular nuclei with unlabelled extravesicular nuclei prior to spectral acquisition (Fig. 8). The extent of diminution of the extravesicular resonance is an indication of the rate of exchange of labelled intravesicular and unlabelled extravesicular nuclei and hence is related to the transmembrane equilibrium exchange rate.

The time-dependence of the intensities (strictly speaking, magnetizations) of the extra- and intravesicular resonances is described by the Bloch-McConnell equations (McConnell, 1958):

$$dI_{int}/dt = -[(I_{int} - I_{inteq})/T_{1,int}] - k_1 I_{int} + k_{-1} I_{ext} \quad (12)$$

$$dI_{ext}/dt = -[(I_{ext} - I_{exteq})/T_{1,ext}] + k_1 I_{int} - k_{-1} I_{ext} \quad (13)$$

where k_1 and k_{-1} are the unitary rate constants describing the equilibrium exchange process, I_{inteq} and I_{exteq} are the equilibrium intensities of the intra- and extravesicular resonances, respectively, and $T_{1,int}$ and $T_{1,ext}$ are the corresponding longitudinal relaxation times. These simultaneous differential equations can be readily solved [e.g., Robinson et al. (1984) used the Laplace-Carson integral transform method] to yield analytical solutions for the time-dependence of I_{int} and I_{ext}. These biexponential solutions can be fitted by nonlinear regression onto plots of I_{ext} and I_{int} versus time to yield the unitary rate constants. Alger and Prestegard (1979) used this approach to determine the dependence of the unitary rate constants for the exchange of acetic acid across vesicle membranes on the pH and acetic acid concentration (Fig. 8); they concluded that neutral acetic acid monomer is the dominant permeant species. Nonlinear regression analysis requires initial estimates of the exponents in the analytical expressions; these are often obtained by exponential peeling of the data (Kuchel, 1985). However, there is a much more convenient method for obtaining initial estimates of k_1 and k_{-1}. Note from the time-dependence of the intensity of the uninverted extravesicular resonance (Fig. 8) that it initially declines to a minimum as the pre-acquisition delay is increased, and then returns to its equilibrium value. If we let I_{int}^{min} and I_{ext}^{min} be the intensities of the resonances arising from the internal and external compartments at the minimum then, by noting that dI_{ext}/dt is zero at this point, Eqn (13) can be rearranged to yield the following expressions for k_1 and k_{-1} (King et al., 1986):

$$k_{-1} = (I_{int}^{min} - I_{int}^{eq})/T_{1,int}(I_{ext}^{min} - K_e I_{int}^{min}) \quad (14)$$

$$k_1 = K_e k_{-1} \quad (15)$$

where the equilibrium constant $K_e = I_{int}^{eq}/I_{ext}^{eq}$, and $T_{1,int}$ can be obtained from a separate experiment.

McCain and Markley (1985) used a different type of magnetization transfer experiment known as saturation transfer to measure the water permeability coefficient for chloroplasts. The ^1H NMR spectrum contains two just barely-resolved resonances, the upfield one due to chloroplast water and the other due to the superimposed resonances of water from all remaining compartments. The chloroplast water signal was partially saturated (i.e. its intensity significantly reduced) by application of an rf pulse of appropriate frequency. A variable delay was allowed prior to spectral acquisition, which enabled the intensities of the saturated and non-saturated water resonances to be monitored as a function of time. Appropriate expressions describing the time dependence (Campbell et al., 1978) were then fitted by nonlinear regression to the intensity-versus-time profiles to yield the water residence time within the chloroplasts, from which the permeability coefficient was calculated. The estimated water permeability coefficient for chloroplasts, 9×10^{-4} cm/s, is slightly lower than values obtained by NMR spectroscopy for human erythrocyte membranes, typically $3\text{-}5 \times 10^{-3}$ cm/s (Benga, 1988; Herbst and Goldstein, 1989), indicating that perhaps the chloroplast double membrane presents a stronger barrier to water diffusion.

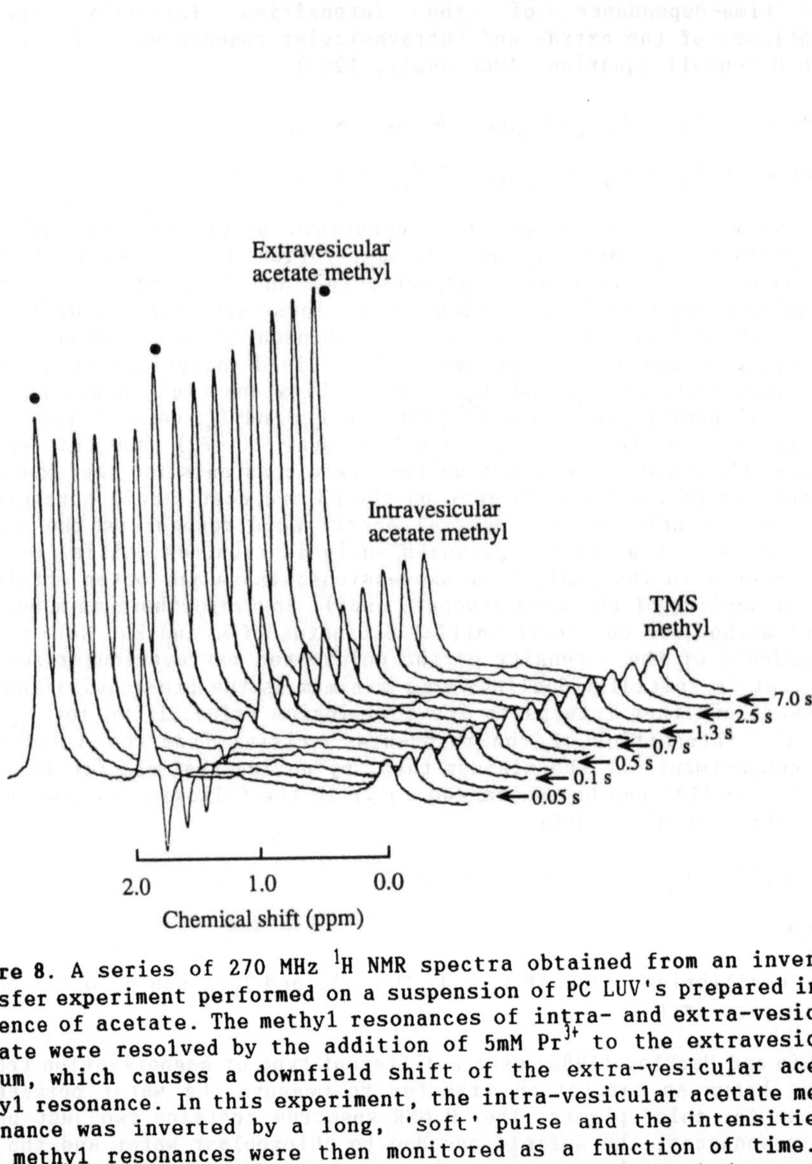

Figure 8. A series of 270 MHz ^{1}H NMR spectra obtained from an inversion transfer experiment performed on a suspension of PC LUV's prepared in the presence of acetate. The methyl resonances of intra- and extra-vesicular acetate were resolved by the addition of 5mM Pr^{3+} to the extravesicular medium, which causes a downfield shift of the extra-vesicular acetate methyl resonance. In this experiment, the intra-vesicular acetate methyl resonance was inverted by a long, 'soft' pulse and the intensities of both methyl resonances were then monitored as a function of time. The delay times are shown adjacent to every second spectrum and those marked with an asterisk were control experiments with delay times sufficient for attainment of equilibrium intensities for both resonances. Note the initial diminution of intensity of the extravesicular acetate methyl resonance due to exchange with 'inverted' intravesicular nuclei. Adapted from Alger and Prestegard (1979).

In cases where it is possible to completely and selectively saturate a resonance from the permeant in a particular compartment, the analysis is considerably simplified. For example if the resonance from the internal compartment is completely saturated then $I_{int} = 0$ and, since the rate of change of I_{int} prior to spectral acquisition will be zero, then $dI_{int}/dt = 0$. Thus, Eqn (12) can be rearranged to yield the following expression for k_1 (Kuchel, 1990):

$$k_1 = [(I_{ext}^{eq} - I_{ext}^{sat})/I_{ext}^{eq}]\ (1/T_{1,int}^{sat}) \qquad (16)$$

where I_{ext}^{eq} and I_{ext}^{sat} are the intensities of the external compartment resonance in the absence and presence respectively, of the additional rf field used for saturation, and $T_{1,int}^{sat}$ is the longitudinal relaxation time of the external compartment resonance measured in the presence of the additional rf field. $T_{1,int}^{sat}$ can be obtained using any standard T_1 measurement technique performed in the presence of the additional rf field; it is related to the intrinsic longitudinal relaxation time ($T_{1,int}$) by the expression (Kuchel, 1990):

$$T_{1,int}^{sat} = T_{1,int}/(1 + k_1 T_{1,int}) \qquad (17)$$

As yet no ^{1}H NMR studies have employed the complete resonance saturation technique to measure transmembrane equilibrium exchange: in both equilibrium exchange experiments described above (Alger and Prestegard, 1979; McCain and Markley, 1985), the compartmental chemical shift difference was insufficient to achieve complete and selective saturation of the resonances from one of the compartments. This is due to instrumental limitations on the bandwith of the saturating rf field, which 'spills over' to resonances within ~50 Hz of the resonance being saturated. Conversely, inversion transfer experiments in which resonance inversion is achieved through manipulation of the exchanging magnetizations, such as the delta-ordered inversion transfer (DOIT) sequence first introduced by Robinson et al (1984), do not have such stringent requirements for transmembrane chemical shift differences; for example, selective inversion was readily achieved using the DOIT sequence in one equilibrium exchange study in which the resonances of the exchanging nuclei were only separated by 34 Hz (King et al., 1986). Furthermore, the extent of resonance inversion achieved using this pulse sequence is considerably better than by simply using long, soft pulses.

No particular type of magnetization transfer experiment can be considered as most appropriate for all situations. Inversion transfer may be the method of choice when the internal and external resonances are not clearly resolved. However, saturation transfer is likely to be preferred in situations where the intensity of either the internal or external resonance is negligible (Kirk and Kuchel, 1986). Although they have not yet been used to measure transmembrane equilibrium exchange rates, two-dimensional NMR magnetization transfer experiments such as EXSY (Mendz et al., 1986) and ACCORDION (Bodenhausen and Ernst, 1982) may prove useful in future transport studies; the relative merits of these techniques are discussed in a recent review by Kuchel (1990).

Kuchel et al. (1987a) have recently introduced a novel NMR technique for measuring transmembrane equilibrium exchange rates which does not rely on compartmental differences in chemical shift, relaxation times, or diffusion properties of the permeant. Rather, the technique utilises a transmembrane difference in the rate of chemical exchange between the permeant and another solute and requires that the chemical shift of at least one nucleus differs between the permeant and the exchanging solute, such that the exchange can be monitored using magnetization transfer techniques. Because there is no stringent requirement for the permeant to exhibit compartmental chemical shift differences, this technique should find wider application to the measurement of transmembrane exchange processes than direct magnetization transfer methods; however to date this technique has only been used in ^{13}C NMR studies to monitor the exchange of ^{13}C-labelled bicarbonate (Kuchel et al., 1987a) and D-[1-^{13}C]glucose (Kuchel et al., 1987b) across erythrocyte membranes.

Relaxation rate profiles If a nucleus on a solute undergoing exchange between two compartments exhibits a significant compartmental difference in T_1 or T_2, then NMR experiments designed to measure the relaxation properties of the nucleus will yield two apparent relaxation rates which contain contributions from both the intrinsic relaxation rates and the rate of transmembrane solute exchange. Examination of the relaxation rate profile enables the transmembrane exchange rate to be extracted.

Although the theoretical considerations which follow are equally applicable to either T_1 or T_2 relaxation rate profiles, T_2 measurements are generally preferred to more time-consuming T_1 experiments (Pirkle et al., 1979). T_2 profiles are generally obtained using the Carr-Purcell-Meiboom-Gill (CPMG) pulse sequence (Carr and Purcell, 1954; Meiboom and Gill, 1958); this sequence, unlike the single spin-echo experiment (see section 3.2.2), eliminates relaxation effects due to diffusion through magnetic field gradients and thus enables measurement of the intrinsic T_2 rather than T_2*. Often, transmembrane exchange rates can be obtained from CPMG experiments performed in just a few seconds [e.g. Pirkle et al. (1979)].

Consider a solute nucleus exchanging between two compartments, and for which there is no significant transmembrane chemical shift difference. The time-dependence of its NMR signal intensity (i.e., the sum of the coincident internal and external resonances) in the CPMG experiment is given by (Pirkle et al., 1979; Hazelwood et al., 1974):

$$I_{tot} = [P_{int}'\exp(-t/T_{2,int}') + P_{ext}'\exp(-t/T_{2,ext}')] \quad (18)$$

$$\text{where} \quad 1/T_{2,int}' = C_1 - C_2 \quad (19)$$

$$1/T_{2,ext}' = C_1 + C_2 \quad (20)$$

$$P_{ext}' = 0.5\text{-}0.25\ [(P_{ext}-P_{int})\ (1/T_{2,int}-1/T_{2,ext}) + k_{-1} + k_1]/C2 \quad (21)$$

$$P_{int}' = 1-P_{ext}' \quad (22)$$

$$C_1 = 0.5(1/T_{2,int} + 1/T_{2,ext} + k_{-1} + k_1) \quad (23)$$

$$C_2 = 0.5[(1/T_{2,ext} + 1/T_{2,int} + k_1 + k_{-1})^2 + 4/k_{-1}k_1]^{0.5} \quad (24)$$

P_{int} and P_{ext} are the fractions of the total signal intensity arising from the internal and external compartments, respectively, $T_{2,int}$ and $T_{2,ext}$ are the corresponding intrinsic transverse relaxation times of the solute nucleus in these compartments, and k_1 and k_{-1} are the unitary rate constants for the influx and efflux of solute, respectively.

Under certain conditions, this relationshp can be considerably simplified. If $T_{2,ext} \ll T_{2,int}$ (such as when a paramagnetic relaxation reagent is added to the external compartment) and $T_{2,ext} \ll 1/k_{-1}$, then Eqn (18) simplifies to (Kirk, 1990):

$$1/T_{2,int}' = 1/T_{2,int} + k_{-1} \quad (25)$$

Conlon and Outred (1972) first used this technique to measure the exchange of water across human erythrocyte membranes; high extracellular concentrations of Mn^{2+} (0.5 -48 mM) were used to ensure that Eqn (25) applied. The strategy for estimating k_{-1} was as follows: First a CMPG experiment was performed on a dilute (haematocrit = 30%) suspension of erythrocytes. The resulting intensity-versus-time profile had, as expected, bi-exponential character; the faster and slower decaying components correspond to $T_{2,ext}'$ and $T_{2,int}'$, respectively (Fig.9). Subsequently, the cells were packed by centrifugation, and a second CPMG experiment performed on the packed cells which in the absence of supernatant yielded the single relaxation component $T_{2,int}$. The values of $T_{2,int}'$ and $T_{2,int}$ were then used in combination with Eqn (25) to yield the mean lifetime for water in the intracellular compartment, $1/k_{-1}$, which ranged from 7.8 to 9.8 ms.

This technique has been used extensively to monitor various aspects of water diffusion across the membranes of human erythrocytes and red cell ghosts (Benga, 1988; Herbst and Goldstein, 1989). However, it is not without inherent problems: Mn^{2+} is slowly permeable to human erythrocytes (and even more permeable in other cells), thus leading to incorrect estimates of $T_{2,int}'$ if significant amounts of this para-magnetic cation accumulate in the intracellular space. In fact, the apparent discontinuity in the Arrhenius plot for water diffusion into human erythrocytes, as measured by NMR, was shown to be an artefact resulting from increased intracellular [Mn^{2+}] when cells were incubated above 20°C for longer than 30 min (Benga, 1988). More recently Ashley and Goldstein (1980) demonstrated that dextran magnetite, a paramagnetic complex of iron and dextran, is impermeant to human erythrocytes and since it has similar effects on $T_{1,ext}$ as Mn^{2+} it should be preferred if exchange measurements are to be performed at temperatures >20°C and/or over prolonged times.

An alternative approach is to fit Eqn (18) directly by non-linear regression to the CPMG data (Ashley and Goldstein, 1980; Pirkle et al., 1979; Fabry and Eisenstadt, 1975). This yields values for P_{int}', P_{ext}', $T_{1,int}'$ and $T_{2,ext}'$ and enables calculation of the unitary rate constants using Eqns (19-24) so long as one of $T_{2,ext}$, $T_{2,int}$, or P_{ext} (= 1-P_{int}) can be estimated independently. A detailed analysis using simulated CPMG data revealed that of these three parameters, error in the independent estimation of $T_{2,int}$ introduced less error into calculation of the unitary rate constants than errors in the independent estimation of either $T_{2,ext}$ or P_{ext} (Pirkle et al., 1979). Thus, like the analysis using Eqn (25), the complete curve-fitting procedure also requires estimation of $T_{2,int}$ from a sample of packed cells. However, the curve-fitting procedure is probably more robust since it is not based on any assumptions and, since it is not strictly required that $T_{2,ext} << T_{2,int}$, more conservative extracellular concentrations of paramagnetic reagents can be used. An analogous curve-fitting procedure using T_1 measurements has also been used to measure water transport in erythrocytes (Fabry and Eisenstadt, 1975, 1978).

Numerous studies have employed T_1 relaxation rate profiles to measure the solute permeabilities of lipid vesicles in which the intra-vesicular space is doped with a paramagnetic relaxation reagent such as Gd^{3+} (Bar-on and Degani, 1985; Brown et al., 1982) or Mn^{2+} (Lipschitz-Farber and Degani, 1980). Under such conditions, $T_{1,int} << T_{1,ext}$, and, so long as $P_{int} << P_{ext}$ (which is generally true) and $T_{1,int} << 1/k_{-1}$, Eqn (18) simplifies to:

$$k_{-1} = (P_{int}/P_{ext})(1/T_{1,ext}' - 1/T_{1,ext}) \qquad (26)$$

T_1 relaxation rate profiles are obtained using the inversion-recovery (IR) pulse sequence (Sanders and Hunter, 1988). The slowly decaying component of the bi-exponential T_1 relaxation rate profile obtained from the vesicle suspension yields $1/T_{1,ext}'$, while $1/T_{1,ext}$ is obtained from an IR experiment performed on the supernatant. This procedure has been used to measure the permeability of lipid vesicles to water (Lipschitz-Farber and Degani, 1980),

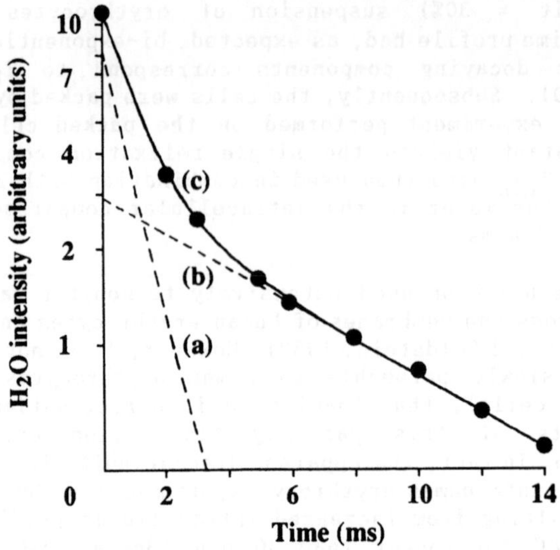

Fig. 9. Logarithmic plot of water resonance intensity-versus-time obtained from a CPMG experiment performed on a sample of whole blood whose plasma had been doped with 24 mM Mn^{2+}. The curve denoted by (c) represents the complete bi-exponential intensity-versus-time profile, while (a) and (b) represent the deconvoluted fast and slow decaying components, respectively, of curve (c). The decay rates of (a) and (b) correspond to $T_{1,ext}$' and $T_{2,int}$', respectively. Adapted from Fig 1 of Conlon and Outred (1972).

glycerol (Brown et al., 1982), and n-alkylamines (Bar-On and Degani, 1985). This technique was also used to measure the water permeability of certain algae by utilising the fact that these cells exhibit an intrinsic compartmental difference in the T_1 of water protons, thus obviating the need for doping one compartment with a paramagnetic reagent (Degani and Avron, 1982).

CONCLUSIONS AND OUTLOOK

It is hoped that this brief review will give the non-expert NMR user an indication of the numerous NMR techniques available for monitoring membrane transport processes. The appropriate NMR experiment for monitoring a particular transport process will depend primarily on the magnitude of the transmembrane solute flux. However, it should be noted that application of NMR techniques to membrane transport studies is generally restricted to circumstances where there is an intrinsic or induced compartmental difference in either the chemical shift, relaxation rate, or diffusion coefficient of a nucleus on the solute undergoing transport. Some recently developed methods have obviated these restrictions (Kuchel et al., 1987a; Wang et al., 1986) and it will be interesting to see if they find general application.

Because there are numerous alternative techniques available for monitoring membrane transport processes NMR spectroscopy will often not be the method of choice. However, some circumstances under which it may offer particular advantages over more conventional techniques include: (i) In proton NMR transport studies all ^1H nuclei are labelled by the radio-frequency pulse; consequently there is absolutely no requirement for the solute to be isotopically labelled. This allows study of the transport of solutes which would be undetectable by any other means or that would be difficult and/or expensive to isotopically label. (ii) ^1H NMR spectroscopy enables several biochemical events to be observed simultaneously, since all ^1H nuclei are concomitantly observed in the NMR spectrum. Thus, it may be possible to monitor simultaneously both solutes in a cotransport process (therefore enabling direct measurement of stoichiometries) or to observe intracellular metabolic events and various physiological properties of the cell (such as cell volume and intracellular pH) during the transport process. (iii) NMR spectrosocpy offers an abundance of methods for monitoring 'fast' membrane transport processes occuring over a timescale of seconds or less.

The possibility of combining a number of these advantages means that NMR will uniquely contribute to our further understanding of membrane solute transport.

ACKNOWLEDGMENTS

The authors would like to thank Drs Kirk and Steward and Professor Kuchel for providing manuscripts prior to publication.

REFERENCES

Alger, J. R. and Prestegard, J. H., 1977, Investigation of peptide bond isomerization by magnetization transfer NMR, J. Magn. Reson. 27:137

Alger, J. R. and Prestegard, J. H. 1979, Nuclear magnetic resonance study of acetic acid permeation of large unilamellar vesicle membranes. Biophys. J. 28:1

Ashley, D. L. and Goldstein, J. H. 1980, The application of dextran magnetite as a relaxation agent in the measurement of water exchange using pulsed nuclear magnetic resonance spectroscopy. Biochem. Biophys. Res. Commun. 97:114

Bar-On, Z. and Degani, H. 1985, Permeability of alkylamines across phosphatidylcholine vesicles as studied by ^1H-NMR. Biochim. Biophys. Acta, 813:207

Beilharz, G. R., Middlehurst, C. R., Kuchel, P. W., Hunt, G. E. and Johnson, G. F. S., 1984, Determination of choline in erythrocytes using high resolution proton nuclear magnetic resonance spectroscopy: comparison with a choline oxidase method. Analyt. Biochem. 127:324

Benga, G. 1988, Water transport in red blood cell membranes. Prog. Biophys. Molec. Biol. 51:193

Bodenhausen, G. and Ernst, R. R. 1982, Direct determination of rate constants of slow dynamic processes by two-dimensional "accordion" spectroscopy in nuclear magnetic resonance. J. Am. Chem. Soc. 104:1304

Boyd, C. A. R., Campbell, I. D., King, G. F. and Quirk, P. G. 1988, Proton NMR studies on epithelial cells isolated from chicken small intestine. J. Physiol. 407:14P

Brindle, K. M. and Campbell, I. D. 1987, NMR studies of kinetics in cells and tissues. Quart. Rev. Biophys. 19:159

Brindle, K. M., Brown, F. F., Campbell, I. D., Grathwohl, C. and Kuchel, P. W. 1979, Application of spin-echo nuclear magnetic resonance to whole-cell systems. Membrane transport. Biochem. J. 180:37

Brown, F. F. 1983, The effect of compartmental location on the T_2* of small molecules in cell suspensions: A cellular field gradient model. J. Magn. Reson. 54:385

Brown, F. F., Campbell, I. D., Kuchel, P. W. and Rabenstein, D. L. 1977, Human erythrocyte metabolism studies by ^1H spin-echo NMR. FEBS Lett. 82:12

Brown, F. F., Sussman, I., Avron, M. and Degani, H. 1982, NMR studies of glycerol permeability in lipid vesicles, erythrocytes and the alga *Dunaliella*. Biochim. Biophys. Acta 690:165

Brown, F. F., Jaroszkiewicz, G. and Jaroszkiewicz, M. 1983, An NMR method for studying the intracellular distribution and transport properties of small molecules in cell suspensions: The chicken erythrocyte system. J. Magn. Reson. 54:400

Campbell, I. D., Dobson, C. M., Ratcliffe, R. G. and Williams, R. J. P. 1978, Fourier transform NMR pulse methods for the measurement of slow exchange rates. J. Magn. Reson. 29:397

Carr, H. Y. and Purcell, E. M. 1954, Effects of diffusion on free precession in nuclear magnetic resonance experiments. Phys. Rev. 94:630

Conlon, T. and Outred, R. 1972, Water diffusion permeability measurements of erythrocytes using an NMR technique. Biochim. Biophys. Acta 288:354

Cramer, J. A. and Prestegard, J. H. 1977, NMR studies of pH induced transport of carboxylic acids across phospholipid vesicle membranes. Biochem. Biophys. Res. Commun. 75:295

Degani, H. 1978, NMR kinetic studies of the ionophore X-537A-mediated transport of manganous ions across phospholipid bilayers. Biochim. Biophys. Acta 509:364

Degani, H. and Avron, M. 1982, The diffusional water permeability in the halotolerant algae *Dunaliella* as measured by nuclear magnetic resonance. Biochim. Biophys. Acta 690:174

Degani, H., Simon, S. and McLaughlin, A. C. 1981, The kinetics of ionophore X-537A-mediated transport of manganese through dipalmitoylphosphatidycholine vesicles. Biochim. Biophys. Acta 646:320

Fabry, M. E. and Eisenstadt, M. 1975, Water exchange between red cells and plasma: measurement by nuclear magnetic resonance. Biophys. J. 15:1101

Fabry, M. E. and Eisenstadt, M. 1978, Water exchange across red cell membranes: II. Measurement by nuclear magnetic resonance T_1, T_2 and T_{12} hybrid relaxation. J. Memb. Biol. 42:375

Fabry, M. E. and San George, R. C. 1983, Effects of magnetic susceptibility on NMR signals arising from red cells: A warning. Biochemistry, 22:4119

Freeman, R. 1988, in A handbook of nuclear magnetic resonance, pp250-258, Longman Scientific and Technical, Avon

Gadian, D. G. 1982, in Nuclear magnetic resonance and its application to living systems. Clarendon Press, Oxford

Glasel, J. A. and Lee, K. H. 1974, On the interpretation of water nuclear magnetic resonance relaxation times in heterogeneous systems. J. Am. Chem. Soc. 96:970

Gupta, R. K. and Gupta, P. 1982, Direct observation of resolved resonances from intra- and extracellular sodium-23 ions in NMR studies of intact cells and tissues using dysprosium (III) tripolyphosphate as paramagnetic shift reagent. J. Magn. Reson. 47:344

Guy, R. D., Razi, M. T. and Rabenstein, D. L. 1986, Measurement of rates of transport across erythrocyte membranes by ^{1}H nuclear magnetic resonance spectroscopy. J. Magn. Reson. 66:434

Hazlewood, C. F., Chang, D. C., Nichols, B. L. and Woessner, D. E. 1974, Nuclear magnetic resonance transverse relaxation times of water protons in skeletal muscle. Biophys. J. 14:583

Herbst, M. D. and Goldstein, J. H., 1989, A review of water diffusion measurements by NMR in human red cells. Am. J. Physiol. 256: C1097

Hockings, P., Healy, P. C. and Rogers, P. J., 1989, Shift and relaxation reagents with preference for lactate in ^{1}H-NMR and ^{13}C-NMR spectroscopy of cell suspensions of *Streptococcus faecalis*. Proc. Aust. Biochem. Soc. 21:SP1

Hunt, G. R. A. 1980, A comparison of Triton X-100 and the bile salt taurocholate as micellar ionophores or fusogens in phospholipid vesicular membranes: A ^{1}H NMR method using the lathanide probe ion Pr^{3+}. FEBS Lett. 119:132

Hunt, G. R. A. and Jones, I. C., 1982, Lathanide-ion transport across phospholipid vesicular membranes: a comparison of alamethicin 30 and A23187 using ^{1}H NMR spectroscopy. Biosci. Rep. 2:921

Hunt, G. R. A., Tipping, L. R. H. and Belmont, M. R., 1978, Rate-determining processes in the transport of Pr^{3+} ions by the ionophore A23187 across phospholipid vesicular membranes. A ^{1}H-NMR and theoretical study. Biophys. Chem. 8:341

Hunt, G. R. A., Jones, I. C. and Veiro, J. A., 1984, Phosphatidic acid regulates the activity of the channel-forming ionophores alamethicin, metlittin and nystatin: A ^{1}H-NMR study using phospholipid membranes. Biosci. Rep. 4:403

Jones, A. J. and Kuchel, P. W., 1980, Measurement of choline concentration and transport in human erythrocytes in ^{1}H NMR: Comparison of normal blood and that from lithium-treated psychiatric patients. Clin. Chim. Acta 104:77

Jones, G. P., Starr, R. K. and Paleg, L. G., 1985, Promotion of cation transport across phospholipid vesicular vesicles by the plant hormone indole-3-acetic acid as studied by ^{1}H-NMR. Biochim. Biophys. Acta 812:77

King, G. F. and Kuchel, P. W., 1984, A proton NMR study of iminodipeptide transport and hydrolysis in the human erythrocyte: Possible physiological roles for the coupled system. Biochem. J. 220:553

King, G. F. and Kuchel, P. W., 1985, Assimilation of α-glutamyl peptides by human erythrocytes: A possible means of glutamate supply for glutathione synthesis. Biochem. J. 227:833

King, G. F., Middlehurst, C. R. and Kuchel, P. W., 1986, Direct NMR evidence that prolidase is specific for the trans isomer of imidodipeptide substrates. Biochemistry 25:1054

King, G. F., York, M. J., Chapman, B. E. and Kuchel, P. W., 1983, Proton NMR spectroscopic studies of dipeptidase in human erythrocytes. Biochem. Biophys. Res. Commun. 110:305

King, G. F., Deegan, J., Smith, R., Matouschek, A. and Campbell, I. D., 1988, An NMR study of alamethicin-, δ-haemolysin-, and melittin-induced leakage from large unilamellar vesicles. Biochem. Soc. Trans. 16:595

Kirk, K., 1990, NMR methods for monitoring membrane transport rates. NMR in Biomed., (in press)

Kirk, K. and Kuchel, P. W., 1986, Equilibrium exchange of dimethyl methylphosphonate across the human red cell membrane measured using NMR spin transfer. J. Magn. Reson. 68:311

Kirk, K. and Kuchel, P. W., 1988, The contribution of magnetic susceptibility effects to transmembrane chemical shift differences in the ^{31}P NMR spectra of oxygenated erythrocyte suspensions. J. Biol. Chem. 263:130

Kuchel, P. W., 1981, Nuclear magnetic resonance of biological samples. CRC Crit. Rev. Analyt. Chem. 12:155

Kuchel, P. W., 1985, Kinetic analysis of multienzyme systems in homogeneous solution. in: Catalytic facilitation in organised multienzyme systems (Welch G R ed.) pp 303-380, Academic Press, New York

Kuchel, P. W., 1989, Biological applications of NMR. in: Analytical NMR (Field L D and Sternhell S eds.) pp 157-219, John Wiley and Sons, Chichester

Kuchel, P. W., 1990, Spin-exchange NMR spectroscopy in studies of the kinetics of enzymes and membrane transport. NMR in Biomed. (in press)

Kuchel, P. W. and Bulliman, B. T., 1990, Perturbation of homogeneous magnetic fields by isolated single and confocal spheroids: Implications for NMR spectroscopy of cells. NMR in Biomed. (in press)

Kuchel, P. W., Chapman, B. E., Endre, Z. H., King, G. F., Thorburn, D. R. and York, M. J., 1984, Monitoring metabolic reactions in erythrocytes using NMR spectroscopy. Biomed. Biochim. Acta 43:719

Kuchel, P. W., King, G. F. and Chapman, B. E., 1987, No evidence of high capacity α-glutamyl-dipeptide transport into human erythrocytes. Biochem. J. 242:311

Kuchel, P. W., Bulliman, B. E., Chapman, B. E. and Kirk, K., 1987a, The use of transmembrane differences in saturation transfer for measuring fast membrane transport: Application to $H^{13}CO_3$-exchange across the human erythrocyte. J. Magn. Reson. 74:1

Kuchel, P. W., Chapman, B. E. and Potts, J. R., 1987b, Glucose transport in human erythrocytes measured using ^{13}C NMR spin transfer. FEBS Lett. 219:5

Labotka, R. J. and Omachi, A., 1987, Erythrocyte anion transport of phosphate analogs. J. Biol. Chem. 262:305

Lau, A. L. Y. and Chan, S. I., 1976, Voltage-induced formation of alamethicin pores in lecithin bilayer vesicles. Biochemistry 5:2551

Lipschitz-Farber, C. and Degani, H., 1980, Kinetics of water diffusion across phospholipid membranes: ^{1}H and ^{17}O-NMR relaxation studies. Biochim. Biophys. Acta 600:291

Lee, Y. and Chan, S. I., 1977, The effect of lysolecithin on the structure and permeability of lecithin bilayer vesicles. Biochemistry 16:1303

McCain, D.C. and Markley, J. L., 1985, Water permeability of chloroplast envelope membranes. *In vivo* measurement by saturation-transfer NMR. FEBS Lett 183:353

McConnell, H. M., 1958, Reaction rates by nuclear magnetic resonance. J. Chem. Phys. 28:430

Meiboom, S. and Gill, D., 1958, Modified spin-echo method for measuring relaxation times. Rev. Sci. Instrum. 29:688

Mendz, G. L., Robinson, G. and Kuchel, P. W., 1986, Direct quantitative analysis analysis of enzyme-catalysed reactions by two-dimensional nuclear magnetic resonance spectroscopy: Adenylate cyclase and phosphoglyceromutase. J. Am.Chem.Soc. 108:169

Middlehurst, C. R., King, G. F., Beilharz, G. R., Hunt, G. E., Johnson, G. F. S. and Kuchel, P. W., 1984, Studies of rat brain metabolism using proton nuclear magnetic resonance spectroscopy: Spectral assignments and monitoring of prolidase, acetylcholinesterase and glutaminase. J. Neurochem. 43:1561

Nicholson, J. K., Buckingham, M. J. and Sadler, P. J., 1983, High resolution ^{1}H n.m.r. studies of vertebrate blood and plasma. Biochem. J. 211:605

Nicholson, J. K., Timbrell, J. A., Bales, J. R. and Sadler, P. J., 1985, A high resolution proton nuclear magnetic resonance approach to the study of hepatocyte and drug metabolism. Application to acetaminophen. Molec. Pharmacol. 27:634

Pirkle, J. L., Ashley, D. L. and Goldstein, J. H., 1979, Pulse nuclear magnetic resonance measurements of water exchange across the erythrocyte membrane at low Mn concentration. Biophys. J. 25:389

Prestegard, J. H., Cramer, J. A. and Viscio, D. B., 1979, Nuclear magnetic resonance determinations of permeation coefficients for maleic acid in phospholipid vesicles. Biophys. J. 26:575

Quirk, P. G., Boyd, C. A. R. and King, G. F., 1990, (unpublished results)

Rabenstein, D. L., 1978, Resolution enhancement with multiple-pulsed techniques in pulsed Fourier transform nuclear magnetic resonance spectroscopy. Analyt. Chem. 50:1265A

Rabenstein, D. L., 1984, ^{1}H NMR methods for the noninvasive study of metabolism and other processes involving small molecules in intact erythrocytes. J. Biochem. Biophys. Methods 9:277

Rabenstein, D. L. and Isab, A. A., 1982, Determination of the intracellular pH of intact erythrocytes by ^{1}H NMR spectroscopy. Analyt. Biochem. 121: 423

Robinson, G., Chapman, B. E. and Kuchel, P. W., 1984, ^{31}P NMR spin-transfer in the phosphoglyceromutase reaction. Eur. J. Biochem. 143:643

Sanders, J. K. M. and Hunter, B. K., 1988, in Modern NMR spectroscopy: A guide

for chemists, pp61-65, Oxford University Press, Oxford
Steward, M. C., Seo, Y., Rawlings, J. M. and Case, R. M., 1990, Water permeability of acinar cell membranes in the isolated perfused rabbit mandibular salivary gland. (in press).
Ting, D.Z., Hagan, P. S., Chan, S. I., Doll, J. D. and Springer, C. S., 1981, Biophys. J. 34:189
Tofts, P. S. and Wray, S., 1988, A critical assessment of methods of measuring metabolite concentrations by NMR spectroscopy. NMR in Biomed 1:1
Vandenberg, J. I., King, G. F. and Kuchel, P. W., 1985, The assimilation of tri- and tetrapeptides by human erythrocytes. Biochim. Biophys. Acta 846:127
Vandenberg, J. I., Kuchel, P. W. and King, G. F., 1986, Application of progress curve analysis to in situ enzyme kinetics using ^{1}H NMR spectroscopy. Analyt. Biochem. 155:38
Wang, J-F, Falke, J. J. and Chan, S. I., 1986, A proton NMR study of the mechanism of the erythrocyte glucose transporter. Proc. Natl. Acad. Sci. USA 83:3277
Westerhoff, H. V., 1982, NMR sheds more light on ion transport. Trends Biochem. Sci. 7:232
Yoon, P. S. and Sharp, R. R., 1985, Ca^{2+} and proton transport in chromaffin granule membranes: A proton NMR study. Biochemistry 24: 7269
York, M. J., Kuchel, P. W. and Chapman, B. E., 1984, A proton nuclear magnetic resonance study of γ-glutamylcyclotransferase in human erythrocytes. J. Biol. Chem. 259:15085
Young, J. D. and Ellory, J. C., 1977, Red cell amino acid transport. in: Membrane transport in red cells (Ellory J C and Lew V L eds.) pp 301-325, Academic Press, London

17

OPTICAL AND NMR SPECTROSCOPY IN THE STUDY OF SECRETION

Tomio Kanno

Department of Physiology, Faculty of Veterinary Medicine

Hokkaido University, Sapporo 060, Japan

INTRODUCTION

The sequence of events in stimulus-secretion coupling in pancreatic acinar cell can be divided into three main steps; activation of receptors by proper stimuli, conduction of excitation, and extrusion of secretory substances. These three main steps occur in acinar cells in three functionally different cellular structures : The receptive portion is basiolateral membrane, and the output portion is the apical membrane. Following eceptor activation, intrtacellular processes, including an increase in cytosolic Ca^{2+}, $[Ca^{2+}]_c$, mediate exocytosis and Cl^- efflux at the apical membrane. Every step in stimulus-secretion coupling may utilize energy, the immediate source of which is ATP. There is evidence that mitochondria supply more than 90 % of the energy needed by the pancreas (Bauduin et al., 1969). The dynamic change in the production and utilization of energy in the exocrine pancreas can be monitored noninvasively with the aid of optical and NMR spectroscopy in the isolated perfused pancreas, which maintains the polarity during every step of stimulus-secretion coupling. The steps that can be minitored by optical and NMR spectroscopy are shown schematically in Fig. 1.

ROLES OF MITOCHONDRIA

During the last four decades, mitochondoria have become an important subject of research within several disciplines of experimental biology. For the biochemist, the identification of mitochondria as the site of cell respiration and respiration-linked phosphorylation implied a decisive step towards the resolution and reconstruction of these processes at a molecular level and the elucidation of their relationship to cellular membranes. For the physiologist, mitochondria afforded the first opportunity for an experimental approach to structure-function relationships, in particular those involved in active transport, vectorial metabolism, and metabolic control mechanism at a subcellular level (Ernster and Schatz, 1981). Activities of mitochondria in the processes of active transport and membrane-mobile transport (endocytosis and exocytosis) in various organs have been investigated by monitoring the redox states of the electron transfer system.

Cell Membrane Transport, Edited by D. L. Yudilevich *et al.*
Plenum Press, New York, 1991

The redox states of pyridine nucleotides in isolated perfused organs have been monitored by recording fluorescence intensities of the amounts of reduced nicotinamide adenine dinucleotide (NADH) and nicotinamide adenine dinucleotide phosphate (NADPH). The redox states of cytochromes in mitochondria have been monitored with a scanning organ spectrophotometer.

RECORDING OF NAD(P)H-FLUORESCENCE

The instrument used for monitoring NAD(P)H-fluorescence is a modification of the microfluorometer introduced by Chance, et al. (1962). It records field of tissues upon light excitation at 366 nm and emission at 460 nm. The fluorescence intensity thus recorded is a function of the amount of reduced form of NADH and NADPH, so that a direct and continuous measurement of the relative value of the intracellular redox state of pyridine nucleotides can be attained. The fluorescence intensity in blood-circulating organs in situ can be disturbed by an optical artifact derived from variations in blood volume in the tissue. Such an optical artifact is compensated by the use of the relation described by Kobayashi, et al. (1971). A microfluorometer used in our laboratory was made by Tateisi Institute of Life Sciences (type HEF-4, Kyoto).

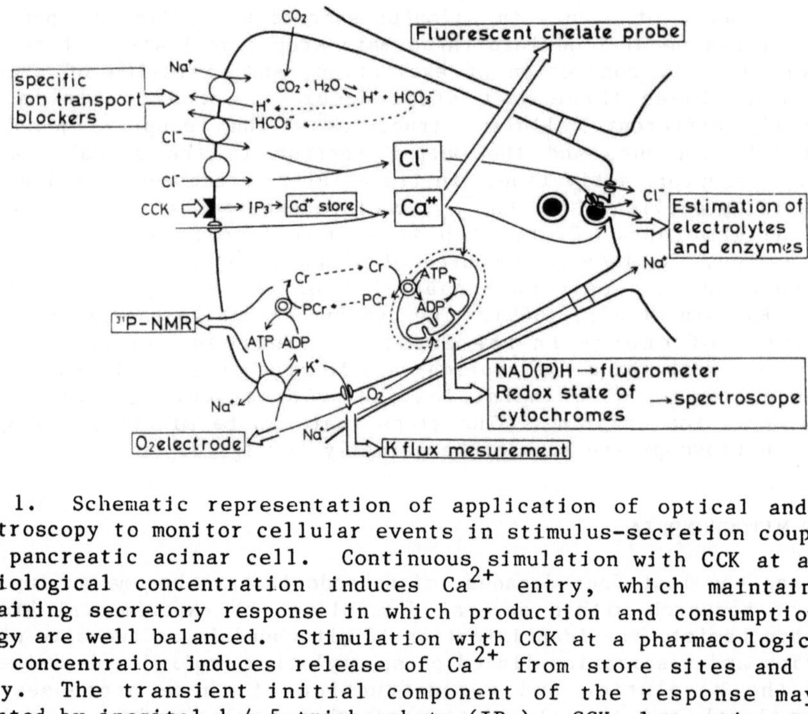

Fig. 1. Schematic representation of application of optical and NMR spectroscopy to monitor cellular events in stimulus-secretion coupling in a pancreatic acinar cell. Continuous simulation with CCK at a low physiological concentration induces Ca^{2+} entry, which maintains a sustaining secretory response in which production and consumption of energy are well balanced. Stimulation with CCK at a pharmacologically high concentraion induces release of Ca^{2+} from store sites and Ca^{2+} entry. The transient initial component of the response may be mediated by inositol 1,4,5-triphosphate (IP_3). CCK also activates the Na^+,K^+-ATPase and other ionic transporters in the basolateral membrane. ATP for the ATPase may be supplied via the cytoplasmic creatine kinase (CPK) system from phosphocreatine (PCr). PCr is supplied by diffusion from the mitochondrial CPK system. The respiratory chain in a mitochondria supplies ATP to the mitochondrial CPK system. Steps that can be monitered by spectroscopy and other methods are shown in the squares in the margin.

The photometric method of observing mitochondrial metabolism through monitoring the redox state of the respiratory enzymes was introduced by Chance and his colleagues, and numerous reports have appeared. The photometry included dual-wavelength spectrophotometry (Chance, 1951) and split-beam spectrophotometry (Yang and Legallais, 1954). These techniques have been applied principally to turbid suspensions of biological particles in vitro. To obtain information concerning dynamic transients, as well as, steady states in the oxidative and reductive responses of the respiratory pigments in a whole organ, several problems more difficult than those of turbid materials required resolution (Kobayashi, et al., 1981). Studies of perfused organs with their integral structure and function are required to elucidate the metabolic correlation. This was attained by developing a special spectrometer capable of recording dynamic changes in wavelength and optical density (Kobayashi et al. 1981). The mechanical and electrical constitutions are as follows.

The light source of scanning organ spectrophotometer we used (Tateishi Institute of Life Science, Kyoto) is a 300 W xenon arc (VIX-300, Varian) powered by a stabilized power supply (PS 300-1J, Evic). The wavelength scanning is realized by repeated vibration of a grating (M 27, Jobin Yvon: 600 grooves/mm). Time for one scan of the 330-660 nm region is 0.8 sec. Each wavelength is read with a rotary encoder (RA-3,600-22-1, Nikon) connected to the driving cam shaft of a synchronous motor that is used for mechanical oscillation of the grating. The encoder generates 3,600 pulses in one rotation of the motor. The monochromatic light from the grating is divided with a beam-splitter into each branch of a pair of two-branched flexible light guides leading to the sample and to the reference in the ratio of 1:1. The reference material is usually MgO, but when measuring a high absorption sample, a cut-off filter with a known spectrum is effective for cancellation especially in the Soret region. The tip of the guide is held facing the tissue at a suitable distance. The light reflected from the sample and the reference passes through the second branch of the guides and then illuminates the respective photomultipliers (R-374, Hamamatsu TV). The light guides are bundles of quartz fibres (Schott). Transmission measurement can be performed by replacing each two-branched guide with two separate guides, one of which introduces the incident light to the sample (or the reference) and the other which receives the transmitted light.

The output analog signals of the photo-detectors are amplified logarithmically in order to get values of optical density:

$$\log(I_{reference}/I_{sample})$$

While the job is interrupted by the wavelength signal, a microprocessor (uPD-753, NEC) sends the photolectric signals to a sample-hold circuit (SHM-LM-2, Datel) and an analog-to-digital converter (ADC-eH12B3, Datel) yielding to make 12-bit digita data, which are then temporarily stored in RAMs and subsequently transferred to a floppy disc (Driven by a JK-880, Matsushita). The disc is able to store data from 166 scans. The data input interval corresponds to 0.2 nm for each wavelength. The data processed in the processor according to the instruction program are converted into analog form using four digital-to-analog converters (DAC-12BGC, Datel) (Kobayashi et al., 1981). Data processing and control of the instrument are carried out with a microprocessor, which contains computing program was for absolute spectrum, difference spectrum, time course of difference between two points in the spectrum, and mean value of data (Kanno and Saito, 1982).

CONTINUOUS RECORDINGS OF REDOX STATES OF ELECTRON TRANSFER SYSTEM IN VARIOUS SECRETORY CELLS

Development of the microfluorometer and the scanning organ spectrophotometer enabled us to record redox states of pyridine nucleotides and cytochromes during continuous monitoring of the other cellular functions in an isolated organ or tissue preparation.

The isolated preparation was first perfused or perifused for 20 min to allow the responses to reach steady levels. After this period, NAD(P)H-fluorescence and absorption spectra of the cytochromes were recorded for 10 min to determine the 'resting levels'. Averages of 16 absorption spectra scans were computed within 1 sec and stored in the built-in microprocessor. The differences between the stored initial average and subsequent averages were continuously computed and displayed as the difference spectra every 16 sec throughout the experiments. The perfusion or perifusion was then switched from the standard solution to a solution containing a secretatogue, and these measurements were continued for a further ten min periods to confirm the responses. The perfusing or perifusing solution was then replaced by a secretagogue-free standard solution, and a third set of measurements were obtained. The perfusion or perifusion was then stopped, and a final set of measurements obtained. Steady levels of these final sets of measurements were considered to be the maximum reduction levels for the pyridine nucleodites and the cytochromes, as no further reduction was observed when preparations were perfused or perifused with anoxic solutions saturated with $Na_2S_2O_4$, a strong reducing agent.

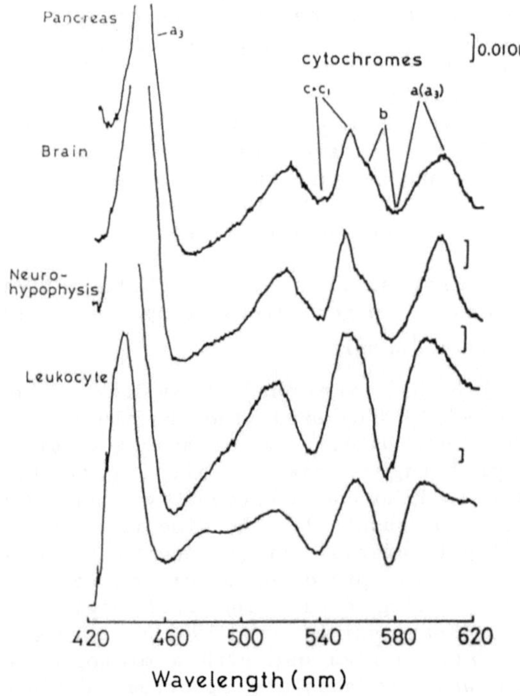

Fig. 2. Examples of the recording of difference spectra at the maximum reduction levels in the isolated perfused rat pancreas, the perfused bullfrog brain, the isolated perfused guinea pig neurohypophysis, and the isolated human leukocytes as indicated.

Figure 2 shows examples of the recordings of difference spectra in the final set of redox state measurements of cytochromes in the isolated perfused rat pancreas, the perfused bullfrog brain, the isolated perifused guinea pig neurohypophysis, and isolated perifused human leukocytes. The patterns of difference spectra at the maximum reduction levels are similar among these preparations. For all of these organs and tissues, a major part of the energy is known to be supplied by mitochondrial oxidative phosphorylation.

At the maximum reduction level, the difference between optical density at 575 nm and that of cytochrome $a(a_3)$ (605 nm); at 575 nm and that of cytochrome b(562 nm); or at 540 nm and that of cytochrome $c+c_1$ (550 nm) was considered to be 100 percent reduction of the respective cytochromes. Cytochromes were assumed to be fully oxidized (0 % reduction) during control perfusion at adequate flow rates, because application of 1 mM sodium amobarbital failed to induce optical absorbance changes during control perfusion.

The time course of NAD(P)H-fluorescence in an isolated preparation is almost identical to that of reduction of the cytochromes.

SIMULTANEOUS MEASUREMENTS OF REDOX STATE OF CYTOCHROMES AND SECRETORY RESPONSE IN ISOLATED PERFUSED PANCREAS

Among the four isolated preparations depicted in Fig.2, the mechanism of transcellular transport of electrolytes and proteins has been studied most intensively in the exocrine pancreas. In our laboratory, correlation between the transcellular transport and the redox state of cytochromes has been studied in the isolated perfused rat pancreas. Pancreatic juice flow

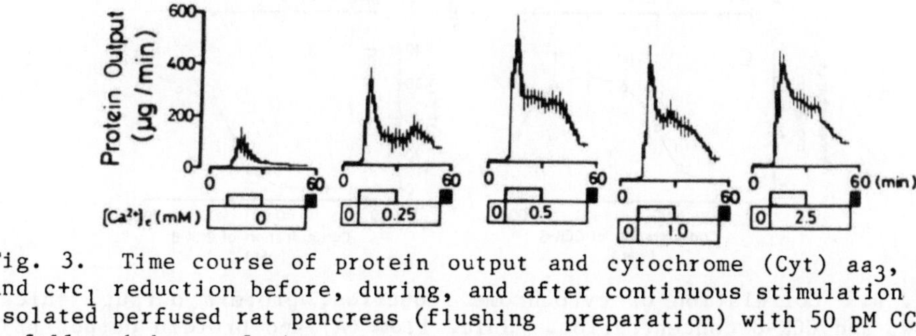

Fig. 3. Time course of protein output and cytochrome (Cyt) aa_3, b, and $c+c_1$ reduction before, during, and after continuous stimulation of isolated perfused rat pancreas (flushing preparation) with 50 pM CCK-8 followed by perfusion stop in environment containing 0 (A), 0.25 (B), 0.5(C), 1.5 (D), or 2.5 (E) mM $CaCl_2$. Stippled horizontal and shaded bars indicate period of perfusion with solution containing 50 pM CCK-8 and period of perfusion stop, respectively. Value are means. ± S. E. of 4 experiments. Reproduced, with permission, from Kanno, and Matsumoto (1986).

and protein output were estimated simultaneously before, during, and after continuous stimulation with different concentration of C-terminal octapeptide of cholecystokinin (CCK-8) in the draining preparation of the isolated perfused rat pancreas, in which the pancreatic juice was collected from the duodenal end after the hepatic end of the duct was ligated (Kanno, et al., 1981; Shibuya and Kanno, 1985; Kanno and Matsumoto, 1986; Kanno, 1987). In secretory responses to different concentrations of CCK-8, the pattern and time course of juice flow was almost identical to that of protein output. Because replacement of the tube attached to the free end of the common duct during continuous measurement of redox state with a scanning organ spectrophotometer is nearly impossible in the dark room, the flushing preparation of isolated perfused pancreas was adopted for simultaneous measurement of redox state and protein output. In the flushing preparation, the common duct was continuously flushed from the hepatic end through the duodenal end at a constant pressure of 5 cmH_2O. The time course of protein output induced by 50 pM CCK-8 in the flushing preparation was identical to that in the draining preparation, but the initial peak of protein output preceded that of reduction of cytochromes (Fig.3). Lowering extracellular Ca^{2+} caused a gradual diminution of the CCK-8-induced responses, protein output and cytochrome reduction.

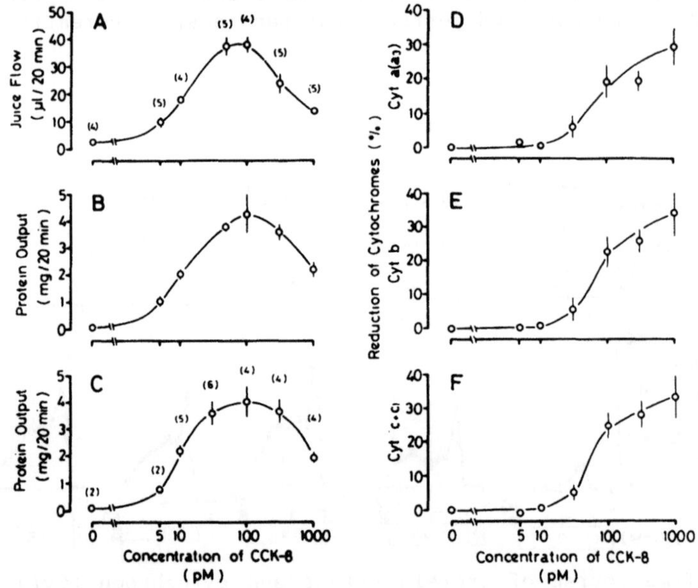

Fig. 4. Correlation of cytochrome reduction, protein output, juice flow, and CCK-8 concentration. Juice flow (A) and protein output (B) were obtained from experiments in draining preparations. Protein output and juice flow are total amount of protein and flow in a sample collected for 20 min of stimulation with CCK-8 of different concentrations. Reduction of cytochromes is shown as the mean of 20-min measurement during respective CCK-8 stimulation. Numbers in parentheses are numbers of experiments. Reproduced, with permission, from Kanno and Matsumoto (1986).

Dose-response relations were also obtained for the redox state, juice flow, and protein output in response to various concentrations of CCK-8 (Fig.3). The dose-response relations for redox state and protein output were obtained from results carried out in flushing preparations, whereas those for protein output and juice flow were obtained from results carried out in draining preparations. There was a distinct discrepancy between the dose-dependent redox response and the dose-dependent secretory responses. A sigmoidal relation was obtained between the concentration of CCK-8 and the level of reduction of cytochromes recorded from the pancreas, whereas a bell-shaped relation was obtained between the concentration of CCK-8 and the secretory responses. Continuous stimulation of the pancreas with a physiological CCK-8 concentration, 10 pM or lower, induced a small but distinct increase in secretory responses and little if any changes in redox state of cytochromes. Continuous stimulation with 30 pM CCK-8 caused near-maximum secretory responses and only a small reduction of cytochromes. Continuous stimulation with 1 nM CCK-8, which has been shown to cause the morphological changes of acute pancreatitis (Kanno, et al., 1984), induced only a slight increase in secretory responses but a large reduction of cytochromes.

Dose-response relations were further obtained for the redox state, the NAD(P)H-fluorescence, O_2 consumption, and the protein output (Kanno, et al., 1983). The dose-response relation for ACh-induced reduction of cytochromes resembled that of ACh-induced O_2 consumption and NAD(P)H-fluorescence, except that there was an excessive large reduction with 0.1 μM ACh. The protein output induced by stimulation with 0.1 μM ACh was the largest compared with secretion induced by 0.01, 1 or 10 μM ACh.

^{31}P-NMR AND SECRETORY RESPONSES IN THE ISOLATED PERFUSED PANCREAS

Phosphorus nuclear magnetic resonance (^{31}P-NMR) was applied to measure phosphorus metabolites in the isolated perfused rat pancreas (Matsumoto, et al., 1988 a,b). Removal of the duodenum from the pancreas in the process of isolating the pancreas from rat made it possible to record ^{31}P-NMR spectra exclusively from the pancreas. Rats were anesthetized by an intraperitoneal injection of pentobarbital sodium. The pancreas was isolated for perfusion by a modification of the methods of Kanno, Suga and Yamamoto (1976). The duodenum was separated surgically from the pancreas. The part of the pancreas that was supplied with blood by the superior mesenteric artery (Kanno, et al., 1985) was removed after ligation of all branches of the superior mesentric artery. The remaining portion of the pancreas, weighing 500-900 mg, was placed in a 10-mm diameter NMR tube and perfused via the celiac artery at the constant flow rate of 2 ml/min with the aid of a peristaltic roller pump. The vascular perfusate from the portal vein was removed with another roller pump. The composition of the standard Ringer solution used for perfusing the preparation was the same as reported previously (Matsumoto et al., 1988 a).

The isolated perfused pancreas was continuously stimulated with ACh or CCK-8. Pancreatic juice was collected from the duodenal end of the pancreatic duct, and its protein concentration was determined by the method of Lowey, et al. (1951) with bovine serum albumin as a standard.

The ^{31}P-NMR spectra were recorded with a WM-360wb NMR spectometer (Bruker) operated at 145.8 MHz using a broad band tunable probe (10 mm in diameter). For kinetic measurements, a ratio frequency pulse of 5 μ sec (15 pulse) was followed by a relaxation delay of 0.3sec, and 1,000 transients of free induction decay were accumulated for 5 min. Proton decoupling was not used. The Fourier transform was applied to free induction decay with 15 Hz line broadening to improve the signal-to-noise ratio of the spectrum. A picture-analyzing system (MOP AM03, Kontron) was used to measure the area beneath the resonant peak corresponding to each phosphorus compound; inorganic phosphate (Pi), creatine phosphate (PCr), ATP, ADP, and phosphomonoesters (PMEs) (Fig. 5) . The tissue levels of phosphorus compounds were expressed as percentages of the resting values. The tissue pH (pHt) was calculated from the chemical shift of inorganic phosphate by the equation shown in the previous paper (Seo, et al., 1983).

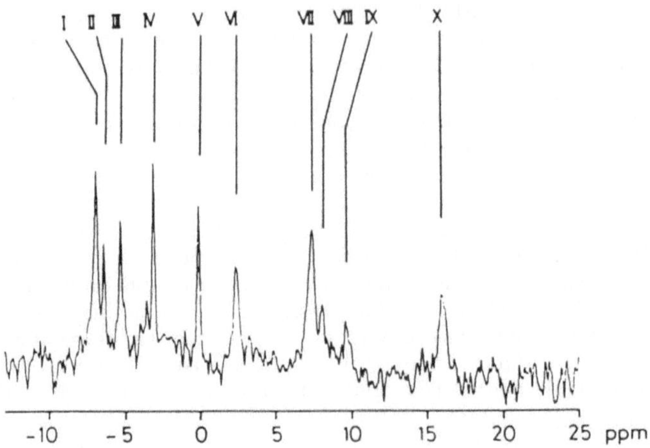

Fig. 5. Assignment of signals in the phosphorus nuclear magnetic responace (^{31}P-NMR) spectrum of perfused rat pancreas. Phosphomonoesters, e.g., sugar phosphates, AMP, etc. (peaks I and II), inorganic phosphate (peak III), phosphodiesters (peak IV), creatine phosphate (peak V), γ-ATP and β-ADP (peak VI), α-ATP and α-ADP(peak VII), NADH and NAD^+ (peak VIII), UDP-glucose (peak IX), and β-ATP (peak X). Reproduced, with permission, from Matsumoto et al. (1988).

In different individual preparations, the resonance of PCr varied, but those of ATP were almost the same. The initial levels of PCr and ATP in individual preparations, however, remained almost unchanged during perfusion with the standard solution for 2 h. When the perfusion was stopped, the levels of ATP and PCr decreased while the levels of PME and Pi increased. At that time, the Pi resonance shifted to a higher magnetic field, indicating that the tissue pH decreased. Upon reperfusion, the tissue levels of phosphorus compounds and the tissue pH were restored to their initial resting levels. Continuous infusion of 0.1 μM ACh caused marked and sustained increases in the flow of pancreatic juice and protein output. During the stimulation the tissue levels of phosphorus compounds remained unchanged, while the tissue pH decreased slightly (0.05 pH unit). Continuous stimulation with CCK-8 at a lower concentration (10 or 100 pM), caused sustained increases in pancreatic secretion, but did not induce any changes in the tissue level of ATP (Matsumoto et al., 1988c). However, continuous stimulation with CCK-8 at a high concentration (1 nM), which was shown to induce morphological changes of acute pancreatitis (Kanno, et al. 1984), caused only a small increase in secretory responses and decreased the tissue level of ATP, whereas it did not change or even increased the tissue level of PCr. These findings suggest that stimulation with high concentrations of secretagogues inhibits creatine phosphokinase (CPK) system near ATP-consuming sites or disturbs PCr transport between a CPK system near ATP-consuming sites and another CPK system near mitochondria in pancreatic acinar cells. Dissociation of secretion and energy supply by stimulation with high concentrations of ACh was also demonstrated in isolated perfused mandibular gland of rats (Murakami et al. 1988). During stimulation with 1 mM ACh, the level of PCr first decreased and then partially recovered, but the level of ATP continued to decrease and the levels of Pi and suger phosphates increased markedly. These findings suggest compartmentalization of CPK systems and the possibility that a high concentration of ACh interferes with the transport of PCr between one CPK system near ATPase and another system near mitochondria in acinar cells in mandibular gland of rat (Murakami et al., 1988).

SIMULTANEOUS MEASUREMENTS OF REDOX STATE OF CYTOCHROME AND ELECTRICAL ACTIVITY IN PERFUSED BULLFROG BRAIN

Under normal conditions, the brain uiliizes glucose as the sole substrate for oxidative phosphorylation, which is maintained by a continuous supply of oxygen. Since the brain has a high rate of oxygen utilization and small reserves of oxygen and energy, any change in oxygen and glucose supply to the brain may result in blockage of electron transfer through the resporatory chain, as well as a net decrease in ATP production. With the aid of a scanning organ spectrophotometer, the redox state of cytochromes was recorded in the perfused bullfrog brain (Kanno et al., 1985).

Male and female bullfrogs were anesthetized by packing in ice until reflex responsiveness was lost and curarized, plastic cannulae were inserted into the internal carotid arteries, into which fluid was perfused at a constant rate of 2 ml/min by means of a roller pump at room temperature. The skull was opened to expose the whole cerebral hemispheres, onto which the tip of the light guide of the scanning organ spectrophotometer (3 or 5 mm in diameter) was placed. The dorsal root of the spinal cord was cut and its afferent end was stimulated electically with a suction electrode.

Changes in the redox state of cytochromes and the frequency of electrical activity (mean $\pm$ S.E.) before, during, and after the stimulation of the dorsal root with a square-wave pulse (25 mA, 20 Hz, 5 msc for 2 min) (Fig.6). The peak of reduction of the cytochromes was in accord with the peak of increasing frequency of the electrical activity. The reduction of the cytochromes and the evoked electrical activity were both abolished when 2 μM tetrodotoxin was added to the perfusing solution. As the electrical stimulation of the dorsal root was prolonged for 5 or 10 min, the reduction of the cytochromes and burst activity recurred for up to several minutes after the cessation of stimulation.

We also demonstrated that perfusion with CCK (Kanno and Saito, 1982), neuropeptide Y (Saito et al., 1984), and thyrotropin releasing hormone (Kanno and Saito, 1988), following perfusion with a hypertonic solution of 300 nM arbinose, known to induce functional opening of the blood-brain barrier, caused marked reduction of cytochromes in the perfused bullfrog brain.

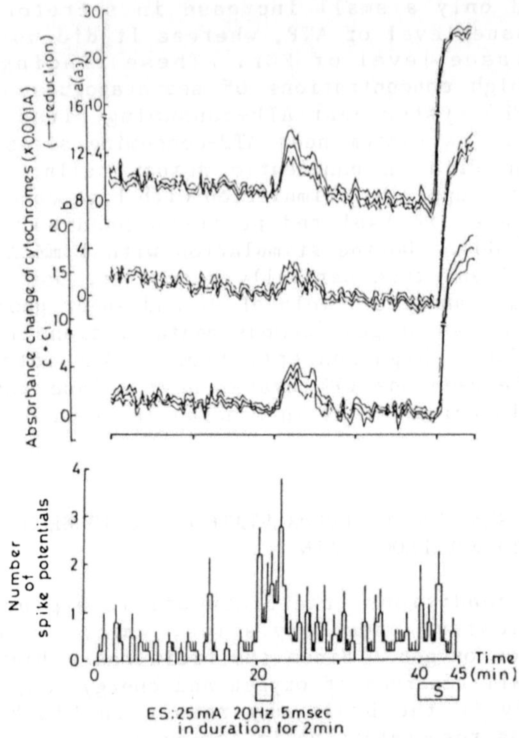

Fig. 6. Time-course of the redox states of cytochromes $a(a_3)$, b and $c+c_1$ and the number of spike potentials (not less than 100 μV in an amplitude) per 39 sec during perfusion with a standard solution in the perfused bullfrog brain. The dorsal root was stimulated electrically during the period shown by ES. Each value is the mean $\pm$ S.E. of 5 experiments. S. Period of stopping flow. Reproduced, with permission, from Kanno et al.(1985)

REDOX STATE AS AN INDICATION OF CELLULAR FUNCTION

The redox state of cytochromes is an indication of the respiratory state of mitochondria. In isolated preparations of mitochondria, Chance and Williams (1956) proposed five redox states of the electron transfer system: fresh, starved, active, resting, and anoxic. Parallel reduction of cytochromes is regarded as a sign of an anoxic state.

We have provided evidence that the parallel reduction of cytochromes is associated with CCK-8-induced secretory responses, protein output and fluid secretion in the isolated perfused rat pancreas, and that the parallel reduction is paralleled by an increase in electrical activity in the perfused bullfrog brain. At least three explanations may be postulated for the genesis of the parallel reduction of cytochromes (Kanno, 1988). First, the parallel reduction may be a sign of the transition from the resting state (state 4) to the anoxic state (state 5) as has been proposed in studies with the isolated mitochondria. Relative hypoxia may occur when the rate of oxidative phosphorylation is accelerated under limited oxygen supply to the mitochondria in the cells of isolated organs or tissues. The rate of mitochondrial respiration is governed by the interplay of four factors, among which the cytosolic [ATP] /[ADP][P_i] serves as the primary sensor of an altered need for energy production (Erecińska and Wilson, 1982). The major consumer of ATP in various secretory cells, including acinar cells and neurons, is (Na^+, K^+)ATPase. Furthermore, a substantial concentration gradient of oxygen is postulated to exist in the vicinity of mitochondoria in cells and tissue (Jones, 1986). Secondly, the functional status of (Na^+, K^+)ATPase activated by CCK or ACh may be regulated by intracellular processes that depend on cytosolic Ca^{2+}, $[Ca^{2+}]_c$, which also play a key role in stimulus-secretion coupling in the pancreatic acinar cell (Hootman, Ochs and Williams, 1985). Third, the increase in $[Ca^{2+}]_c$ may also be a factor in the parallel reduction of cytochromes, since it may also accelerate the intramitochondrial metabolic state (Denton and McCormack, 1986; Hansford, 1985).

REFERENCES

Chance, B. (1951) Rapid and sensitive spectrophotometry. III. A double beam apparatus. *Rev.Sci.Instrum.*, 22:634-638.

Chance, B. and Williams, G.R. (1956) The respiratory chain and oxidative phosphorylation. In: *Advance in Enzymology and Related Subjects of Biochemistry*.,ed. Nord, F.F., vol.17, Interscience Pub. Inc., New York, 65-134.

Chance, B., Cohen, P., Jobsis, F. and Schoener, B. (1962) Intracellular oxidation-reduction state in vivo. *Science*, 137:499-508.

Denton, R.M. and McCormack, J.G. (1985) Ca^{2+} transport by mammalian mitochondria and its rolein hormone action. *Am.J.Physiol.*, 249:E543-E554.

Ernster, L. and Schatz, S. (1981) Mitochondria: A Historical Review. *J.Cell Biol.*, 91:227-255.

Erecińska, M. and Wilson, D.F. (1982) Regulation of cellular energy metabolism. *Membrane Biol.*, 70:1-14.

Hansford, R.G. (1985) Relation between mitochondrial calcium transport and control of energy metabolism. *Rev.Physiol.Biochem.Pharmacol.*, 102:1-72.

Hootman, S.R., Ochs, D.L. and Williams, A. (1985) Intracellular mediators of Na^+-K^+ pump activity in guinea pig pancreatic acinar cells. *Am.J.Physiol.*, 12:G470-G478.

Kanno, T. (1987) The perfused pancreas for studying exocrine secretion. In: *In Vitro Methods for Studying Secretion*, eds. Poisner, A.M. and Trifaró, J.M., Elsevier, Amsterdam, 45-61.

Kanno, T. (1988) Roles of Mitochondria in Stimulus-Secretion Coupling in Pancreatic Acinar Cells. In: *Molecular Mechanisms in Secretion*, ed. Thorn, N.A., Treiman, M., Petersen, O.H., Muksgaard, Copenhagen, 315-323.

Kanno, T. and Matsumoto, T. (1986) Influence of extracellular Ca^{2+} on secretory and redox responses to CCK-8 in perfused rat pancreas. *Am.J.Physiol.*, 251:C10-C16.

Kanno, T. and Saito, A. (1982) Direct effect of CCK-8 stimulating respiratory chain in bullfrog brain. *Biomed.Res.*, 3:313-318.

Kanno, T. and Saito, T. (1988) Dose-related effects of TRH on electrical activity and redox response in the perfused bullfrog brain. *Abstracts of 8th International Congress of Endocrinology*, 163.

Kanno, T., Ikei, N. and Nakase, Y. (1981) Simultaneous measurements of cholecystokinin-induced changes in absorption spectrum of cytochromes, NAD(P)H-fluorescence, and enzyme output in isolated perfused rat pancreas. *Biomed.Res.*, 2:381-389.

Kanno, T., Saito, A. and Ikei, N. (1983) Dose-dependent effects of acetylcholine stimulating respiratory chain and secretion of isolated perfused rat pancreas. *Biomed.Res.*,4:175-186.

Kanno, T., Suga, T. and Yamamoto, M. (1976) Effects of oxygen supply on electrical and secretory responses of humorally stimulated acinar cells in isolated rat pancreas. *Jpn.J.Physiol.*, 26:101-115.

Kanno, T., Saito, T. and Saito, A. (1985) Cytochrome reduction coincides with electrical activity in perfused bullfrog brain. *Brain Res.*, 338:237-242.

Kanno, T., Matsumoto, T., Mori, M., Oyamada, M. and Nevalainen, T.J. (1984) Secretin prevents hyporeactive and morphological responses of rat pancreatic acinar cells to stimulation with supraoptimal concentration of cholecystokinin-octapeptide. *Biomed.Res.*, 5:355-370.

Kobayashi, S., Kaede, K., Nishiki, K. and Ogata, E. (1971) Microfluorometry of oxidation-reduction state of the rat kidney in situ. *J.Appl.Physiol.*, 31:693-696.

Kobayashi, S., Yoshimura, M., Shibahara, T., Nakase, T. and Yaono, T. (1981) Scanning spectrophotometry for dynamic study of the organ oxidative metabolism. *Biomed.Res.*,2:390-397.

Lowry, O.H., Rosebrough, N.J., Farr, A.L. and Randall, R.G. (1951) Protein measurement with the Folin phenol reagent. *J.Biol.Chem.*, 193:265-275.

Matsumoto, T., Kanno, T., Seo, Y., Murakami, M. and Watari, H. (1988a) Phosphorus nuclear magnetic resonance in isolated perfused rat pancreas. *Am.J.Physiol.*, 17:G575-G579.

Matsumoto, T., Kanno, T., Seo, Y., Murakami, M. and Watari, H. (1986b) ^{31}P NMR studies of the isolated perfused pancreas of the rat. *Biomed.Res.*, 7,suppl,2:29-31.

Matsumoto, T., Kanno, T., Seo, Y., Murakami, M. and Watari, H. (1988c) Dissociation between exocrine secretion and energy supply in rat pancreas during stimulation with high doses of secretagogues. *J.Physiol.Soc.Jpn.*, 50:394.

Murakami, M., Seo, Y. and Watari, H. (1988) Dissociation of fluid secretion and energy supply in rat mandibular gland by high dose of ACh. *Am.J.Physiol.*, 17:G781-G787.

Saito, T., Kanno, T., Tatemoto, K. and Mutt, V. (1983) Dose-related effect of neuropeptide Y stimulating respiratory chain in the bullfrog brain. *Biomed.Res.*, Suppl., 4:173-178.

Seo, Y., Murakami, H., Watari, Y., Imai, Y., Yoshizaki, K., Nishikawa, H. and Morimoto, T. (1983) Intracellular pH detection by a ^{31}P NMR technique. The second dissociation constant of phosphoric acid in biological system. *J.Biochem.*, 94:729-734.

Shibuya, I. and Kanno, T. (1985) Calcium-dependent secretory and redox response to CCK-8 in isolated perfused rat pancreas. *Am.J.Physiol.*, 248:C228-C234.

Yang, C.-C. and Legallais, V. (1954) A rapid and sensitive recording spectrophotometer for the visible and ultraviolet region. I. Description and performance. *Rev.Sci.Instrum.*, 25:801-807.

Part 5: MOLECULAR APPROACHES TO STRUCTURE AND FUNCTION OF TRANSPORTERS

Chapter 18

Anion transport systems: continuous monitoring of transport by fluorescence (CMTF) in cells and vesicles
Ioav Cabantchik

Chapter 19

The molecular biology of sugar transport proteins
Peter Henderson, Elaine Davis, Brian McKeown and Martin Maiden

Chapter 20

Chemical and molecular probes of nucleoside transport mechanisms in mammalian tissues
Simon Jarvis

18

ANION TRANSPORT SYSTEMS: CONTINUOUS MONITORING OF TRANSPORT BY FLUORESCENCE (CMTF) IN CELLS AND VESICLES

Z. Ioav Cabantchik and Ofer Eidelman

Department of Biological Chemistry
The Hebrew University of Jerusalem
Jerusalem, 91904, Israel

INTRODUCTION

Anion transport mechanisms are found in mammalian cell membranes in the form of exchangers or antiporters (Passow, 1987; Aronson, 1989), cation-anion cotransporters or symporters (Hoffman, 1986; O'Grady *et al.*, 1987) and Cl channels (Greger, 1985, 1988; Gogelein, 1988; Frizzell *et al.*, 1986). The transport of the anions across both biological and artificial membranes are studied by following the transcompartmental movement of physiologically relevant substrates such as Cl^-, HCO_3^-, SO_4^{-2} or $H_2PO_4^-$. For electroneutral mechanisms, the anion flux can be traced by a variety of physical or chemical techniques, the most widely used being tracing the movement of a radiolabeled anion by separating the compartments at various time points, sampling their contents, and measuring the amount of tracer in one of them. This method is, however, limited by the space available to the substrate, its specific activity and total radioactivity, the efficiency and speed of separation relative to the actual transport rates, as well as by the capacity of the biological system to retain its structural integrity during separation steps. Since this method is discrete in nature, the amount of information it can yield is usually limited and often not sufficiently precise for a thorough kinetic evaluation. For charge-conducting transport systems, the classical and most straightforward approach is the electrophysiological one, which is used in various forms according to the level of information required and the nature of the biological system. Electrophysiological techniques are of an invasive nature and most commonly inaccessible to small organelles. On the other hand for conductive systems operating in cells and vesicles, isotopic techniques can still be very useful when used under conditions where a large <u>chemical</u> gradient of the ion is established across the membrane and a radio-labeled tracer of the ion is placed at the low concentration side (Garty *et al.*, 1983). In such conditions, the diffusion potential created across the

Cell Membrane Transport, Edited by D. L. Yudilevich *et al.*
Plenum Press, New York, 1991

membrane by the transportable ion will drive the isotopic species into the vesicles at a rate commensurate with the potential and the intrinsic conductivity of the channel tested (Landry *et al.*, 1987; Breuer, 1989).

Fluorescence can be considered as a suitable alternative to the classical approaches used for studying anion transport for various reasons:

(a) it has the requisite sensitivity and temporal resolution for following movement of either fluorescent-substrates or fluorescence-quencher-substrates across membranes of relatively small compartments;
(b) it permits continuous monitoring of transport (CMTF) without need for periodic separation of compartments;
(c) it allows selection of a discrete area of a given compartment (cellular, organellar or extracellular) for monitoring localized transport events (*e.g.*, by using physical-optical focusing, chemical-impermeant light quenchers, or biochemical-processing of a permeable dye precursor by cellular or organellar enzymes so as to produce the dye indicator at the site required (for recent reviews see Loew, 1989; Eidelman and Cabantchik, 1989a,b).

An additional advantage of CMTF is its modest demand for biological material such as cells or proteoliposomes, both because of its high signal and temporal sensitivity and because it uses the same biological sample for the entire flux measurement. The limitations inherent in "sampling" methodologies, which are especially manifested when studying transport with radiotracers in vesicles, can largely be overcome by using CMTF. Typical vesicles isolated from cell membranes or reconstituted proteoliposomes are of 200 nm diameter, and can accommodate about 100,000 molecules or ions per vesicle when using 100 mM solutions. Thus two serious problems become apparent: First, the time required for the solute contents of the vesicles to be exchanged with that of the medium (t_{eq}) is dictated by the transfer rate of the transporter, the surface density of transporters and the area to volume ratio of the vesicles. Even for vesicles harboring in their membrane a single functional unit of a channel (with a transfer rate of 10^6 ions per sec) or a fast carrier-like entity (with turnover rates of 10^4-10^5 molecules per sec) t_{eq} is relatively short (in the range of 100 msec for channel-containing and 1-10 sec for carrier-containing vesicles). Second, the amount of labeled material which can be enclosed in cells or vesicles determines the amount of biological material required for flux measurement. Typical vesicles isolated from animal cell membranes have 0.3 mg phospholipid/mg protein and have an internal volume of 1 pl/mg protein or 3 μl/mg phospholipid. Working concentrations of substrates enclosable in vesicles are in the order 10^8 cpm/ml radiolabeled material. Therefore, the amount of vesicles required for obtaining <u>a single time point sample</u> of 10^3 cpm/ml is in the order of 10 nl/μl (or 10 μg protein). In contrast, an equivalent amount of vesicles loaded with 10 mM fluorescent probe can yield <u>a complete flux profile</u> (or even a full kinetic profile of a transport system) using standard fluorescence equipment.

In this chapter we describe the basic principles of CMTF as applied to anion transport of various cell systems, the methodologies and approaches used for the selection of substrate or of fluorescence modulators, the acquisition, calibration, and data analysis, and various applications of CMTF methodologies to selected anion transport transporters, primarily to anion exchangers and Cl channels. For a more detailed description and applications of CMTF the reader is referred to recent reviews (Eidelman and Cabantchik, 1989a,b; Loew, 1988).

PRINCIPLES

Monitoring of anion transport by fluorescence is based on variations in fluorescence caused by translocation of anions between compartments. The anionic substrate can in principle be a fluorophore, a quencher of another fluorophore or even a modulator of a physical property which can be monitored by fluorescence, such as membrane potential or pH. An anionic substrate, specific for a given anion transport system (the permeant) is designed or selected together with a suitable impermeant substance, such that (a) one of them is fluorescent or potentially fluorescent and (b) interactions between the two will lead to a swift and sizeable change in the fluorescence signal. Transport is followed as either an increase or a decrease in fluorescence signal depending on the characteristics of the permeant-impermeant couple.

Several factors can cause compartment-dependent variations in optical properties of fluorescent probes affected by substrate concentration:

(a) quenching by collisions, energy transfer or complexation with a suitable substance (*e.g.*, anti-fluorophore antibody, a chelator, or a low molecular weight collision quencher);
(b) absorption of the exciting or the emitted light by a substance confined to the internal compartment such as in pigment containing cells or organelles (*e.g.*, hemoglobin in erythrocytes, chlorophyll in chloroplasts or cytochromes in chromaffin granules) or exogenously added impermeant chromophores (in reconstituted vesicles);
(c) transmembrane redistribution of dyes as a result of membrane potential changes generated by the movement of a permeant ion;
(d) physical selection of a single compartment (*i.e.*, a single cell or organelle) by using a light detector mounted on a fluorescence microscope focused on the given compartment.

Some methods have general applicability to all membranous enclosed compartments, while others are more suitable to cells because of the chemical nature of agents or the feasibility of confining the impermeant partner in a given compartment. For example, fluorescent substrates are useful for measuring transport in the efflux mode from a cellular or vesicular compartment into a medium containing an impermeant quencher, such as antifluorophore antibody or from a pigment-containing compartment in which fluorescence is almost null into an optically clear medium. The CMTF method depends entirely on the possibility of having one member of the couple mostly confined to one compartment throughout the entire transport

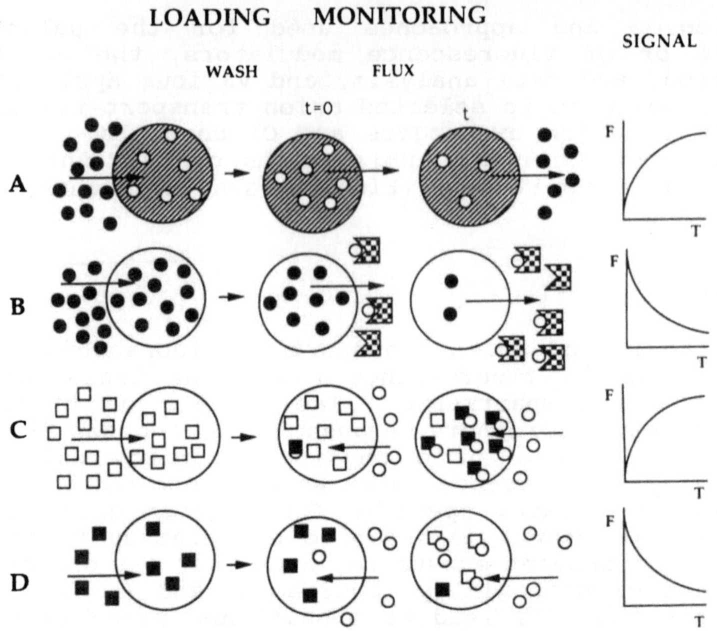

Figure 1. MODES OF CMTF IN BIOLOGICAL SYSTEMS (adapted from Eidelman and Cabantchik, 1989a, with permission). *The basic steps of measuring transport (loading, washing and monitoring) are depicted in four different modes of CMTF. Fluorescent substrate molecules are shown as solid circles and fluorescent indicators as solid squares, and their respective non-fluorescent forms as empty circles or squares. The time (T) dependent changes in fluorescence intensity (F) in the four modes described are shown schematically in the right side. A. Efflux of a* *__fluorescent substrate__* *from pigmented cells, vesicles or organelles (shadowed backgrounds). Substrate fluorescence is quenched by light absorption. B. Efflux of* *__fluorescent substrate__* *from non-pigmented cells or vesicles (clear background) into quencher (checkered forms) containing medium. Substrate fluorescence is quenched upon interaction with quencher. C. Influx of substrate into* *__indicator__* *containing cells or vesicles. Indicator fluorescence changes upon complexation with substrate. D. Influx of* *__quenching substrate__* *into cells or vesicles loaded with a suitable fluorophore. The signal decreases due to collisions with substrate.*

measurement and on the selective traffic of substrate between defined compartments.

The change in fluorescence associated with solute movement across membranes is usually attained by optical modulation of fluorophore properties as a result of interactions between the fluorophore and the modulator.

Quenching is an extreme case of modulation which can be either of a dynamic (collisional) or of a static (complex forming) nature. Dynamic quenching is given by the classical Stern-Volmer equation (Lakowicz, 1983):

$$\frac{F_0}{F} = 1 + \frac{[Q]}{K_{SV}} \qquad \text{(Eq. 1)}$$

where F_0 and F represent the fluorescence emitted in the absence or the presence of a given concentration [Q] of quencher, respectively, and K_{SV} is the Stern-Volmer constant for quenching, *i.e.*, the concentration needed for attaining half maximal quenching. This equation can be used to calculate the instantaneous concentration of substrate in a given compartment. Moreover, when applied to a continuous trace of fluorescence intensity, it enables the determination of the instantaneous rate of transport over the entire flux profile. Under appropriate experimental conditions, that is using CMTF and starting with a sufficiently high concentration of the anionic substrate ($[S]>3*K_s$, a single sample can be used for assessing the kinetic parameters of the transport system (see below) and with present day state of the art technology a similar approach is applicable for studying transport kinetics in single cells.

An example of complexation-type quenching is the use of anti-fluorophore antibodies as efficient (K_d < 1 nM) quenchers of fluorescence (Darmon *et al.*, 1982). The quenching mechanism is usually attributed to the formation of electron donor- acceptor or charge transfer complexes. Fluorescent dyes (non-substrate) or fluorescence quenchers (non-substrate) can be loaded into cells or vesicles by prolonged incubations at relatively high temperatures, while fluxes are conducted at temperatures at which probe leakage is minimal during the time scale of substrate flux measurement and does not compromise it. Incorporation of impermeant molecules into vesicles can be accomplished by freeze-thaw manipulations or during detergent removal for reconstituted proteoliposomes (Kasahara and Hinkle, 1976; Cabantchik and Darmon, 1985). For intact cells, incorporation of impermeant molecules can be attained by the cell scraping technique (McNeil *et al.*, 1984) or by endocytosis of material encapsulated in liposomes in hyperosmotic media and dye release into cytoplasm after restoration of isotonicity (Rothenberg *et al.*, 1983). Ideally, impermeant dyes or indicators could be chemically modified into permeant precursors which are rendered impermeant by intracellular enzymatic hydrolysis, such as in the case of pH and Ca indicators. Pertinent examples include fluorophores such as -COOH containing molecules which were modified into their acetoxy derivatives (see Molecular Probes Inc. catalogue for list of materials)). A free bonus acquired by using these relatively permeable precursors is their low fluorescence efficiency, thus minimizing the step normally required for eliminating the contaminating extracellular fluorescent material. Nonetheless, in all these examples removal of extracellular or extravesicular probe (quencher, fluorescent indicator or substrate) prior to transport measurements is essential.

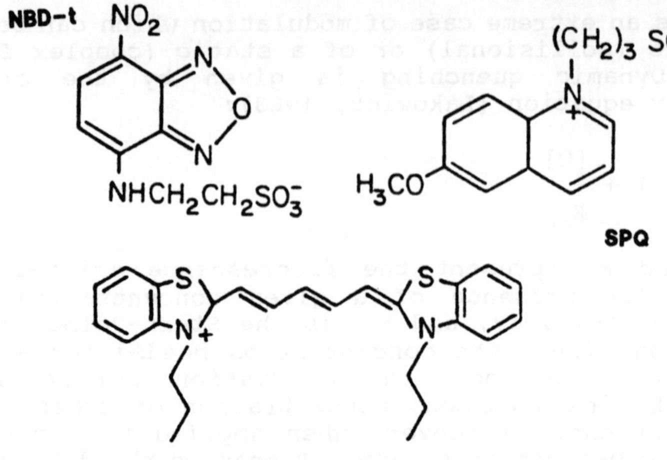

Figure 2: CHEMICAL STRUCTURES OF FLUORESCENT PROBES FOR ANION TRANSPORT SYSTEMS: *N-(-2-aminoethylsulfonate)-7-nitrobenz-2-oxa-1,3-diazole (NBD-taurine, Eidelman and Cabantchik, 1989 a,b) a fluorescent substrate for the CBE in red blood cells; 6-methoxy-n-(-3-sulfopropyl) quinolinium (SPQ, Illsley and Verkman, 1987) a fluorescent indicator for Cl (and other halides) and DiS-C_3-(5) a cationic potentiometric dye (Loew, 1989; Cabrini and Verkman, 1986)*

An indirect method for following anion in cells and in vesicles is based on the use of fluorescent dyes which can sense membrane (diffusion) potential changes generated by the net movement of the anions across the cell or vesicle membrane (Breuer, 1989). The dye added exogenously to the test system either adsorbs or partitions into the membranous systems and the diffusion potentials generated by fluxes of ions via particular conductive pathways affect the transmembrane distribution of the various molecular forms of the dye and the associated fluorescence signals (Loew, 1988). The method is economical, but demands careful calibrations and controls, as the amphiphilic fluorescent dyes tend to adsorb to charged and hydrophobic surfaces and to various biological materials.

ANALYSIS OF CMTF DATA

Since the fluorescence record is well resolved from the fractional second to the hour range even with standard fluorescence equipment, the data can be used to their fullest extent for calculating the kinetic parameters of the transport system. The analysis of CMTF kinetic data is based on classical transport (enzyme) kinetics schemes. The first step in the analysis is the conversion of fluorescence intensity record to substrate concentration units. This part of the calculation is where knowledge of the factors affecting fluorescence signal is needed. These factors include on the one hand, substrate concentration-dependent effects which have to be calibrated, and on the other hand, "environmental" (*i.e.*, artifactual) factors such as light scattering, autofluorescence, effects of pH and other solutes, etc.) all of which have to be subtracted as background. The case of the fluorescent substrate is the simplest, as the fluorescence intensity is usually linearly related to the concentration of the fluorophore in the compartment where quenching is lowest. However, even in this case, it needs to be established that fluorescence intensity is linear with dye concentration and is not affected by factors such as inner filter effect etc. In the different case where the fluorophore is impermeant and the substrate acts as a modulator, the relationship between fluorescence intensity and substrate concentration depends on the dynamics of the interaction between the two partners, the quencher and the fluorophore. In the simplest case of collisional quenching, the concentrations of free dye and free substrate virtually do not change following their interaction, but in the case of complexation both these concentrations are constantly changing.

Data acquisition

The raw data from a CMTF experiment is obtained as a continuous trace of fluorescence intensity F(t) versus time (t). The initial fluorescence intensity signal F(0), generated immediately after addition of cells or vesicles, is usually obtained from the recorded fluorescence trace by extrapolation to t=0. This signal is composed of the background signal F_{bg} due to light scattering, to some external probe remaining even after the wash, and from instrumental noise and stray light. For the particular case of pigment-free cells or vesicles used in conjunction with anti-fluorophore antibodies, F(0) can also be obtained by addition of excess of a non-fluorescent hapten, thus leading to restoration of the original fluorescence, such as dinitrophenyl-taurine added to NBD-taurine-anti-NBD complexes formed in the assay mixture at the end of the flux (Darmon *et al.*, 1982). The background (mainly due to light scattering) is normally assumed to be constant when no major volume changes of cellular or vesicular compartments occur, otherwise the data should be corrected by subtracting the measured light scattering intensity of a blank system (*i.e.*, with fluorophore omitted). The fluorescence signal attained at equilibrium, commonly referred to as F_∞, or F_{eq} is obtained either by waiting for the time required to reach equilibrium or by addition of external factors, such as ionophores or detergent to lyse the membranes, depending on the system in question.

Calibration of fluorescence and determination of substrate concentration

The relationship between the substrate concentration and the associated fluorescence signal is of great practical importance in quantitative analysis of CMTF data. It is only after the determination of this relationship that the working range of substrate/dye/quencher concentrations should be chosen while giving attention to the following points:

a) the range of substrate concentrations at which signal to noise values are adequate for calculating flux rates;
b) the range of substrate concentration suitable for

estimating the apparent K_s ($0.3K_s < [S] < 3K_s$); and

c) possible direct or indirect interference of the substrate or the impermeant dye with the biological system.

Thus, the optimal dye concentration and the optimal range of substrate concentrations have to be selected according to the particular details of the permeant-impermeant couple used, as well as those of the transport system under consideration. Calibration curves of F(t) versus [S] for the fluorescent substrate or indicator have to be done with the probes in the appropriate compartments. In the external medium, the calibration is usually done with fluorophores in the micromolar range. Deviation from linearity occurs mainly because of inner filter effects. Caution has therefore to be exercised to limit the total amount of dye in a flux assay so that the final concentration attained is in the linear range of F versus [S]. For the internal compartment, samples are prepared by loading the cells with different concentrations of probe. The calibration is done by measuring the initial signal either in the presence of flux inhibitors, or by extrapolation of the signal intensity to the time of mixing, t = 0. When impermeant quenchers, such as anti-fluorophore antibodies (Darmon *et al.*, 1982), are used in the external medium, the minimal amount of quencher required should be determined before flux measurements. This is important since quenchers might interfere with the transport activity (collisional quenchers) or cause excessive background signal (antibodies). The amount of quencher required is found by adding the anticipated final concentration of fluorescent substrate in a cuvette and adding increasing concentrations of quencher needed until fluorescence signal is substantially reduced. However, in the case of antibodies, where excessive amounts might contribute significantly to the fluorescent signal, we have found it useful to determine the quenching capacity of the antibodies, that is the amount of fluorophore which can be maximally quenched by a given amount of Ab. This is done by placing a known amount of Ab in the cuvette and adding increasing concentrations of fluorophore and the quenching capacity is determined by extrapolating the increasing part of the signal vs. [fluorophore] plot back to the background (*i.e.* with Ab alone) level.

For the case of CMTF based on an impermeant fluorescent indicator and a fluorescence modulating substrate, the fluorescence signal is not directly proportional to the substrate concentration in either compartment, but follows a more complex function. For a collisional quencher it is given by the Stern-Volmer equation (see above Eq. 1), and for a titratable indicator, the dependence is that of a bimolecular chemical reaction. However, for either case, the dependence can be expressed as:

$$F(t) = \frac{F_{max}}{(1+[S_q]/K_{SV})} + F_{bg} \qquad \text{(Eq. 2)}$$

where K is either K_{SV}, the Stern-Volmer constant of the collision quenching substrate, or K_s, the dissociation constant for the binding of the substrate to the titratable indicator. In order to determine the appropriate K constant of interaction between the substrate and the indicator and the concentration range where this relation is applicable, it is necessary to obtain a reliable calibration curve. This is not a trivial point, since the fluorescence signal is affected on the one hand by self quenching and inner filter effects which arise when the fluorophore is present at relatively high concentrations or on the other hand, at relatively low concentrations of dye, a relatively large fraction of the dye is bound to substrate, so that Eq. 2 no longer holds.

DATA ANALYSIS

Fluorescent substrate: CMTF is commonly used in a mode whereby the fluorescent substrate is present in trace concentrations. In these conditions ($[S]<<K_s$) the probe does not interfere with the normal performance of the transport function (translocation of the primary substrate) but only reports about the kinetic status of the system (*i.e.*, the transport rate of the probe is proportional to the concentration of free transport sites at the cis side of probe addition (Eidelman and Cabantchik, 1989a,b). A more versatile and kinetic informative mode of CMTF is based on the use of substrate-analog concentrations higher than the apparent K_s ($[S(0)]>>K_s$). In this mode, a single fluorescence trace can provide the information required for obtaining the values of (apparent) K_s and V_{max} of probe transport. An example of such a trace taken from a typical efflux experiment and several analytical procedures used for obtaining the kinetic parameters are shown in Fig. 3. In the kinetic analysis it is assumed that either the fluorescence signal is linear with substrate concentration or that the original data record has been corrected accordingly.

The fluorescence intensity originating from probe present either in the intracellular (i) or the external (e) compartment (*e.g.*, red cells) is given by:

$$F(t) = F_{bg}+\alpha_i*N_i(t)+\alpha_e*N_e(t) \qquad \text{(Eq. 3a)}$$

where N_i and N_e are the amounts of probe in the respective internal and external compartments and α_i and α_e are the respective proportionality coefficients of fluorescence signal associated with the amount of probe present in a given compartment i or e. In the efflux mode all the probe is in the inner compartment at t=0 and as $t \rightarrow \infty$ all probe is in the external one, so that

$$F(0) = F_{bg}+\alpha_i*N_T \qquad \text{(Eq. 3b)}$$

and

$$F(\infty) = F_{bg}+\alpha_e*N_T \qquad \text{(Eq. 3c)}$$

where $N_T = N_i + N_e$ is the total amount of probe. Combining equations 3a-3c, we obtain

$$\frac{N_i(t)}{N_T} = \frac{F(\infty)-F(t)}{F(\infty)-F(0)} \qquad \text{(Eq. 4a)}$$

and

$$\frac{N_e(t)}{N_T} = \frac{F(t)-F(0)}{F(\infty)-F(0)} \quad \text{(Eq. 4b)}$$

Thus, the probe concentration in the internal compartment is:

$$S_i(t) = \frac{N_i}{V_i} = \frac{N_T}{V_i} * \frac{F(\infty)-F(t)}{F(\infty)-F(0)} = S_0 * \frac{F(\infty)-F(t)}{F(\infty)-F(0)} \quad \text{(Eq. 5)}$$

where [S(O)] is the initial internal concentration of probe, and V_i is the effective volume of cells or vesicles present in the assay system. It should be stressed that [S(0)] inside the cells is not necessarily equal to the concentration of substrate used in the loading medium, since Donnan effects or membrane potentials have marked effect on the transmembrane distribution of charged species.

<u>Trace mode</u>: ([S(O]<<K_s) In this mode the efflux of probe follows either a single exponential decay (for pigment-free system) or a single exponential increase (for pigmented system) with a characteristic rate constant, k. This constant can be evaluated by a variety of methods as shown in Fig. 3:

(a) By obtaining the tangent to the curve at the start of the flux and dividing it by the total fluorescence:

$$k = \text{initial slope}/[F(\infty)- F(0)] \quad \text{(Eq. 6a)}$$

Though straightforward, this method is rather subjective and often unsuitable for data obtained with either relatively fast or very slow transport systems.

(b) By logarithmic conversion of the fluorescence intensity:

$$\log[F(\infty)- F(t)]= -k*t \quad \text{(Eq. 6b)}$$

which, when plotted against t, yields a straight line with a slope of k.

(c) The fluorescence data can be fitted either directly to an exponential function and analyzed by a nonlinear least squares regression program or, alternatively, it call be linearized as in (b) and analyzed by a linear least squares regression.

(d) Calculating the instantaneous rate constant k(t) for each point along the entire fluorescence trace according to Eidelman and Cabantchik, 1989a,b):

$$k(t) = \frac{1}{F(\infty)-F(t)} * \frac{dF(t)}{dt} \quad \text{(Eq. 6c)}$$

For a simple exponential curve k(t) should be constant throughout the entire profile (though the region near the plateau level shows excessive calculation noise).

In the trace mode, the calculated k values are useful for assessing and comparing the effects of external conditions such as temperature, pH, substrate composition, inhibitors, etc. The k values can also be used to study the transport system in nonequilibrium conditions, that is, when k changes with time as a result of changes in physical or chemical factors which affect the system. For such a case, calculation of the instantaneous rate parameters (Eq. 6c.) was found to be very useful.

Michaelian mode: (Fig. 3) $\{[S(O)]>>K_s\}$ The unidirectional flux of fluorescent substrate can be used to determine the kinetic parameters of the system, K_s, and V_{max} when the initial concentration of probe is considerably larger than K_s. The simplest and most common means to obtain the kinetic parameters is by calculating both the S(t) at selected time points (t) and the corresponding velocities or flows (v=dS/dt) or fluxes (J=flow/surface area), plot the values either according to the Michaelis-Menten formulation (Fig 3B) and fit the data by non-linear regression or use one of the more familiar linearized versions and fit the data by linear regression (Fig. 3D). The experimental data can also be fitted to the integrated form of the Michaelis-Menten equation, using time as the dependent variable:

$$t = \frac{N_T}{V_{max}} * f_t - \frac{K_s * N_T}{V_{max} * [S(0)]} * \log(1-f_t) \qquad \text{(Eq. 7)}$$

where $f_t = [F(t)-F(0)]/[F(\infty)-F(0)]$ and the data are analyzed by nonlinear least squares regression.

Analogously, one can linearize the equation and obtain K_s (or apparent K_s) and V_{max} by linear least square analysis of t/f_t versus $[\log(1-f_t)/f_t]$, which is based on:

$$\frac{t}{f_t} = \frac{N_T}{V_{max}} - \frac{K_s * N_T}{V_{max}} * [S(0)] * \frac{\log(1-f_t)}{f_t} \qquad \text{(Eq. 8)}$$

The value of N_T/V_{max} is given by the y-intercept and that of $[S(O)]/K_s$ by the x-intercept (Fig. 3C). N_T can be determined from the fluorescence signal at t=0 or t=∞ and the proper calibration curve.

Fluorescence modulating substrate: For a collision quencher, the fluorescence signal of an impermeant indicator affected by collision-quenching with permeant substrate S_q follows the Stern Volmer equation (Eqs. 1&2), which can be rearranged to give:

$$F(t) = F_{bg} + \frac{F_{max}}{1+[S_q(t)]/K_{SV}} \qquad \text{(Eq. 9)}$$

where F_{bg} is the background and F_{max} is the fluorescence emitted by the indicators in the absence of quencher, *i.e.* when substrate concentration is zero. These two parameters can usually be estimated from the measured fluorescence intensity values at t=o and t->∞:

$$F_{max} = (1 + \frac{K_{SV}}{[S_q(\infty)]}) * [F(0)-F(\infty)] \qquad \text{(Eq. 10)}$$

$$F_{bg} = F(0) - F_{max} \qquad \text{(Eq. 11)}$$

and thus,

$$F(t) = F(0) - F_{max} * \frac{[S_q(t)]}{K_{SV}+[S_q(t)]} \qquad \text{(Eq. 12)}$$

where F(0), F(∞) and F(t) are the fluorescence intensities at times t=0, t->∞ and t, respectively, while $[S_q(0)]$, $[S_q(\infty)]$ and $[S_q(t)]$ are the corresponding concentrations of the quenching permeant.

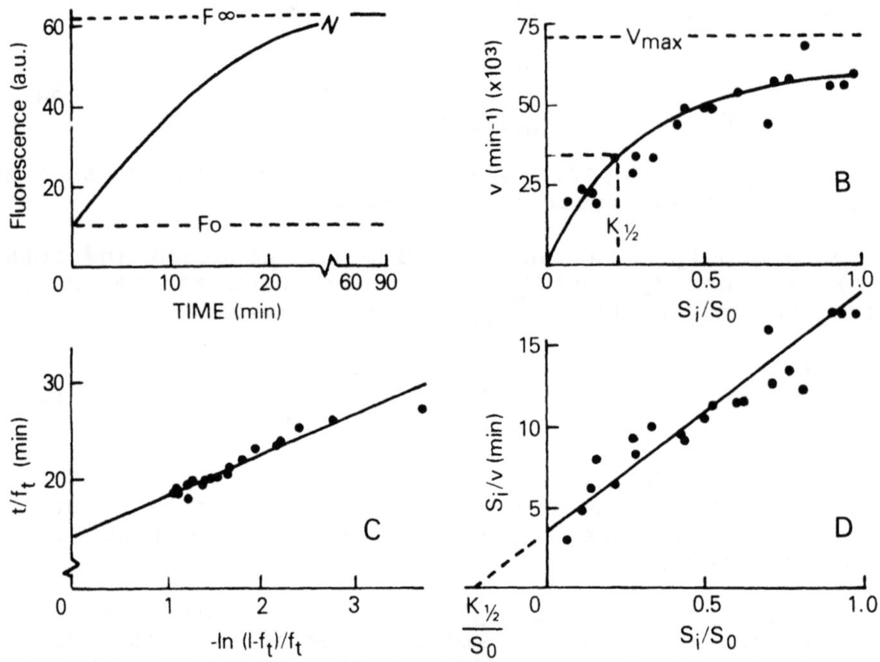

Figure 3. KINETIC ANALYSIS OF EFFLUX IN MICHAELIAN MODE: *The fluorescence (F) profile of NBD-t efflux from red cells is depicted in A and the calculated flow of NBD-t v=dF/dt versus f_t with the respective non-linear least squares (based on the Michaelian formulation) are given in B, and according to eq. 8 in C (solid line is the linear least squares fit). In D the data are analyzed according to the classical Hanes-Woolf representation (adapted with permission from Eidelman and Cabantchik, 1989).*

The relationship between the fluorescence intensity and permeant concentration deserves some further considerations, since it determines the range of $[S_q]$ which is useful for estimating the instantaneous concentration of substrate and the corresponding fluxes. Rearranging Eq. 10 we obtain:

$$[S_q] = \frac{K_{SV}}{\frac{F_{max}}{F(0)-F(t)} - 1} \qquad \text{(Eq. 13)}$$

and the error in the determination of $[S_q]$ is given by

$$\delta[S_q] \approx (K_{SV}+S_q) * \frac{2\delta F}{F(t)-F_{bg}} \qquad \text{(Eq. 14)}$$

or,

$$\delta[S_q] \approx \frac{2\delta F}{F_{max}} * \frac{(K_{SV}+S_q)^2}{K_{SV}} \qquad \text{(Eq. 15)}$$

where δF represents the measurement error (noise) in determining F, and $\delta[S_q]$ is the calculation error in determining the corresponding substrate concentration $[S_q]$. Thus, the accuracy of the calculations is reduced as S increases and F decreases, and at concentrations considerably higher than K_{SV}, the noise increases as $[S_q]^2$. An important corollary emerging from Eq. 14b is that the noise in determining $[S_q]$ is inversely proportional to F_{max}, or the amount of dye indicator present in the cuvette, so that the measurements are equally sensitive whether performed with, for example, 10 µg protein sample loaded with 1 mM of the Cl-sensitive indicator SPQ, or a 1 µg protein sample loaded with 10 mM dye. If dye can be loaded at higher concentrations without affecting the system, it is better to decrease the amount of cells or vesicles and increase dye concentration, since higher amounts of vesicles, in addition to being more "wasteful", also increase the light scattering background level.

The rate of quencher entry into the cells can be calculated from the time dependent changes in fluorescence signal since:

$$\frac{d[S_q]}{dt} = -K_{SV} * \frac{F_{max}}{[F(t)-F_{bg}]^2} * \frac{dF}{dt} \qquad \text{(Eq. 16)}$$

In this case, as in the fluorescent substrate case, measurements can be done in the Michaelian mode for determining the kinetic parameters (K_s and V_{max}) from an experiment with a single sample. When measuring substrate efflux from cells or vesicles, Eq. 8 can be used by substituting

$$f_t = 1 - \frac{[S_q(t)]}{[S_q(0)]} \qquad \text{(Eq. 17)}$$

The uncertainty in determining $d[S_q]/dt$ is directly related to the uncertainty in $[S_q]$ (Eq. 14) and so the most reliable range for calculating fluxes is the when $[S_q]<<K_{SV}$. Thus, when employing the method in the Michaelian mode (Eq. 8), an indicator has to be found such that $K_{SV}>K_S$, since the reliable region is that for which $[S_q]>K_s$ (see above).

The transport systems for which CMTF has been successfully applied are the anion exchangers and Cl-channels of mammalian cell membranes, both of which have been studied in native systems as well as in isolated and reconstituted forms. The Cl-HCO_3 exchanger (CBE) is the most prominent transport system of red blood cells (RBC) and has been widely explored in terms of its physiological role (Cabantchik *et al.*, 1978; Knauf, 1979; Gunn, 1979; Jennings, 1985; Frohlich and Gunn, 1986; Passow, 1987), kinetic mechanisms (Cabantchik *et al.*, 1978; Knauf, 1979; Gunn, 1979; Jennings, 1985; Frohlich and Gunn, 1986; Passow, 1987) chemical architecture (Cabantchik *et al.*, 1978; Jay and Cantley, 1986; Passow, 1987) and encoding gene structure (Kopito and Lodish, 1985; Demuth *et al.*, 1986; Kopito *et al.*, 1987; Cox and Lazarides,1988; Alper *et al.*, 1988; Tanner *et al.*, 1988) and it has provided one of the best conceptual and methodological models for understanding protein mediated transport. In red cells, CBE subserves the systemic transport of CO_2 from tissues to lungs, while in other cells, CBE activity might be involved directly in regulating cellular pH homeostasis (Reinertsen *et al.*, 1988; Ganz *et al.*, 1989) and, in conjunction with other transport systems, it is assumed to participate in cell volume regulation (Finn, 1985; Cala, 1985; Hoffman, 1986; Spring and Ericson, 1983) and epithelial salt secretory or absorptive processes (Greger, 1988; Muallem, 1988; Tago *et al.*, 1986; Schuster *et al.*, 1987; Wagner *et al.*, 1987; Schwartz *et al.*, 1985). The two other major mammalian cell membrane Cl-transporters whose functions have been identified are the (Na,K,Cl) cotransporter (SPCC) and the Cl channel (ClCh). The relative activities of these systems vary among cells, and in the case of epithelial cells, they vary even between luminal and contraluminal membranes. The SPCC has been recognized (Hoffman, 1986; O'Grady *et al.*, 1987) as an essential component of net salt and water movement in both absorptive and secretory epithelia (Greger, 1988) and down-regulation of cell volume upon hyperosmotic shock and hormonal stimulation (Hoffman, 1986). Various types of ClCh have been described in electrophysiological terms, primarily in salt secretory and absorptive epithelia (Greger, 1987; Frizzell *et al.*, 1986; Gogelein, 1988) nerve (Hamill *et al.*, 1983) and muscle (Blatz and Magleby, 1985) lymphocytes (Chen *et al.*, 1989) and intracellular membranes (Glickman *et al.*, 1983; Xie *et al.*, 1983). One particular type of ClCh activity which is normally hormone-regulated (Donowitz and Welsh, 1986) was found to be defective in salt secretory tissues of cystic fibrosis patients, as evidenced by an apparent lack of response to hormonal activation (Frizzell *et al.*, 1986; ; Welsh and Liedtke, 1986).

Specific inhibitor of Cl transport systems

In addition to physiological and mechanistic differences, in practice the three major Cl transport systems are identified by their differential susceptibilities to pharmacological agents with demonstrable inhibitory activity (Fig. 4):

CBE is highly susceptible to disulfonic stilbene derivatives, the most prominent representatives of which are the covalently reactive derivative DIDS

(4,4'-diisothiocyano-2,2'- disulfonic stilbene) and the reversibly acting derivatives DNDS and DBDS (the 4,4'-dinitro and the 4,4'-dibenzamido analogs of DIDS, respectively). In RBC, these derivatives inhibit CBE competitively in the lower μM range (Cabantchik *et al.*,1978; Barzilay *et al.*, 1979; Passow, 1987).

SPCC is specifically inhibited by nM-μM concentrations of loop diuretics such as furosemide, piretanide, benzmetanide and bumetanide (O'Grady, 1987; Schlatter *et al.*, 1983) particularly in the presence of the substrates.

ClCh is particularly sensitive (in the nM-μM concentration range) to phenyl-aminoalkyl-phenylcarboxylates, of which the nitro-benzoate derivatives are among the most potent (Wangemann *et al.*, 1986) and to indanyloxyacetic derivatives (Landry *et al.*, 1987).

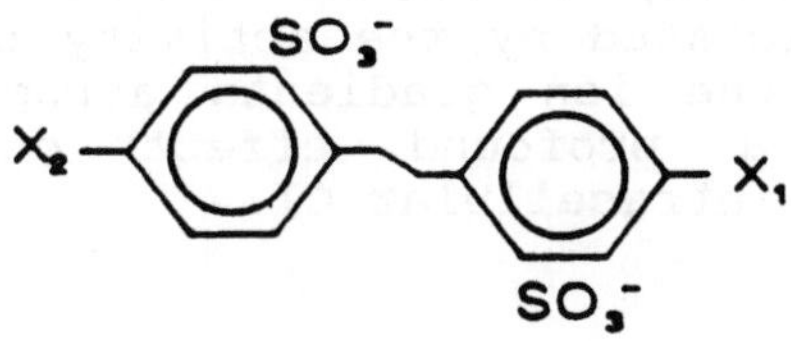

1

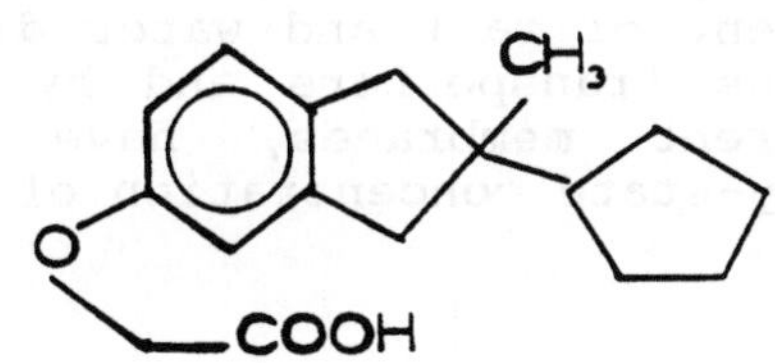

2

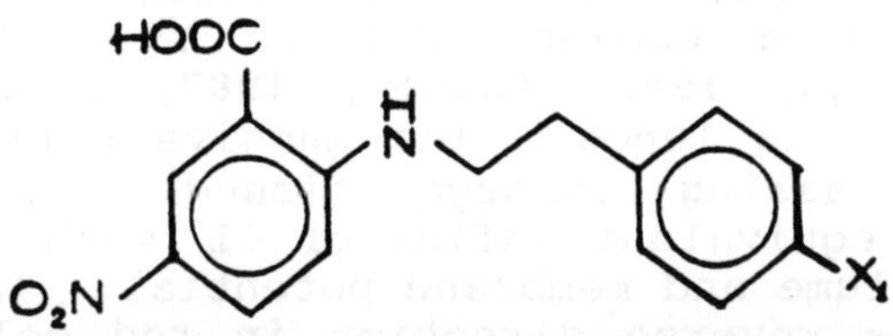

3

Figure 4. CHEMICAL STRUCTURES OF BLOCKERS OF ANION TRANSPORTERS: *(1) Disulfonic stilbene derivatives (IC_{50} 1-50 M for CBE); (2) indanyloyloxyacetic acid; (3) nitrophenylanthranilates (IC_{50} 1-50 M for ClCh).* $X_{1,2}$*=-NCS,-NN$^+$,-NHCOpheNNN,-NO$_2$,-NHCOphe,-NNN.*

These inhibitors, used as biochemical, physiological and pharmacological tools, have contributed enormously to our present understanding of anion transport. In fact, the inhibitory action of various "specific" blockers has frequently been employed as the major criterion for identification of particular transport activities in intact cells or tissues. However, the inhibitory efficacy of the various agents for a particular transport system varies considerably among cell types and organisms, and cross inhibitory effects are not unusual. Therefore, the spectrum of specificities would need to be determined on an individual basis for each probe, transport system and cell type, before any conclusion can be drawn as to the operation of a particular system in a given cell.

With the exception of CBE in red blood cells, our knowledge of the pathways for Cl transport and the forces involved in Cl translocation in mammalian cell membranes is very limited. The fact that various Cl transporters are present in the same cells and that the spectrum of specificities of the above inhibitors is variable, poses some additional problems for analyzing transmembrane anion movements. The red cell is a rather unique example of a cellular system in which Cl^- as well as H^+ are in thermodynamic equilibrium (Cabantchik *et al.*, 1978; Knauf, 1979; Gunn, 1979; Jennings, 1985; Passow, 1987). Due to the Donnan effect, the δpH of RBC determines to a large extent the cellular Cl concentration ($[Cl]_i$). In other cells, $[Cl]_i$ is markedly above equilibrium level, the deviation determined by the weighed activities of the various Cl transporters as well as (a) by the intracellular pH (for cells with relatively high CBE activity), (b) by the Na gradient generated by the Na-pump (for cells with relatively high SPCC activity), or (c) by the membrane potential (for cells with relatively high ClCh activity). In epithelial cells, vectorial transcellular movement of salt and water dictated by the activity of the various transporters and by the ion gradients across the different membranes, have a profound effect on the steady-state concentration of intracellular Cl.

The Cl-HCO_3 exchanger (CBE) of red blood cell

This is the transport function for which the red cell is most highly specialized (Cabantchik *et al.*,1978; Knauf, 1979; Gunn, 1979; Jennings, 1985; Passow, 1987) subserving CO_2 removal from tissue to lungs. The massive efflux of HCO_3^- from red cells passing through tissue capillaries is accompanied by an equivalent influx of Cl^- with no apparent changes in cell volume and membrane potential. This process takes place in the reverse direction in red cells passing through lung capillaries. The combined processes, namely CO_2 diffusion, HCO_3^- formation and HCO_3^-/Cl^- exchange, has to be completed within the passage time of red cells through capillaries (about 1 sec, see Klocke, 1976) which is consistent with experimental observations showing CBE transport rates to be fast enough to export 80% of HCO_3^- produced within 1/3 of a second (Klocke, 1976).

The anion exchange protein of red blood cells is a simple carrier system which operates as an electroneutral, obligatory 1:1 exchanger of anions with broad substrate admission properties yet highly specialized for the fast translocation of halides and HCO_3^- (Cabantchik *et al.*, 1978; Knauf, 1979; Gunn, 1979; Jennings, 1985; Passow, 1987). It's main characteristics are (Frohlich and Gunn, 1986)):

1. All halides as well as HCO_3^-, NO_3^-, SO_4^{-2} and HPO_4^{-2} display saturation kinetics, compete with each other, have similar K_s values (in the 20-50 mM range), but differ considerably in their transport rates ($Cl^- > Br^- > I^- >> SO_4^{=}$). The list of anionic compounds, both inorganic and organic, which can be translocated by the exchanger is remarkable (*e.g.*, aliphatic monocarboxylic and dicarboxylic acids, aliphatic, aromatic and heterocyclic phosphates and sulfonates, cAMP, etc). However their transport rates span an enormous range

of values, which, for most of them, are orders of magnitude slower than those of Cl or bicarbonate, the two physiologically important substrates (*e.g.*, Cl is transported about 10^4 fold faster than sulfate). The list of anionic compounds which can be admitted, but not necessarily translocated by the system (*i.e.* impermeant competitive inhibitors) is even larger. The structure-activity relationship of these inhibitors (Barzilay *et al.*, 1978; Motais and Cousin, 1978) provided the basis for the design of affinity labels for the exchanger which led to the identification of band 3 as the relevant functional component of the red cell membrane anion exchange (Cabantchik *et al.*, 1978).

2. All transportable anions, monovalent and divalent alike, show a similar value of energy of activation (E_a of 30-40 Kcal/mol or 120-160 KJ/mol or Q_{10} of 5-6), which is considerably higher than that of free diffusion (2-4 Kcal/mol), diffusion through lipid bilayers (10-12 Kcal/mol) and even higher than some active transport mechanisms (20-30 Kcal/mol). For the particular case of halides, the temperature profiles of self exchange show a break at a temperature which differs for the various anions (for halides, below the break point, the E_a is 20 Kcal/mol; Passow, 1987).

3. While halide exchange rates have pH optima in the 7.5-8.5 pH range, the divalent sulfate and phosphate have pH optima at 6.4. This, as well as other properties, were best explained by the mechanism of a titratable carrier which can convert from a Cl^- transporting form (E^+) into a SO_4^{-2} one (E^{++}) by protonation of a single group (Knauf, 1979; Gunn, 1979; Jennings, 1985; Frohlich and Gunn, 1986; Passow, 1987):

$$E^+ + H^+ \longleftrightarrow E^{++}$$
$$E^+ + Cl^- \longleftrightarrow E\text{-}Cl$$
$$E^{++} + SO_4^{--} \longleftrightarrow E\text{-}SO_4$$

thus, translocation in this model is always electroneutral. In the E^+ conformation, the exchanger can also transport anionic ion-pairs (such as $NaCO_3^-$ and $LiCO_3^-$, both analogs of HCO_3^-) in an electroneutral fashion. Additional effects of pH on the transport mechanism demanding modifications of the titratable carrier model can not be ruled out at present (Knauf, 1986).

4. Halides and sulfate also interact with an additional site on AEP, the so called modifier site. This interaction, which causes a non-competitive self inhibition of anion exchange, is of no obvious physiological relevance, as it is of a relatively low affinity character (K_I of 200-300 mM) (Frohlich and Gunn, 1986).

5. The high degree of coupling between inward and outward translocation of anions (net Cl permeability is about four orders of magnitude slower than exchange permeability) could be explained by simultaneous as well as by sequential models. However, detailed kinetic analysis demonstrated that the basic mechanism of anion translocation could be best explained by a ping-pong model, with a single site alternating between an outward facing conformation and an inward facing one (Knauf, 1979; Gunn, 1979; Frohlich and Gunn, 1986).

6. In addition to the highly coupled influx and efflux of anions, the RBC has an additional mode of Cl translocation, which is an electrically operative (*i.e.* charge conductive) mechanism, it is about four orders of magnitude slower than Cl self exchange and is partially sensitive to disulfonic

stilbenes (Knauf, 1979). This mode of translocation has been associated with the same CBE system, except that the uncoupled anion flux was attributed to either slippage (Knauf, 1979) or tunneling (Frohlich, 1988) through the CBE carrier. However, the possibility that part of the stilbene disulfonate inhibitable conductive fluxes in erythrocytes could be attributed to Cl^- or K^+ channels which have been reported to occur in RBC (Hamill, 1983) cannot be excluded at this stage.

Cl-channels of mammalian epithelial cell membranes

Current models of transepithelial NaCl and fluid reabsorption invoke the concerted operation of two anion transport systems: a sulfonamide-sensitive SPCC and a ClCh located on opposite membrane (Greger, 1988). Hitherto, the characterization of the channels has been limited to analysis of their electrophysiological properties such as single-channel conductance, voltage dependence and the effects of second messengers on channel activation. Several types of ClCh have been identified in electrophysiological terms, primarily in the apical membrane of secretory epithelia (Gogelein, 1988): highly voltage-dependent *maxi-channels* (g=200-400 pS) which are activated at relatively high voltages (*e.g.*, in urinary bladder, A_6-cells, alveoli, apical membrane of salt secretory cells); outwardly rectifying *midi-channels* (40-60 pS) which are c-AMP stimulated and modulated by various kinases (*e.g.*, in apical membrane of various salt secretory cells including airway epithelial cells) and *mini-channels* (5-30 pS), which show no rectification (*e.g.*, urinary bladder, thick ascending limb of Henle's loop). Some attempts have been done at identifying and reconstituting ClCh both from kidney and trachea (Landry *et al.*, 1989; Breuer, 1990) using isotopic techniques for measurements of conductive fluxes. CMTF was used to study Cl fluxes in vesicles isolated from human placenta (Illsley and Verkman, 1987) and basolateral membranes from pig renal outer medulla (Breuer, 1989).

CMTF of the anion exchanger using a fluorescent substrate analog

NBD-taurine (Figure 2, for experimental details see Eidelman and Cabantchik, 1989a,b) was designed on the basis of structure-activity relationship studies of CBE (Barzilay *et al.*, 1979) and the spectroscopic properties of the NBD group: excitation and emission wavelengths which are overlap with the absorption spectrum of hemoglobin so that intracellular probe fluorescence will be largely quenched by absorption, relatively large Stokes shift (80 nm) and the availability of anti-NBD antibodies which efficiently quench its fluorescence (Darmon *et al.*, 1982). NBD-taurine has been shown to be a substrate which can serve as "bona fide" reporter of the CBE of RBC based on the findings that it competes with halides for common transport sites and that it is inhibitable by all the blockers of CBE and it displays mechanistic features and energy of activations similar to other monovalent anions (Eidelman and Cabantchik, 1983a,b).

NBD-taurine can be synthesized as described elsewhere (Eidelman *et al.*, 1981) or procured from a commercial source (Molecular Probes,Inc.). Anti-NBD antibodies are prepared by standard immunological procedures (Darmon *et al.*,1983) or obtained commercially (Molecular Probes,Inc.)

TRANSPORT ASSAYS

Human red blood cells: The number of cells required for a single CMTF assay using a regular spectrofluorometer is $1\text{-}5 \times 10^6$ (ca. 1 μl whole blood). Whole blood is washed in buffered isotonic saline solutions pH 7.4, 25°C, using repeated centrifugation and aspiration of supernate and buffy coat for removal of white cells. Other methods of removal of white cells are also adequate.

Loading of NBD-taurine: cells at 10 to 20% hematocrit are incubated with 1 mM NBD-taurine for 1-3 hrs at 37°C and after equilibration with the probe, cells are kept on ice or at 5°C.

Removal of external probe: the cells can be washed in isotonic sulfate buffer (5°C), by centrifugation or by filtration through a Dowex AG (1 x 8, 25-50 mesh, sulfate or citrate form) minicolumn pre-cooled to 5°C.

Flux measurement: an aliquot of the NBD-taurine-loaded cells free of extracellular probe ($0.5\text{-}5 \times 10^6$ cells) is jetted into a spectrofluorometer cuvette containing 0.2-2 ml buffer pre-equilibrated at the desired temperature. Continuous stirring is recommended when fluxes are followed for more than 20 min to avoid settling of RBC. For most cases, however, periodic mixing of cells by pipetation is adequate. The value of the fluorescence signal F_∞, attained when 99% the probe has exited the cells, is needed for quantitative data analysis. In most cases, F_∞ can be obtained by permeabilizing the cells with detergent (*e.g.*, Nonidet P-40, Triton X-100 or Na-lauryl sulfate at 0.05 mg/ml final concentration; $C_{12}E_9$ (Calbiochem) at 0.5 mg/ml final concentration). With Nonidet P-40, care has to be exercised to avoid clouding of the detergent at temperatures above 30°C (this occurs generally 1 to 2 min after addition of the detergent). Since the amount of detergent to be added may vary with cell type and cell number, it is recommended that the amount be determined in a preliminary experiment. When applied to RBC, NBD-taurine CMTF is useful for assessing the mechanism of transport and the mode of inhibition of inhibitors (Eidelman and Cabantchik, 1983; 1989a,b).

Pigment-free cells, ghosts, or vesicles

The amount of biological material and the concentration of fluorophore required for a transport assay are in the same range as those of erythrocytes, although the loading times may vary considerably. For ghosts, resealing in the presence of probe circumvents subsequent loading. Similarly, for vesicles, freezing and thawing commonly used to seal the membranous systems, also facilitate entrapment of probe.

Removal of external probe: Initial removal of external probe after loading can be done by centrifugation. Since the cells, or vesicles, are still in the loading solution, there are no limitations on time of centrifugation. However, there might be time limitations on the second separation step, that is the time of centrifugation has to be much shorter than the half time for probe efflux in the washing solution. The latter can be increased by lowering the temperature, by addition of transport inhibitors, or by removing a required factor for the flux. For cells or resealed red cell ghosts, quick centrifugation (Eppendorf, 1 sec to 5 min) in the cold (5°C) in isotonic sulfate medium can serve the purpose. However, for vesicles as well as ghosts, chromatography through a Dowex AG 1x8 column (100-200 mesh) as shown before for erythrocytes, is faster and more convenient. For fast transporting systems (*e.g.*, reconstituted band 3 vesicles) separation can be avoided by quenching all external probe (after centrifugation and resuspension) with excess anti-NBD antibodies. However, determination of the requisite amount of anti-fluorophore antibodies has to be assessed independently (*e.g.*, by using a sample permeabilized with non-ionic detergent).

Flux measurements are done essentially as described above except that the membrane material and antibody are added in appropriate amounts. Using this approach, it was possible to study the function vesicles isolated from RBC membranes and reconstituted band 3 proteoliposomes (Cabantchik and Darmon, 1985).

Measuring Cl fluxes using the fluorescent indicator SPQ

The Cl-sensitive fluorescent probe SPQ (Figure 2) is a zwitterionic molecule which has been applied in CMTF of various biological systems (Illsley *et al.*,1988; Krapf *et al.*, 1988a,b; Verkman *et al.*, 1989). The probe can be loaded into pigment-free cells (red cell ghosts) or into vesicles isolated from epithelial cell membranes (Illsley and Verkman, 1987) by incubation for 30 min. at 37°C with 1 mM probe and into vesicles by overnight incubation at 4°C in a solution containing 10 mM dye (Illsley and Verkman, 1987). The internal contents of the vesicles as well as the loading medium should be devoid of Cl, but should be provided with a non-quenching permeant anion as bicarbonate for CBE. As with other low molecular weight fluorophores or quenchers, incorporation of SPQ (and also the requisite ion components) can be attained by incorporating them in the vesicular compartment during vesicle preparation or by rapid freezing at liquid N_2 temperature and thawing at room temperature and subsequent removal of extravesicular probe at ice-cold temperatures by: (a) gel filtration (wet); (b) gel filtration-centrifugation (Sephadex G-25 placed in 5 ml syringes held on centrifuge tubes are drained of solvent by a 1,500 g x 5 min spin, loaded with vesicle suspension and spun again, the liquid collected in the centrifuge contains the extravesicular probe-free vesicles usually after 2-3 spins); (c) ion exchange chromatography on a DOWEX AG 1X8 (100-1200 mesh) minicolumn or (d) centrifugation and washing at ice cold temperatures.

Fluxes are carried out after rapid mixing of an aliquot of SPQ-loaded vesicles into an isotonic medium containing Cl, usually with the aid of a rapid mixing device. As the probe might leak from vesicles or cells at significant rates at temperatures above 20°C, the fluxes should be conducted at the appropriate (low) temperatures. The fluorimeter settings should be: 350 nm for excitation and 420 nm for emission.

CMTF of conductive pathways using a potentiometric dye

This method is suited for measuring conductive pathways in systems amenable to manipulation of the ion composition of the internal (*i.e.* cellular or vesicular) compartment (Cabrini and Verkman, 1986, Chen *et al.*, 1988, Breuer, 1989, 1990). This usually requires placement of impermeant ions to compensate for the primary permeating ionic species to be tested. Gluconic acid and N-methyl-glucamine are among the most preferred species used for that purpose. The test ion is usually added to the external medium at the start of the experiment and after assessing the relationship between fluorescent changes and a membrane potential artificially established with K-valinomycin.

The technique for preparing apical membrane vesicles from epithelial cells has been thoroughly described (Illsley and Verkman, 1987; Mircheff, 1989). It is important to emphasize that the internal composition of the vesicles can be determined during the step of cell homogenization or during

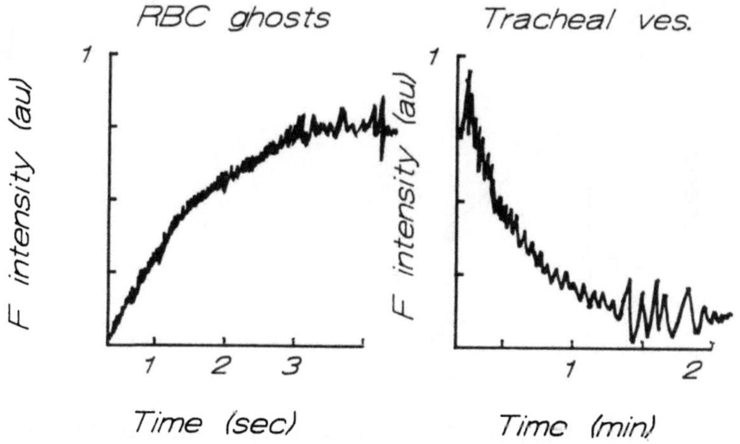

Figure 5. CMTF USING SPQ AS AN INDICATOR AND Cl AS A COLLISIONAL QUENCHER: *Human erythrocyte ghosts resealed with 120 mM NaCl and 1 mM SPQ, HEPES 10 mM, pH 7.4 were washed with isotonic Na-gluconate and jetted into a flow cuvette of a rapid mixing device. The medium contained 25 mM Na-bicarbonate and 110 mM Na-gluconate, pH 7.4 at 10 C. F was followed with time as indicated in the text. Bovine tracheal (apical) vesicles containing isotonic Na-gluconate, pH 7.4 were loaded with SPQ (1mM) overnight in the cold, and F was followed after mixing with isotonic Na-gluconate solution containing 5 mM NaCl.*

freezing of the vesicles, provided the latter is done at sufficiently high dilution so as to nullify the components carried over from previous isolation steps. A 100 mM K-gluconate, and 150 mM sucrose pH 7.4 HEPES or MOPS buffered solution could well serve the purpose. Isolation of sealed vesicles is accomplished by centrifugation over a 30% Ficoll cushion (or equivalent metrizamide cushion) and a 100,000 g x 20 min spin. The vesicles at the upper layer are washed in the same medium as that present in the vesicles. The protein concentration of the vesicle suspension can be determined by the BCA method (Pierce), the Bradford method (BioRad) or the more conventional Lowry method in its numerous variations. Working concentrations in the 5-15 mg/ml range are usually obtained after centrifugation at the above centrifugal forces.

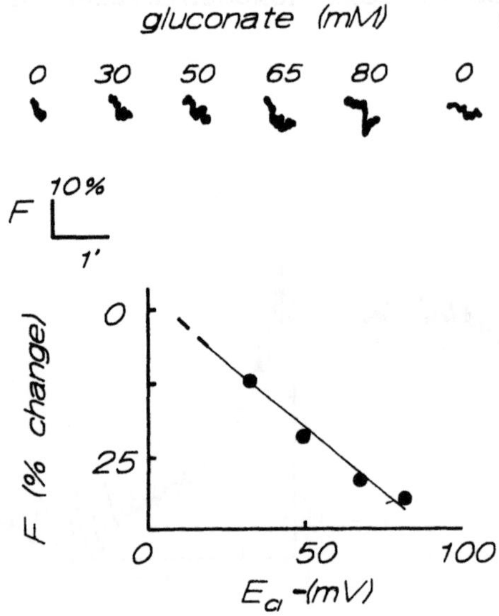

Figure 6: DIFFUSION POTENTIALS GENERATED BY Cl PERMEATION THROUGH CHANNELS IN MEMBRANE VESICLES: *A preparation of human placenta membrane vesicles containing 100 mM K-gluconate, 10 mM Hepes, 150 mM sucrose, pH 7.4 was suspended in isotonic Na-gluconate medium and treated with 2 μM dye (added from DMSO solution). Upper traces show fluorescence changes (in percent of original signal) after addition of the indicated final concentration of either NaCl or Na-gluconate. The lower graph represents the percent fluorescence change vs. the Cl-equilibrium potentials as calibrated with valinomycin added to vesicles suspended in isotonic sucrose gluconate solutions with various potassium concentrations.*

<u>Flux measurements</u>: A 1-2 ml solution of buffered Na-gluconate (100 mM), 1mM NaCl and sucrose (100mM) containing vesicles (50-100 μg/ml protein) is placed in a well-stirred cuvette (polystyrene or acrylate) sitting in a thermostated (25°C) cuvette holder of a spectrofluorometer (excitation 620 nm and emission 670 nm). In these conditions $[Cl^-]_i=[Cl^-]_o$ and the external concentration of K^+ carried over from the vesicle suspension depends on the dilution of the latter in the flux medium (usually a 1:100 dilution will yield a 1 mM final $[K^+]$;

if necessary, the K-gluconate solution can be replaced with Na-gluconate so as to attain the requisite final concentration of both external and internal K+). di-$SC_3(5)$ is administered from a 1 mM solution (in DMSO) so as to reach a 1-2 μM final concentration. A transient rise in fluorescence signal is followed by a relatively slower signal decay to a stable level due to binding of the dye to the vesicles. Generation of E_m (negative inside) by addition of valinomycin (5 μM) is manifested in signal decrease (membrane hyperpolarization) the extent of which is governed by [K+] present in the medium (the $[K_i]/[K_o]$ gradient). In order to obtain a relationship between the magnitude of the fluorescence signal and the membrane potential, aliquots of K are added from a concentrated K-gluconate or K_2SO_4 solution and the fluorescence signals of the depolarization are recorded (the membrane potential E_m is calculated from the Nernst potential given (at 20°C, in mV) by $-58*\log([K^+]_i/[K^+]_o)$. The determination of Cl channel activity is accomplished in a parallel experiment by stepwise addition of Cl from an NaCl solution. A decrease in fluorescence signal indicates quenching by dye accumulation due to hyperpolarization of the membranes.

The extent of the diffusion potentials created by Cl-channel activity depends on the tightness of the vesicles, usually the fluorescence signal levels are not sustained, probably because of the dissipation of the gradients by anion exchange activity (which is either omnipresent or from contaminating cell membranes). Unfortunately, this method does not lend itself to inhibition studies, because inhibitors of anion transport were found to either interact directly with the cyanine dyes or, when covalently bound to membranes, to alter the partition profiles of the dye. A typical experiment conducted with diS-$C_3(5)$ and apical membrane vesicles from bovine human placenta (Figure 6), depicts the Cl generated changes in fluorescence (upper half) and the change in fluorescence as a function of Cl Nernst potential.

ACKNOWLEDGEMENTS

Supported in part by a grant from the US National Institutes of Health HL-40158 and by the Germany-Israel Binational Science Foundation (GIF).

REFERENCES

Alper, S. L., Kopito, R., Libresco, S. M., and Lodish, H. F., 1988: *Cloning and characterization of murine band 3 related cDNA from kidney and from a lymphoid cell line,* J. Biol. Chem., 263:17902-17909.

Aronson, P., 1989: *The renal proximal tubule: a model for diversity of anion exchangers and stilbene sensitive anion transporters,* Ann. Rev. Physiol., 51:419-441.

Barzilay, M., Jones, D., and Cabantchik, Z. I., 1978: *Sidedness of inhibitory effects as evidence for asymmetric location of the anion transport system of red blood cell membranes,* Fed. Proc., 37:1295.

Barzilay, M., Ship, S., and Cabantchik, Z. I., 1979: *Anion transport in red blood cells: 1. Chemical properties of anion recognition sites as revealed by structure activity relationships of aromatic sulfonic acids,* Membr. Biochem., 2:227-254.

Blatz, A. E., and Magleby, K. L., 1985: *Single chloride-selective channels active at resting membrane potentials in cultured rat skeletal muscle,* Biophys. J., 47:119-123.

Breuer, V. W., 1989: *Characterization of Cl channels in membrane vesicles from the kidney outer medulla,* J. Membr. Biol., 107:35-42.

Breuer, V. W., 1990: *Selective solubilization and reconstitution of a kidney chloride channel,* Biochim. Biophys. Acta, In press.

Cabantchik, Z. I., and Darmon, A., 1985: *Reconstitution of Membrane Transport Systems,* in: "Structure and Properties of Membranes" (Benga, G., ed.), pp. 123-165, CRC Press.

Cabantchik, Z. I., Knauf, P. A., and Rothstein, A., 1978: *The anion transport system of the red blood cell. The role of membrane protein evaluated by use of "probes",* Biochim. Biophys. Acta, 515: 239-303.

Cabrini, G. and Verkman A. S., 1986: *Mechanism of interaction of the cyanine dye diS-C_3-(5) with renal brush-border vesicles,* J. Membr. Biol., 90:163-175.

Cala, P., 1985: *Volume regulation by Amphiuma red blood cells: strategies for identifying alkali-/H transport,* Fed. Proc., 44:2500-2507.

Chen, J. H., Schulman, H., and Gardner, P., 1989: *A c-AMP regulated chloride channel in lymphocytes that is affected in cystic fibrosis,* Science, 243:657-660.

Chen, P. Y., Illsley, N. P., and Verkman, A. S., 1988: *Renal brush border chloride transport mechanisms characterized using a fluorescent indicator,* Am. J. Physiol., 254:F114-F120

Cox, J. V., and Lazarides, E., 1988: *Alternative primary structures in the transmembrane domain of the chicken erythroid anion transporter,* Mol. Cell. Biol., 8:1327-1335.

Darmon, A., Eidelman, O., and Cabantchik, Z. I., 1982: *A method for measuring anion transfer across membranes of hemoglobin free cells and vesicles by continuous monitoring of fluorescence,* Anal. Biochem., 119:313-321.

Darmon, A., Zangvill, M., and Cabantchik, Z. I., 1983: *New approaches for the reconstitution and functional assay of membrane transport proteins. Application to the anion transporter of human erythrocytes,* Biochim. Biophys. Acta, 727:77-88.

Demuth, D. R., Showe, L. C., Ballantine, M., Palumbo, A., Fraser, P. J., Cioe, L., Rovera, R., and Curtis, R. J., 1986: *Cloning and structural characterization of a human non-erythroid band 3-like protein,* EMBO J., 5:1205-1214.

Donowitz, M., and Welsh, M. J., 1986: *Ca and cAMP in regulation of intestinal Na, K and Cl transport,* Ann. Rev. Physiol., 48:135-150.

Eidelman, O., Zangvil, M., Razin, M., Ginsburg, H., and Cabantchik, Z. I., 1981: *The anion transfer system of erythrocyte membranes: N-(7-nitrobenzofurazan-4-yl) taurine, a fluorescent substrate-analogue of the system,* Biochem. J., 195:503-513.

Eidelman, O., and Cabantchik, Z. I., 1983a: *The mechanism of anion transport across human red blood cell membranes as revealed by a fluorescent substrate: I. Kinetic properties of NBD-taurine transfer in symmetric conditions,* J. Membr. Biol., 71:141-148.

Biol., 71:141-148.
Eidelman, O., and Cabantchik, Z. I., 1983b: *The mechanism of anion transport across human red blood cells membranes as revealed by a fluorescent substrate: II. Kinetic properties of NBD-taurine transfer in asymmetric conditions.*, J. Membr. Biol., 71:149-161.
Eidelman, O., and Cabantchik, Z. I., 1989a: *Continuous monitoring of transport by fluorescence,* Biochim. Biophys. Acta (Reviews in Biomembranes) 988:319-334.
Eidelman, O., and Cabantchik, Z. I., 1989b: *Fluorescent methods for monitoring transport in cells and vesicles,* Meth. Enzymol., 172:122-135.
Finn, A. L., 1985: *Volume dependent pathways in animal cells,* Fed. Proc., 44:2599.
Frizzell, R. A., Halm, D. R., Rechkemmer, G., and Shoemaker, R. L., 1986: *Chloride channel regulation in secretory epithelia,* Fed. Proc., 45:2727-2731.
Frohlich, O., 1988: *The "tunneling" mode of biological carrier mediated transport,* J. Membr. Biol., 101:189-198.
Frohlich, O., and Gunn, R. B., 1986: *Erythrocyte anion transport: the kinetics of a single site obligatory exchange system,* Biochim. Biophys. Acta, 864:169-194.
Ganz, M. B., Boyarsky, G., Sterzel, R.B., and Boron, W.F., 1989: *Arginine vasopressin enhances pH_i regulation in the presence of HCO_3^- by stimulating three acid-base transport systems,* Nature, 332:648-651.
Garty, H., Rudy, B. and Karlish, S. J. D., 1983: *A simple and sensitive procedure for measuring isotopic fluxes through ion-specific channels in heterogeneous populations of membrane vesicles,* J. Biol. Chem., 258: 13054-13059.
Glickman, J., Croen, K., Kelly, S., and Al-Awqati, Q., 1983: *Golgi membranes contains an electrogenic H-pump in parallel with a chloride conductance,* J. Cell. Biol., 97:1303-1308.
Gogelein, H., 1988: *Chloride channels in epithelia,* Biochim. Biophys. Acta, 947:521-547.
Greger, R., 1987: *Ion transport mechanisms in thick ascending limb of Henle's loop of mammalian nephron,* Physiol. Rev., 65:755-797.
Greger, R., 1988: *Chloride transport in thick ascending limb, distal convolution, and collecting duct,* Ann. Rev. Physiol., 50:111-122.
Gunn, R. B., 1979: *Anion transport in red cells: an asymmetric, ping pong mechanism,* in: "Mechanisms of Intestinal Secretion" (Binder, H. J., ed.), pp. 25-43, Alan Liss, Inc., New York.
Hamill, O. P., 1983: *Potassium and chloride channels in red blood cells,* In: "Single channel recording" (Sackman B. and Neher E., Eds.) pp. Plenum Press, New York.
Hamill, O. P., Bormann, J., and Sackmann, 1983: *Activation of multiple conductance state chloride channels in spinal neurones by glycine and GABA,* Nature, 305:805-808.
Hoffman, E. K., 1986: *Anion transport systems in the plasma membrane of vertebrate cells,* Biochim. Biophys. Acta, 864:1-32.
Jay, D., and Cantley, L. C., 1986: *Structural aspects of the red cell anion exchange protein,* Ann. Rev. Biochem., 55:511-538
Jennings, M. L., 1985: *Kinetics and mechanism of anion transport in red blood cells,* Ann. Rev. Physiol., 47:519-533.
Illsley, N. P., and Verkman A. S., 1987: *Membrane chloride*

transport measured using a chloride-sensitive fluorescent probe, Biochemistry, 26:1215-1219.

Illsley, N. P., Glaubensklee, C., Davis, B., and Verkman, A. S., 1988: *Chloride transport across placental microvillous membranes measured by fluorescence,* Am. J. Physiol., 255:C789-C797.

Kasahara, M. and Hinkle, P. C., 1976: *Demonstration of D-glucose transport catalyzed by a protein fraction from human erythrocyte sonicated liposomes,* Proc. Natl. Acad. Sci., 73:396-400.

Klocke, R. A., 1976: *Rate of bicarbonate-chloride exchange in human red cells at 37 C,* J. Appl. Physiol., 40:707-714.

Knauf, P. A., 1979: *Erythrocyte anion exchange and the band 3 protein: transport kinetics and molecular structure,* in: "Current Topics in Membrane Transport" (Bronner, F., and Kleinzeller, A., eds.), pp. 249-363, Academic Press, New York.

Knauf, P. A., 1986: *Anion transport in erythrocytes,* in: "Membrane Transport Disorders" (Andreoli, T., Hoffman, J. F., Schultz, S. G., and Fanestil, D. D., eds.), pp. 191-220, Plenum Press, Inc., New York.

Kopito, R. R., and Lodish, H. F., 1985: *Structure of the murine anion exchange protein,* J. Cell. Bioch., 29:1-17.

Kopito, R. R., Andersson, M., and Lodish, H. V., 1987: *Structure and organization of the murine band 3 gene.* J. Biol. Chem., 262:8035-8040.

Krapf, R., Berry, C. A. and Verkman, A. S., 1988a: *Estimation of intracellular chloride activity in isolated perfused rabbit proximal convoluted tubules using a fluorescent indicator,* Biophys. J., 53:955-962.

Krapf, R., Illsley. N. P., Tseng, H. C. and Verkman, A. S., 1988b: *Structure-activity relationships of chloride sensitive fluorescent indicators for biological application,* Anal. Biochem., 169:142-150.

Lakowicz, J. R., 1983: "Principles of fluorescence spectroscopy". Plenum Press, New York.

Landry, D. W., Reitman, M., Cragoe, E. J., and Al-Awqati, Q., 1987: *Epithelial chloride channel: Development of inhibitory ligands,* J. Gen. Physiol., 90:779-798.

Loew, L. M., 1988: "Spectroscopic membrane probes". CRC Press, Boca Raton Florida.

McNeil, P. F., Murphy, R. F., Lanni, F. and Taylor, D. L., 1984: *A method for incorporating macromolecules into adherent cells,* J. Cell Biol., 98:1556-1564.

Mircheff, A. K., 1989: *Isolation of plasma membranes from polar cells and tissues: apical/basolateral separation, purity, function,* Meth. Enzymol., 172:18-34.

Motais, R., and Cousin, J. L., 1978: *A structure activity study of some drugs acting as reversible inhibitors of chloride permeability in red cell membranes: Influence of ring substituents,* in: "Cell membrane receptors for drugs and hormones: A multidisciplinary study" (Straub, R. W., and Bolis, L., Eds.) pp. 219-225, Raven Press, New York.

Muallem, S., Blissard, D., Cragoe, E. J., and Sachs, G., 1988: *Activation of Na/H and Cl/HCO$_3$ exchange by stimulation of acid secretion in the parietal cell,* J.Biol. Chem., 263:14703-14711.

O'Grady, Z. M., Palfrey, H. C., and Field, M., 1987: *Characteristics and functions of Na-K-Cl cotransport in epithelial tissues,* Am. J. Physiol., 253:C177-C192.

Passow, H., 1987: *Molecular aspects of band 3 protein-*

mediated anion transport across the red blood cell membrane, Rev. Physiol. Biochem. Pharmacol., 103:62-217.

Reinertsen, K. V., Tonnessen, T. I., Jacobsen, J., Sandvig, K., and Olsnes, S., 1988: *Role of chloride/bicarbonate antiport in control of cytosolic pH. Cell-line differences in activity and regulation of antiport,* J. Biol. Chem., 263:11117-11125.

Rothenberg, P., Glaser, L., Schlesinger, P. and Cassel, D., 1983: *Activation of Na/H exchange by epidermal growth factor elevated intracellular pH in A431 cells,* J. Biol. Chem., 258:12644-12653.

Schlatter, E., Greger, R., and Weidtke, C., 1983: *Effect of high ceiling diuretics on active salt transport in the cortical thick ascending limb of Henle's loop of rabbit kidney: Correlations of chemical structure and inhibitory potency,* Pflug. Arch., 396:210-217.

Schuster, V. L., Bonsib, S. M., and Jennings, M. L., 1987: *Two types of collecting duct mitochondria-rich (intercalated) cells: lectin and band 3 cytochemistry,* Am. J. Physiol., 251:C347-C355.

Schwartz, G. J., Barasch, J., and Al-Awqati, Q., 1985: *Plasticity of functional epithelial polarity,* Nature, 318:368-371.

Spring, K. R., and Ericson A. C., 1982: *Epithelial cell volume modulation and regulation* J. Membr. Biol., 69:167-176.

Tago, K., Schuster, V. L., and Stokes, J. B., 1986: *Regulation of Cl self exchange by cAMP in cortical collecting tubule,* Am. J. Physiol., 251:F40-F48.

Tanner, M. J. A., Marti, P. G., and High, S., 1988: *The complete amino acid sequence of the human erythrocyte membrane anion-transport protein deduced from the cDNA sequence,* Biochem. J., 256:703-712.

Verkman, A. S., Sellers, M. C., Chao, A. C., Leung, T. and Ketcham, R., 1989: *Synthesis and characterization of improved chloride-sensitive fluorescent indicators for biological applications,* Anal. Biochem., 178:355-361.

Wagner, S., Vogel, R., Lietzke, R., Koob, R., and Drenckhahn, D., 1987: *Immunochemical characterization of band 3-like anion exchanger in collecting duct of human kidney,* Am. J. Physiol., 253:F213-F221.

Wangemann, P., Wittner, M., DiStefano, A., Englert, H. C., Lang, H. J., Schlatter, E., and Greger, R., 1986: *Cl-channel blockers in the TALH: Structure activity relationship,* Pflug. Arch., 407:S128-S141.

Welsh, M. J., and Liedtke, C. M., 1986: *Chloride and potassium channels in cystic fibrosis airway epithelia,* Nature, 322:467-470.

Xie, X. S., Stone, D. K., and Racker, E., 1983: *Determinants of clathrin coated vesicle acidification,* J. Biol. Chem., 258:14834-14838.

19

THE MOLECULAR BIOLOGY OF SUGAR TRANSPORT PROTEINS

Peter J.F. Henderson, Elaine O. Davis,
Brian J. McKeown, and Martin C.J. Maiden

Department of Biochemistry, University of Cambridge
Tennis Court Road, Cambridge CB2 1QW, U.K.

INTRODUCTION

Many species of bacteria live in environments where nutrients are in short supply. Individual species must therefore accumulate nutrients at rates commensurate with rapid growth and successful competition. In order to increase the uptake of a given nutrient metabolic energy is used in its transport across the cell membrane (Button, 1985). In *Escherichia coli,* the ubiquitous model microorganism, transport processes are energised by several mechanisms (Fig.1). Energy sources used include: trans-membrane electrochemical gradient, of protons or sodium; adenosine triphosphate (ATP); or phosphoenol pyruvate (PEP) (illustrated in Fig.1; reviewed in Button, 1985; Saier, 1985; Henderson, 1986).

The present article will review recent advances in the study of some of the proton-electrochemical gradient-linked transporters that occur in *E. coli.* These proteins catalyse the obligatory translocation of protons with a substrate molecule, "symport", or the experimentally indistinguishable "antiport" of hydroxyl ions (processes envisaged by Mitchell, 1961, 1963, 1973). There are six proton-symport proteins known to transport sugars into *E. coli,* one each for L-arabinose, D-xylose, D-galactose, L-rhamnose, L-fucose, and lactose (Fig. 1; West and Mitchell, 1972, 1973; Henderson, 1986; Henderson and Macpherson, 1986; Bradley et al., 1987; Muiry, 1989). The disaccharide transporters for lactose/proton symport (Saier, 1985; Kaback, 1986, 1990), or melibiose/sodium symport (Wilson et al., 1986; Leblanc et al., 1989; Pourcher et al., 1990), will not be considered in detail here, though they may operate by a similar mechanism.

Most of the active transporters of *E.coli,* which are located in the inner membrane, are of low abundance (typically less than 0.5% of the total membrane proteins).. This contrasts with the situation for the non-energised "porins" in the outer membrane (Nikaido and Vaara, 1987). The low abundance of these proteins, together with the problems which their hydrophobic nature presents the biochemist, can be overcome by recombinant DNA and genetical techniques. As antibodies and peptide sequence are often not available for membrane proteins, partly because of the problems outlined above, it is generally not possible to utilise cloning techniques that rely on these. The work described in the present article was developed to overcome these problems by exploiting the highly developed genetical techniques available in *E. coli* to study two of its membrane transport proteins, the arabinose (AraE protein) and xylose (XylE protein) proton symporters. As genetical systems for other organisms become more developed approaches similar to the ones described should have broader applicability.

Recently, recombinant DNA technologies have contributed spectacularly to our understanding of membrane proteins, and have enabled the discovery of unexpected interrelationships between transport systems. For example, the arabinose and xylose transporters of *E. coli* are found to be related to glucose transporters in other bacteria, algae, yeasts, and man (see below). It is anticipated that the application of ever more developed recombinant DNA techniques (for example in the protein engineering of transport molecules, Kaback, 1987) will greatly enhance our understanding of the molecular mechanisms of these membrane proteins.

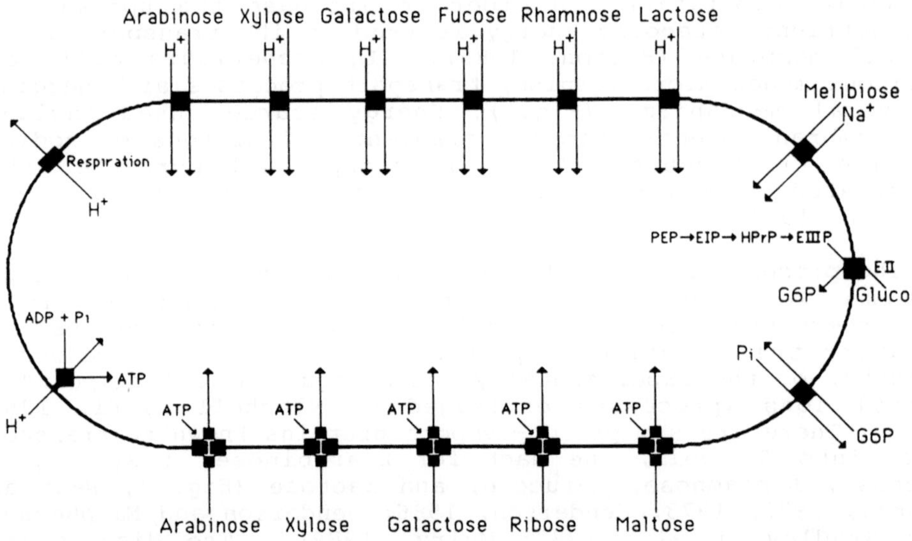

Fig. 1. Energisation of sugar transport in *E. coli*. The large oval represents the cytoplasmic membrane of the microorganism. A trans-membrane electrochemical gradient of protons is generated by respiration or ATP hydrolysis (Mitchell, 1961, 1963; reviewed by Henderson, 1986), depicted on the left. This can be utilised by the different sugar/H^+ symport systems shown along the top. Some sugars can be accumulated by an alternative mechanism involving ATP, a binding protein, and two or three other proteins (Quiocho, 1986, 1990; Higgins et al., 1990; Furlong, 1987) shown along the bottom, or by a phosphotransferase mechanism involving PEP and two or three proteins (Saier, 1985; Postma and Lengeler, 1985; Lengeler et al., 1990) shown on the right. Melibiose transport occurs by symport with sodium ions (top right) (Wilson et al., 1986; Leblanc et al., 1989), and hexose phosphates utilise an antiport mechanism with phosphate ions (bottom right; Eiglmeier et al., 1987; Maloney, 1990).

THE DESIGN OF A SUITABLE CLONING STRATEGY

Although the use of recombinant DNA and genetical techniques solves many problems, care is needed in the rational design of a cloning strategy for a given transport system. Some knowledge of the genetical organisation of the transport genes is essential, as is the availability of stable mutants of the transporter and a reliable assay system (we have found that it is often difficult or impossible to determine the transport phenotype for organisms with mutant sugar-proton symporters without the use of the direct proton symport assay, Jones-Mortimer and Henderson, 1986; Henderson and Macpherson, 1986).

In some cases, where genetic organisation is relatively simple, it is possible to clone a whole operon, including the genes for transport and metabolism of a given substrate, by complementation. This has been done in our laboratory for the rhamnose operon of *Salmonella typhimurium* (Muiry, 1989). In other cases, however, complementation of mutants by cloned genes on multicopy plasmids does not work. This is true of the genes coding for the AraE protein (*araE*) and XylE protein (*xylE*), probably because they occur in operons separate from the operons containing the genes for the positive regulatory elements (the presence of an additional transport system for each of these sugars further complicates the situation, see below). This problem can sometimes be overcome by the use of an adjacent gene that is closely linked to the desired gene and which has an easily selected phenotype. For example, it was possible to clone the *araE* gene of *S. typhimurium* by selecting cosmids containing the adjacent *lysA* gene in a *lysA*$^-$ mutant.

For cloning the *araE* and *xylE* genes a strategy was devised that did not rely on complementation or on the use of antibodies or peptide sequence. Firstly, their position on the *E. coli* linkage map (Bachmann, 1983) was established. Secondly, the phage lambda p*lac*Mu was inserted into the gene and used to establish a restriction map of the *E. coli* genome around the gene. Finally, a specialised transducing phage carrying DNA adjacent to and including the gene was obtained and a specific restriction fragment that contained the gene was subcloned into a multicopy plasmid.

It was then possible to sequence the DNA and construct vectors for the amplified expression of the gene products. Now, with the availability of an ordered library of the *E. coli* genome (Kohara et al., 1987; Medigue et al., 1990) the production of specialised transducing phage would be unnecessary.

A final consideration is the choice of cloning vector (see Sambrook et al, 1989, for review). We have tended to use plasmid pBR322 and derivatives (Balbas et al., 1986) for most of our work rather than the very high copy number pUC series of plasmids (Yanisch-Perron et al., 1985), which can present problems with gene products that are lethal when produced at elevated levels. Cosmids (Sambrook et al., 1989) have the disadvantage of deleting to smaller sizes, although we have used them successfully as mentioned above. Vectors based on phage lambda (Hendrix et al., 1983; Sambrook et al., 1989) are slightly less convenient to handle, but have given us very satisfactory results. A plasmid vector, pAD284, constructed specifically for the amplification of gene expression is discussed further below.

THE LOCATION OF GENES ENCODING THE SUGAR TRANSPORT PROTEINS ON THE *Escherichia coli* LINKAGE MAP

Random mutagenesis of the *E. coli* chromosome is conveniently accomplished by the use of transposable genetic elements. There are a variety of such elements available, some of them specially constructed for certain types of mutagenesis (Silhavy et al., 1984; Jones-Mortimer and Henderson, 1986). Often they contain an antibiotic resistance marker and/or some other conveniently scored marker gene to facilitate the selection of insertion mutants. The mutants obtained are then further scored or selected to obtain bacterial strains with insertions in the desired transport gene. Examples of selections useful in obtaining transport mutants include the appearance of sugar resistant phenotypes in *fda* or *ppc* strains and the acquisition of resistance to toxic sugar analogues. The phenotypes of the strains obtained can be confirmed by transport assays (Kornberg and Henderson, 1974; Henderson, 1986; Jones-Mortimer and Henderson, 1986).

To mutate the *xylE* gene, the MudAp*lac* phages (Casadaban and Cohen, 1979; Silhavy et al., 1984; Jones-Mortimer and Henderson, 1986) were used; these insert randomly in the bacterial chromosome and allow the selection of ampicillin resistance and xylose-dependence of lactose utilisation in addition to changes in xylose transport phenotype (for details of the selection used, see Davis et al., 1984). The genetical techniques of interrupted conjugation and phage P1-mediated transduction (Miller, 1972) were then used to locate the inserted phage (and hence the *xylE* gene) at 91.4 min on the linkage map of the chromosome (Davis et al., 1984; Davis, 1985).

By similar types of experiment, a number of sugar transport genes have been located on the linkage map (Fig. 2). As mentioned above, some occur in operons together with genes encoding enzymes for the metabolism of the transported sugar and also with the genetical elements that regulate gene expression. Examples are the *lac* operon (Beckwith and Zipser, 1970) and the genes for the utilisation of rhamnose (Badia et al., 1989; Tobin and Schleif, 1990) and fucose (Lu and Lin, 1989). In other cases the transport and metabolic genes are separate. In the case of arabinose utilisation, the *araE* gene is located at 61 minutes, the metabolic and control genes at 0 min and the binding protein dependent transport system at 45 min on the linkage map (Fig 2). Xylose and galactose utilisation genes show a similar dispersion about the linkage map (Fig.2, Bachmann, 1983; Jones-Mortimer and Henderson, 1986; Riordan and Kornberg, 1977; Sumiya, 1989; Roberts et al., 1990).

The expression of the various sugar utilisation genes (for metabolism and sometimes for more than one transporter) are usually coordinated and the appearance of the sugar in the environment often increases the levels of gene expression by factors of tens to several hundreds. So far, only the expression of the *ara* operons has been studied in any detail (Kolodrubetz and Schleif, 1981; Kosiba and Schleif, 1982; Stoner and Schleif, 1983; Hendrickson and Schleif, 1984), but it is still unclear how the transcription of the *araE* gene is coordinated with the expression of the other *ara* genes.

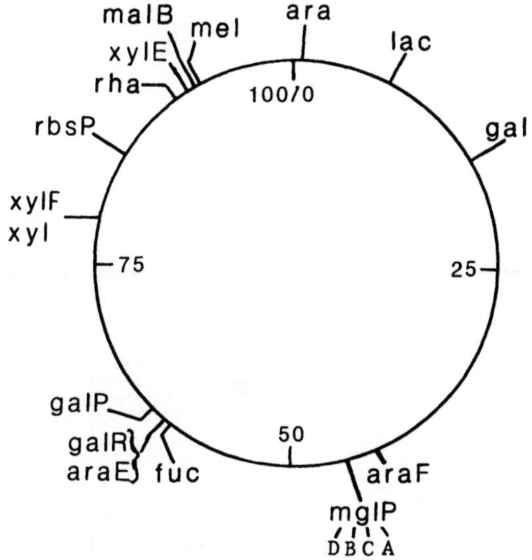

Fig. 2. The location of some *E. coli* genes involved in sugar transport and metabolism. The 100min map (Bachmann, 1983) starts at the mark inside the circle and proceeds clockwise. *ara* = operons for regulation and metabolic enzymes for L-arabinose; *lac* = operon for regulation, metabolism and transport of lactose; *gal* = operon for metabolism of D-galactose; *araF* = genes for binding protein system of L-arabinose transport; *mglP* = genes for binding protein system of D-galactose transport; *fuc* = operon for regulation, metabolism and transport of L-fucose; *araE* = gene for L-arabinose/H^+ symport; *galR* = gene for regulator of gal operons; *galP* = gene for D-galactose/H^+ symport; *xyl* = operon for regulation and metabolism of D xylose; *xylF* = genes for binding protein system of D-xylose transport; *rbsP* = genes for regulation, metabolism and transport of D-ribose; *rha* = genes for regulation, metabolism and transport of L-rhamnose; *xylE* = gene for D-xylose/H^+ symport; *malB* = genes for binding protein system of maltose transport; *mel* = gene for melibiose/Na^+ symport.

ISOLATING DNA CONTAINING THE TRANSPORT GENES

A further transposable element, lambda p*lac* Mu, was used to isolate parts of the transport genes. This phage has many of the properties of MudAp*lac* but in addition can be excised from the chromosome to obtain specialised transducing phage (that is phage that carry a specific part of the *E. coli* chromosome; Bremer et al., 1984; Silhavy et al., 1984). The maps of restriction endonuclease sites of phage DNA derived from insertions known to be in the target gene were compared with those from phage inserted close to the target gene on the second phage (Davis and Henderson, 1987; Maiden et al., 1988). In each case, a restriction endonuclease fragment containing the required gene was subcloned into a multicopy plasmid. The presence of transport activity associated with these plasmids was then checked in a transport-negative mutant strain tranformed with the relevant plasmid. One of these subcloning procedures is illustrated in Figs. 3, 4, 5 (for more detail see Davis and Henderson, 1987, and Maiden et al., 1988).

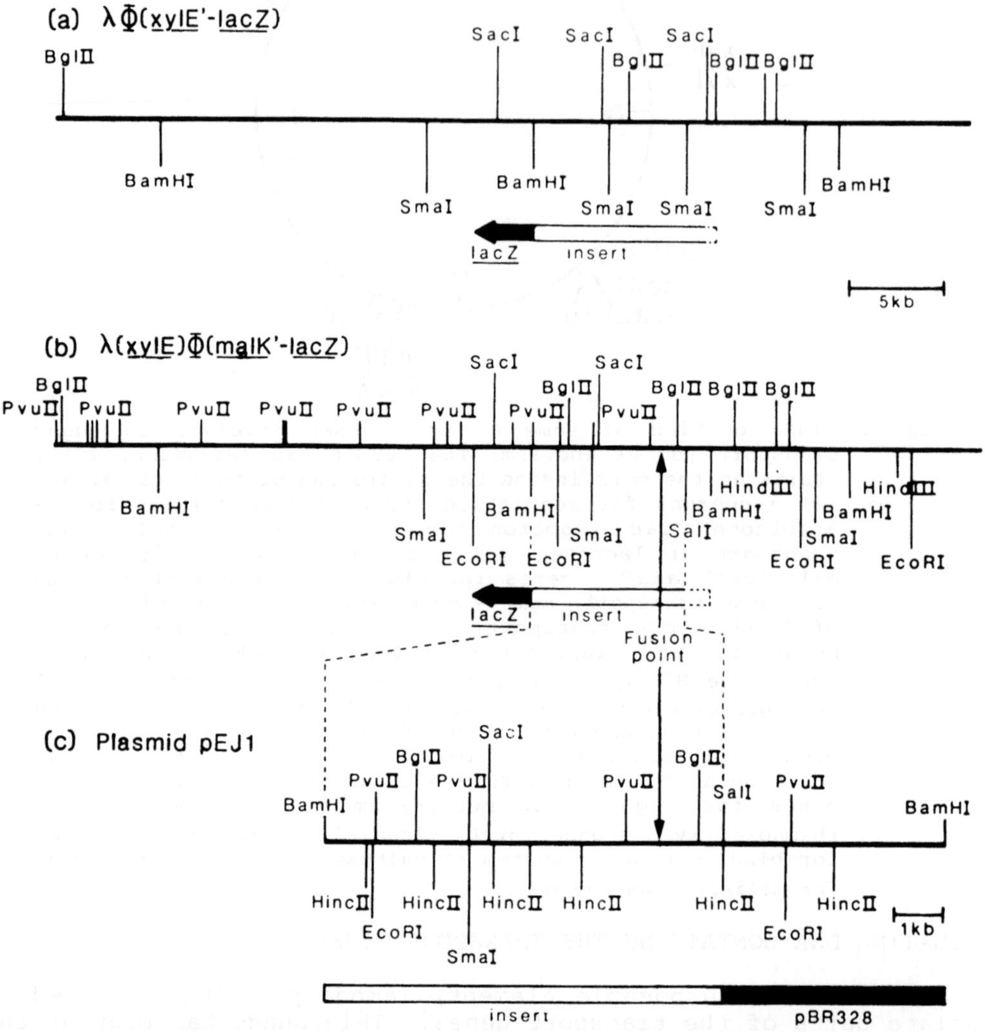

Fig 3. Restriction maps of lambda phages and a plasmid carrying the *xylE* gene. (a) phage lambda (*xylE'-lacZ*) containing the N-terminal region of *xylE* fused to the *lacZ* gene; (b) phage lambda (*xylE*) Φ (*malK'-lacZ*) containing intact *xylE* and the *malK* gene fused to the *lacZ* gene; (c) plasmid pEJ1 derived by subcloning the 8kB *Bam*HI-*Sal*I fragment into plasmid pBR328. The region of overlapping DNA in the two phage clones was identified from the *Sac*I *Sma*I *Bgl*II pattern of restriction sites, which is in the opposite arrangement in the two clones. The arrow indicates the probable insertion site of lambda p*lac* Mu in *xylE*. Details are given in the text and in Davis and Henderson (1987)

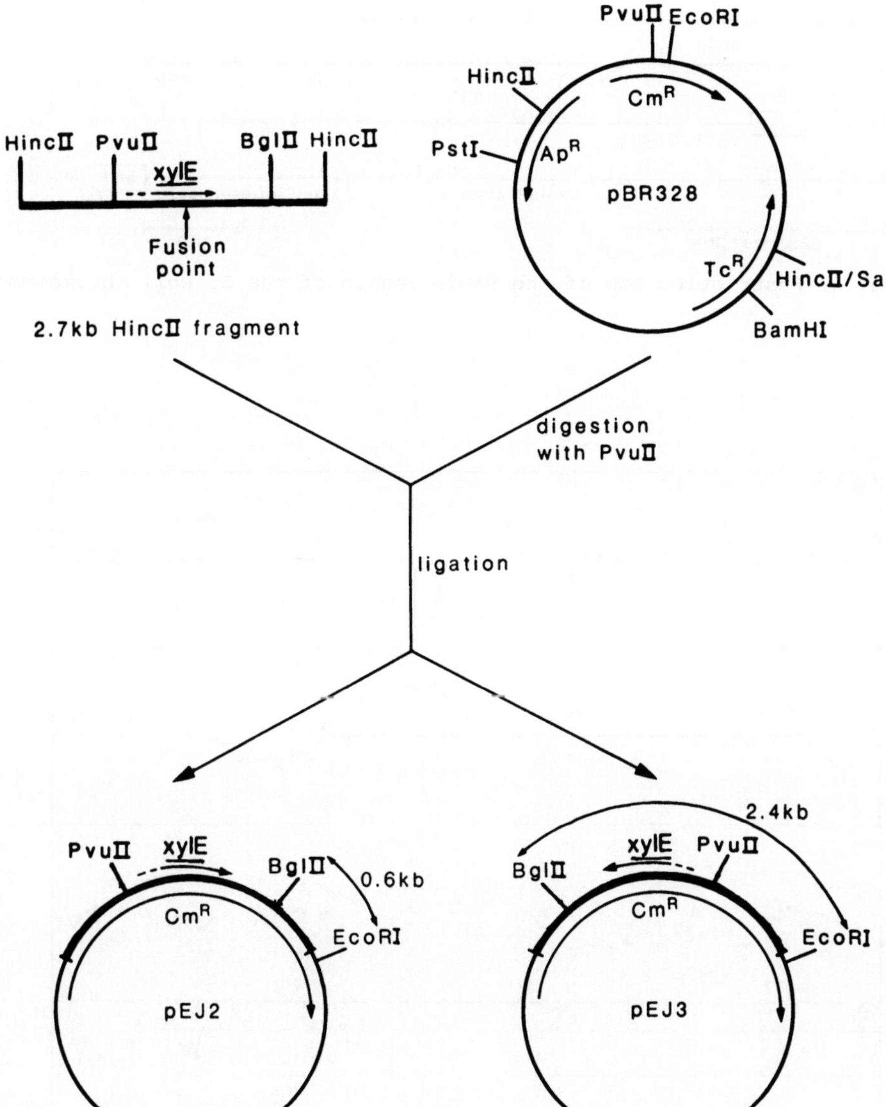

Fig. 4. Sub-cloning of a HincII DNA fragment containing the intact *xylE* gene. Details are given in the text and in Davis and Henderson (1987).

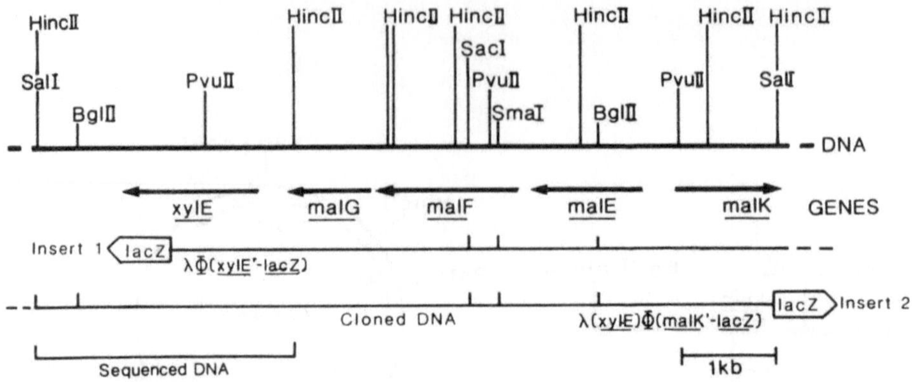

Fig. 5. Restriction map of the 91min region of the *E. coli* chromosome

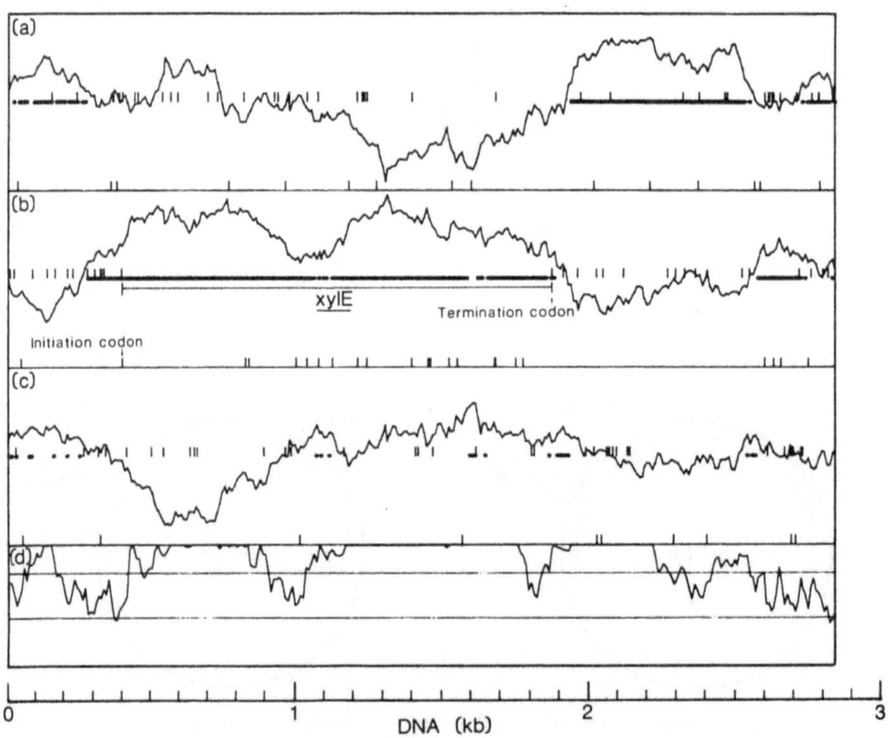

Fig. 6. Identification of the reading frame of the *xylE* gene. The actual distribution of codons in each of the three possible reading frames (a), (b), (c) was compared with the expected distribution (Staden, 1984a,b; von Heijne, 1987). The additional absence of termination codons and the length identified the *xylE* reading frame in (b).

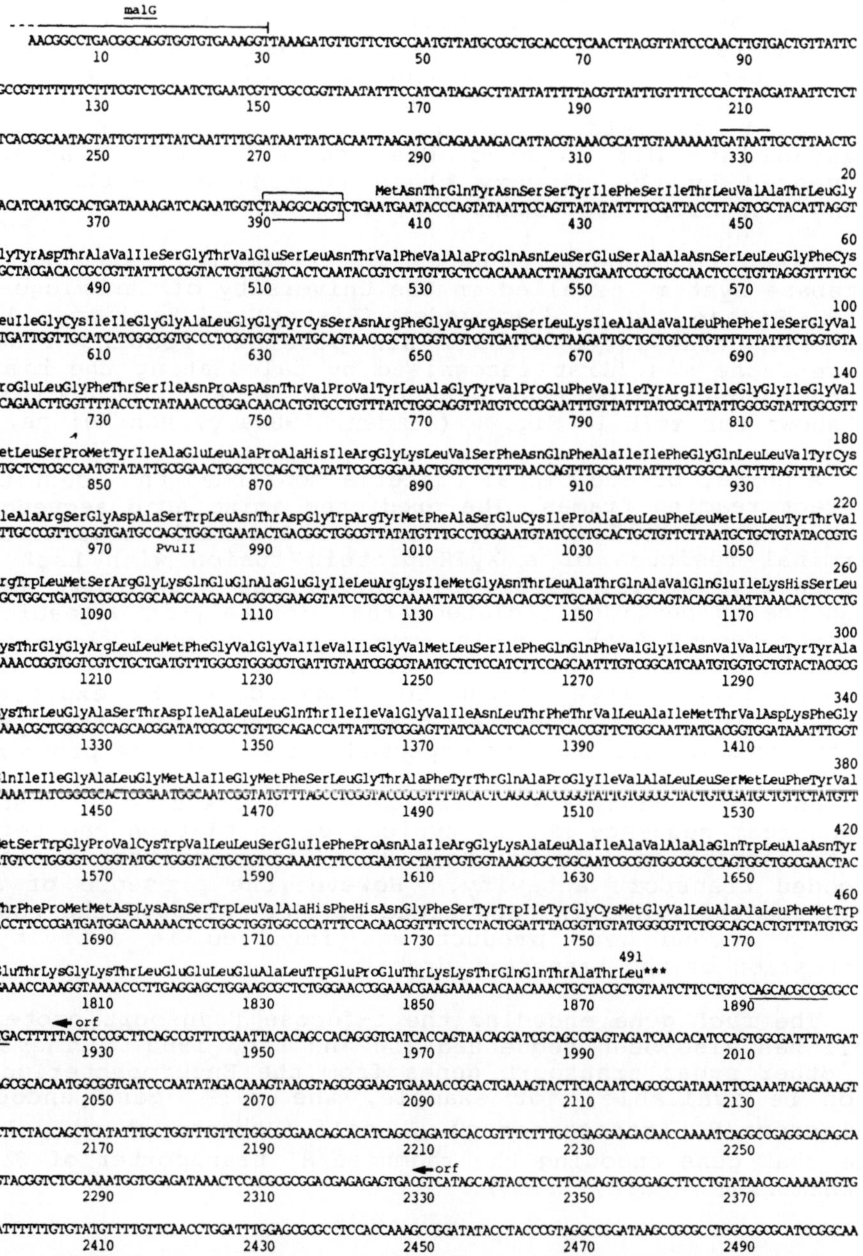

Fig. 7. The Sequence of a *Hinc*II DNA fragment containing the *xylE* gene.

THE SEQUENCING OF GENES ENCODING SUGAR TRANSPORTERS OF PROKARYOTES

The *araE* and *xylE* genes of *E. coli* have been transferred to plasmid vectors in which their activities were expressed (Davis and Henderson, 1987; Maiden et al., 1988). The DNA of each was then sequenced (Davis and Henderson, 1987; Maiden et al., 1988). In each case a 2.7-2.8kb piece of DNA (see e.g. Figs. 4,5) was sonicated to yield random fragments of 0.2-0.7kb. These were ligated into M13 phage clones and individual fragments were sequenced by the dideoxy chain termination method of Sanger (Sanger et al., 1980; Bankier et al., 1987).

Each contiguous sequence was assembled using R. Staden's Database system installed on the University of Cambridge IBM 3080 by M. Bishop and P. Alefounder (Bishop and Rawlings, 1987). It was then translated in all six possible reading frames. The correct one was first recognised by calculating the bias of the codon usage towards that expected for an *E. coli* coding region, as shown for *xylE* in Fig. 6 (Staden, 1984a,b; von Heijne, 1987).

A number of additional criteria confirmed the identity of the correct reading frames. The predicted amino acid sequence of the N-terminus of XylE (Fig. 7) matched the actual sequence of six N-terminal residues of a XylE protein fusion with LacZ isolated separately (Davis, 1985). Similarly, the N-terminal amino acid sequence of peptides isolated after proteolytic digestion of the over-expressed AraE protein (see below) matched those predicted from the DNA sequence of the *araE* gene. Each was the only open reading frame large enough to accommodate the expected M_r of >36000. Also, each predicted sequence contained a high proportion of hydrophobic residues as expected for a membrane protein.

For both the *xylE* and *araE* genes the presence of upstream and downstream sequence motifs typical of initiation and termination of transcription (von Heijne, 1987) indicated that a single gene encoded transport activity. However,the presence of a second open reading frame in each piece of DNA allowed the possibility that a second gene product was involved in activity and/or expression of the transport protein.

The *fucP* gene encoding the L-fucose transport protein of *E. coli* has also been sequenced (Lu and Lin, 1989). The sequences of other sugar transport genes from the Enterobacteriaceae will soon be available, for example, the *galP* gene encoding the galactose/H^+ transporter of *E. coli* (Roberts et al., 1990) and the *rhaT* gene encoding the rhamnose/H^+ transporter of *Salmonella typhimurium* (Muiry, 1989).

THE SEQUENCING OF GENES ENCODING SUGAR TRANSPORTERS OF MAMMALS

The isolation of genes encoding sugar transport proteins from mammals required different strategies. Mueckler et al. (1985) exploited antibodies raised to the human erythrocyte glucose transporter to screen a lambda gt11 expression library from a human hepatoma HepG2 cell line and identify cDNA that encoded a glucose transport protein. This was subsequently sub-cloned and sequenced. The cDNA for the rat brain glucose transporter was isolated by a similar technique and then sequenced (Birnbaum et al., 1986). Once this DNA was available it could be exploited as a hybridisation probe to detect similar genes. This led to the cloning and sequencing of related, but not identical, glucose transport proteins characteristic of either liver (Thorens et al., 1988) or insulin-responsive tissues (James et al., 1989; Birnbaum, 1989). By a novel strategy screening for the expression of mRNA in oocytes, the sequence of DNA encoding the glucose/Na^+ transporter of rabbit intestine was also established (Hediger et al., 1987). The rabbit gene DNA could then be used to probe a library of human DNA, and isolate the gene encoding the human transporter (Hediger et al., 1989)

An important feature of these separate studies was their prediction of the primary sequence of the amino acid residues that comprised each sugar transport protein. The following sections discuss the valuable information obtained from comparisons of these sequences.

ANALYSES OF THE PREDICTED AMINO ACID SEQUENCES OF THE SUGAR TRANSPORT PROTEINS

Computerised analyses (von Heijne, 1987; Bishop and Rawlings, 1987) of the predicted amino acid sequences of the *E. coli* XylE (Fig. 7) and AraE proteins revealed many interesting features (Maiden et al., 1988). For each protein, twelve hydrophobic regions of 21+ amino acids were predicted to loop through the membrane as alpha-helices. Certain parts of the primary sequence were therefore predicted to occur on the outside of the cell membrane and other parts on the inside (Fig. 8). The central hydrophilic portion of the transport protein may be alpha-helical in structure (Fig. 9). The positions of some beta-turns are predicted (Fig. 9). The XylE protein was homologous to the *E. coli* AraE protein with about 28% identical residues and an additional 12% conservative substitutions. The protein was equally similar to glucose transport proteins from a diversity of other organisms (see below). Residues conserved in identical locations in all the aligned proteins probably play a critical role in the maintenance of structure and function.

The structural model of each protein (e.g. Fig. 8) generated from the sequence provides the database and design features of numerous future experiments to elucidate the structure and molecular mechanism of the transport process. However, such experiments often require amounts of protein greater than is possible or convenient to obtain from wild-type *E. coli* strains. Recombinant DNA techniques may then be exploited to increase the expression of the appropriate genes.

AMPLIFIED EXPRESSION OF GENES ENCODING TRANSPORT PROTEINS

To enhance the expression of a procaryotic gene in *E. coli* it is usually sufficient simply to increase the level of messenger RNA produced by that gene (with non-procaryotic messages additional work may be necessary to stabilise the mRNA once it has been produced). This is best achieved on multicopy plasmids; the increase in the number of copies of the gene within the cell may be sufficient to amplify the level of expression to the desired level. With most bacterial transporters it is usually necessary to introduce a new promoter upstream of the gene to be expressed. A particularly useful

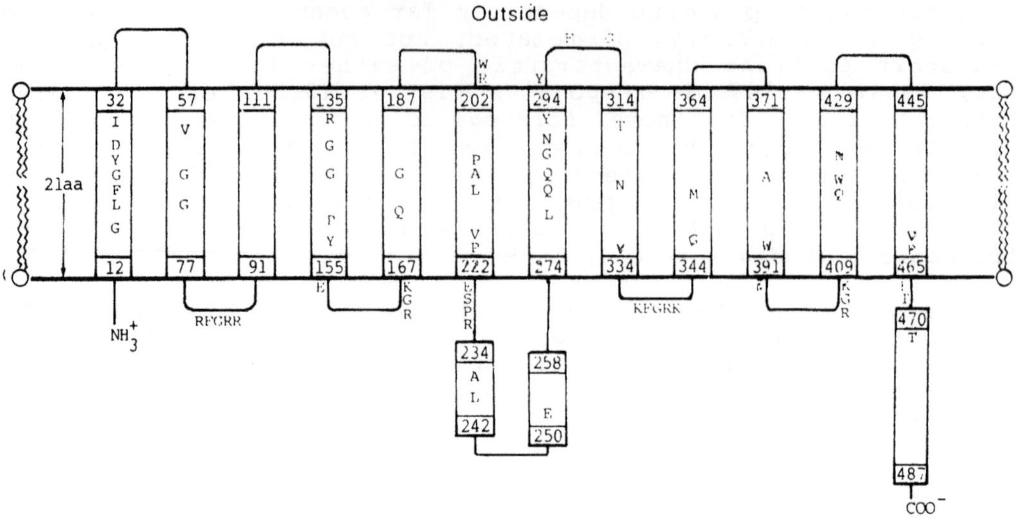

Fig. 8. Proposed orientation of the xylose/H^+ transport protein in the membrane. The model is based on predictions of membrane-spanning regions (Garnier et al., 1978; Kyte and Doolittle, 1982) and alpha-helix content (Garnier et al., 1978; Eisenberg et al., 1984; von Heijne, 1988) and the alignment with the glucose transporter (Mueckler et al., 1985; Baldwin and Henderson, 1989; Fig. 13a,b). The 12 membrane-spanning segments are represented as rectangles with the end residues numbered from the N-terminus. The rectangles outside the membrane represent predicted alpha-helices. The single residues marked are usually conserved in all the aligned sugar transporter sequences (Fig. 13a,b), and some recur at equivalent positions in each half of the molecule.

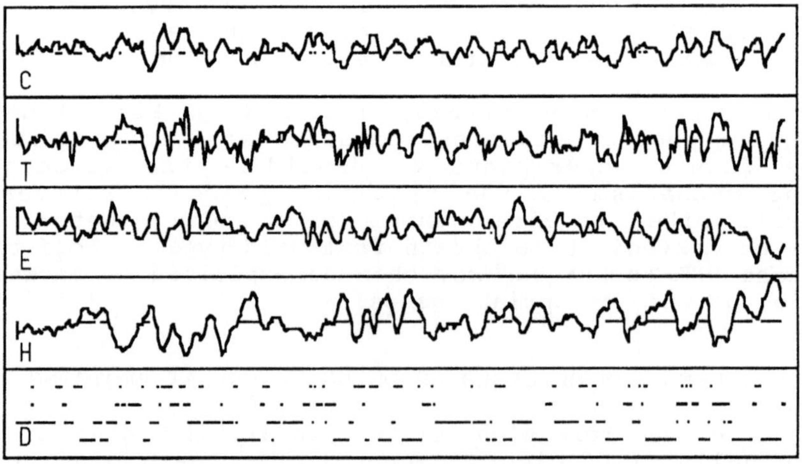

Fig. 9. A Robson plot of the secondary structure of the XylE protein. C,T,E and H represent the probabalistic profiles for forming random coil, reverse turn, ß-sheet and α-helix, respectively (Garnier et al., 1978). The program selects sequentially a 17-residue segment starting at the N-terminus and predicts the contribution of the 9th residue to formation of one of the four states in that region. This process is repeated and D (decision) represents the most likely structure at each point throughout the sequence.

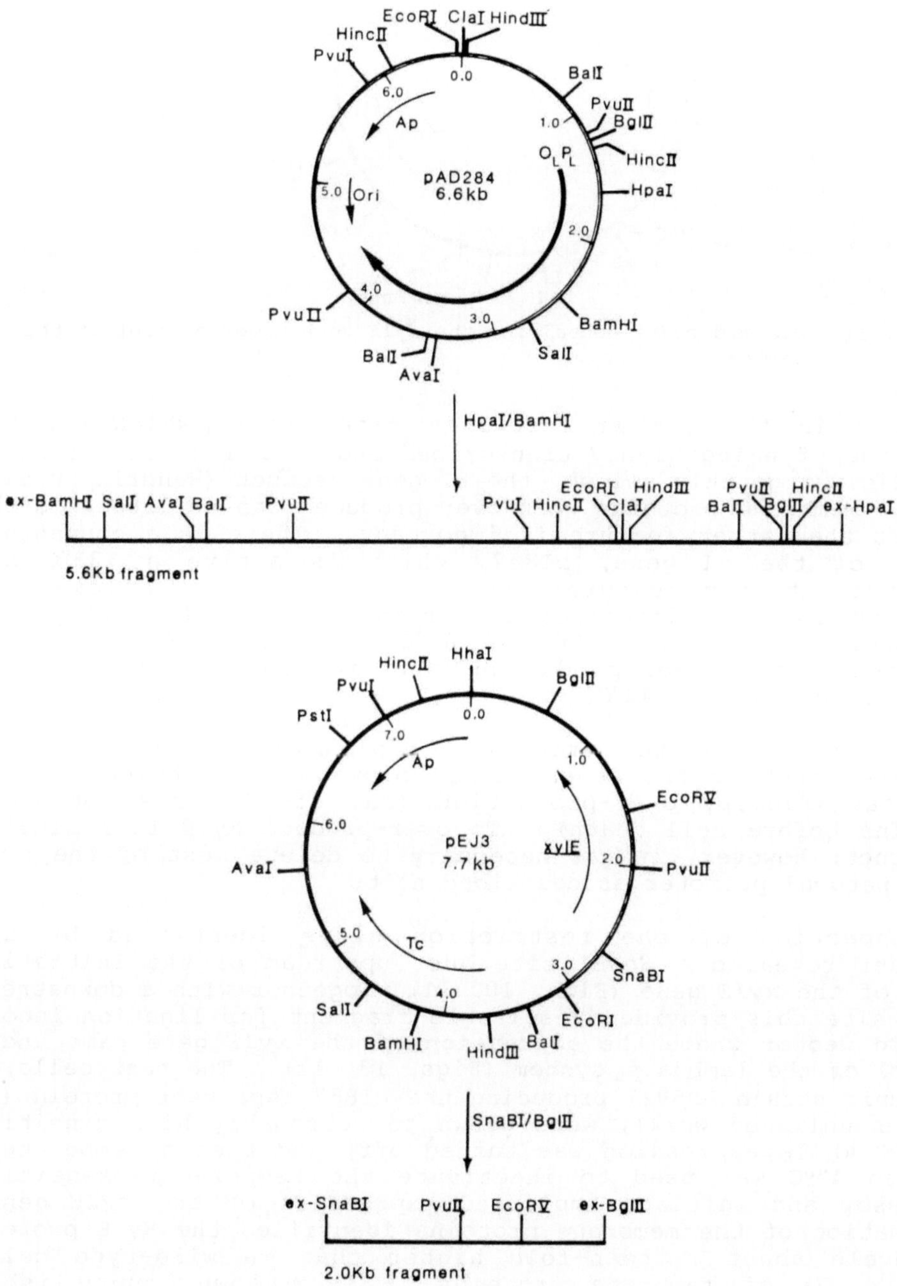

Fig. 10. Restriction digestion of plasmids pEJ3 and pAD284 prior to construction of pBJM1.

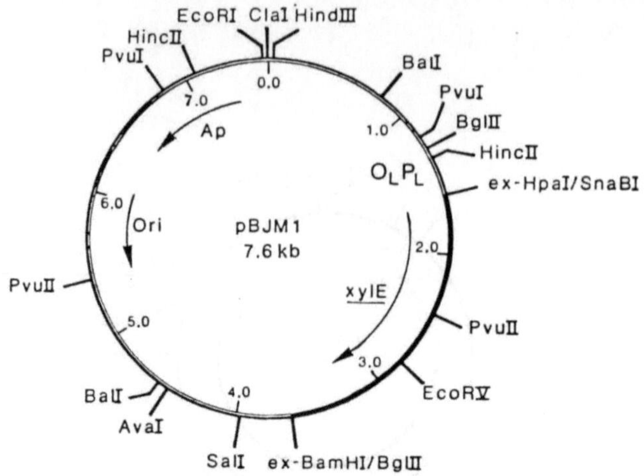

Fig. 11. Plasmid pBJM1 containing the *xylE* gene under control of the lambda promoter.

promoter is the lambda leftwards promoter p_L, which has the advantage of being highly transcribed and of being controlled by an efficient genetic switch, the *c*I gene product (Hendrix et al., 1983). Membrane proteins when over produced can be lethal to the cell so the latter feature is important. There is a convenient mutant of the *c*I gene, *c*I857, which is active at 33°C but inactive at temperatures above 37°C. This allows the construction and maintenance of strains capable of growing with the promoter switched off (at 33°C) and then of turning on expression and over-producing the membrane protein by a temperature shift to 42°C.

In the case of the *araE* gene a plasmid was produced that contained much of the genes normal promoter (see below). This gave satisfactory over-production (ca. 15-20% total membrane proteins before cell death). To over-produce XylE in a similar construct, however, it was necessary to delete most of the *xylE* genes natural promoter as described next.

Inspection of the restriction sites identified by the computer revealed a *Sna*BI site just upstream of the initiation codon of the *xylE* gene (Figs. 10, 11). Together with a downstream *Bgl*II site this provided a suitable fragment for ligation into a plasmid vector where the expression of the *xylE* gene came under control of the lambda p_L system (Figs. 10, 11). The host cells, a lysogenic strain AD5827 producing the *c*I857 repressor protein (A. Das, unpublished work), were grown to reasonably high densities at 33°C while expression was turned off, and then a temperature jump to 42°C was used to inactivate the temperature-sensitive repressor and initiate amplified expression of the *xylE* gene. Examination of the membrane proteins identified the XylE protein at levels about fifteen-fold higher than in wild-type cells (approx. 7% of membrane proteins, B.J. McKeown, unpublished data).

A similar strategy was successful for amplifying expression of the *araE* gene. It is illustrated in Fig. 12 and described in detail by Maiden et al. (1988). The amounts of AraE protein produced, 15-20% of the total membrane proteins, proved to be sufficient for raising antibodies and for structural studies (Smith and Lucas, unpublished data).

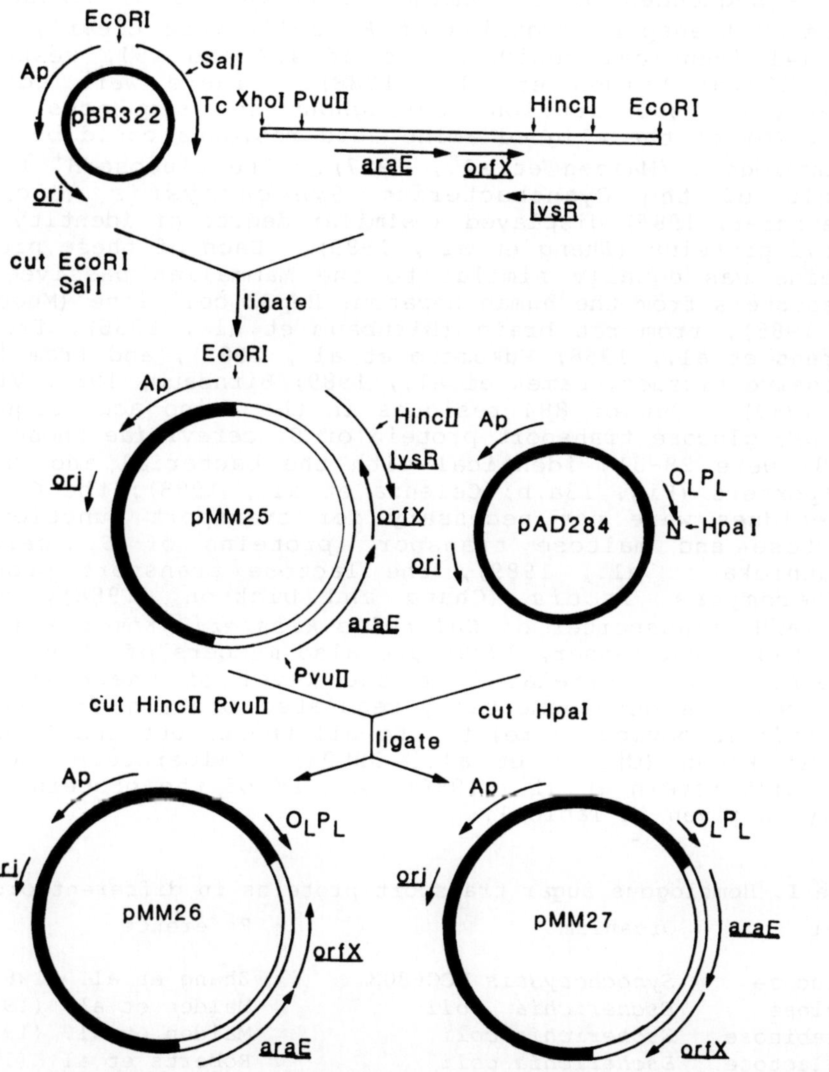

Fig. 12. Strategy for cloning and amplifying expression of the *araE* gene. A *Xho*I-*Eco*RI fragment of 4.8kb was sub-cloned from lambda d(*galR*$^+$*lysA*$^+$*araE*$^+$) phage into the *Eco*RI and *Sal*I sites of pBR322. This plasmid (pMM25) contained *araE orfX lysR* and part of *lysA*. *E. coli* strain MM23(*araE araF*) acquired arabinose/H$^+$ symport activity when transformed with pMM25, demonstrating the presence of an intact *araE* gene on the latter. A smaller HincII-PvuII fragment of 2.8kb also carried arabinose/H$^+$ symport activity. This was used to generate sequence data, and was cloned into a lambda p_L expression vector (pAD284, from Asis Das). In plasmid pMM27 *araE* was orientated to allow its expression from the lambda promoter; in plasmid pMM26 *araE* was in the opposite orientation.

AMINO ACID SEQUENCE HOMOLOGIES BETWEEN SUGAR TRANSPORT PROTEINS FROM PROKARYOTES AND EUKARYOTES

The sequences of the amino acids in the arabinose/H^+ and xylose/H^+ transport proteins of *E. coli* were clearly related, with 141 identical residues out of 472 and 491, respectively (Fig. 13a,b; Maiden et al., 1988). There were additional conservative substitutions throughout the sequences, so that nearly 40% of the aligned amino acid residues could be regarded as homologous (Maiden et al., 1987). The glucose/H^+ transport protein of the Cyanobacterium *Synechocystis* (Flores and Schmetterer, 1986) displayed a similar degree of identity to both *E. coli* proteins (Zhang et al., 1989). Each of these prokaryote proteins was equally similar to the mammalian passive glucose transporters from the human hepatoma HepG2 cell line (Mueckler et al., 1985), from rat brain (Birnbaum et al., 1986), from liver (Thorens et al., 1988; Fukumoto et al., 1988), and from insulin-responsive tissues (James et al., 1989; Birnbaum, 1989; Vilaro et al., 1989). Out of 884 residues in the amino acid sequence of the *SNF3* glucose transport protein of *S. cerevisiae* those from 86 to 581 were 28-31% identical with the bacterial and mammalian transporters (Fig. 13a,b; Celenza et al., 1988); the C-terminal 87 residues were not necessary for transport function. The galactose and maltose transport proteins of *S. cerevisiae* (Szkutnicka et al., 1989), the lactose transport protein of *Kluyveromyces lactis* (Chang and Dickson, 1988), and the glucose/H^+ transporter of *Chlorella kessleri* (Komor and Tanner, 1974; Sauer and Tanner, 1989) are also members of this family of proteins that are related at the level of their amino acid sequences. A developmentally-regulated protein of *Leishmania enriettii* is obviously related to all these, but its function is not yet known (Cairns et al., 1989a; Zilberstein and Dwyer, 1985; Zilberstein et al., 1986). A list of the proteins in this family is given in Table 1.

Table I. Homologous sugar transport proteins in different organisms

Sugar	Organism	Reference
D-Glucose	*Synechocystis* PCC6803	Zhang et al. (1989)
D-Xylose	*Escherichia coli*	Maiden et al. (1987)
L-Arabinose	*Escherichia coli*	Maiden et al. (1987)
D-Galactose	*Escherichia coli*	Roberts et al. (1990)
D-Galactose	*Saccharomyces cerevisiae*	Szkutnicka et al. (1989)
D-Glucose	*Saccharomyces cerevisiae*	Celenza et al. (1988)
Maltose	*Saccharomyces carlsbergensis*	Yao et al. (1989)
	Saccharomyces cerevisiae	Szkutnicka et al. (1989)
Lactose	*Kluyveromyces lactis*	Chang and Dickson (1988)
Quinate	*Neurospora crassa*	Geever et al. (1989)
D-Glucose	*Chlorella kessleri*	Sauer and Tanner (1989)
D-Glucose	*Leishmania donovanii*	Zilberstein et al.(1986)
	Leishmania enriettii	Cairns et al. (1989a)
D-Glucose	Human hepatoma/erythrocyte	Mueckler et al. (1985)
	Rat brain	Birnbaum et al. (1986)
D-Glucose	Rat liver	Thorens et al. (1988)
	Human liver	Fukumoto et al. (1988)
D-Glucose	Rat adipocyte/heart/muscle	James et al. (1989)
(Insulin-		Birnbaum (1989)
regulated)	Mouse adipocyte/heart/muscle	Kaestner et al. (1989)
D-Glucose	Human small intestine	Gould and Bell (1990)
D-Glucose	Human foetal muscle	Gould and Bell (1990)

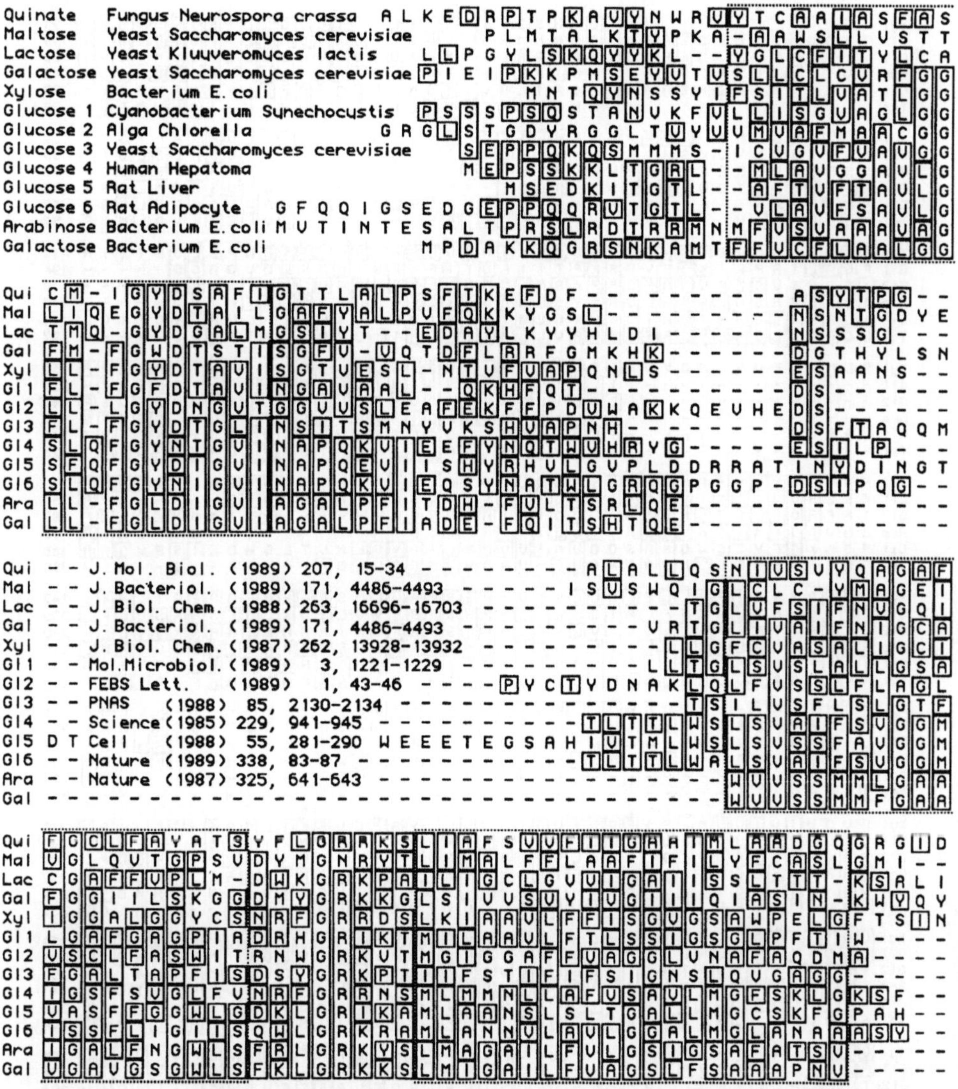

Fig. 13a. The aligned amino acid sequences of sugar transport proteins from prokaryotes and eukaryotes. The single letter code for amino acids is employed for the indicated transport proteins. The residues are boxed if they occur in more than one sequence. The dotted rectangles enclose hydrophobic regions predicted to span the membrane. For reasons of space only the N-terminal regions of the proteins are shown. However, in Fig. 13b the complete sequences of the xylose/H^+ symporter of *E. coli* (Xyl), the glucose transporter of *Saccharomyces cerevisiae* (Gl1), the glucose transporter of a human hepatoma cell line (Gl2), the arabinose/H^+ symporter of *E. coli* (Ara), the plasmid-encoded citrate/H^+ symporter (Cit), and the plasmid pBR322-encoded tetracycline/H^+ antiporter, are aligned. References are given in the text.

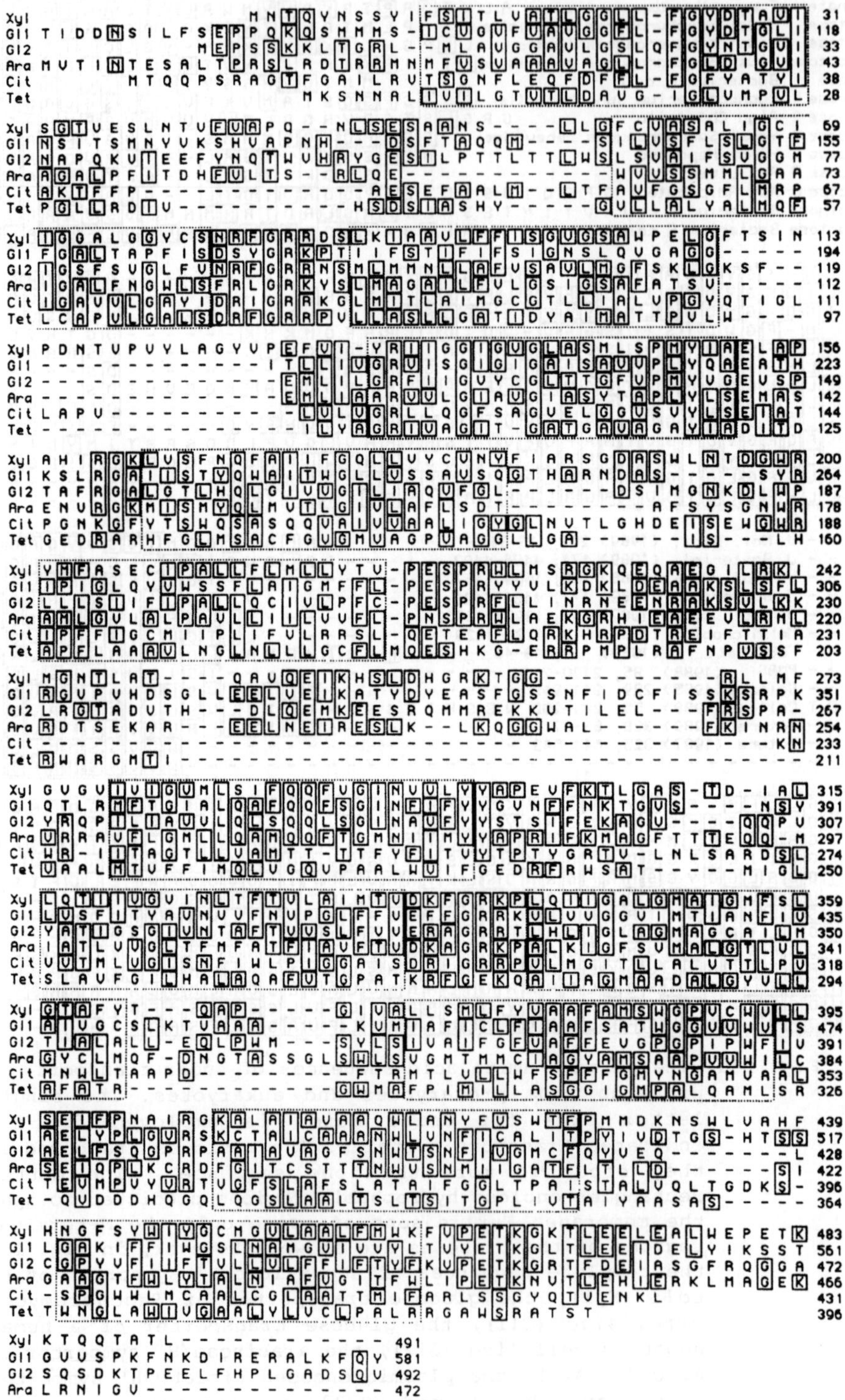

Fig. 13b. The aligned amino acid sequences of transport proteins. Details are given in the legend to Fig. 13a.

It was curious that none of these proteins was detectably homologous to either the lactose/H^+ or melibiose/Na^+ transporters of *E. coli* (Buchel et al., 1980; Yazyu et al., 1984), or to the glucose/Na^+ transporter of rabbit intestine (Hediger et al., 1987). Perhaps their three-dimensional structures are more similar than their sequences would indicate. Intriguingly, the mammalian glucose/Na^+ transporter is related to the proline/Na^+ transporter of *E. coli* (Hediger et al., 1989).

The N-terminal ends of thirteen of the homologous sugar transporters are aligned in Fig. 13a. The conservation of a single amino acid in all the sequences may reveal a feature critical for maintenance of structure and function. The most highly conserved residues throughout the whole sequences were glutamate, glycine, arginine and proline. Cysteine and histidine residues were not conserved, indicating that their occurrence at specific locations is not essential for activity.

Several of the related sugar transporters do not involve cation symport. Thus, the basic structures, and presumably molecular mechanism, of proteins that catalyse 'active' versus 'passive' transport are not substantially different.

Some transport proteins were apparently not related to the aligned transporters in Fig. 13a, but might nevertheless contain isolated regions of identity that enable subsequent alignment. For example, a sequence motif RXGRR found in two places (Figs. 8, 13b), was also found in the citrate/H^+ transporter encoded on a prokaryote plasmid (Ishiguro and Sato, 1985; Sasatsu et al., 1985) and in tetracycline/proton(potassium) antiporters encoded on a transposon and several plasmids (Hillen and Schollmeier, 1983; Levy, 1988). This 'landmark' enabled alignment of the whole citrate and tetracycline transporters with the sugar transporters (Maiden et al., 1987; Henderson and Maiden, 1990; Fig. 13b), and identified 45 conserved residues in the citrate transporter (out of a total of 431) that may be vital for the basic structure and common transport function of all the proteins.

A UNIFYING MODEL FOR THE STRUCTURE OF THE SUGAR TRANSPORT PROTEINS

The homologous sugar/H^+ symporters and passive sugar transporters from prokaryotes and eukaryotes are each independently predicted to contain twelve membrane-spanning segments (Mueckler et al., 1985; Maiden et al., 1987; Celenza et al., 1988; Birnbaum et al., 1986; Szkutnicka et al., 1989), although these predictions must be regarded with caution (Lodish, 1988). Nevertheless, by combining them with the aligned sequences (Fig. 13a,b), with computer predictions of secondary structure (Fig. 9; Garnier et al., 1978; Eisenberg et al., 1984; von Heijne, 1988), and with experiments on the human erythrocyte glucose transporter (reviewed by Baldwin and Henderson, 1989), a model of the folding of each transport protein in the membrane can be generated (Fig.8). Three more examples are shown in Fig. 14. The AraE protein (Maiden, 1986), the citrate transporter (Reynolds and Silver, 1983), and the tetracycline transporter (McMurry et al., 1980; Dosch et al., 1984; Levy, 1984; Eckert and Beck, 1989). They include the twelve membrane-spanning segments, the central hydrophilic region, the positions of residues conserved in the transporters, and some sequence motifs

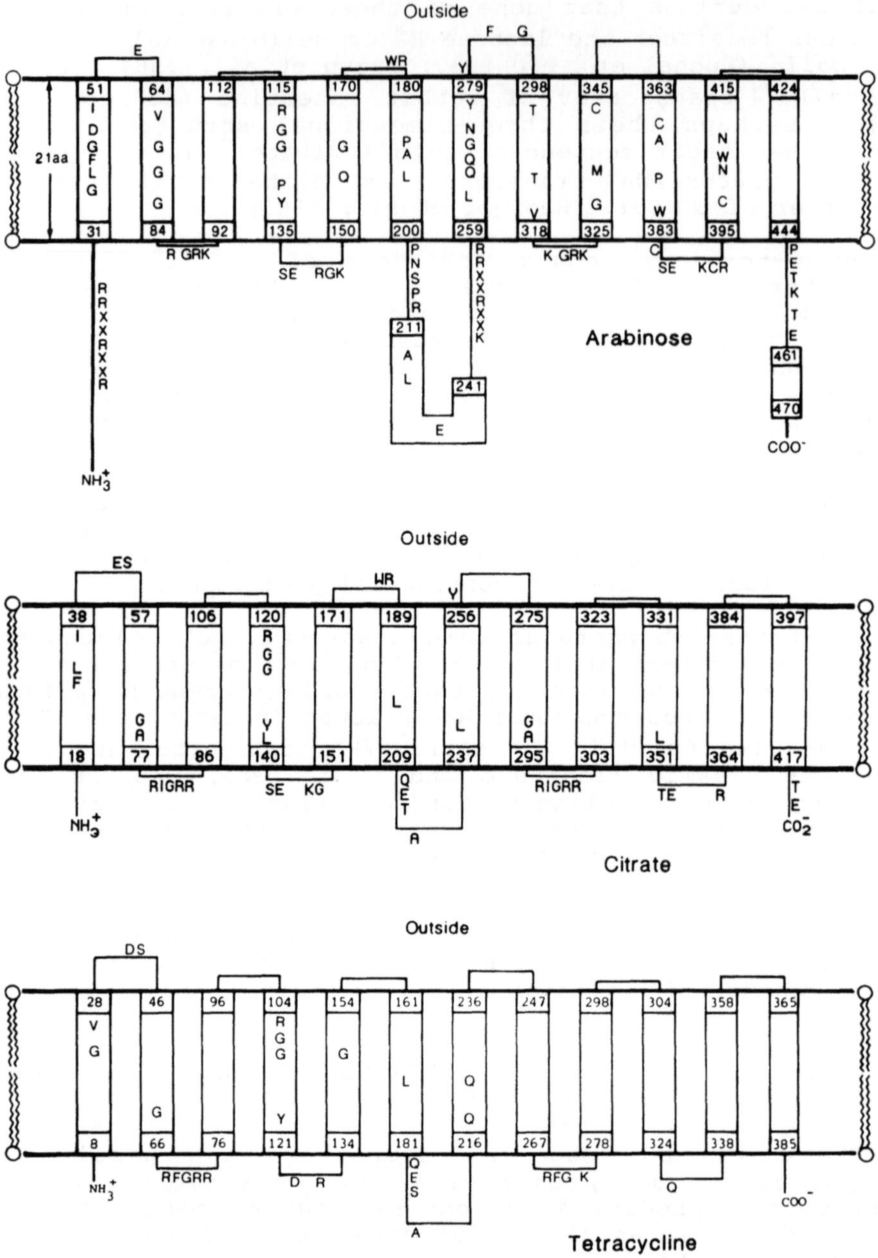

Fig. 14. Proposed orientation of the arabinose, citrate and tetracycline transporter proteins in the membrane. The models are based on predictions of membrane-spanning regions (Garnier et al., 1978; Kyte and Doolittle, 1982) and alpha-helix content (Garnier et al., 1978; Eisenberg et al., 1984; von Heijne, 1988), the alignment with the glucose transporter (Mueckler et al., 1985; Baldwin and Henderson, 1989; Fig. 13a,b) and experimental data (Cairns et al., 1987; Holman and Rees, 1987; Karim et al., 1987, Haspel et al., 1988; Walmsley, 1988; Baldwin and Henderson, 1989). The 12 membrane-spanning segments are represented as rectangles with the end residues numbered from the N-terminus. The rectangles outside the membrane represent predicted alpha-helices. The single residues marked are usually conserved in all the aligned sugar transporter sequences (Fig. 13a,b).

apparently repeated in similar positions in each half of the molecule. It seems possible that these proteins arose from an ancestral gene duplication event (Doolittle, 1981). Such models form the basis for future experiments in which 'topological' reagents and antibodies (e.g Cairns et al., 1987), and gene fusions (Broome-Smith and Spratt, 1986; Manoil and Beckwith, 1986), are used to explore the actual folding of the protein in the membrane.

These transport proteins are not the only ones predicted to have twelve membrane-spanning regions. Others are the lactose/H+ (Foster et al., 1983), proline/Na^+ (Nakao et al., 1987), hexosephosphate-phosphate (Friedrich and Kadner, 1987), and glycerolphosphate-phosphate (Friedrich and Kadner, 1987; Eiglmeier et al., 1987) transporters of *E. coli,* the uracil transporter of yeast (Jund et al., 1988), and possibly the human glucose/Na^+ transporter (Hediger et al., 1989).

CONCLUSIONS

The family of sugar transport proteins described in this article has been found in such diverse organisms so far that it seems likely to occur in most forms of life. This implies that investigations of the molecular mechanism of their transport activities in microorganisms, with the attendant conveniences of experimental manipulation, will uncover features equally valid for their operation in higher organisms, including man. It seems that insight into the mechanism of transport of substrates other than sugars will also accrue, if the apparent similarities to the citrate and tetracycline (and other?) transporters described here are real. The homologies also imply a fundamental similarity between proteins that catalyse facilitated diffusion ('passive transport'), substrate/H^+ symport ('secondary active transport') and substrate/H^+ antiport - a perceived difference in mechanism seems not to require a profound difference in structure. This similarity is reinforced by the sensitivity of several of the sugar transporters from *E. coli, Leishmania* and *Homo sapiens* to the antibiotic, cytochalasin B (Shanahan, 1982; Zilberstein et al. 1986; Walmsley, 1988; Baldwin and Henderson, 1989; Cairns et al., 1989b; Smith et al., 1990). The availability of a variety of sugars and their structural analogues, and the known sequence homologies of the L-arabinose, D-xylose, D-galactose and D-glucose transporters, make this a particularly promising series of proteins in which to establish the characteristics of molecular recognition and their evolution (Fersht, 1985; Quiocho, 1986, 1989). There also exist transport proteins with the same function of sugar/cation symport, but without any detectable similarity in primary sequence, i.e. the lactose/H^+ (Kaback, 1990), L-rhamnose/H^+ (Muiry, 1989) and melibiose/Na+ (Leblanc et al., 1989) transporters of Enterobacteria and the D-glucose/Na^+ transporter of mammalian intestine (Hediger et al., 1989); perhaps these arose by convergent evolution.

The sugar transporters are not the only membrane proteins that have homologues in bacteria and mammals. The proline/Na^+ transporter of *E. coli* is homologous to the glucose/Na^+ transporter of mammalian intestine (Hediger et al., 1989); the ATP-energised KdpB K^+ transport protein of *E. coli* has regions of primary sequence similar to the K^+/Na^+, Ca^{++}, and $H^+(K^+)$ ATPases of mammalian tissues (Epstein et al., 1990); the H^+ATPase of

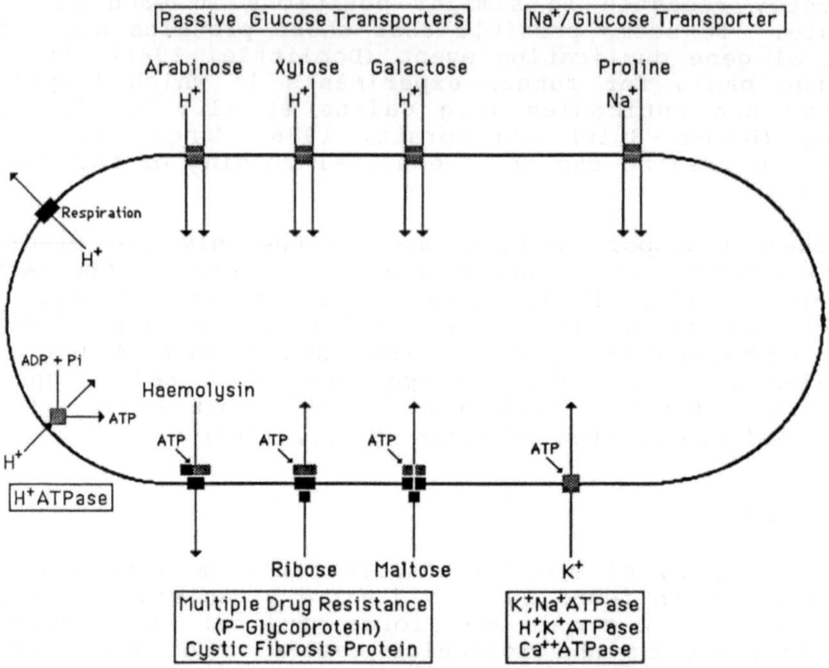

Fig. 15. An illustration of transport systems in *Escherichia coli* that have homologous proteins in mammals (boxed). For details see Fig.1.

bacteria is homologous to the H^+ATPase of mammalian mitochondria; and the ATP-binding protein component of numerous transporters for sugars, amino acids, vitamins, peptides and proteins in bacteria has its equivalents in the multiple drug resistance protein and the cystic fibrosis protein of mammals (Higgins, 1989; Higgins et al., 1990). These relationships are illustrated in Fig. 15.

A real understanding of all these transport phenomena is critically dependent upon determination of the three-dimensional structure of the proteins at the level of atomic resolution. Tentative models of transport protein structure such as those in Figs. 8 and 14 are the basis of present and future experiments in which 'topological' protein reagents and antibodies, and gene fusions (Cairns et al., 1987; Haspel et al., 1988; Manoil and Beckwith, 1986; Broome-Smith and Spratt, 1986), and other techniques, are used to explore the actual folding of the protein in the membrane. The structural and functional roles of individual amino acids may also be evaluated by *in vivo* and *in vitro* mutagenesis. (e.g. Kaback, 1987; Menick et al., 1987; Botfield and Wilson, 1988). Such approaches are the most practicable at the present time, but they are unlikely to yield a molecular model of sufficient detail to understand the different substrate specificities, the process of cation recognition, the translocation mechanism, and the evolution of the related transport proteins described in this article. The determination of a more detailed three-dimensional structure probably requires crystallisation and X-ray and/or electron diffraction analysis (Diesenhofer et al., 1985; Henderson and Schertler, 1990). The first step is the purification of sufficient stable, undenatured protein. Recombinant DNA techniques, such as those illustrated in this article, are an essential prerequisite for achieving this with membrane transport proteins.

ACKNOWLEDGEMENTS

This research was supported by SERC, MRC, The SmithKline Foundation and The Wellcome Trust. We are indebted to Dr. S.A. Baldwin, Dr. M.T. Cairns, Dr. B.M. Charalambous, Dr. V.A. Lucas, Mr. T. McDonald, Dr. D.C.M. Moore, Dr. J.A.R. Muiry, Dr. C.G. Tate, Ms. K. Petro, Mr. P.E. Roberts, and Dr. G. Smith for unpublished information and helpful discussions. MCJM was the recipient of a MRC postdoctoral training Fellowship. Figs. 3,4,5,7,12 are reproduced with the permission of the Journal of Biological Chemistry, and Fig. 13b with the permission of the Royal Society.

REFERENCES

Bachmann, B.J., 1983, Linkage map of *Escherichia coli* K12, edition 7, Microbiol. Rev., 47:180-230.

Badia, J., Baldoma, L., Aguilar, J., and Boronat, A., 1989, Identification of the *rhaA*, *rhaB* and *rhaD* gene products from *Escherichia coli* K12, FEMS Microbiol. Lett., 65:253-258.

Balbas, P., Soberon, X., Merino, E., Zurita, M., Lomeli, H., Valle, F., Flores, N., and Bolivar, F., 1986, Plasmid vector pBR322 and its special purpose derivatives - a review, Gene, 50:3-40.

Baldwin, S.A., and Henderson P.J.F., 1989, Homologies between sugar transporters from eukaryotes and prokaryotes, Annu. Rev. Physiol., 51:459-471.

Bankier, A.T., Weston, K.M.G., and Barrell, B., 1987, Random cloning and sequencing by the M13/dideoxynucleotide chain termination method, Meth. Enzymol., 155, 51-93.

Beckwith, and Zipser, 1970, "The lactose operon", Cold Spring Harbor, Cold Spring Harbor.

Birnbaum, M.J., 1989, Identification of a novel gene encoding an insulin-responsive glucose transporter protein. Cell, 57:305-315

Birnbaum, M.J., Haspel, H.C., and Rosen, O.M., 1986, Cloning and characterisation of a cDNA encoding the rat brain glucose-transporter protein, Proc. Natl. Acad. Sci. USA, 83:5784-5788.

Bishop, M.J., and Rawlings, C.J., 1987, "Nucleic acid and protein sequence analysis - a practical approach", IRL Press, Oxford.

Botfield M.C., and Wilson, T.H., 1988, Mutations that simultaneously alter both sugar and cation specificity in the melibiose carrier of *Escherichia coli*, J. Biol. Chem., 263:12909-12915.

Bradley, S.A., Tinsley, C.R., Muiry J.A.R., and Henderson, P.J.F., 1987, Proton-linked L-fucose transport in *Escherichia coli*, Biochem. J., 248:495-500.

Bremer, E., Silhavy, T.J., Wiesemann, J.M., and Wienstock, G.M., 1984, Lambda *plac* Mu: a transposable derivative of phage lambda for creating *lacZ* protein fusions in a single step, J. Bacteriol., 158:1084-1093.

Broome-Smith, J.K., and Spratt, B.G., 1986, A vector for the construction of translational fusions to TEM ß-lactamase and the analysis of protein export signals and membrane protein topology, Gene, 49:341-349.

Buchel, D.E., Gronenborn, B., and Muller-Hill, B., 1980, Sequence of the lactose permease gene, Nature, 283:541-545.

Button, D.K., 1985, Kinetics of nutrient-limited transport and microbial growth, Microbiol. Rev., 49:270-297.

Cairns, M.T., Alvarez, J., Panico, M., Gibbs, A.F., Morris, H.R., Chapman, D., and Baldwin, S.A., 1987, Investigation of the structure and function of the human erythrocyte glucose transporter by proteolytic dissection, Biochim. Biophys. Acta, 905:295-310.

Cairns, B.R., Collard, M.W., and Landfear, S.M., 1989a, Developmentally regulated gene from Leishmania encodes a putative membrane transport protein, Proc. Natl. Acad. Sci. USA, 86:7682-7686.

Cairns, M.T., Smith, G., Henderson, P.J.F., and Baldwin, S.A., 1989b, Photoaffinity labelling of the GalP D-galactose transport protein of *Escherichia coli* with cytochalasin B, Biochem.Soc.Trans., 17:552-553.

Casadaban, M.J., and Cohen, 1979, Lactose genes fused to exogenous promoters in one step using a Mu-*lac* bacteriophage: *In vivo* probes for transcriptional control, Proc. Natl. Acad. Sci. USA,76:4530-4533.

Celenza J.L., Marshall-Carlson, L., and Carlson, M., 1988, The yeast SNF3 gene encodes a glucose transporter homologous to the mammalian protein, Proc. Natl. Acad. Sci. USA, 85:2130-2134.

Chang, Y-D,. and Dickson, R.L., 1988, Primary structure of the lactose permease from the yeast *Kluyveromyces lactis*, J. Biol.Chem., 263:16696-16703.

Davis, E.O., Jones-Mortimer, M.C.J., and Henderson, P.J.F., 1984, Location of a structural gene for xylose/H+ symport at 91min on the linkage map of *Escherichia coli*, J. Biol. Chem., 259:1520-1525.

Davis, E. O., 1985, "Xylose transport in *Escherichia coli*", Ph.D. dissertation, University of Cambridge.

Davis, E.O., and Henderson, P.J.F., 1987, The cloning and DNA sequence of the gene *xylE* for xylose-proton symport in *E. coli* K12, J. Biol. Chem., 262:13928-13932.

Diesenhofer, J., Epp, O., Miki, K., Huber, R., and Michel, H., 1985, Structure of the protein subunits in the photosynthetic reaction centre of *Rhodopseudomonas viridis* at 3A resolution, Nature, 318:618-624.

Doolittle, R.F., 1981, Similar amino acid sequences: chance or common ancestry, Science, 214:149-159.

Dosch, D., Salvacion, F. and Epstein, W., 1984, Tetracycline resistance element of pBR322 mediates potassium transport, J.Bacteriol., 160:1188-1190.

Eckert, B., and Beck, C.F., 1989, Topology of the transposon Tn10-encoded tetracycline resistance protein within the inner membrane of *Escherichia coli*. J. Biol. Chem., 264: 11663-11670.

Eiglmeier, K., Boos, W., and Cole, S.T., 1987, Nucleotide sequence and transcriptional startpoint of the *glpT* gene of *Escherichia coli*, Molec. Microbiol., 1:251-258.

Eisenberg, D., Schwarz, E., Komaromy, M. and Wall, R., 1984, Analysis of membrane and surface protein sequences with the hydrophobic moment plot, J. Mol. Biol., 179:125-142.

Epstein, W., Walderhaug, M.O., Polarek, J.W., Hesse, J.E., Dorus, E., and Daniel, J.M., 1990, The bacterial Kdp K^+-ATPase and its relation to other transport ATPases, Phil. Trans. R. Soc. Lond. B, 326:479-487.

Fersht, A.R., 1985, "Enzyme Structure and Mechanism", pp.317-331, Wiley, New York.

Flores, E., and Schmetterer, G., 1986, Interaction of fructose with the glucose permease of the cyanobacterium *Synechocystis* sp. strain PCC 6803, J. Bacteriol., 166: 693-696.

Foster, D.L., Boublik, M., and Kaback, H.R., 1983, Structure of the *lac* carrier protein of *Escherichia coli,* J. Biol.Chem., 258:31-34.

Friedrich, M.J., and Kadner, R.J., 1987, Nucleotide sequence of the *uhp* region of *Escherichia coli*, J. Bacteriol., 169: 3556-3563.

Fukumoto, H., Seino, S., Imura, H., Seino, Y., Eddy, R.L., Fukushima, Y., Byers, M., Shows, B.T., and Bell, G.I., 1988, Sequence, tissue distribution, and chromosomal localisation of mRNA encoding a human glucose transporter-like protein, Proc. Natl.Acad. Sci. USA, 85:5434-5438.

Furlong, 1987, Osmotic-shock-sensitive transport systems, in "*Escherichia coli* and *Salmonella typhimurium*", Niedhardt, ed., pp. 768-796, American Society for Microbiology, Washington.

Garnier, J., Osguthorpe, D.J., and Robson, B., 1978, Analysis of the accuracy and implications of simple methods for predicting the secondary structure of globular proteins, J. Mol. Biol., 120:97-120.

Geever, R.F., Huiet, L., Baum, J.A., Tyler, B.M., Patel, V.B., Rutledge, B.J., Case, M.E. and Giles, N.H., 1989, DNA sequence, organisation and regulation of the *qa* gene cluster of *Neurospora crassa*, J. Mol. Biol., 207:15-34

Gould, G., and Bell, G.I., 1990, Facilitative glucose transporters: an expanding family, TIBS, 15:18-23.

Haspel, H.C., Rosenfeld, M., and Rosen, O.M., 1988, Characterisation of antisera to a synthetic carboxyl-terminal peptide of the glucose transporter protein, J. Biol. Chem., 263:398-403.

Hediger, M.A., Coady M.J., Ikeda T.S., and Wright E.M., 1987, Expression cloning and cDNA sequencing of the glucose/Na^+ co-transporter, Nature, 330:379-381.

Hediger, M.A., Turk, E., and Wright E.M., 1989, Homology of the human intestinal glucose/Na^+ and *Escherichia coli* proline/Na^+ cotransporters, Proc. Natl. Acad. Sci. USA, 86: 5748-5752.

Henderson, P.J.F., 1986, Active transport of sugars into *Escherichia coli*, in "Carbohydrate Metabolism in Cultured Cells", M.J. Morgan, ed., pp. 409-460, Plenum Press, London.

Henderson, P.J.F., and Macpherson, A.J.S., 1986, Assay, genetics, proteins and reconstitution of proton-linked galactose, arabinose, and xylose transport systems of *Escherichia coli,* Methods Enzymol., 125: 387-429.

Henderson, P.J.F., and Maiden, M.C.J., 1990, Homologous sugar transport proteins in *Escherichia coli* and their relatives in both prokaryotes and eukaryotes, Phil. Trans. R. Soc. Lond. B, 326:391-410.

Henderson, R., and Schertler, G., 1990, The structure of bacterio-rhodopsin and its relevance to the visual opsins and other seven helix G-protein coupled receptors, Phil. Trans. R. Soc.Lond. B, 326:379-389.

Hendrickson, W., and Schleif, R.F., 1984, Regulation of the *Escherichia coli* L-arabinose operon studied by gel elecrophoresis DNA binding assay, J. Mol. Biol., 174: 611-628.

Hendrix, R.W., Roberts, J.W., Stahl, F.W., and Weisberg, R.A., 1983, "Lambda II", Cold Spring Harbor, Cold Spring Harbor.

Higgins, C.F., 1989, Protein joins transport family, Nature, 341:103.

Higgins, C.F., Gallagher, M.P., Hyde, S.C., Mimmack, M.L., and Pearce, S.R., 1990, The high affinity active transport systems in bacteria: the membrane-associated components, Phil. Trans.R. Soc. Lond. B, 326:353-365.

Hillen, W., and Schollmeier, K., 1983, Nucleotide sequence of the Tn10 encoded tetracycline resistance gene, Nucleic Acids Res., 11:525-539.

Holman, G.D., and Rees, W.D., 1987, Photolabelling of the hexose transporter at external and internal sites, Biochim. Biophys. Acta, 897:395-405.

Ishiguro, N. and Sato, G. 1985. Nucleotide sequence of the gene determining plasmid-mediated citrate utilisation, J. Bacteriol., 164:977-982.

James,D.E., Strube, M. and Mueckler, M., 1989, Molecular cloning and characterisation of an insulin-regulatable glucose transporter, Nature, 338:83-87.

Jones-Mortimer, M.C.J., and Henderson, P.J.F., 1986, Use of transposons to isolate and characterise mutants lacking membrane proteins, illustrated by the sugar transport systems of *Escherichia coli*, Methods Enzymol., 125:157-180.

Jund, R., Weber, E. and Chevallier M-R, 1988, Primary structure of the uracil transport protein of *Saccharomyces cerevisiae*, Eur.J. Biochem.,171:417-424.

Kaback, H.R., 1986, Proton electrochemical gradients and active transport: the saga of *lac* permease. Annals N.Y. Acad. Sci., 456:291-304.

Kaback, H.R., 1987, Use of site-directed mutagenesis to study the mechanism of a membrane transport protein, Biochemistry, 26: 2071-2076.

Kaback, H.R., 1990, Lac permease of *Escherichia coli*: on the path of the proton, Phil. Trans. R. Soc. Lond. B, 326:425-436.

Karim, A.R., Rees, W.D., and Holman, G.D., 1987, Binding of cytochalasin B to trypsin and thermolysin fragments of the human erythrocyte hexose transporter, Biochim. Biophys. Acta, 902:402-405.

Kohara, Y., Akiyama, K., and Isono, K., 1987, The physical map of the whole *E. coli* chromosome, Cell, 50:495-508.

Kolodrubetz, D., and Schleif, R., 1981, Regulation of the L-arabinose transport operons in *Escherichia coli*, J. Mol. Biol., 151:215-227.

Komor, E., and Tanner, W., 1974, The hexose-proton symport system of *Chlorella vulgaris*, Eur. J. Biochem., 44:219-223.

Kornberg, H.L., and Henderson, P.J.F., 1976, Mutant methodology in the study of transport, in "Horizons in biochemistry", vol. 2, Palmieri, F., and Singer, T.P., eds., pp. 1-31, Addison Wesley, London.

Kosiba, B.E., and Schleif, R., 1982, Arabinose-inducible promoter from *Escherichia coli*, J. Mol. Biol., 156:53-56.

Kyte, J. and Doolittle, R.F., 1982, A simple method for displaying the hydropathic character of a protein, J. Mol. Biol., 157:105-132.
Leblanc, G., Pourcher, T., and Bassilana, M., 1989, Molecular biology and bacterial secondary transporters. Biochimie, 71; 969-979.
Lengeler, J.W., Titgemeyer, F., Vogler, A.P., and Wohrl, B.M., 1990, Structures and homologies of carbohydrate:phosphotransferase system (PTS) proteins, Phil. Trans. R. Soc. Lond. B, 326:489-504.
Levy, S., 1984, Resistance of the tetracyclines, in "Antimicrobial drug resistance", L.E. Brian, ed., pp.191-240, Academic Press, New York.
Levy, S. 1988, Tetracycline resistance determinants are widespread, ASM News, 54:418-421.
Lodish, H.F., 1988, Multi-spanning membrane proteins: how accurate are the models?, Trends in Biochem. Sci., 13: 332-334.
Lu, Z., and Lin, E.C.C., 1989, The nucleotide sequence of *Escherichia coli* genes for L-fucose dissimilation, Nucl. Acids Res., 17:4883-4884.
Maiden, M.C.J., 1986, "Arabinose-proton symport in *Escherichia coli*", Ph.D. dissertation, University of Cambridge.
Maiden, M.C.J., Davis, E.O., Baldwin, S.A., Moore, D.C.M., and Henderson, P.J.F., 1987, Mammalian and bacterial sugar transport proteins are homologous, Nature, 325:641-643.
Maiden M.C.J., Jones-Mortimer M.C., and Henderson P.J.F., 1988, The cloning, DNA sequence and overexpression of the gene *araE* coding for arabinose-proton symport in *Escherichia coli* K12, J. Biol. Chem., 263:8003-8010.
Maloney, P.C., 1990, Resolution and reconstitution of anion exchange systems, Phil. Trans. R. Soc. Lond. B, 326:437-454.
Manoil C., and Beckwith, J., 1986, A genetic approach to analysing membrane topology, Science, 233:1403-1408.
McMurry, L., Petrucci, R.E. and Levy, S.B., 1980, Active efflux of tetracycline encoded by four genetically different tetracycline resistance determinants in *Escherichia coli*, Proc. Natl. Acad.Sci. USA, 77:3974-3977.
Medigue, C., Bouche, J.P., Henaut, A., and Danchin, A., 1990, Mapping of sequenced genes (700kbp) in the restriction map of the *Escherichia coli* chromosome, Molec. Microbiol., 4:169-187.
Menick, D., Lee, J.A., Brooker, R. J., Wilson, T. H. and Kaback, H.R., 1987, Role of cysteine residues in the Lac permease of *Escherichia coli*, Biochemistry, 26:1132-1136.
Miller, J.H., 1972, "Experiments in molecular genetics", Cold Spring Harbor, Cold Spring Harbor.
Mitchell, P. 1961, Coupling of phosphorylation to electron and hydrogen transfer by a chemiosmotic type of mechanism. Nature, 191:144-148.
Mitchell, P., 1963, Molecule, group and electron transfer through natural membranes, Biochem. Soc. Symp., 22:142-169.
Mitchell, P. 1973, Performance and conservation of osmotic work by proton-coupled solute porter systems. Bioenergetics, 4:63-91.
Mueckler, M., Caruso, C., Baldwin, S.A., Panico, M., Blench, I., Morris, H.R., Allard, W.J., Lienhard, G.E., Lodish, H., 1985, Sequence and structure of a human glucose transporter, Science, 229:941-945.
Muiry, J.A.R., 1989, "The bacterial transport systems for L-rhamnose and L-fucose", Ph.D.dissertation, University of Cambridge.

Nakao, F, Yamato, I., and Anraku, Y., 1987, Nucleotide sequence of *putP,* the proline carrier gene of *Escherichia coli K12*, Mol.Gen. Genet., 208:70-75.

Nehlin, J.O., Carlberg, M., and Ronne, H., 1989, Yeast galactose permease is related to yeast and mammalian glucose transporters, Gene, 85:313-319.

Nikaido, H., and Vaara, M., 1987, Outer Membrane, in "*Escherichia coli* and *Salmonella typhimurium*", Niedhardt, ed., pp., American Society for Microbiology, Washington.

Postma, P.W., and Lengeler, J.W., 1985, Phosphoenolpyruvate: carbohydrate phosphotransferase system of bacteria, Microbiol.Rev., 49:232-269.

Pourcher, T., Bassilana, M., Sarkar, H.K., Kaback, H.R., and Leblanc, G., 1990, The melibiose /Na^+ symporter of *Escherichia coli*: kinetic and molecular properties, Phil. Trans. R. Soc. Lond. B, 326:411-423.

Quiocho, F.A., 1986, Carbohydrate binding proteins: tertiary structures and protein-sugar interactions, Annu. Rev. Biochem., 55:287-315.

Quiocho, F.A., 1990, Atomic structures of periplasmic binding proteins and the high affinity active transport systems in bacteria, Phil. Trans. R. Soc. Lond. B, 326:341-351.

Reynolds, C.H., and Silver, S., 1983, Citrate utilisation by *Escherichia coli*: plasmid- and chromosome-encoded systems, J.Bacteriol., 156:1019-1024.

Riordan, C., and Kornberg, H.L., 1977, Location of *galP*, a gene which specifies galactose permease activity, on the *Escherichia coli* linkage map, Proc. R. Soc. Lond. Ser. B, 198:401-410.

Roberts, P.E.R., Moore, D.C.M., Jones-Mortimer, M.C.J., and Henderson, P.J.F., 1990, Location of the *galP* gene at 63.7min on the *Escherichia coli* linkage map, next to *metK*, in preparation.

Saier, M.H., 1985, "Mechanisms and regulation of carbohydrate transport in bacteria", Academic Press, New York.

Sambrook, J., Fritsch, E.F., and Maniatis, T., 1989, "Molecular cloning, a laboratory manual", Cold Spring Harbor, Cold Spring Harbor.

Sanger, F., Coulson, A.R., Barrell, B.G., Smith, A.J.H., and Roe, B.A., 1980, Cloning in single-stranded bacteriophage as an aid to rapid DNA sequencing, J. Mol. Biol., 143:161-178.

Sasatsu, M., Misra, T.K., Chu, L., Ladagu, R., and Silver, S., 1985, Cloning and DNA sequence of a plasmid-determined citrate utilisation system in *Escherichia coli*, J. Bacteriol., 164:983-993.

Sauer, N., and Tanner, W., 1989, The hexose carrier from *Chlorella* - cDNA cloning of a eucaryotic H^+-cotransporter, FEBS Lett., 259:43-46.

Shanahan, M.F., 1982, Cytochalasin B. A natural photoaffinity ligand for photolabelling the human erythrocyte glucose transporter, J. Biol. Chem., 257:7290-7293.

Silhavy, T.J., Berman, M.L., and Enquist, L.W., 1984, "Experiments with gene fusions", Cold Spring Harbor, Cold Spring Harbor.

Smith, G., Petro, K.P., Cairns, M.T., Baldwin, S.A., Maiden, M.C.J., and Henderson, P.J.F., 1990, Interaction of bacterial transport proteins with cytochalasin B, in preparation.

Staden, R., 1984a, Computer methods to locate signals in nucleic acid sequences, Nucl. Acids Res., 12:505-519.

Staden, R., 1984b, Measurements of the effects that coding for a protein has on a DNA sequence and their use for finding genes, Nucl. Acids Res., 12:5551-567.

Stoner, C., and Schleif, R., 1983, The *araE* low affinity L-arabinose transport promoter, J. Mol. Biol., 171:369-381.

Sumiya, M., 1989, "The XylF D-xylose binding protein transport system in *Escherichia coli*", Ph.D. dissertation, University of Cambridge.

Szkutnicka, K., Tschopp, J.F., Andrews, L., and Cirillo, V.P., 1989, Sequence and structure of the yeast galactose transporter, J.Bacteriol., 171:4486-4493.

Thorens, B., Sarkar, H.K., Kaback, H.R., and Lodish, H.F., 1988, Cloning and functional expression in bacteria of a novel glucose transporter present in liver, intestine, kidney and ß-pancreatic islet cells, Cell, 55:281-290.

Tobin, J.F., and Schleif, R.F., 1990, Transcription from the *rha* operon p_{sr} promoter, J. Mol. Biol., 211:1-4.

Vilaro, S., Palacin, M., Pilch, P.F., Testar, X., and Zorzano, A., 1989, Expression of an insulin-regulatable glucose carrier in muscle and fat epithelial cells, Nature, 342:798-800.

von Heijne, G., 1987, "Sequence analysis in molecular biology - treasure trove or trivial pursuit", Academic press, London.

Von Heijne, G., 1988, Transcending the impenetrable: how proteins come to terms with membranes, Biochim. Biophys. Acta, 947:307-333.

Walmsley, A.R., 1988, The dynamics of the glucose transporter, Trends in Biochem. Sci., 13:226-231.

West, I.C. and Mitchell, P., 1972, Proton-coupled ß-galactoside translocation in non-metabolising *Escherichia coli*, Bioenergetics, 3:445-462.

West, I.C., and Mitchell, P., 1973, Stoichiometry of lactose-proton symport across the plasma membrane of *Escherichia coli*, Biochem. J., 132:587-592.

Wilson, D.M., Tsuchiya, T., and Wilson, T.H., 1986, Methods for the study of the melibiose carrier of *Escherichia coli*, Methods Enzymol., 125:377-387.

Yanisch-Perron, C., Vieira, J., and Messing, J., 1985, Improved M13 phage cloning vectors and host strains: Nucleotide sequences of the M13mp18 and pUC19 vectors, Gene, 33: 103-119.

Yao, B., Sollitti, P., and Marmur, J., 1989, Primary structure of the maltose-permease-encoding gene of *Saccharomyces carlsbergensis*, Gene, 79:189-197.

Yazyu, H., Shiota-Niiya, S., Shimamoto, T., Kanazawa, H., Futai, M., and Tsuchiya, T., 1984, Nucleotide sequence of the *melB* gene and characteristics of deduced amino acid sequence of the melibiose carrier in *Escherichia coli*, J. Biol. Chem., 259:4320-4326.

Zhang, C.-C., Durand, M.-C., Jeanjean, R., and Joset, F., 1989, Molecular and genetical analysis of the fructose-glucose transport system in the cyanobacterium *Synechocystis* PCC6803, Molec. Microbiol., 3:1221-1229.

Zilberstein, D., and Dwyer, D.M., 1985, Protonmotive force-driven active transport of D-glucose and L-proline in the protozoan parasite *Leishmania donovani*. Proc. Natl. Acad. Sci. USA, 82:1716-1720.

Zilberstein, D., Dwyer, D.M., Matthei, S., and Horuk, R., 1986, Identification and biochemical characterisation of the plasma membrane glucose transporter of *Leishmania donovani*, J. Biol.Chem., 261:15053-15057.

20

CHEMICAL AND MOLECULAR PROBES OF NUCLEOSIDE TRANSPORT MECHANISMS IN MAMMALIAN TISSUES

Simon M. Jarvis

Biological Laboratory
University of Kent, Canterbury
Kent CT2 7NJ, U.K.

INTRODUCTION

The passage of nucleosides and synthetic nucleoside analogues across plasma membranes of animal cells may occur by simple diffusion or more frequently by carrier-mediated processes. Two types of facilitated-diffusion nucleoside transporters that differ in their sensitivity to inhibition by nitrobenzylthioinosine (NBMPR) have been recognized (Jarvis and Young, 1987; Jarvis, 1988a; Plagemann et al., 1988). More recently, active nucleoside transport systems have been described in brush-border membrane vesicles and cells derived from renal and intestinal epithelia (Jarvis et al., 1989; Vijayalakshmi and Belt, 1988; Jakobs and Paterson, 1986). This article reviews results of recent studies of nucleoside transport in mammalian cells with particular emphasis on the methodologies employed in using chemical and molecular probes to explore the properties of nucleoside transporters.

FACILITATED-DIFFUSION NUCLEOSIDE TRANSPORT

Transport Measurements - Estimates of Initial Rates

A critical factor for valid analysis of the kinetic properties of a transport process is to measure true initial uptake velocities. For many types of cells nucleoside transport is rapid, and hence determination of the initial velocities of radioactive permeant transport requires rapid sampling techniques (assay times down to a few seconds). A number of different methodologies have been used.

Oil-stop: For suspended cells, a common sampling technique is to terminate nucleoside transport by centrifuging the cells and extracellular medium through an oil layer contained in a microcentrifuge tube. The density of the oil is chosen such that cells pass through while the extracellular medium remains on top of the oil layer. Useful oils are *n*-dibutyl phthalate for erythrocytes and a 84:16 mixture of Dow Corning DC 555: Dow Corning DC 200 for most cultured cells (specific gravity 1.03g/ml) (Jarvis et al., 1980; Harley et al., 1982). The time taken to sediment the cells through the oil is approximately 2-5s and must be taken into account during analysis of the initial rates of transport (see Plagemann and Wohlhueter, 1980; Paterson et al., 1985).

Cell Membrane Transport, Edited by D. L. Yudilevich *et al.*
Plenum Press, New York, 1991

Inhibitor-stop: An added refinement to the oil-stop procedure is to add an inhibitor to block nucleoside influx and efflux before centrifugation. NBMPR and dilazep have both been employed for nucleoside transport studies (Jarvis et al., 1980; Paterson et al., 1984). Such inhibitors must act instantaneously (NBMPR and dilazep fulfil this criterion at high concentrations, > 10μM) and also inhibit transport. This latter point needs to be emphasised since some cells possess facilitated-diffusion nucleoside transporters that are resistant to NBMPR inhibition (see section on inhibitor studies). Dilazep (100μM) is thus the inhibitor of choice if nothing is known about the nucleoside carrier mechanism. Moreover, it is readily soluble in buffer. Some workers have also used dipyridamole as a nucleoside transport stopper but it should be noted that nucleoside transport in rat and mouse derived cells is much less sensitive to inhibition by dipyridamole compared to other species (compare Figs. 4 and 5) (Jarvis and Young, 1987; Jarvis, 1988a; Plagemann et al., 1988). With practice and using a metronome, assay times as short as 1 sec. can be achieved and should be chosen such that the intracellular substrate concentration does not exceed 10-20% of the extracellular substrate concentration. Alternatively, the entire time course of substrate equilibration across the membrane can be determined and initial transport velocities calculated as the zero-time slopes of the progress curves (for detailed discussion see Plagemann et al., 1988; Plagemann and Wohlhueter, 1980). Recently, a quenched-flow method has been described that provided time courses of cellular uptake of adenosine during intervals of 0.05 to 0.5 sec. (Paterson et al., 1984).

Filtration. For membrane vesicles and synaptosomes, rapid filtration through either glass fibre or nitrocellulose filters, following the addition of ice-cold stop buffer (buffer containing a transport inhibitor) has been used to measure nucleoside uptake (Lee and Jarvis, 1988a; Lee et al., 1988). Control experiments must be performed to ensure that the filters retain all the membrane vesicles. More detailed reviews of methods for measuring nucleoside transport have been published (Plagemann and Wohlhueter, 1980; Paterson et al., 1985).

Effect of Metabolism on Initial Rates of Transport

An additional problem in estimating transport velocities is rapid metabolism of nucleosides. Indeed, much of the early work (pre-1977) resulted in measurements of nucleoside metabolism rather than nucleoside transport. Subsequent investigations have used modified cells that are incapable of nucleoside metabolism to resolve transport and metabolic components of nucleoside uptake, e.g. ATP depletion by incubating cells in a medium lacking D-glucose and glutamine and containing inhibitors of energy production, such as sodium cyanate and iodoacetate, cells with appropriate enzyme lesions, and specific inhibitors of enzymes that do not interfere with the transport measurements, such as deoxycoformycin, a potent inhibitor of adenosine deaminase (Plagemann et al., 1985; Wohlhueter et al., 1979; Plagemann et al., 1976). Non-metabolized nucleoside substrates have also been used for this purpose, e.g. uridine and thymidine in mammalian erythrocytes and formycin B, a C-nucleoside analog of inosine (Plagemann and Woffendin, 1989).

Kinetic Properties of Facilitated-Diffusion Nucleoside Transport

Nucleoside transport measured under these initial rate conditions has been found to be adequately described by a simple carrier model (Jarvis et al., 1983; Wohlhueter and Plagemann, 1982). The nucleoside transporters of cultured cell lines exhibit directional symmetry (influx = efflux) and equal "mobility" (conformational change) of the loaded and empty carrier (Wohlhueter and Plagemann, 1982). Similarly, the transporters of fresh human, pig and guinea-pig erythrocytes also exhibit directional symmetry but differential "mobility" of loaded and empty carrier when assayed with uridine as substrate (Table 1) (Jarvis and Young, 1987; Jarvis, 1988a; Plagemann et al., 1988; Jarvis et al., 1983; Jarvis and Martin, 1986). This differential mobility of loaded and empty carrier is accentuated during blood storage and at low temperature (Jarvis, et al., 1983; Jarvis and Martin, 1986). Moreover, uridine transport in erythrocytes from outdated blood is asymmetric with respect to direction (Jarvis et al., 1983). The basis for this change is unknown but does not appear to be related to ATP depletion during storage. However, conclusions obtained with uridine may not be directly applicable to other nucleoside permeants. For example, the mobility of the adenosine-loaded and empty carrier in ATP-depleted, deoxycoformycin treated human erythrocytes is about the same (Table 1) and adenosine fails to stimulate the zero-trans efflux of uridine (referred to as trans acceleration) from preloaded human erythrocytes (Plagemann et al., 1985; Jarvis 1986a). 2-Chloroadenosine inhibits rather than accelerates uridine efflux (Jarvis, 1986a). The kinetic parameters in Table 1 suggest an explanation for these differences, namely that the mobility of the adenosine and 2-chloroadenosine-loaded carrier is lower than the uridine-loaded carrier. These results demonstrate that trans-acceleration between two nucleosides is an unreliable criterion for determining whether nucleosides are permeants for the same transporter.

Table 1. Kinetic parameters for nucleoside transport across fresh human erythrocytes at 22°C

Kinetic Parameter	Uridine	Substrate Adenosine	2-Chloroadenosine
K_{21}^{zt} (μM)	170	61	33
V_{21}^{zt} (μM/s)	32	28	7
V_{21}^{zt}/K_{21}^{zt} (s^{-1})	0.19	0.45	0.21
K^{ee} (μM)	760	60	ND
V^{ee} (μM/s)	128	29	ND
V^{ee}/K^{ee} (s^{-1})	0.17	0.49	ND
R_{ee} (s/mM)	8	35	240
R_{oo} (s/mM)	54	38	≈40
R_{oo}/R_{ee}	6.7	1.1	0.17

The kinetic parameters have been calculated from previously published experimental values (Plagemann et al., 1985; Jarvis et al., 1983, 1985; Jarvis, 1986a) where the resistance parameters, R, are defined as R_{12} = $1/V_{max}$ zero-trans efflux, $1/V_{12}^{zt}$; $R_{21} = 1/V_{max}$ zero-trans influx, $1/V_{21}^{zt}$; $R_{ee} = 1/V_{max}$ equilibrium exchange, $1/V^{ee}$ and $R_{oo} + R_{ee} = R_{12} + R_{21}$. ($K_{21}^{zt}$, K_m zero-trans influx; K^{ee}, K_m equilibrium exchange; R_{oo}, parameter for the 'unloaded' carrier; ND, not determined).

More reliable evidence for the transport of two nucleosides by the same carrier is the mutual inhibition of their transport or the counter-transport by each other. From such studies, the facilitated-diffusion nucleoside transporters in animal cells have been shown to have considerable flexibility for the aglycone portion of nucleoside molecules, accepting a broad range of purine and pyrimidine nucleosides (e.g. K_m values for adenosine and uridine influx are approximately 30-100μM and 200μM at 25°C, respectively). One notable exception is

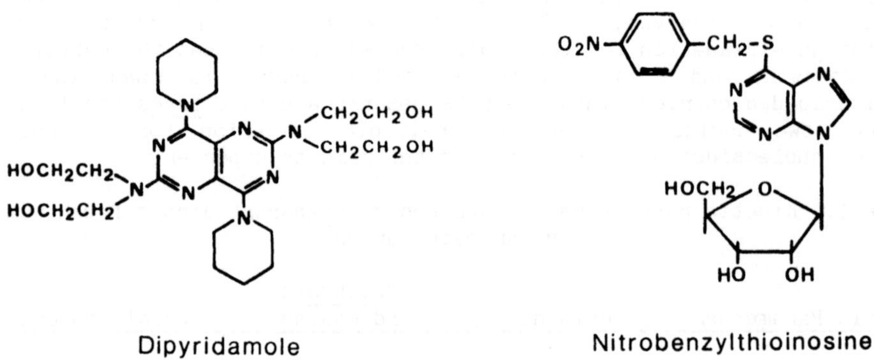

Fig. 1. Structures of dipyridamole and nitrobenzylthioinosine.

3'-azido-3'-deoxythymidine which has been reported not to be a permeant for the human erythrocyte nucleoside transporter (Zimmerman et al., 1987). The absence of the 3'OH group impedes the binding and transport of the nucleoside via the erythrocyte carrier.

INHIBITORS OF FACILITATED-DIFFUSION NUCLEOSIDE TRANSPORT

Nitrobenzylthioinosine and Dipyridamole - Mechanism of Inhibition

NBMPR (Fig. 1) is a potent inhibitor (K_i 0.1 to 1nM) of nucleoside transport in many types of cells (for reviews see Jarvis and Young, 1987; Plagemann et al., 1988). Inhibition is associated with binding of NBMPR to the cell membrane and a direct correlation between inhibition of transport and NBMPR binding has been demonstrated (Cass et al., 1974).

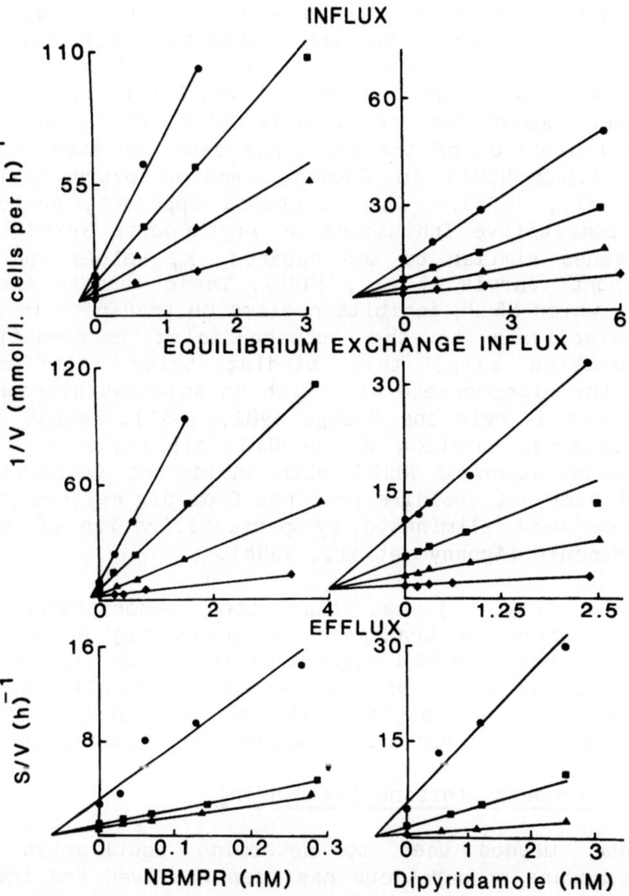

Fig. 2. Effect of $[^3H]$nitrobenzylthioinosine and $[^3H]$dipyridamole on zero-<u>trans</u> uridine influx, efflux and equilibrium exchange influx by guinea-pig erythrocytes. Initial rates of $[^{14}C]$uridine transport are plotted against the final free equilibrium concentrations of $[^3H]$inhibitor. Uridine concentrations (mM) for zero-trans influx were 1.0 (♦), 0.5 (▲), 0.1 (■) and 0.05 (●) and 0.85 (♦), 0.21 (▲), 0.085 (■) and 0.043 (●) for NBMPR and dipyridamole, respectively, and 1.0 (♦), 0.5 (▲), 0.1 (■) and 0.05 (●) for equilibrium exchange influx. Initial intracellular uridine concentrations (mM) for the efflux experiments were 0.026 (▲), 0.055 (■) and 0.3 (●) for NBMPR and 0.033 (▲), 0.24 (■) and 0.85 (●) for dipyridamole. Part of the data taken (with permission) from Jarvis 1986b.

The mechanism of interaction of NBMPR with the nucleoside transporter has been investigated by a number of different but complementary approaches. Kinetically, NBMPR is a competitive inhibitor of zero-trans uridine influx and equilibrium exchange uridine influx in guinea-pig and nucleoside-permeable sheep erythrocytes but a noncompetitive inhibitor of zero-trans uridine efflux (Fig.2). It should be noted that these experiments were conducted with cell types that slowly transport uridine thereby ensuring that the rate of uridine transport is much slower than the rate of NBMPR binding. In cells that rapidly transport nucleosides considerable uptake of nucleoside can occur before maximum inhibition of transport has been achieved, e.g. $t^{1}/_{2}$ for inhibition at 1.5nM NBMPR in Chinese hamster ovary cells is ≈ 5s (Wohlheuter et al., 1978). In the second approach, nucleosides were shown to be competitive inhibitors of erythrocyte NBMPR binding with apparent K_i values similar to the apparent K_m values for equilibrium exchange transport (Jarvis et al., 1982). These results are consistent with a model in which NBMPR inhibits nucleoside transport in erythrocytes by binding selectively to the outward-facing conformation of the substrate permeation site, this binding being stabilized by the interaction of the nitrobenzene ring with an adjacent hydrophobic domain on the transporter (Jarvis and Young, 1982; 1987). Recent studies have confirmed the external location of the NBMPR binding site by demonstrating that the association of NBMPR with inside-out vesicles was slower than with right-side-out vesicles prepared from pig erythrocytes and that such differences were eliminated by permeabilization of the membrane vesicles with saponin (Agbanyo et al., 1988).

Dipyridamole (Fig. 1) has also been demonstrated to inhibit NBMPR-sensitive nucleoside transport in guinea-pig erythrocytes in a similar manner to that of NBMPR suggesting that both ligands interact at common or overlapping sites on the carrier (Fig. 2) (Jarvis, 1986b; 1988a). These data do not exclude the possibility of NBMPR and dipyridamole binding to a separate allosteric inhibitor site.

Methodology of Nitrobenzylthioinosine Binding

The general method used to determine equilibrium binding of [^{3}H]NBMPR to membrane preparations has been reviewed and involves rapid filtration through glass fibre filters to separate free inhibitor from membrane-bound inhibitor (Young and Jarvis, 1985). However such filters do not retain 100% small membrane vesicles or solubilized membranes. Nitrocellulose filters with a smaller pore size are not suitable since considerable binding of free [^{3}H]NBMPR to these filters occurs. Thus, we have modified our existing protocol. Using the methodology developed by Bruns et al., 1983, Whatman GF/B glass fibre filters are presoaked for 1 hour in 0.3% (v/v) polyethylenimine and placed on the filtration manifold without washing. Aliquots of the incubation are poured onto a filter under suction and washed twice with 5ml of ice-cold buffer. The filters are placed in scintillation vials with 4ml of Optiphase T (LKB Scintillation Products), left overnight and counted for radioactivity.

The binding of [^{3}H]NBMPR to human placental brush-border and basal membrane vesicles using this protocol is shown in Fig. 3, where membrane-associated binding is plotted against the equilibrium free concentration of the inhibitor. The concentration dependence of [^{3}H]NBMPR binding was resolved into saturable and linear components. For brush-border and basal membrane vesicles, respectively, the saturable parameters of binding were estimated: B_{max} (maximal binding) 1.35 and 1.20pmol/mg protein, apparent K_d 1.0 and 1.5nM. Saturable binding was abolished in the presence of 10μM NBMPR and the difference in binding in the presence and absence of NBMPR is defined as specific binding. This filtration methodology should also be applicable to reconstituted protein fractions in liposomes and soluble transporter protein, e.g. detergent solubilised. One problem that needs to be overcome is that detergents such as Triton X-100 are inhibitors of [^{3}H]NBMPR binding (Jarvis and Young, 1981). Preliminary experiments suggest that [^{3}H]NBMPR binding can be detected in the presence of CHAPS.

Nitrobenzylthioinosine-Insensitive Transporters

The specificity and potency of NBMPR has led to the use of NBMPR for the detection and localization of nucleoside transporters in cells and tissues, to estimate the turnover number for the nucleoside transporter and to purify the carrier (Jarvis, 1988a; Jarvis and Young, 1987; Plagemann et al., 1988). However, it is now apparent that many cells also possess facilitated diffusion transporters that are resistant to inhibition by concentrations of NBMPR as high as 1μM (Belt and Noel, 1985; Plagemann and Wohlhueter, 1984). Some cultured cells, e.g. Walker 256 rat carcinoma, only express NBMPR-insensitive transporters which correlates with an absence of high-affinity NBMPR binding sites. Other cells only possess NBMPR-sensitive transporters, e.g. human, pig and guinea-pig erythrocytes, but most cells investigated express simultaneously both NBMPR-sensitive and -insensitive transport but in different proportions. The proportion of NBMPR-sensitive transporters ranges from about 99% for S49 mouse lymphoma cells to 20% for rat erythrocytes (see Fig. 4).

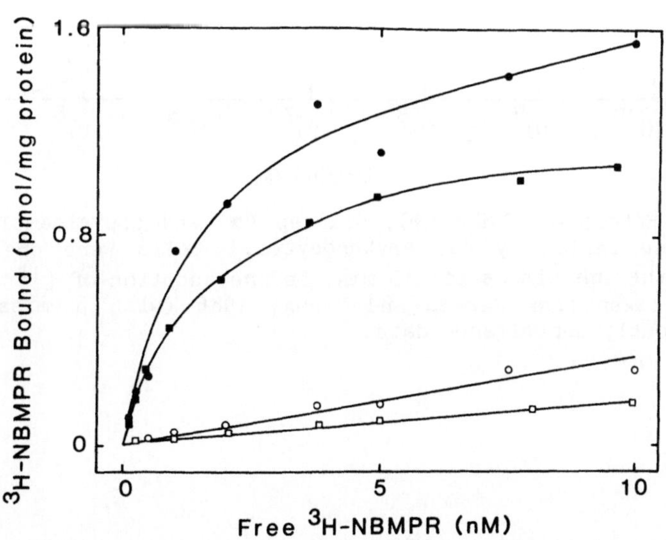

Fig. 3. Binding of [^{3}H]nitrobenzylthioinosine to human placental brush-border (●,○) and basal (■,□) membrane vesicles. [^{3}H]NBMPR binding to the membrane vesicles in the presence (open symbols) and absence (solid symbols) of 10μM NBMPR was measured as described in the text. Values are the means of duplicate estimates (L.F. Barros, J.C. Bustamante, D.L. Yudilevich and S.M. Jarvis, unpublished data).

The two transporters are apparently kinetically similar although in some cells the affinity for certain nucleosides may differ as much as 3-fold (Jarvis and Young 1986; Lee and Jarvis, 1988b). However, NBMPR-sensitive and -insensitive nucleoside transporters do differ with respect to sensitivity to p-chloromecuriphenyl sulphonate (PCMBS). This organomecurial inhibits NBMPR-sensitive transport in mammalian erythrocytes at the cytoplasmic surface of the inward-facing permeation site (Jarvis and Young, 1982). In contrast, the PCMBS-reactive thiol(s) in NBMPR-insensitive transporters are accessible from the external membrane surface (Belt and Noel, 1985; Jarvis and Young, 1986). This

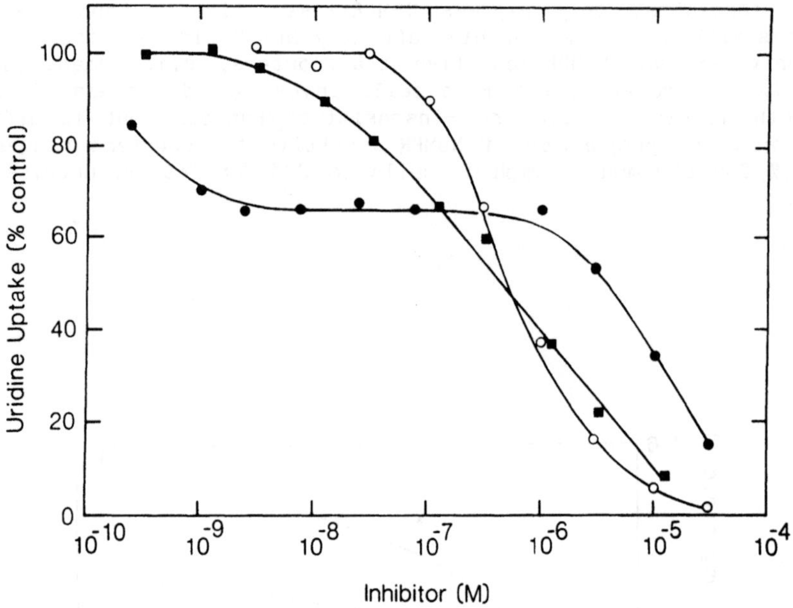

Fig. 4. Effect of NBMPR (●), dilazep (■) and dipyridamole (□) on uridine influx by rat erythrocytes. Cells were preincubated with the inhibitors for 15 min. before addition of [^{14}C]uridine. Data taken from Jarvis and Young, 1986 (with permission) and previously unpublished data.

differential sensitivity to PCMBS between the two facilitated-diffusion nucleoside transporters may possibly form the basis of an approach to isolate the NBMPR-insensitive nucleoside carrier. Nothing is known about the molecular properties of the NBMPR-insensitive transporter or its relationship to the NBMPR-sensitive system. For example, the two transporters may represent different forms of the same protein or different gene products. Support for this latter view has recently come from the isolation of nucleoside transport mutants (Belt and Noel, 1988).

The existence of NBMPR-insensitive nucleoside transport has been exploited in experimental chemotherapy strategies. For example, the coadministration of NBMPR with several cytotoxic nucleosides can dramatically increase the therapeutic index of the nucleoside drugs in the treatment of neoplasms and parasitic diseases, achieving cures in some instances (Lynch et al., 1981; El Kouni et al., 1983). The basis of this selective protection of host tissues appears to be the insensitivity of the tumor cell or parasite transport system to NBMPR. Thus, there is much interest in whether other potent nucleoside transport inhibitors that are available for use in the clinic can discriminate between NBMPR-sensitive and NBMPR-insensitive nucleoside transport. We have demonstrated that NBMPR-insensitive nucleoside transport in guinea-pig synaptosomes is also resistant to inhibition by dilazep and dipyridamole, but both compounds could block NBMPR-sensitive adenosine influx in the nanomolar range (Fig. 5) (Lee and Jarvis, 1988b). High µmolar concentrations of dilazep and dipyridamole were effective at inhibiting the total adenosine flux and thus these results support the use of dilazep as a transport 'stopper' of total facilitated-diffusion nucleoside transport. Other workers have reported no difference in the sensitivity of the two transporters to inhibition by dipyridamole and dilazep (Plagemann and Woffendin, 1988). One complication in many of these latter studies is that rat cells and tissues were used and rat is notable in exhibiting a 100-1000-fold lower sensitivity to dipyridamole inhibition than guinea-pig cells (Jarvis, 1988a; Jarvis and Young, 1987). Indeed, the results of Fig. 4 indicate that the NBMPR-sensitive and -insensitive transporters in rat erythrocytes are equally sensitive to dipyridamole inhibition as judged from the monophasic inhibition curve. In contrast, inhibition of uridine influx in rat erythrocytes by dilazep (Fig. 4) was shallow consistent with at least two different components.

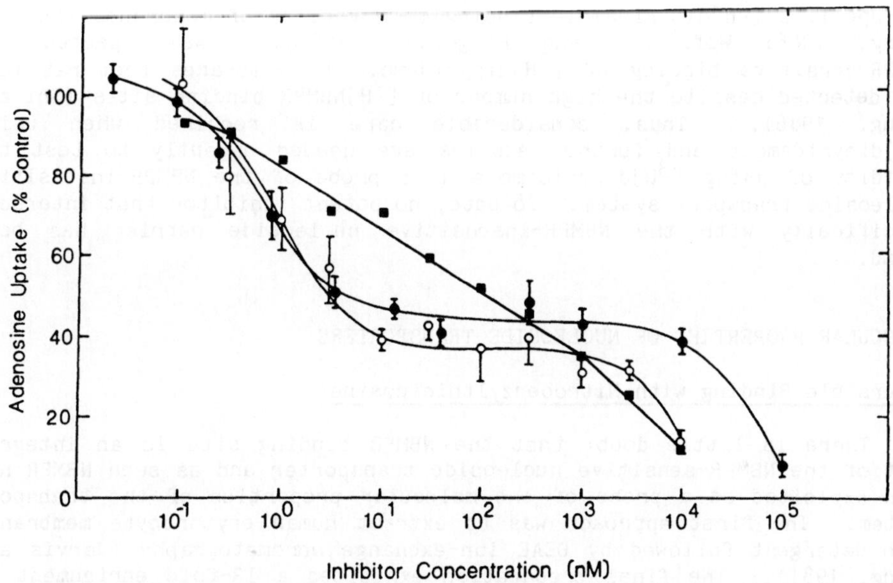

Fig. 5. Effect of nitrobenzylthioinosine (●), dilazep (○) and dipyridamole (■) on adenosine influx by guinea pig cerebral cortical synaptosomes. Taken (with permission) from Lee and Jarvis 1988b. The error bars for the dipyridamole data have been omitted for clarity.

Dipyridamole Binding - Uses and Problems

The finding of NBMPR-insensitive nucleoside transport has clearly raised doubts about the usefulness of [^{3}H]NBMPR as a marker for nucleoside uptake sites, especially its ability to label all of them. As a result, other probes have been suggested, especially [^{3}H]dipyridamole. In human erythrocytes and guinea-pig lung membranes, a single high-affinity [^{3}H]dipyridamole binding site has been detected that is blocked by nucleosides and other nucleoside transport inhibitors (Jarvis, 1986b; Shi and Young, 1986). The maximum number of high-affinity [^{3}H]dipyridamole binding sites on human erythrocytes and guinea-pig lung membranes is similar to that estimated using [^{3}H]NBMPR. However, in guinea-pig brain membranes approximately 4-times more [^{3}H]dipyridamole binding sites (K_d 3.5nM) than [^{3}H]NBMPR binding sites were detected (Marangos and Deckert, 1987; Deckert et al., 1987).Moreover, NBMPR only inhibited 50% of dipyridamole binding (Marangos and Deckert, 1987). Thus, Marangos and co-workers have suggested that the remaining 50% of the dipyridamole binding sites resistant to inhibition by NBMPR represent the NBMPR-insensitive nucleoside transporter in guinea-pig brain membranes. However, the nature of these [^{3}H]dipyridamole binding sites are uncertain. For example, nucleosides exhibit K_i values for inhibition of dipyridamole binding that are 10 to 100-fold higher than the K_m values for transport of these nucleosides (Lee and Jarvis, 1988a,b; Marangos and Deckert, 1987; Morgan and Marangos, 1987). The inhibition profile of adenosine influx by dipyridamole (Fig. 5) is inconsistent with a high-affinity [^{3}H]dipyridamole binding site to the NBMPR-insensitive nucleoside carrier.

Technical problems also exist with the use of [^{3}H]dipyridamole that many workers have failed to address. These include the chemical instability of [^{3}H]dipyridamole, the tendency of [^{3}H]dipyridamole to stick to plastic, the binding of [^{3}H]dipyridamole to glass fibre filters and the finding that unlabelled dipyridamole underestimates the level of non-specific [^{3}H]dipyridamole binding to a variety of membranes (Shi and Young, 1986; Woffendin and Plagemann, 1987). As expected, no NBMPR-sensitive binding of [^{3}H]dipyridamole to membranes from rat lung was detected despite the high number of [^{3}H]NBMPR binding sites (Shi and Young, 1986). Thus, considerable care is required when using [^{3}H]dipyridamole and further studies are needed urgently to test the validity of using [^{3}H]dipyridamole as a probe of the NBMPR-insensitive nucleoside transport system. To date, no potent inhibitor that interacts specifically with the NBMPR-insensitive nucleoside carrier has been found.

MOLECULAR PROPERTIES OF NUCLEOSIDE TRANSPORTERS

Reversible Binding with Nitrobenzylthioinosine

There is little doubt that the NBMPR binding site is an integral part of the NBMPR-sensitive nucleoside transporter and as such NBMPR has been exploited as a probe of the molecular properties of the transport system. The first approach was to extract human erythrocyte membranes with detergent followed by DEAE ion-exchange chromatography (Jarvis and Young, 1981). The final preparation exhibited a 13-fold enrichment of NBMPR binding activity and consisted mainly ($\approx$ 95%) of band 4.5 polypeptides. The preparation when reconstituted into soybean phospholipid vesicles is capable of catalyzing NBMPR-sensitive saturable uridine transport (Tse et al., 1985). A complication with the use of NBMPR is that NBMPR binding activity is reversible and some detergents are potent inhibitors of binding activity necessitating the removal of the detergent. This has led to the covalent labelling of the NBMPR binding site by exposing [^{3}H]NBMPR and membranes to UV light (Wu et al., 1983).

Photoaffinity Labelling with Nitrobenzylthioinosine - Experimental Techniques

The methodology employed to study the photoaffinity labelling of [^{3}H]NBMPR to plasma membranes from a wide variety of sources is as follows. Plasma membranes suspended in an appropriate buffer (~ 2mg protein per ml plus protease inhibitors) are equilibrated with a saturating concentration of [^{3}H]NBMPR (~ final free concentration 5nM) in the presence and absence of 20μM NBMPR or nitribenzylthioguanosine as competing ligand used to estimate non-specific binding. After 30 min., incubations are cooled to 4°C, supplemented with 10mM dithiothreitol which acts as a free radical scavenger and transferred to silica spectrophotometer cuvettes. Samples are exposed to UV light at a distance of 6.5cm from a 450W mercury arc lamp. (Conrad-Hanovia Inc., Newark, New Jersey, U.S.A.) for 20-120 sec. Samples are diluted 10-75 fold with buffer containing a nucleoside transport inhibitor, e.g. NBMPR (10μM) and allowed to stand at room temperature for 10 min. before recovery of the membrane fraction by centrifugation. In this way non-covalently bound [^{3}H]NBMPR and unreacted ligand is removed. Further centrifugation and washing may be required to remove the remaining traces of non-covalently bound [^{3}H]NBMPR. The membrane pellets are dissolved in gel electrophoresis buffer or utilized for further studies, e.g. further purification of the [^{3}H] NBMPR labelled protein. SDS-polyacrylamide gel electrophoresis is carried out by standard procedures with the exception that the samples are not heated as this causes protein aggregation (Jarvis and Young, 1987; Wu et al., 1983). Radioactivity in the various regions of the gel is determined either by slicing the gel into 2mm fractions and measuring the ^{3}H content of the slices by liquid scintillation counting using Econofluor containing 3% (v/v) Protosol (New England Nuclear) or by fluorography after treatment of the Coomassie Blue-stained gels with Englightning autoradiography enhancer solution (New England Nuclear). For a more detailed review of the methodology, potential problems and artifacts that may arise during the course of NBMPR photoaffinity labelling the reader is referred to a recent review article (Jarvis and Young, 1987).

Identification of Transporters by Photoaffinity Labelling with Nitrobenzylthioinosine

Using the above protocol selective incorporation of radioactivity into band 4.5 polypeptides of mammalian erythrocytes (apparent molecular weight 45,000-65,000) is observed (Fig. 6) (Jarvis and Young, 1987; Wu et al., 1983). Minor high- and low-molecular-weight peaks are also apparent and these correspond to aggregates of the transporter and degradation products, respectively. Covalent labelling of membrane protein is inhibited by nucleosides, nitrobenzylthioguanosine and dipyridamole (Jarvis and Young, 1987; Wu et al., 1983). Moreover, no radiolabelling occurred in membranes from transport deficient cells (Wu et al., 1983; Young et al., 1984). These experiments strongly implicate band 4.5 membrane proteins in NBMPR-sensitive erythrocyte nucleoside permeation. In the native membrane the size of the functional transporter as determined by radiation-inactivation is estimated to be 120,000 (Jarvis et al., 1984).

The NBMPR photoaffinity labelling experiments have been extended to a variety of other cell types and tissues providing evidence that NBMPR-sensitive nucleoside transporters from different sources have broadly similar molecular weights to that of the human erythrocyte system (Jarvis and Young, 1987; Shi et al., 1984; Kwan and Jarvis, 1984; Jarvis and Ng, 1985; Young et al., 1984; Wu and Young, 1984). Rat nucleoside transporters have a consistently higher apparent M_r (average) in the region of 60,000 (Shi et al., 1984; Wu and Young, 1984). Treatment of rat and guinea-pig liver nucleoside transporters with endoglycosidase F abolished this difference (Jarvis and Young, 1987).

Enzyme Cleavage Studies

The availability of a specific covalent probe has allowed the transmembrane structure of the human erythrocyte nucleoside transporter to be explored by limited proteolysis. Treatment of unsealed membranes with trypsin cleaves [^{3}H]NBMPR labelled band 4.5 to yield two radioactive fragments with apparent M_r 38,000 and 24,000 (Janmohamed et al., 1985). Similar treatment of intact cells has no effect on the radiolabelled band 4.5, suggesting that at least one of the trypsin cleavage sites is on the cytoplasmic surface. The affinity for the reversible binding of [^{3}H]NBMPR to the nucleoside transporter was lowered two-fold following trypsin treatment of unsealed membranes, but the maximum number of inhibitor binding sites was unaffected despite the cleavage of band 4.5 to lower-M_r fragments. These results suggest that either a low-M_r fragment of band 4.5 can bind [^{3}H]NBMPR or alternatively the three-dimensional structure of the polypeptide has not been drastically disrupted following trypsin treatment.

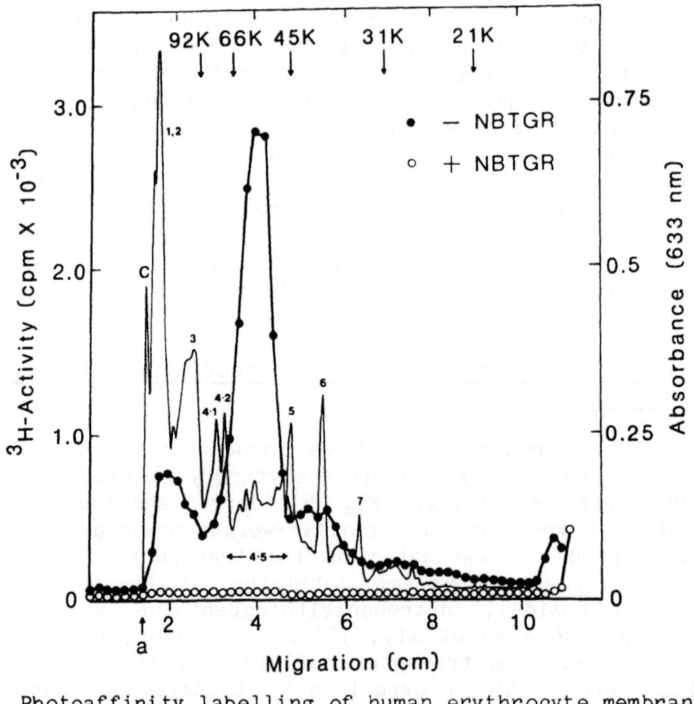

Fig. 6. Photoaffinity labelling of human erythrocyte membranes with [^{3}H]NBMPR. Membranes were equilibrated with 75nM [^{3}H]NBMPR in the presence and absence of 20μM nitrobenzylthioguanosine (NBTGR) and exposed to UV light. Membrane samples were then dissolved in gel sample buffer and subjected to SDS-gel electrophoresis. The absorbance scan (633nm) of Coomassie blue-stained membrane proteins, ^{3}H profiles and positions of the molecular weight standards are from the same slab gel. Data taken from Wu et al., 1983 with permission.

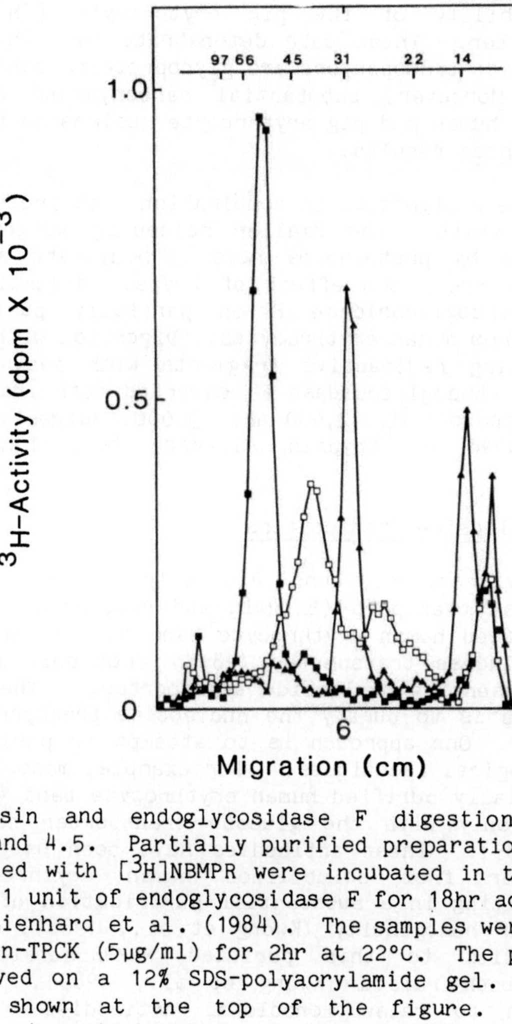

Fig. 7. Trypsin and endoglycosidase F digestion of [^{3}H]NBMPR labelled band 4.5. Partially purified preparations of band 4.5 photolabelled with [^{3}H]NBMPR were incubated in the presence or absence of 1 unit of endoglycosidase F for 18hr according to the method of Lienhard et al. (1984). The samples were then treated with trypsin-TPCK (5μg/ml) for 2hr at 22°C. The protein samples were resolved on a 12% SDS-polyacrylamide gel. M_r values (x 10^{-3}) are shown at the top of the figure. Symbols ■, control; □, trypsin treated; ▲, trypsin plus endoglycosidase F treated.

The broadness of the [^{3}H]NBMPR-labelled band 4.5 (Fig. 6) suggested that the polypeptide(s) is extensively glycosylated. Endo-β-galactosidase and endoglycosidase F digestion of the partially purified [^{3}H]NBMPR labelled band 4.5 from human erythrocytes converted the average apparent M_r of the nucleoside transporter from 55,000 to 47,000 and 44,000, respectively, following digestion and detection of the polypeptides by SDS-polyacrylamide gel electrophoresis (Kwong et al., 1986). The nucleoside transport protein of pig erythrocytes exhibited a higher M_r than the human protein with endoglycosidase-F treatment decreasing its apparent M_r (average) from 64,000 to 57,000 (Kwong et al., 1986). Interestingly, endo-β-galactosidase had no effect on the electrophoretic mobility of the pig erythrocyte [^{3}H]NBMPR labelled nucleoside transporter. These data demonstrate that the human and pig erythrocyte nucleoside transporters are glycoproteins containing N-linked oligosaccharide. Moreover, substantial carbohydrate and polypeptide differences between human and pig erythrocyte nucleoside transporters are also suggested by these results.

Endoglycosidase F digestion in combination with proteolysis has been used to ascertain whether the smaller molecular weight radiolabelled fragments generated by proteinases have carbohydrate attached to the fragments. Figure 7 shows the effect of trypsin digestion alone or in combination with endoglycosidase F on partially purified [^{3}H]NBMPR labelled band 4.5 from human erythrocytes. Digestion with trypsin yields two broadly migrating radioactive fragments with peak apparent M_r of 41,000 and 25,000. Endoglycosidase F converted both fragments into two sharper bands of apparent M_r 32,000 and 13,000, suggesting that the two polypeptides produced by trypsin cleavage both have carbohydrate attached.

Purification of Nucleoside Transporters

In human erythrocytes, band 4.5 polypeptides have also been implicated in hexose transport (Baldwin and Henderson, 1989). Indeed, the partially purified human erythrocyte band 4.5 preparation is mainly composed of the glucose transporter (~80%) with only about 3% of the total protein representing nucleoside transporter. Therefore, a major aim of many studies is to purify the nucleoside transporter free of the sugar transporter. One approach is to attempt to purify the membrane protein by immunological techniques. For example, monoclonal antibodies derived using partially purified human erythrocyte band 4.5 polypeptides that cross react only with the glucose transporter have been raised (Boyle et al., 1985). These antibodies have been used to remove the glucose transporter from solubilized human erythrocyte band 4.5 preparations, resulting in a substantial purification of the nucleoside transporter to near homogeneity (Kwong et al., 1988). Very recently, polyclonal antibodies to this purified NBMPR-sensitive nucleoside transporter have been raised. (Kwong et al., 1989). An alternative approach has been to raise monoclonal antibodies directly to the nucleoside transporter using partially purified antigen (Good et al., 1987). Both the monoclonal and polyclonal antibodies should be useful in screening cDNA libraries.

In a complementary approach, the nucleoside transporter from pig erythrocytes is being purified since these cells lack a functional glucose carrier but express an NBMPR-sensitive nucleoside transport system that is kinetically similar to that of the human erythrocyte system (Kwong et al., 1987). In contrast to the human erythrocyte nucleoside transporter, pig erythrocyte nucleoside transporter can be absorbed by DEAE-cellulose at low ionic strength and subsequently eluted by NaCl. The overall purification of NBMPR binding activity was 140-fold. The final preparation contained two protein staining regions on SDS-gel electropherograms - one in the band 4.5 region of the gel and migrating in the same position as [^{3}H]NBMPR labelled pig transporter (average apparent M_r 64,000), the other with a lower apparent molecular weight of 43,000. The estimated purification of the radiolabelled NBMPR-binding protein with respect to the M_r 64,000 band is estimated to be in the region of 350-fold. This value compares with a theoretical maximum purification of ~ 650-fold and thus the NBMPR-binding protein is one of the major polypeptide species present in the M_r 64,000 band. This highly-enriched nucleoside transporter preparation should also prove to be extremely useful in molecular studies of the NBMPR-sensitive nucleoside transporter function.

Our knowledge of the molecular properties of nucleoside transporters in cells other than erythrocytes is limited and we have hardly any information available on the NBMPR-insensitive carrier. Photoaffinity analogues of transported nucleosides have been developed and tested as possible covalent probes for the human erythrocyte nucleoside transporter before being applied to the NBMPR-insensitive system. 8-Azidoadenosine is transported by the human erythrocyte nucleoside transporter but unexpectedly was shown to photolabel the sugar rather than the nucleoside transporter (Jarvis et al., 1986). This result, together with the known molecular similarities of the nucleoside and sugar transporters, i.e. both carriers are band 4.5 glycoproteins, raises the possibility that nucleosides may interact with the glucose transporter. Indeed, a recent study (Jarvis, 1988b) has shown that nucleosides are capable of inhibiting glucose transport activity by interacting at the cytoplasmic surface of the glucose transporter. In contrast, sugars have no effect on nucleoside transport.

ACTIVE NUCLEOSIDE TRANSPORT

Kinetic Properties

In addition to the non-concentrative, facilitated-diffusion systems that have been described, active nucleoside transport has been detected in a number of specialized cells. For example, brush-border membrane vesicles from rabbit and rat kidney and intestine exhibit Na^+-cotransport systems for nucleosides that exhibit a transient overshoot in the presence of a sodium gradient (extravesicular > intravesicular)(Lee et al., 1988; Williams and Jarvis, 1987; Le Hir and Dubach, 1985; Darnowski et al., 1987; Jarvis, 1989). The affinity of nucleosides for influx in the presence of Na^+ is remarkably high (K_m 1-10μM) and is one to two orders of magnitude higher than that of the facilitated-diffusion systems. NBMPR, dilazep and dipyridamole have no effect on Na^+-dependent nucleoside transport. A sodium:uridine coupling stoichiometry of 1:1 has been found for Na^+-dependent uridine transport in rabbit and rat renal brush-border membrane vesicles.

Permeant Specificity

Studies on the substrate specificity of Na^+-dependent nucleoside transport have shown a wide range of specificities that vary with cell type and tissue. For example, in our laboratory we have demonstrated that Na^+-dependent uridine uptake by bovine renal brush-border membrane vesicles occurs by two distinct systems. Figure 8 shows the dose response curves for the inhibition of Na^+-dependent uridine uptake by nucleosides. Thymidine only inhibits about 50% of the uptake of uridine. No further inhibition is observed if other pyrimidine nucleosides are added in addition to thymidine. In contrast, low concentrations of purine nucleosides, e.g. guanosine and inosine, will inhibit the remaining uridine flux component. These studies and direct transport measurements have demonstrated that bovine renal brush-border membrane vesicles have two Na^+-dependent nucleoside transport mechanisms with differing, but overlapping, substrate specificities. One system has a specificity for purine nucleosides but uridine is also a substrate for this carrier. The other system accepts pyrimidine nucleosides but adenosine and some adenosine analogues are also transported on this carrier. Recent studies with mouse intestinal epithelial cells also indicate the presence of two Na^+-dependent nucleoside transporters with substrate specificities similar to that in the kidney (Vijayalakshmi and Belt, 1988). In contrast, Na^+-dependent uridine transport by murine splenocytes was inhibited by purine nucleosides, but not by other pyrimidine nucleosides (Darnowski et al., 1987). Furthermore, no Na^+-dependent thymidine uptake was detected. These results would suggest that murine splenocytes only possess the Na^+-dependent "purine nucleoside" transporter. Therefore, it seems likely that there exists a number of Na^+-dependent nucleoside transporters and further work will be required to fully characterize the properties of these different active transport systems.

Recent studies have also suggested the presence of a K^+-cotransporter for nucleosides in rat renal cortical brush-border membrane vesicles (Lee et al., 1988). This system has not been detected in rabbit and bovine kidney (Jarvis et al., 1989). At the present time we have no information on the molecular properties of these active nucleoside transporters.

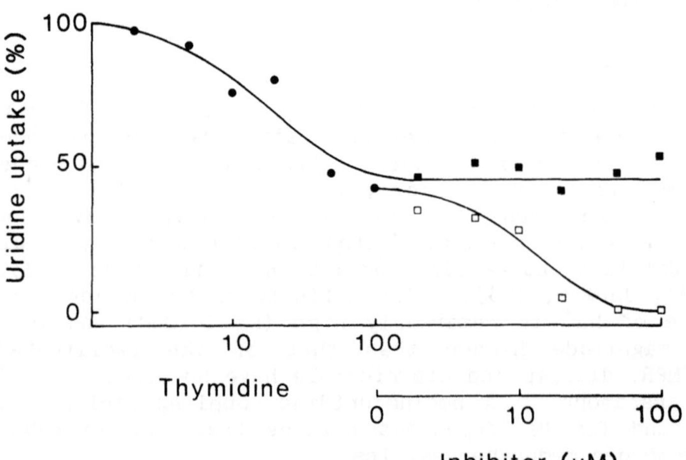

Fig. 8. Definition of Na^+-dependent nucleoside transport systems. Na^+-dependent uptake of 10μM [^{3}H] uridine into bovine renal brush-border membrane vesicles at 5 sec was measured in the presence of increasing concentrations of unlabelled thymidine (●). Substantial inhibition of uridine uptake by thymidine was observed. Uridine uptake in the presence of 100μM thymidine was further inhibited by increasing concentrations of guanosine (□). In contrast, cytidine caused no further inhibition of uridine uptake (■). Each point represents the mean of triplicate estimates (T.C. Williams and S.M. Jarvis, unpublished data).

Table 2. Properties of nucleoside transporters in animal cells

	Nucleoside Transporter			
	Equilibrative		Concentrative	
	NBMPR-sensitive	NBMPR-insensitive	"Purine"	"Pyrimidine"
Na^+-dependent	–	–	+	+
Inhibited by NBMPR at less than 1 μmolar	+	–	–	–
Substrate specificity				
Uridine	+	+	+	+
Adenosine	+	+	+	+
Thymidine	+	+	–	+
Formycin B	+	+	+	–
Guanosine	+	+	+	–
Tissue distribution	Most tissues	Most tissues	Kidney Intestine Mouse splenocytes	Kidney Intestine

CONCLUSIONS

There are at least four separate nucleoside transport systems in mammalian cells: an NBMPR-sensitive facilitated diffusion system, an NBMPR-insensitive facilitated diffusion process and two Na^+-dependent cotransporters. The general properties of these important transporters are summarized in Table 2 and further kinetic studies will provide additional insights into the characteristics of these systems, especially the Na^+ cotransporters. Moreover, it is anticipated that in the next few years the genes for these carriers will be isolated and sequenced thereby permitting comparison of the amino acid sequence of the transporters and possible identification of structural features important to function.

ACKNOWLEDGEMENTS

The work done in this laboratory was supported by grants from the MRC, The Nuffield Foundation and The National Kidney Research Fund.

REFERENCES

Agbanyo, F.R., Cass, C.E. and Paterson, A.R., 1988, External location of sites on pig erythrocyte membranes that bind nitrobenzylthioinosine, Mol. Pharmacol., 33 : 332-337.

Baldwin, S.A. and Henderson, P.J.F., 1989, Homologies between sugar transporters from eukaryotes and prokaryotes, Annu. Rev. Physiol., 51: 459-471.

Belt, J.A. and Noel, L.D., 1985, Nucleoside transport in Walker 256 rat carcinosarcoma and S49 mouse lymphoma cells. Differences in sensitivity to nitrobenzylthioinosine and thiol reagents, Biochem. J. 232 : 681-688.

Belt, J.A. and Noel, L.D., 1988, Isolation and characterization of a mutant of L1210 murine leukemia deficient in nitrobenzylthioinosine- insensitive nucleoside transport, J. Biol. Chem., 263 : 13819-13822.

Boyle, J.M., Whetton, A.D., Dexter, T.M., Meeran, K. and Baldwin, S.A., 1985, Characterization of monoclonal antibodies which specifically recognize the human erythrocyte glucose transport protein, Embo. J., 4 : 3093-3098.

Bruns, R.F., Lawson-Wendling, K. and Pugsley, T.A., 1983, A rapid filtration assay for soluble receptors using polyethylenimine-treated filters, Anal. Biochem., 132 : 74-81.

Cass, C.E., Gaudette, L.A. and Paterson A.R.P., 1974, Mediated transport of nucleosides in human erythrocytes. Specific binding of the inhibitor nitrobenzylthioinosine, to nucleoside transport sites in the erythrocyte membrane. Biochim. Biophys. Acta, 345 : 1-10.

Darnowski, J.W., Holdridge, C. and Handschumacher, R.E., 1987, Concentrative uridine transport by murine splenocytes: kinetics, substrate specificity and sodium dependency, Cancer Res., 47 : 2614-2619.

Deckert, J., Bisserbe, J.C. and Marangos, P.J., 1987, Quantitative [^3H]dipyridamole autoradiography: evidence for adenosine transporter heterogeneity in guinea pig brain, Naunyn-Schmiedeberg's Arch. Pharmacol., 335 : 660-666.

El Kouni, M.H., Diop, D. and Cha, S., 1983, Combination therapy of schistosomiasis by tubercidin and nitrobenzylthioinosine 5'-monophosphate, Proc. Natl. Acad. Sci., 80 : 6667-6670.

Good, H.A., Craik, J.D., Jarvis, S.M., Kwong, F.Y.P., Young, J.D., Paterson, A.R.P. and Cass, C.E., 1987, Characterization of monoclonal antibodies that recognize band 4.5 polypeptides associated with nucleoside transport in pig erythrocytes, Biochem. J., 244: 749-755.

Harley, E.R., Paterson, A.R.P. and Cass, C.E., 1982, Initial rate kinetics of the transport of adenosine and 4-amino-7-(β-D-ribofuranosyl) pyrrolo [2, 3-d] pyrimidine (tubercidin) in cultured cells, Cancer Res., 42: 1289-1295.

Jakobs, E.S. and Paterson, A.R.P., 1986, Sodium-dependent, concentrative nucleoside transport in cultured intestinal epithelial cells, Biochem. Biophys. Res. Commun., 140 : 1028-1036.

Janmohamed, N.S., Young, J.D. and Jarvis, S.M., 1985, Proteolytic cleavage of the [^{3}H]nitrobenzylthioinosine labelled nucleoside transporter in human erythrocytes, Biochem. J., 230 : 777-784.

Jarvis, S.M., 1986a, Trans-stimulation and trans-inhibition of uridine efflux from human erythrocytes by permeant nucleosides. Biochem.J., 233 : 295-297.

Jarvis, S.M., 1986b, Nitrobenzylthioinosine-sensitive nucleoside transport system: mechanism of inhibition by dipyridamole, Mol. Pharmacol., 30 : 659-665.

Jarvis,S.M., 1988a, Adenosine Transporters in "Receptor Biochemistry and Methodology", Vol.11, pp.113-123, D.M.F. Cooper & C. Londos, ed., A.R.Liss, New York.

Jarvis, S.M., 1988b, Inhibition of glucose transport activity in human erythrocytes by nucleosides, Biochem. J., 249 : 383-389.

Jarvis, S.M., 1989, Characterization of sodium-dependent nucleoside transport in rabbit intestinal brush-border membrane vesicles, Biochim. Biophys. Acta, 979 : 132-138.

Jarvis, S.M. and Martin, B.W., 1986, Effects of temperature on the transport of nucleosides in guinea-pig erythrocytes, Can. J. Physiol. Pharmacol., 64 : 193-198.

Jarvis, S.M. and Ng, A.S., 1985, Identification of the adenosine uptake sites in guinea pig brain, J. Neurochem., 44 : 183-188.

Jarvis, S.M. and Young, J.D., 1981, Extraction and partial purification of the nucleoside transport system from human erythrocytes based on the assay of nitrobenzylthioinosine binding activity, Biochem. J., 194 : 331-339.

Jarvis, S.M. and Young, J.D., 1982, Nucleoside translocation in sheep reticulocytes and in erythrocytes from newborn lambs. A proposed model for the nucleoside transporter, J. Physiol. (Lond), 324 : 47-66.

Jarvis, S.M. and Young, J.D., 1986, Nucleoside transport in rat erythrocytes: two components with differences in sensitivity to inhibition by nitrobenzylthioinosine and p-chloromecuriphenylsulphonate, J. Membr. Biol., 93 1:10.

Jarvis, S.M. and Young, J.D., 1987, Photoaffinity labelling of nucleoside transporter polypeptides, Pharmacol. Ther., 32 : 339-359 (1987).

Jarvis, S.M., Young, J.D., Ansay, M., Archibald, A.L., Harkness, R.A. and Simmonds, R.J., 1980, Is inosine the physiological energy source of pig erythrocytes?, Biochim. Biophys. Acta, 597 : 183-188.

Jarvis, S.M., McBride, D., and Young, J.D., 1982, Erythrocyte nucleoside transport: asymmetrical binding of nitrobenzylthioinosine to nucleoside permeation sites, J. Physiol.(Lond), 324 : 31-46

Jarvis, S.M., Hammond, J.R, Paterson, A.R.P. and Clanachan, A.S., 1983, Nucleoside transport in human erythrocytes. A simple carrier with directional symmetry in fresh cells but with directional asymmetry in cells from outdated blood, Biochem. J., 210 : 457-461.

Jarvis, S.M., Finchan, D.A., Ellory, J.C., Paterson, A.R.P. and Young, J.D, 1984, Nucleoside transport in human erythrocytes. Nitrobenzylthioinosine binding and uridine transport activities have similar radiation target sizes, Biochim. Biophys. Acta, 772 : 227-230.

Jarvis, S.M., Martin, B.W. and Ng, A.S., 1985, 2-Chloroadenosine, a permeant for the nucleoside transporter, Biochem. Pharmacol., 34: 3237-3241.

Jarvis, S.M., Young, J.D., Wu, J-S.R., Belt, J.A. and Paterson, A.R.P., 1986, Photoaffinity labelling of the human erythrocyte glucose transporter with 8-azidoadenosine, J. Biol. Chem., 261: 11077-11085.

Jarvis, S.M., Williams, T.C., Lee, C.-W. and C.I. Cheeseman, 1989, Active transport of nucleosides and nucleoside drugs, Biochem. Soc. Trans., 17 : 448-450.

Kwan, K.F. and Jarvis, S.M.,1984, Photoaffinity labelling of adenosine transporter in cardiac membranes with nitrobenzylthioinosine, Am. J. Physiol., 246 : H710-H715.

Kwong, F.Y.P., Baldwin, S.A., Scudder, P.R., Jarvis, S.M., Choy, M.Y.M. and Young, J.D., 1986, Erythrocyte nucleoside and sugar transport. Endo-β-galactosidase and endoglycosidase-F digestion of partially purified human and pig transporter proteins, Biochem. J., 240 : 349-356.

Kwong, F.Y.P., Tse, C.M., Jarvis, S.M., Choy, M.Y.M. and Young, J.D., 1987, Purification and reconstitution of the nucleoside transporter from pig erythrocytes, Biochim. Biophys. Acta, 904: 105-116.

Kwong, F.Y.P., Davies, A., Tse, C.M., Young, J.D., Henderson, P.J.F. and Baldwin, S.A., 1988, Purification of the human erythrocyte nucleoside transporter by immunoaffinity chromatography, Biochem. J., 255 : 243-249.

Kwong, F.Y.P., Fincham, H.B., Young, J.D., and Baldwin, S.A, 1989, Characterization of polyclonal antibodies against the human erythrocyte nucleoside transporter, Biochem. Soc. Trans., 17 : 743-744.

Lee, C.W. and Jarvis, S.M., 1988a, Nucleoside transport in rat cerebral cortical synaptosomes. Evidence for two types of nucleoside transporters, Biochem. J., 249 : 557-564.

Lee, C.W. and Jarvis, S.M., 1988b, Kinetic and inhibitor specificity of adenosine transport in guinea pig cerebral cortical synaptosomes: evidence for two nucleoside transporters, Neurochem. Int., 12 : 483-492.

Lee, C.W., Cheeseman, C.I. and Jarvis, S.M., 1988, Na^+- and K^+-dependent uridine transport in rat renal brush border membrane vesicles, Biochim. Biophys. Acta, 942: 139-149.

Le Hir, M. and Dubach, U.C., 1985, Concentrative transport of purine nucleosides in brush-border vesicles of the rat kidney, Eur. J. Clin. Invest., 15 : 121-127.

Lienhard, G.E., Crabb, J.H. and Ransome, K.J. 1984, Endoglycosidase F cleaves the oligosaccharides from the glucose transporter of the human erythrocyte, Biochim. Biophys. Acta, 769 : 404-410.

Lynch,T.P., Jakobs, E.S., Paran, J.H. and Paterson, A.R.P., 1981, Treatment of mouse neoplasms with high doses of tubercidin, Cancer Res., 41 : 3200-3204.

Marangos, P.J. and Deckert, J., 1987, $[^3H]$Dipyridamole binding to guinea pig brain membranes: possible heterogeneity of central adenosine uptake sites, J. Neurochem., 48: 1231-1236.

Morgan, P.F. and Marangos, P.J., 1987, Comparative aspects of nitrobenylthioinosine and dipyridamole inhibition of adenosine accumulation in rat and guinea pig synaptoneurosomes, Neurochem. Int., 11 : 339-346.

Paterson, A.R.P., Harley, E.R. and Cass, C.E., 1984, Inward fluxes of adenosine in erythrocytes and cultured cells measured by a quenched-flow method, Biochem. J., 244 : 1001-1008.

Paterson, A.R.P., Harley, E.P. and Cass, C.E., 1985, Measurement and inhibition of membrane transport of adenosine, Methods Pharmacol., 6 : 165-180.

Plagemann, P.G.W. and Woffendin, C., 1988, Species differences in sensitivity of nucleoside transport in erythrocytes and cultured cells to inhibition by nitrobenzylthioinosine, dipyridamole, dilazep and lidoflavine, Biochim. Biophys. Acta, 969 : 1-8.

Plagemann, P.G.W. and Woffendin, C., 1989, Use of formycin B as a general substrate for measuring facilitated nucleoside transport in mammalian cells, Biochim. Biophys. Acta, 1010 : 7-15.

Plagemann, P.G.W. and Wohlhueter, R.W., 1980, Permeation of nucleosides and nucleic acid bases and nucleotides in animal cells, Curr. Top. Membr. Transp., 14: 225-330.

Plagemann, P.G.W. and Wohlhueter, R.M., 1984, Nucleoside transport in cultured mammalian cells. Multiple forms with different sensitivity to inhibition by nitrobenzylthioinosine or hypoxanthine, Biochim. Biophys. Acta, 773 : 39-52.

Plagemann, P.G.W., Marz, R. and Erbe, J., 1976, Transport and counter-transport of thymidine in ATP-depleted and thymidine kinase deficient Novikoff rat hepatoma and mouse L cells. Evidence for a high Km facilitated diffusion system with wide nucleoside specificity, J. Cell. Physiol., 89 : 1-18.

Plagemann, P.G.W. Wohlhueter, R.M. and Kraupp,M., 1985, Adenosine uptake, transport, and metabolism in human erythrocytes. J.Cell Physiol., 125 : 330-336.

Plagemann, P.G.W., Wohlhueter, R.W. and Woffendin, C., 1988, Nucleoside transport in animal cells, Biochim. Biophys. Acta, 947: 405-443.

Shi, M.M. and Young, J.D., 1986, [^{3}H]Dipyridamole binding to nucleoside transporters from guinea-pig and rat lung, Biochem. J., 240 : 879-883 (1986).

Shi, M.M., Wu, J-S.R., Lee, C.M. and Young, J.D., 1984, Nucleoside transport. Photoaffinity labelling of high-affinity nitrobenzylt-thioinosine binding sites in rat and guinea pig lung, Biochem. Biophys. Res. Commun., 118 : 594-600.

Tse, C.M., Belt, J.A, Jarvis, S.M., Paterson, A.R.P., Wu, J-S.R. and Young,J.D., 1985, Reconstitution studies of the human erythrocyte nucleoside transporter, J. Biol. Chem., 260 : 3506-3511.

Vijayalakshmi, D. and Belt, J.A., 1988, Sodium-dependent nucleoside transport in mouse intestinal epithelial cells. Two transport systems with differing substrate specificities, J. Biol. Chem., 263: 19419-19423.

Williams, T.C. and Jarvis, S.M., 1987, Sodium-dependent uridine transport in renal cortical brush border membrane vesicles, Biochem. Soc. Trans., 15, 1134-1135.

Woffendin, C. and Plagemann, P.G.W., 1987, Interaction of [^{3}H] dipyridamole with the nucleoside transporters of human erythrocytes and cultured animal cells, J. Membr. Biol., 98 : 89-100.

Wohlhueter, R.W. and Plagemann, P.G.W., 1982, On the functional symmetry of nucleoside transport in mammalian cells. Biochim. Biophys. Acta., 689 : 249-260.

Wohlhueter, R.M., Marz, R. and Plagemann, P.G.W., 1978, Properties of the thymidine transport system of Chinese hamster ovary cells as probed by nitrobenzylthioinosine, J. Membr. Biol., 42 : 247-264.

Wohlhueter, R.M., Marz, R. and Plagemann, P.G.W., 1979, Thymidine transport in cultured mammalian cells. Kinetic analysis, temperature dependence and specificity of the transport system, Biochim. Biophys. Acta, 553 : 262-283.

Wu, J-S.R. and Young, J.D., 1984, Photoaffinity labelling of nucleoside-transport proteins in plasma membranes isolated from rat and guinea-pig liver, Biochem. J., 220 : 499-506.

Wu, J-S.R., Kwong, F.Y.P., Jarvis, S.M., and Young, J.D., 1983, Identification of the erythrocyte nucleoside transporter as a Band 4.5

polypeptide. Photoaffinity labelling studies using nitrobenzylthioinosine, J. Biol. Chem., 258 : 13745-13751.

Young, J.D. and Jarvis, S.M., 1985, The use of ligands in the study of the nucleoside transport complex - nitrobenzylthioinosine, Methods Pharmacol., 6 : 181-190.

Young, J.D., Jarvis, S.M., Belt, J.A., Gati, W.P. and Paterson, A.R.P., 1984, Identification of the nucleoside transporter in cultured mouse lymphoma cells. Photoaffinity labelling of plasma enriched fractions from nucleoside transport competent (S49) and nucleoside-deficient (AE_1) cells with [^{3}H]nitrobenzylthioinosine, J. Biol. Chem., 259 : 8363-8365.

Zimmerman, T.P., Mahony, W.B. and Pruss, K.L., 1987, 3'-Azido-3'-deoxythymidine. An unusual nucleoside analogue that permeates the membrane of human erythrocytes and lymphocytes by non-facilitated diffusion. J.Biol.Chem., 262 : 5748-5754.

Part 6: TRANSPORT REGULATION

Chapter 21

Regulation of junctional intercellular communication by tyrosine-protein kinases. Role of the cellular SRC gene and growth factor receptors
Werner Lowenstein

Chapter 22

Single-channel and whole-cell patch-clamp experiments on gland cells: activation of ion channels via second messengers
Ole Petersen

Chapter 23

Regulation of amino acid transporters in cultured vascular endothelial cells perfused in a microcarrier column
Giovanni Mann, Cherry-Jane Sherrif, Valerie Toothill and Jeremy Pearson

21

REGULATION OF JUNCTIONAL INTERCELLULAR COMMUNICATION BY TYROSINE-PROTEIN KINASES. ROLE OF THE CELLULAR *SRC* GENE AND GROWTH FACTOR RECEPTORS

Werner R. Loewenstein

Department of Physiology and Biophysics
University of Miami School of Medicine
P.O. Box 016430
Miami, Florida 33101

INTRODUCTION

I deal with a membrane channel of a very different sort from the channels of the preceding speakers at this workshop. It is a channel that spans not one, but two cell membranes. The channel is an old friend of mine; I ushered it in twenty-two years ago.

The channel was a solution to an ancient evolutionary problem: how to link the cyberneting loops of information of different cells, and how to do so without mixing the information sources. What finally evolved was a channel 16-20 Å in diameter, wide enough to let through the cytoplasmic inorganic ions, the metabolites, the building blocks of biomolecular synthesis, and the high energy phosphates, but not the macromolecules. The channel is made of two protein halves, one contributed by each membrane, tightly joined, forming a rather leakproof aqueous tunnel between cells (Fig. 1).

This cell-to-cell channel is present in nearly all organized tissues throughout the phylogenetic scale, constituting one of the most basic forms of cellular communication. As a rule, a cell is connected by these channels with several neighbors in a tissue, and so whole organs or large organ parts are continuous from within. The channels tend to cluster in the junctional membrane regions where they are seen electron-microscopically as clusters of membrane particles, called "gap junction". The particles are hexameric: six slanted subunits form a channel halve (Caspar et al., 1977; Unwin & Zampighi, 1980).

Several regulatory mechanism of the channel have been identified. One regulator is Ca^{2+}. The channel closes rapidly when the concentration of this ion rises above 10^{-6}M in the junctional locale (Rose & Loewenstein, 1976). The subunits then slide radially obliterating the channel bore--a shutter mechanism of the iris diaphragm sort (Unwin & Ennis, 1983). This closure mechanism is activated when a cell springs a leak (Oliveira-Castro & Loewenstein, 1971) or its ion pumps stand still (Rose & Loewenstein, 1976). It is a basic survival mechanism allowing the cell community to disconnect an unhealthy member and to seal itself off.

Seven years ago, my colleagues and I came across another regulatory

Cell Membrane Transport, Edited by D. L. Yudilevich *et al.*
Plenum Press, New York, 1991

mechanism in which a cyclic AMP-dependent protein kinase plays a key role. Here, the common isoenzyme form I, which phosphorylates on serine and threonine residues, has an upregulatory role in junctional communication of epithelial and fibroblast cells, (Flagg-Newton et al., 1981; Wiener and Loewenstein, 1983) following this scheme:

$$\text{c-AMP} \rightarrow PK_I \rightarrow \uparrow \text{ junctional permeability}$$

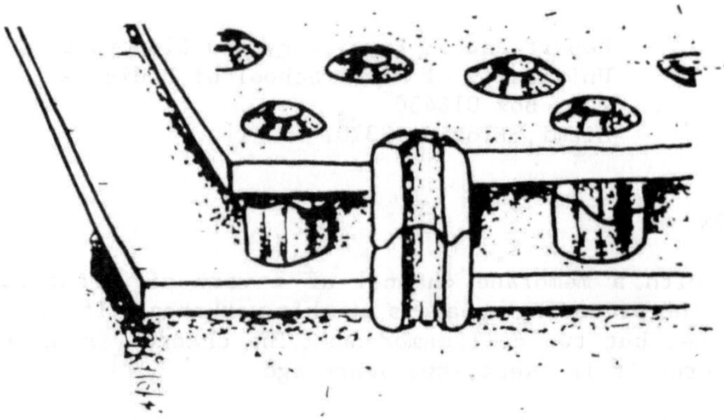

Figure 1. Cell-to-cell channel. The channel is made of two tightly joined transmembrane proteins, one from each apposing cell membrane. The aqueous permeation-limiting bore of the channel is 1.6-2 nm in mammalian cells. (From Loewenstein, 1974.)

Here I give an account of a protein kinase which phosphorylates on tyrosine and acts in the opposite direction on communication:

$$PK_{Tyr} \rightarrow \downarrow \text{ junctional permeability}$$

We were able to trace this regulation to a single gene: the cellular *src* gene. This lucky strike happened just recently, and I tell what Roobik Azarnia, David Shalloway and I have found out so far.

The protein kinase is the membrane-bound $pp60^{c\text{-}src}$, the product of the cellular *src* gene (c-*src*) discovered twelve years ago (Stehelin et al., 1976). The gene is present in all animal species and is highly conserved in evolution. Its expression level is relatively low in ordinary cultured cells and its functions were heretofore unknown.

C-*src* Overexpression Inhibits Junctional Communication

Our approach to function was straightforward. We raised the gene expression level by incorporating highly transcribed c-*src* recombinant DNAs into mammalian cells, and examined the junctional transfer and growth. The expression vectors contained the chicken c-*src* sequences ligated to retroviral long terminal repeats providing promoter/enhancer sequences and polyadenylation sites (Fig. 2). Mouse NIH 3T3 cells were transfected with these plasmids, and cell lines were cloned from foci (suffix *foc*) or co-selected for expression of *Eco-gpt* (suffix *cos*) or *neo*. Thus we obtained cell clones whose $pp60^{c\text{-}src}$ levels were elevated 10 to 20 times over that of the controls.

The effect of this overexpression on junctional communication was striking: whereas normally the NIH 3T3 cells transferred Lucifer Yellow or lissamine rhodamine B-labeled glutamic acid (LRB-Glu) extensively to first- and higher-order neighbors, the overexpressor cells transferred these tracers rarely and only to first-order neighbors. The incidence of cell interfaces permeable to the tracers was reduced by 57 to 73%, the cells with the highest $pp60^{c\text{-}src}$ expression level exhibiting the greatest reduction. Nonjunctional membrane permeability was not affected by the overexpression; the rates of tracer loss from the cells were negligible and not different from those of the control cells (Azarnia et al., 1988).

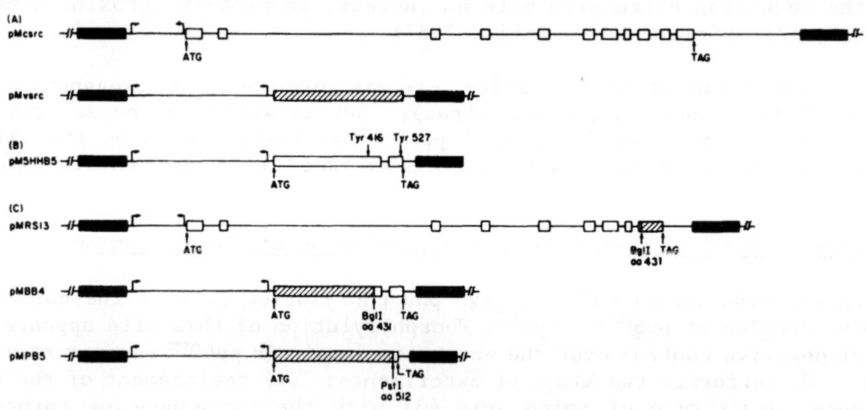

Figure 2. The c-*src* and v-*src* expression plasmids. The plasmids contain the chicken c-*src*, v-*src*, mutant and chimeric *src* genes ligated to transcriptional enhancers, promoters and polyadenylation signals provided by flanking Moloney murine leukemia virus long terminal repeats (filled boxes) from the vector pEVX. The locations of Moloney splice donors and first *src* splice acceptors are denoted by bent arrows. Coding regions are indicated by open (c-*src*) or hatched boxes (v-*src*) and the locations of the initiating (ATG) and terminating (TAG) codons are marked. Only the functionally relevant eukaryotic regions of the plasmids are shown. (A), pMcsrc, the plasmid used in the experiments on c-*src* overexpression, and the structurally similar v-*src* expression plasmid pMvsrc; (B), pM5HHB5, the plasmid used for site-directed mutagenesis to produce Tyr → Phe codon exchange at amino acids 416, 527 or both, generating plasmids pcsrc416, pcsrc527 and pc416•527. (C), Plasmids pMRS13 and pMBB4 are mirror-image chimeric constructions with junctions at site *BglI* at amino acid (aa) codon 431. pMPB5 is similar to pMBB4, except that the junction is at *PstI* site at amino acid 512 (From Azarnia et al., 1988a).

These results showed that pp60$^{c\text{-}src}$ down-regulates junctional permeability and pointed up a simple mode of genetic control: regulation of communication by modification of c-*src* transcription rate.

Middle T Antigen Inhibits Junctional Communication

We explored the possibility of further controls: regulation through post-translational modification of pp60$^{c\text{-}src}$. We were guided by the knowledge that the polyoma middle T antigen --a protein which binds to pp60$^{c\text{-}src}$-- enhances the enzyme and transforming activities of pp60$^{c\text{-}src}$ (Bolen et al., 1984) and that these activities are inhibited by phosphorylation of Tyr^{527}, a site in the carboxyl terminal region of pp60$^{c\text{-}src}$ (Cooper et al., 1986; Kmiecik and Shalloway, 1987; Piwnica-Worms et al., 1987; Cartwright et al, 1987). Thus, we studied the action on communication of middle T and of mutations of the pp60$^{c\text{-}src}$ carboxyl terminal region.

For the study of middle-T action we used an expression vector containing the sequences for middle T ligated to a dexamethasone-inducible promoter (Raptis et al, 1985). The vector was incorporated into rat F-111 cells, and transcription was turned on and off by adding the hormone to the cell medium and removing it.

The effect of middle T was as pronounced as that of c-*src* overexpression. Junctional permeability was reversibly reduced (Fig. 3) and the reduction correlated with an increase in protein tyrosine kinase activity (Azarnia and Loewenstein, 1987).

The depression of communication here was governed by an exogenous gene (middle T is encoded by polyoma virus), but it would not be surprising if the post-translational trick of pp60$^{c\text{-}src}$ activation used by the virus turned out to be a variation, albeit extreme, of older physiological themes.

Mutation of Tyr^{527} Enhances the c-*src* Inhibition of Communication

We centered then on the *in vivo* phosphorylation site of the carboxyl terminal region of pp60$^{c\text{-}src}$, Tyr^{527}. Phosphorylation of this site appears to excert negative control over the enzymatic domain in pp60^{src} (Cooper et al., 1986). We performed two kinds of experiments: (i) replacement of the DNA sequences downstream of amino acid 431 with the corresponding carboxyl terminal sequences of the viral *src* gene (v-*src*) which lacks Tyr^{527} (c-*src*/v-*src* chimeras); and (ii) point mutation of Tyr^{527}. Either mutation led to enhancement of the ability of c-*src* to down-regulate communication.

The c-*src*/v-*src* chimeras induced marked losses of junctional permeability, even at low gene expression. The losses were comparable to those induced by the full viral gene (Atkinson et al., 1981; Azarnia and Loewenstein, 1984). Conversely, mirror-image chimeric constructions (v-*src*/c-*src*) with the c-*src* carboxyl terminal sequences spliced downstream of the amino acid codon 431 of v-*src*, or chimeric constructions with such a splice at codon 512, produced no significant loss of junctional permeability.

For point mutation we used a site-directed mutant c-*src* vector (pcsrc527) that encodes a substitution of Phe (which cannot be phosphorylated) for Tyr^{527}. This mutation was nearly as effective as the c-*src*/v-*src* chimeras; most cells exhibited no junctional transfer at all. The incidence of permeable interfaces in the cells containing the Tyr^{527} mutation was much lower than that of the controls containing the unmodified c-*src* vector (pM5HHB5), even though the pp60$^{c\text{-}src}$ expression level was lower than in the controls. The mutation evidently had caused a reduction of junctional permeability on its own.

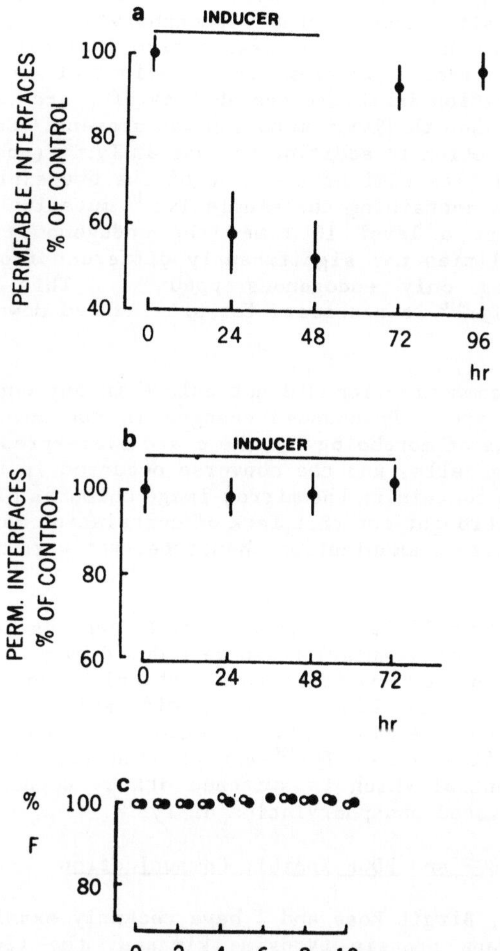

Figure 3. (a) Reversible reduction of junctional LRB-Glu transfer in F-111 cells containing a glucocorticoid-sensitive middle T recombinant DNA (mT). Dexamethasone (1μM), the inducer, was applied at time zero and removed at 48 h. Mean incidence of permeable interfaces (±SE) normalized with respect to uninduced controls (100%). (b) A control run in which dexamethasone (1μM) was applied to F-111 cells not containing mT. (c) Photometric measurement of intracellular LRB-Glu fluorescence loss from cells containing mT induced with 1μM dexamethasone (•) and uninduced (o). (From Azarnia & Loewenstein, 1987.)

As in the case of c-*src* overexpression, the expression of Tyr^{527} mutants or c-*src*/v-*src* chimeras did not sensibly affect the permeability of nonjunctional cell membrane.

Mutation of Tyr^{416} Suppresses the c-*src* Inhibition of Communication

Another phosphorylation site on $pp60^{c\text{-}src}$ is of the immediate interest here: Tyr^{416}. This site becomes stably phosphorylated upon mutation of Tyr^{527} and transiently upon c-*src* overexpression (Kmiecik and Shalloway, 1987; Kmiecik et al., 1987). We examined the effect of a Phe substitution for Tyr^{416}. This mutation inhibited the ability of c-*src* to down-regulate communication, even when the Tyr^{527} mutation was present. Cells containing genes with such a mutation in addition to that at Tyr^{527} (pcsrc416/527) had junctional permeabilities similar to those of the overexpressor controls (pM5HHB5); and cells containing the single Tyr^{416} mutation (pcsrc416) and expressing $pp60^{c\text{-}src}$ at a level 10 times the endogenous one, exhibited junctional permeabilities not significantly different from the controls (pSV2neo) containing only endogenous $pp60^{c\text{-}src}$. This suggests that phosphorylation of Tyr^{416} is necessary for *src*-induced down-regulation of communication.

The changes of communication did not relate in any consistent way to morphological phenotype. Pronounced changes in communication occurred without major changes of morphology in the c-*src* overexpressors and in the middle T- containing cells, and the converse occurred in the v-*src*/c-*src* chimeras. The clones containing the mirror-image chimeras NIH(pMRS13/foc)C and NIH(pMBB4/foc)I brought out this lack of correlation most clearly: the two clones had opposite communication phenotype, yet were morphologically indistinguishable.

In conclusion, the cellular *src* gene product down-regulates junctional communication. This down-regulation can be induced by an increase in gene transcription rate or by modifications enhancing the *src* product's activity (a tyrosine protein kinase). The modification of a single codon, Tyr^{527}, in the carboxyl terminal coding region can activate this down-regulation and modification of Tyr^{416} can abolish it. Tyr^{527} appears to provide negative control which is switched off by dephosphorylation of this site and associated phosphorylation of Tyr^{416}.

The Growth Factors EGF and PDGF Inhibit Communication

Pedro Maldonado, Birgit Rose and I have recently examined the action of two other membrane protein tyrosine kinases: the receptors of the epidermal growth factor, EGF (Carpenter and Cohen, 1979) and the receptor of the platelet-derived growth factor, PDGF (Helding and Westermark, 1984). In these cases, the tyrosine kinase is a trans-membrane protein rather than membrane-bound as in the case of *src* protein. These receptors are expressed in many cell types. As their tyrosine kinase activities are turned on by the binding of their ligands, one can readily manipulate the enzyme activity by external application of the growth factors to cells in culture.

The effect of these growth factors on junctional communication was quite striking: within 2 min the conductance between contiguous NRK cells fell, reaching a minimum level within 10 min (Fig. 4). The effect also is seen by measuring junctional transfer of fluorescent tracer molecules, and Table 1 summarizes the results obtained with various mammalian cell lines. The inhibition of junctional permeability is reversible and dose-dependent, half-maximal effects are obtained at 10^{-9} and 10^{-11} M concentrations of EGF and PDGF, respectively (Fig. 5). This inhibition is among the earliest cellular responses to this growth factor and, like the other effects I discussed before, may be significant in the regulation of growth.

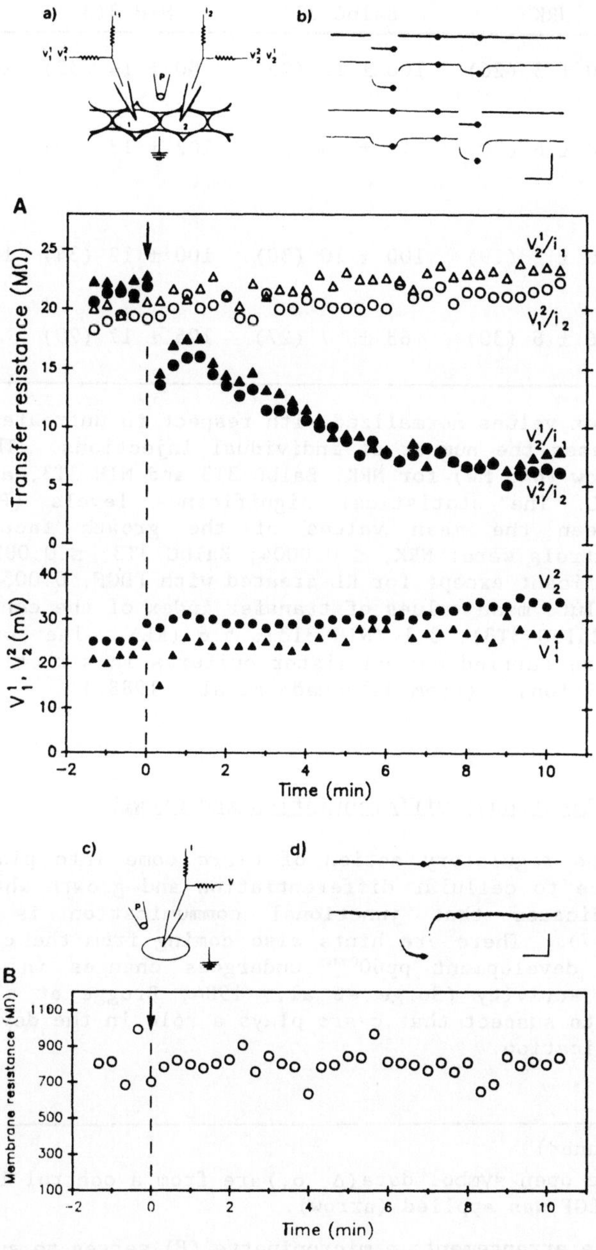

Figure 4. EGF reduces junctional conductance. EGF (80 nM) was applied (arrow) while the electrical coupling between two contiguous NRK cells was measured. The main figure shows the computer plots of the steady-state values of transfer resistance (V_2^1/i_1, ▲; V_1^2/i_2, •) and input voltage (V_1^1; ▲; V_2^2; •), as current was pulsed alternately into the two cells. The transfer resistances measured from either side are nearly identical, and the fall in transfer resistance in response to EGF is concomitant with a rise in input voltage, demonstrating a fall in junctional

(continued)

TABLE 1. EFFECTS OF EGF AND PDGF ON JUNCTIONAL TRANSFER

	NRK	BalbC 3T3	NIH 3T3	RL
Control (30)	100 ± 3 (26)	100 ± 12 (29)	100 ± 14 (33)	100 ± 4
EGF (16 nM) (27)	48 ± 6 (30)	48 ± 8 (33)	102 ± 13 (30)	95 ± 3
Control (30)	100 ± 2 (19)	100 ± 10 (30)	100 ± 12 (34)	100 ± 3
PDGF (0.6 nM) (27)	76 ± 6 (30)	63 ± 7 (27)	105 ± 17 (29)	89 ± 2

Mean transfer index values normalized with respect to untreated controls ± SE; in parentheses the number of individual injections. The tracers were Lucifer Yellow (443 MW) for NRK, BalbC 3T3 and NIH 3T3, and LRB-Glu (688 MW) for RL. The statistical significance levels (P) of the differences between the mean values of the growth factor-treated condition and controls were: NRK, ≤ 0.0004; BalbC 3T3, ≤ 0.001; NIH 3T3 and RL, not significant except for RL treated with PDGF, 0.003 (standard T test). The absolute mean values of transfer index of the controls were 7.4 (NRK), 8.5 (BalbC 3T3), 3.1 (NIH 3T3), 5.6 (RL). The controls for each treatment were carried out on sister cultures in strictly parallel experimental conditions. (From Maldonado et al., 1988.)

Physiological Implications: Differentiation and Growth

When might the regulatory action of c-*src* come into play? We are drawn here at once to cellular differentiation and growth where a body of evidence indicates that junctional communication is important (Loewenstein, 1987). There are hints also coming from the c-*src* side. During embryonic development $pp60^{c\text{-}src}$ undergoes changes in expression level and enzyme activity (Sorge et al., 1984; Brugge et al., 1985). Thus, we are led to suspect that c-*src* plays a role in the developmental program of communication.

Figure 4.(Continued)

conductance. The open-symbol data (Δ, o,) are from a control in which medium without EGF was applied (arrow).

Insets. (*a*). The arrangement: a micropipette (*P*) serves to superfuse the test cells with growth factor-containing medium. Microelectrodes connected to balanced bridge circuits were inserted into cells *1* and *2* to pulse current ($i = 5 \times 10^{-10}$ A) between cell inside and outside (grounded) and to record the resulting input voltages (V_1^1, V_2^2) and transfer voltages (V_2^1, V_1^2) (superscripts denote the cells into which i was injected and subscripts, the cells in which V was recorded). (*b*) Sample of an oscilloscope record of the currents i_1, i_2 (1st and 3rd trace, respectively) and voltages V_1^1, V_2^2 (2nd and 4th trace) from both cells. The computer sampled the steady-state current and voltages at the times marked by dots. Calibration: vertical, 20 mV and 1nA; horizontal, 100 msec. (From Maldonado et al., 1988.)

As to a role in cellular growth, we have the following to report: the growth *in vitro* of the cells, namely their saturation density, correlates with loss of communication. In general in the various c-*src* conditions examined, the larger the loss of communication, the higher was the saturation density (Fig. 6).

A further correlation became apparent when we focussed on cell transformation as defined by the criteria of growth in agarose and tumorigenicity: transformation correlated with loss of communication; but only the highest losses, those produced by the c-*src*/v-*src* chimeras, Tyr^{527} mutation and v-*src*, not those by c-*src* overexpression, were associated with full transformation. We have shown a similar correlation between growth and junctional permeability in the case of the junctional response to retinoids (Mehta et al., 1989).

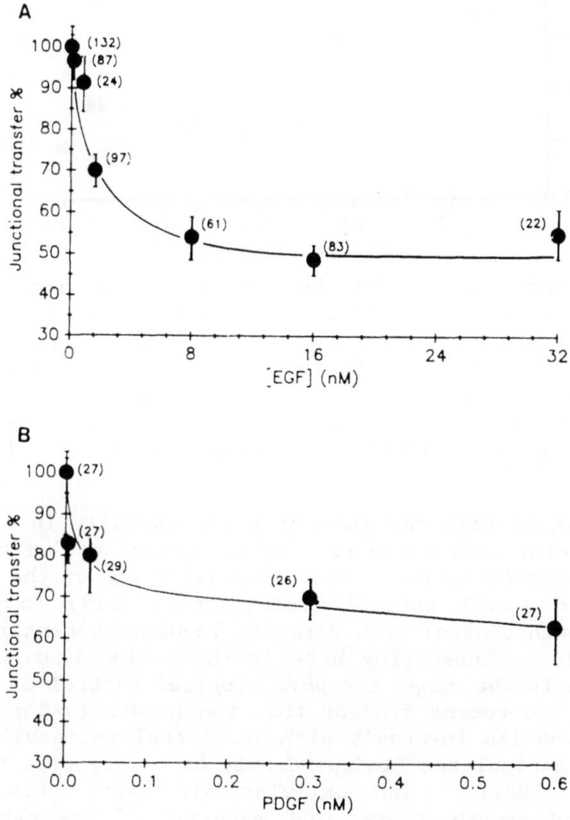

Figure 5. Dose-response curves for EGF and PDGF. The junctional transfer, as indexed by the number of cells to which fluorescent tracer (Lucifer Yellow) is transferred 1 min after its microinjection, at various growth-factor concentrations. Values are means ± SE, normalized with respect to the corresponding untreated controls. In parentheses the number of trials, i.e., the number of individually injected cells. BalbC 3T3 treated (*A*) with EGF and (*B*) with PDGF. Junctional transfer was tested 20 min after growth-factor application (or control medium application). The absolute mean values of the controls were 8.6 cells for (*B*) and ranged 6.8-7.3 cells for (*A*). (From Maldonado et al., 1988.)

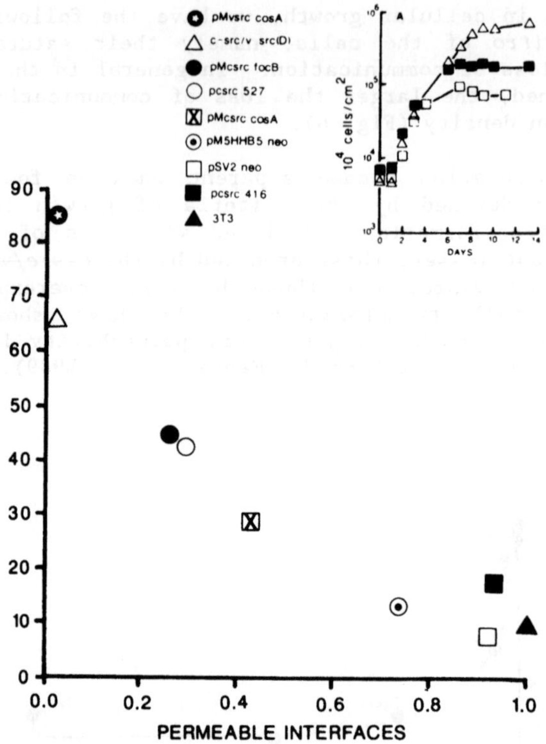

Figure 6. Mean saturation density *vs.* mean incidence of permeable interfaces in cells containing endogenous c-*src*, overexpressed c-*src*, and mutations at the c-*src* carboxyl terminal coding region (key of symbols is given on top). *Inset A*, three sample growth curves --cell density (log scale) *vs.* time-- representative of cells with the highest, intermediate and lowest saturation densities. (From Azarnia et al., 1988b.)

Thus, we suggest that the rule of c-*src* normally is limited to the high range of communication controlling the normal cell growth, that is, growth without transformation. This proposal leans on the hypothesis of junction-mediated growth control (Loewenstein, 1979) in particular on the model of growth control with discrete regulatory centers (Nonner and Loewenstein, 1989). Underlying here is the notion that the high range of communication is the range for *physiological* control of growth, which is supported by the recent finding that the growth (saturation density) of normal cells varies inversely with junctional permeability when this permeability is manipulated (independently of c-*src*) over the high range (Mehta et al., 1989). The cellular *src* gene, thus, would rule communication and growth without the excesses of its pathogenic viral counterpart.

REFERENCES

Atkinson, M. M., Menko, A. S., Johnson, R. G., Sheppard, J. R., and Sheridan, J. D., 1981, Rapid and reversible reduction of junctional permeability in cells infected with a temperature-sensitive mutant of avian sarcoma virus, *J. Cell Biol.* 91:573-578.

Azarnia, R., and Loewenstein, W. R., 1984, Intercellular communication and the control of growth. X. Alteration of junctional permeability by the *src* gene. A study with temperature-sensitive mutant Rous sarcoma virus, *J. Membr. Biol.* 82:191-205.

Azarnia, R., and Loewenstein, W. R., 1987, Polyomavirus middle T antigen downregulates junctional cell-to-cell communication, *Mol. Cell. Biol.* 7:946-950.
Azarnia, R., Reddy, S., Kmiecik, T. E., Shalloway, D., and Loewenstein, W. R., 1988a, The cellular *src* gene product regulates junctional cell-to-cell communication, *Science* 239:398-401.
Azarnia, R., Reddy, S., Kmiecik, T. E., Shalloway, D., and Loewenstein, W. R., 1988b, The cellular *src* gene: Regulation of communication and growth. *In: Gap Junctions.* Modern Cell Biology Series. Editors E. Hertzberg and R. Johnson. A. R. Liss, New York 7:423-433.
Bolen, J. B., Thiele, C. J., Israel, M. A., Yonemoto, W., Lipsich, L. A., and Brugge, J. S., 1984, Enhancement of cellular *src* gene product associated tyrosyl kinase activity following polyoma virus infection and transformation, *Cell* 38:767-777.
Brugge, J. S., Cotton, P. C., Queral, A. E., Barrett, J. N., Nonner, D., and Keane, R. W., 1985, Neurones express high levels of a structurally modified activated form of $pp60^{c\text{-}src}$, *Nature* 316:554-557.
Carpenter, G., and Cohen, S., 1979, Epidermal growth factor, *Ann. Rev. Biochem.* 48:193-216.
Cartwright, C. A., Eckhart, W., Simon, S., and Kaplan, P. L., 1987, Cell transformation by $pp60^{c\text{-}src}$ mutated in the carboxy-terminal regulatory domain, *Cell* 49:83-91.
Caspar, D. L. D., Goodenough, D. A., Makowski, L., and Phillips, W. C., 1977, Gap junction structures. I. Correlated electron microscopy and x-ray diffraction. *J. Cell Biol.* 74:605-628.
Cooper, J. A., Gould, K. L., Cartwright, C. A., and Hunter, T., 1986, Tyr^{527} is phosphorylated in $pp60^{c\text{-}src}$: implications for regulation, *Science* 231:1431-1434.
Flagg-Newton, J. L., Dahl, G., and Loewenstein, W. R., 1981, Cell junction and cyclic AMP. I. Upregulation of junctional membrane permeability and junctional membrane particles by administration of cyclic nucleotide or phosphodiesterase inhibitor, *J. Membr. Biol.* 63:105-121.
Helding, C. H., and Westermark, B., 1984, Growth factors: Mechanism of action and relation to oncogenes, *Cell* 37:9-20.
Kmiecik, T. E., and Shalloway, D., 1987, Activation and suppression of $pp60^{src}$ transforming ability by mutation of its primary sites of tyrosine phosphorylation, *Cell* 49:65-73.
Loewenstein, W. R., 1974, Cellular communication by permeable junctions. *In*: *Cell Membranes: Biochemistry, Cell Biology and Pathology* (Weissmann, G. & Claiborne, R., eds.) pp. 105-114, H.P. Publishing Co. Inc., New York.
Loewenstein, W. R., 1979, Junctional intercellular communication and the control of growth, *Biochem. Biophys. Acta Cancer Rev.* 560:1-65.
Loewenstein, W. R., 1987, The cell-to-cell channels of gap junctions, *Cell* 48:725-726.
Maldonado, P.E., Rose, B., and Loewenstein, W. R., 1988, Growth factors modulate junctional cell-to-cell communication, *J. Membr. Biol.* 106:203-210.
Mehta, P. P., Bertram, J., and Loewenstein, W. R., 1989, The actions of retinoids on cellular growth correlate with their actions on gap junctional communication, *J. Cell Biol.* 108:1053-1065.
Nonner, W. F., and Loewenstein, W. R., 1989. Appendix: A growth control model with discrete regulatory centers, *J. Cell Biol.* 108:1063-1065.
Oliveira-Castro, G. M., and Loewenstein, W. R., 1971, Junctional membrane permeability: effects of divalent cations, *J. Membr. Biol.* 5:51-57.

Piwnica-Worms, H., Saunders, K. B., Roberts, T. M., Smith, A. E., and Cheng, S. H., 1987, Tyrosine phosphorylation regulates the biochemical and biological properties of $pp60^{c\text{-}src}$, *Cell* 49:75-82.

Raptis, L., Lamfrom, H., and Benjamin, T. L., 1985, Regulation of cellular phenotype and expression of polyoma virus middle T antigen in rat fibroblasts, *Mol. Cell. Biol.* 5:2476-2485.

Rose, B., and Loewenstein, W. R., 1976, Permeability of a cell junction and the local cytoplasmic free ionized calcium concentration. A study with aequorin, *J. Membr. Biol.* 28:87-119.

Sorge, L. K., Levy, B. T., Gilmer, T. M., and Maness, P. F., 1984, $pp60^{c\text{-}src}$ expression in embryonic nervous tissue: immunocytochemical localization in the developing chick retina, *Cancer Cells* 1:117-122.

Stehelin, D., Varmus, H. E., Bishop, J. M., and Vogt, P. K., 1976, DNA related to the transforming gene(s) of avian sarcoma viruses is present in normal avian DNA, *Nature* 260:170-173.

Unwin, P. N. T., and Ennis, P. D., 1983, Calcium-mediated changes in gap junction structure: evidence from the low angle X-ray pattern, *J. Cell Biol.* 97:1459-1466.

Unwin, P. N. T., and Zampighi, G., 1980, Structure of the junction between communicating cells, *Nature* 283:545-549.

Wiener, E. C., and Loewenstein, W. R., 1983, Correction of cell-cell communication defect by introduction of a protein kinase into mutant cells, *Nature* 305:433-435.

22

SINGLE-CHANNEL AND WHOLE-CELL PATCH-CLAMP EXPERIMENTS ON GLAND CELLS: ACTIVATION OF ION CHANNELS VIA INTERNAL MESSENGERS

Ole Petersen

M.R.C. Secretory Control Research Group, The Physiological Laboratory
University of Liverpool, P.O. Box 147, Liverpool, L69 3BX, U.K.

INTRODUCTION

Electrophysiological studies on gland cells have contributed much to our understanding of the various membrane transport events and the stimulus-secretion coupling sequences. The earliest microelectrode studies on exocrine glands were carried out by Lundberg (1958) and later extended by Petersen (1980). The small size of mammalian gland cells made proper voltage clamp studies of transmembrane ionic currents difficult, but the first such studies with intracellular microelectrodes were made on pancreatic acini by McCandless et al. (1981) and Maruyama & Petersen (1983). At this time, however, the improved patch-clamp technique (Hamill et al. (1981) had become available and the first studies using this technique on exocrine glands were made by Maruyama and Petersen (1982a,b) on membrane fragments and intact single pancreatic acinar cells. This review aims to explain how the various patch-clamp recording configurations have been used to investigate the mechanisms by which receptor-activation switches on ion channels in exocrine gland cells.

THE PATCH-CLAMP METHOD

Electrical currents in cells are mediated by a single class of proteins in the plasma membrane called ion channels or pores. The first direct observation of current flowing through single-ion channels was a result of studies in bimolecular lipid membranes in the presence of antibiotics. In intact cells the problem is to detect single-channel currents in the presence of background electrical noise. Conventional intracellular microelectrode methods for current measurement are associated with a background noise of at least 100 pA, whereas the current flowing when a single channel opens is only a small fraction of this background noise.

Neher and Sakmann (1976) solved this problem by the patch-clamp method. Instead of inserting a microelectrode into a cell, they pressed a microelectrode tip onto the surface of a cell, effectively isolating a patch of membrane. The intrinsic noise increases with the area of the membrane under study, and when a small area (1 - 10 μm^2) is isolated, the extraneous noise levels are so low that the picoampere currents flowing through single channels can be measured directly.

Cell Membrane Transport, Edited by D. L. Yudilevich *et al.*
Plenum Press, New York, 1991

The seal between the tip of the microelectrode and the outer surface of the cell membrane has, under suitable conditions (fire-polished and clean micropipette tip and clean membrane surface), a high electrical resistance (in the order of gigaohms (10^9 ohm) and is mechanically surprisingly stable. The discovery in 1980 of this high-resistance seal by Neher, Sakmann and co-workers (Hamill et al., 1981) turned out to be very important, as it made entirely new types of experiments possible (Fig. 1).

The electrically isolated membrane patch can be pulled off the cell (excised in such a way that the inside of the plasma membrane faces the bath solution (inside-out) or alternatively so that the inside faces the solution in the

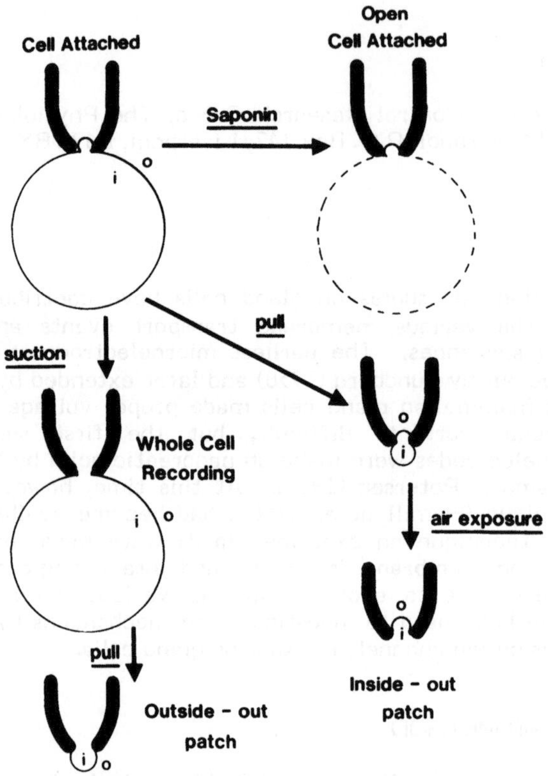

Fig. 1. Schematic representation of different patch-clamp recording configurations and procedures used to establish them. A patch-clamp experiment starts in the cell-attached configuration (top left). The tip of the fire-polished recording pipette is gently brought into contact with the cell surface, and slight suction is applied to the pipette resulting in a high-resistance seal between tip of the glass pipette and the cell membrane. The detergent saponin can be introduced briefly into the bath to permeabilize the cell membrane (outside the isolated patch area), in this way allowing equilibration between cell interior and exterior (open cell-attached configuration). Starting from the original cell-attached situation, the patch membrane can be mechanically pulled off, leading to formation of a closed membrane vesicle in the pipette tip. The outer surface of this vesicle can be disrupted by passing the pipette tip briefly through the air-water interface of the bath, leading to the excised inside-out membrane configuration. If a short pulse of suction is applied to the pipette in the cell-attached mode, the membrane patch can be broken and direct continuity between pipette and cell interior established. Currents flowing across the entire cell membrane can now be recorded (whole-cell recording). If the pipette is thereafter pulled away from the cell, excised membrane fragments will reseal so that an excised outside-out membrane patch is obtained. (Adapted from Petersen and Findlay, 1987).

micropipette (outside-out). By breaking the patch membrane in the cell-attached recording conformation, the solution in the pipette interior gains direct access to the cell interior and cell dialysis is carried out under conditions in which the currents across the whole-cell membrane can be measured. Equilibration of the cell interior with the bath solution can be done while single-channel currents are recorded by making holes in the plasma membrane outside the isolated patch area with the help of detergents such as saponin (Maruyama & Petersen, 1984b).

USE OF PATCH-CLAMP SINGLE-CHANNEL CURRENT RECORDING EXPERIMENTS TO DEMONSTRATE MESSENGER-MEDIATED CHANNEL OPENING

The first experiments in which hormonal activation of single-channel currents via an internal (second) messenger were demonstrated were carried out on mouse pancreatic acinar cells in which the Ca^{2+}-activated nonselective cation channel is dominant (Maruyama & Petersen, 1982a,b). In such experiments, cholecystokinin octapeptide (CCK_8) is applied by leakage from an extracellularly placed micropipette close to the cell from which unit recording was made (Fig. 2). When the tip of the CCK-delivery pipette is far away from the acinar unit under investigation, the patch is always silent. When the pipette is brought close to the recording site, unit activity begins to appear, and after less than 1 min, vigorous activity can be observed (Fig. 3). The cell-attached configuration is very stable, and currents can be recorded at many different pipette potentials. The I-V relationship obtained in these experiments is linear, with a single-channel conductance in the range 30-35 pS. Reversal of current polarity occurs at a pipette potential of about -25 mV. After the I-V relationship has been established in situ, the patch can be excised and the membrane inside exposed to a solution

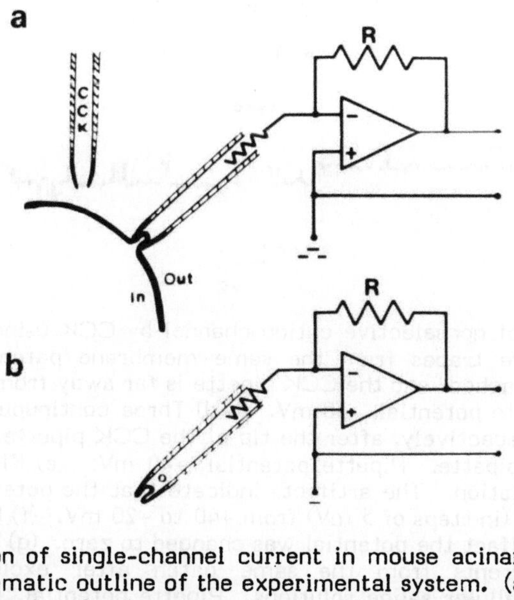

Fig. 2. Activation of single-channel current in mouse acinar cell by local CCK application. Schematic outline of the experimental system. (a) Recording from in situ membrane patch. In these experiments the recording pipette was always filled with extracellular bath solution. No agonist was present in the pipette solution. The pipette labelled CCK contained the octapeptide in a concentration of 5 μM. The spontaneous diffusion of CCK from the tip of the micropipette was used as a means of stimulation. (b) After the in situ recording, the patch-clamp pipette was withdrawn from the acinus to obtain an excised inside-out membrane patch. In these experiments single-channel recordings were made in symmetrical saline solutions. In other experiments the bath fluid was changed to one having·an intracellular composion. (Adapted from Maruyama and Petersen (1982b).

containing a high Ca^{2+} concentration (for example, 10 μM) (Fig. 3). Single-channel currents can be recorded again, and a new I - V relationship (linear) obtained. This again represents a single-channel conductance of 30 - 35 pS, but this time reversal of current polarity occurs at 0 mV pipette potential (same extracellular solution in pipette and bath). It would therefore appear that a membrane potential of about -25 mV is present in the acinar cell during CCK stimulation.

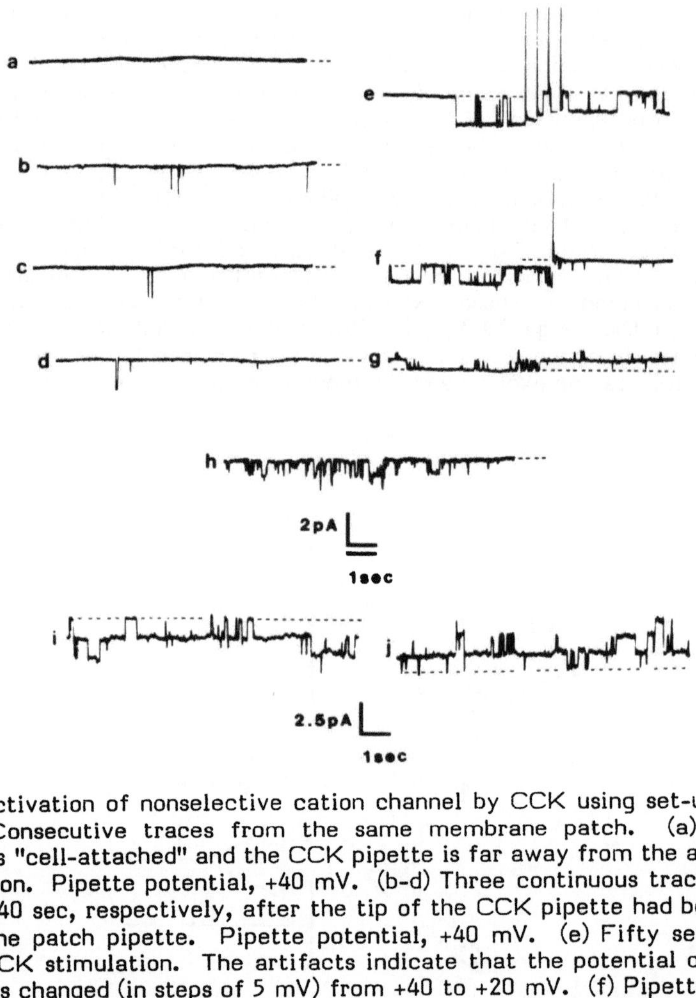

Fig. 3. Activation of nonselective cation channel by CCK using set-up shown in Fig. 2. Consecutive traces from the same membrane patch. (a) Recording situation is "cell-attached" and the CCK pipette is far away from the acinus under investigation. Pipette potential, +40 mV. (b-d) Three continuous traces obtained 20,30 and 40 sec, respectively, after the tip of the CCK pipette had been brought close to the patch pipette. Pipette potential, +40 mV. (e) Fifty seconds after start of CCK stimulation. The artifacts indicate that the potential of the patch pipette was changed (in steps of 5 mV) from +40 to +20 mV. (f) Pipette potential, +20 mV. At the artifact the potential was changed to zero. (g) Pipette potential, -40 mV. (h) Currents from the same patch after excision (inside-out). Symmetrical extracellular saline solutions. Pipette potential, +50 mV. (i and j) Currents recorded from another excised (inside-out) membrane patch with intracellular solution (high K^+, low Na^+) in the bath and the normal extracellular solution (high Na^+, low K^+) in the pipette. Pipette potential, +60 mV in i and -60 mV in j. Dashed lines represent the current level when all channels are closed. Downward deflections represent the current from extracellular to intracellular side of membrane. (Adapted from Maruyama and Petersen, 1982b).

The effect of CCK has also been investigated in the presence of the CCK antagonist dibutyryl cyclic guanosine monophosphate (Philpott & Petersen, 1979). In this case CCK always fails to activate the unitary inward currents, but after excision of the patch and exposure of the inner aspect of the membrane to the high Ca^{2+} concentration of the bath, single-channel currents representative of the nonselective cation channel invariably appear (Maruyama & Petersen, 1982b).

These experimental results show that specific CCK receptor interaction can activate the 30- to 35-pS cation channel localized in the basolateral plasma membrane by a mechanism that would appear to involve an internal messenger, since channels localized in an area to which CCK has no access (the patch area) are controlled by the agonist.

In a separate series of experiments with a similar type of protocol, the action of acetylcholine (ACh) has been investigated. ACh acts in a manner indistinguishable from that of CCK. Thus, it appears that different agonists interacting with two different receptor types control the activation of the same channels (Maruyama & Petersen, 1982b).

The simplest interpretation of the experiments of the type represented by Fig. 3 would be that binding of hormones or neurotransmitters to specific receptor sites evokes an increase in intracellular free Ca^{2+} concentration $[Ca^{2+}]_i$ and that Ca^{2+}, by interacting directly with the channel protein or a closely associated regulatory subunit, causes a conformational change in the channel, resulting in ion flux.

This simple interpretation is, however, too simple. In experiments on pig pancreatic acinar cells, which are dominated electrically by the high-conductance Ca^{2+}- and voltage-activated K^+ channel (Maruyama et al. 1983) it became clear that there must be involvement of at least one other messenger in addition to Ca^{2+} (Suzuki et al., 1985). The experimental approach shown in Fig. 2 was used again, but this time experiments were carried out both in the presence and absence of Ca^{2+} in the recording patch-clamp pipette. The experiments showed that sustained activation of the K^+ channels could only be evoked by excitation of CCK receptors if Ca^{2+} was present in the solution within the recording pipette (Suzuki et al., 1985).

The large Ca^{2+}- and voltage-activated K^+ channel has been extensively characterized in excised membrane patches from a variety of cell types (Petersen & Maruyama, 1984). Channel opening is controlled exclusively by the membrane potential and $[Ca^{2+}]_i$, whereas changes in $[Ca^{2+}]_o$ have no effect. The external Ca^{2+} requirement in the isolated membrane patch area in the experiments demonstrating indirect CCK activation of K^+ channels cannot, therefore, be explained by the action of Ca^{2+} on the external site of the channel, but must be due to an influence on $[Ca^{2+}]_i$. Hormones and transmitters seem to mobilize intracellular Ca^{2+} via the intracellular messenger inositol 1,4,5-trisphosphate (IP_3) and this substance acts on the membrane of Ca^{2+}-storing organelles by opening a Ca^{2+} pathway (Berridge & Irvine, 1989). However, many hormone- and transmitter-evoked responses in exocrine gland cells are dependent on extracellular Ca^{2+}, as documented for the stimulant-evoked hyperpolarization and increase in outward K^+ current in pig pancreatic acinar cells (Maruyama et al. 1983). The experimental procedure in patch-clamp experiments of the type carried out on cell-attached pig pancreatic acinar cells may emphasize the external Ca^{2+} dependence by physically moving the isolated membrane patch away from intracellular Ca^{2+} stores, and Ca^{2+} released in the pancreatic cell by the messenger following stimulation may not easily penetrate into the patch area due to restrictions of intracellular Ca^{2+} diffusion. IP_3, or another messenger may diffuse more easily than Ca^{2+} and could perhaps mediate the CCK-evoked Ca^{2+}-controlled channel opening by allowing Ca^{2+} influx, raising $[Ca^{2+}]_i$ locally (Suzuki et al., 1985).

In other words, the patch-clamp experiments have demonstrated an external Ca^{2+} requirement for sustained K^+ channel opening evoked by receptor (cholecystokinin) activation of pancreatic acinar cells. This Ca^{2+} requirement can be localized at a site separated from the hormone receptor site, which indicates messenger-mediated rather than direct receptor-operated Ca^{2+} entry.

USE OF PATCH-CLAMP WHOLE-CELL CURRENT RECORDING EXPERIMENTS TO DEMONSTRATE MESSENGER-MEDIATED CHANNEL OPENING

The tight-seal whole-cell current recording configuration has been used extensively to characterize currents mediated by Ca^{2+}-activated K^+- and or Cl^- selective channels in exocrine acinar cells.

K^+ Channel Currents

The whole-cell voltage-clamp current can be regarded as the sum of many small elementary currents flowing through individual high-conductance K^+ channels of the type characterized in single-channel studies. There are a number of strong arguments which support this conclusion. (1) The whole-cell K^+ current is Ca^{2+} activated, and the Ca^{2+} sensitivity is similar to that described for the single channels (Maruyama & Petersen, 1984a). (2) The activation of the whole-cell currents occurs within the same voltage range as described for single-channel currents (Maruyama et al., 1983; Maruyama & Petersen, 1984a). (3) Ca^{2+}-activated K^+ currents can be blocked by low concentrations (2 mM) of tetraethylammonium acting specfically from the outside of the plasma membrane (Iwatsuki & Petersen, 1985). (4) Ensemble fluctuation analysis of the whole-cell currents indicates that the outward current can be entirely accounted for by a relatively small number (50 - 150) of high-conductance channels (Maruyama et al., 1986).

The action of CCK-8 has been tested in the whole-cell recording configuration. In experiments of this type, the intracellular ionic composition is determined by the solution used to fill the recording pipette. When the pipette and therefore the cell, is filled with a solution containing a high concentration (10 mM) of Ca^{2+} chelator (ethyleneglycol bis(aminoethyl ether)tetra-acetic acid, EGTA), and no Ca^{2+} is added, CCK has no effect on the whole-cell currents. In this situation $[Ca^{2+}]_i$ is clamped at very low levels (probably $<10^{-9}$M) and this result is, therefore in complete agreement with the hypothesis that the CCK-evoked conductance increase under physiologic conditions is mediated via an increase in $[Ca^{2+}]_i$. If, however, a low EGTA concentration is used in the pipette, CCK is able to increase the K^+ current. The effect of CCK is reversible on removal of the agonist, but can also be reversed by the removal of Ca^{2+} from the bathing solution in contact with the outside of the isolated cell. The effect of CCK added to the outside solution on the whole-cell currents is very similar to the action of internal Ca^{2+}, and taken together, these two types of experiments provide very strong evidence that the secretagogue-evoked increase in K^+ conductance is mediated by internal Ca^{2+} activating voltage-gated K^+-selective channels (Suzuki et al., 1985; Maruyama & Petersen, 1984a).

Sustained stimulant-evoked activation of the K^+ conductance pathway requires the continued presence of extracellular Ca^{2+}, and this result suggests that stimulation, in addition to causing intracellular Ca^{2+} release, also promotes Ca^{2+} uptake from the extracellular fluid (Maruyama & Petersen, 1984a).

Cl^- Channel Currents

In classical intracellular microelectrode studies on mouse pancreatic acinar cells it has been shown that ACh evokes depolarization by increasing the membrane conductance. This ACh controlled conductance pathway is mainly permeable to Cl^- (Petersen & Philpott, 1980). The ACh-evoked depolarization and conductance increase can be mimicked by intracellular Ca^{2+} injection (Iwatsuki & Petersen, 1977) and prevented by intracellular injection of the Ca^{2+} chelator EGTA (Laugier & Petersen, 1980). Patch-clamp whole-cell recording studies have confirmed that ACh evokes opening of Cl^- conductance pathways in mouse and rat pancreatic acinar cell membranes and that this effect is mediated by an increase in $[Ca^{2+}]_i$ (Wakui et al., 1989). In the lacrimal acinar cells a Ca^{2+}-dependent Cl^- current has also been demonstrated (Marty et al., 1984; Findlay & Petersen, 1985; Evans & Marty, 1986) and single channel recording show that it is carried by Cl^- movement through a large number of anion channels with a very low unit conductance (1-2 pS) (Marty et al., 1984). In the lacrimal acinar cells Ca^{2+}-dependent K^+ and Cl^- channels co-exist although there is a tendency for the Cl^- currents to run down in the whole-cell current configuration whereas in the mouse pancreatic acinar cells there are only Ca^{2+}-dependent Cl^- channels which are very stable even during prolonged internal cell perfusion.

Internal Cell Perfusion

Internal cell perfusion with exchange of solutions during individual patch-clamp experiments can be used to demonstrate the importance of inositol polyphosphates. The tight-seal whole-cell current recording configuration (Hamill et al., 1981) has been very useful, but it has the important limitation that only one intracellular soltion can be used in an individual experiment. By combining patch-clamp whole-cell current recording with internal perfusion of patch pipette (Fig. 4) it is possible to carry out experiments in which many different intracellular solutions can be tested in individual experiments (Jauch et al., 1986).

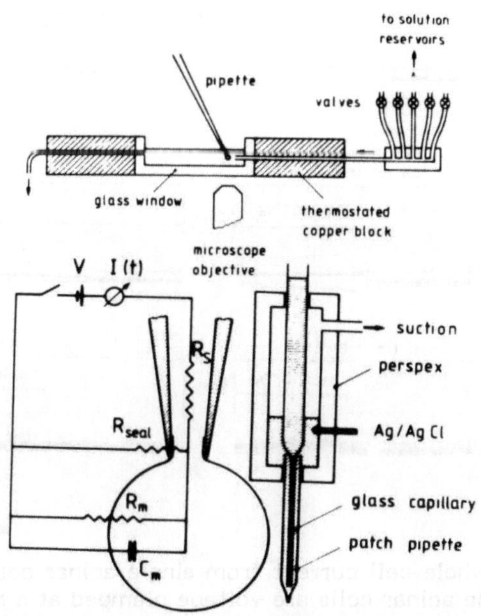

Fig. 4. The experimental arrangements for whole-cell current recording with internal perfusion of the patch pipette. (Upper part) Experimental set-up for whole-cell recordings. (Lower right part) Internal perfusion of the patch pipette (not drawn to scale). The glass capillary had a length of about 5 cm and an internal diameter of 20-40 µm. The distance between the tip of the capillary and the tip of the patch pipette was 100 - 200 µm. (Lower left part) Circuit parameters of a whole-cell recording experiment; R_s, series resistance of the pipette, R_{seal}, pipette-membrane seal resistance; R_m and C_m, resistance and capacitance of the cell membrane. (Adapted from Jauch et al., 1986).

In the lacrimal acinar cells used for such a study, ACh evokes sustained membrane hyperpolarization due to a marked and sustained increase in outward K^+ current, mediated by high-conductance Ca^{2+} and voltage-activated K^+ channels. The activation of K^+ current by ACh is only sustained in the presence of extracellular Ca^{2+}. Figure 5 shows recordings of whole-cell currents from single isolated mouse lacrimal acinar cells. At the holding potential of -40 mV (in some experiments -35 or -50 mV), which is close to the normal resting potential, ACh evokes an outward current response which is substantially enhanced at a membrane potential of 0 mV. At more negative potentials (-75 to -90 mV), an inward current response is observed. These currents have been well characterized: the ACh-evoked outward current is carried by K^+ through high-conductance (200 - 300 pS) voltage-activated and Ca^{2+}-activated K^+ channels

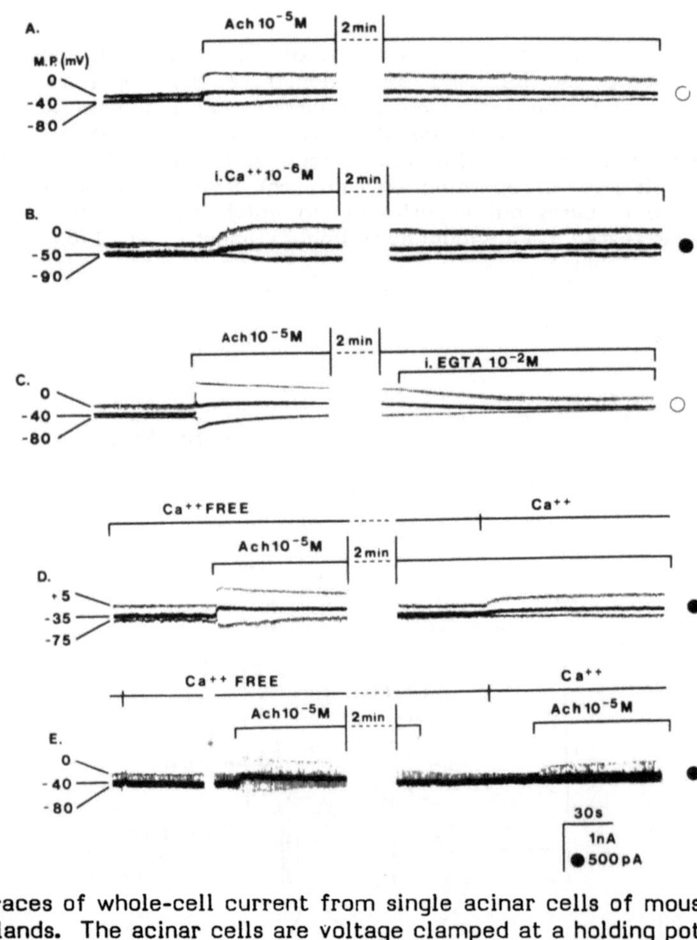

Fig. 5. Traces of whole-cell current from single acinar cells of mouse exorbital lacrimal glands. The acinar cells are voltage clamped at a holding potential very close (within 3 mV) to the zero current potential, namely, the resting membrane potential of the internally perfused cells. The internal solution contains 0.5 mM EGTA and no added calcium. From this holding potential a series of depolarizing and hyperpolarizing voltage jumps are superimposed (100 - 200 msec in duration) evoking outward K^+ (upward deflections) and inward Cl^- (downward deflections) currents, respectively. Membrane potential (MP) (millivolts) indicates the holding potential and the membrane potential during the superimposed voltage jumps. (A) Currents in the unstimulated cell and changes on exposing the cell to ACh, 10^{-5}M. ACh stimulation is associated with a sustained increase in the amplitude of the outward current and a largely transient increase in the inward currents. (B) At

whereas the inward current is carried by Cl^- through Ca^{2+}-dependent Cl^- channels with a low unit conductance (1 - 2 pS). Figure 5 shows that the effect of continued ACh stimulation is reversible and can be blocked either by introducing a high concentration (10 mM) of EGTA into the cell or by removing external Ca^{2+}. Readmission of extracellular Ca^{2+} during sustained ACh exposure brings back the full response. On the other hand, removal and readmission of external Ca^{2+} in the absence of ACh stimulation has no effect on the magnitude of the transmembrane currents. The effect of ACh can also be reversibly mimicked by introducing a solution with a high Ca^{2+} concentration (about 1 µM) into the cell. The results shown in Fig. 5 are very similar to those discussed in the previous section dealing with cholecystokinin-evoked Ca^{2+}-dependent K^+ currents in pig pancreatic acinar cells, but are more direct and convincing as a result of being able to change the intracellular ion composition during individual experiments.

It is now generally recognized that receptor-activated Ca^{2+} mobilization is mediated by inositol (1,4,5) trisphosphate (IP_3). One would therefore expect that internal cell perfusion with IP_3 should mimick the effect of external ACh application. This is the case (Fig. 6). As shown in Fig. 6, ACh can evoke a pulsatile Ca^{2+}-dependent Cl^- current in mouse pancreatic acinar cells and internal application of both IP_3 and the non-metabolizable analogue inositol (1,4,5) trisphosphorothioate (IPS_3) can also evoke pulsatile Cl^- currents. Since these effects are independent of external Ca^{2+} (Fig. 6) it is clear that IP_3 evokes repetitive release of Ca^{2+} from internal stores (Wakui et al., 1989).

It is possible to combine whole-cell current recording with single-cell fluorimetric measurements of cytoplasmic Ca^{2+} concentration. An example of this is shown in Fig. 7. Essentially we have here a method to simultaneously assess $[Ca^{2+}]_i$ in the cell at large (microfluorimetry using Fura-2) and $[Ca^{2+}]_i$ very close to the cell membrane (whole-cell Cl^- current). As seen in Fig. 7 a low dose of ACh evokes Ca^{2+} spikes near the membrane, but has little effect on $[Ca^{2+}]_i$ in general whereas a somewhat higher dose evokes synchronous change in $[Ca^{2+}]_i$ as seen with both methods (Osipchuk et al., 1990). Similar findings have been obtained with intracellular infusion of IP_3 and GTP-g-S.

The method of intracellular perfusion has also been valuable in giving novel information on the mechanism underlying the receptor-activated cytoplasmic Ca^{2+} oscillations. The general hypothesis is now that IP_3 evokes a small steady flow of Ca^{2+} from the IP_3-sensitive Ca^{2+} pool and that this Ca^{2+} then triggers pulses of Ca^{2+} release from a separate intracellular store. It is this Ca^{2+}-induced Ca^{2+} release that is quantitatively dominant and gives rise to the oscillations in

---

the point indicated, the internal perfusion fluid has been changed to a solution with a free calcium concentration of 10^{-6}M (i, intracellular). The rise in intracellular free calcium is associated with a sustained increase in the amplitude of the outward current. (C) Effects of extracellular application of ACh (10^{-5}M), as in (A). At the point indicated, the control intracellular solution is replaced by one containing 10 mM EGTA and no added calcium. Dialysis of the cell with this calcium chelator reduces the ACh-activated currents to prestimulus levels. (D) The acinar cell is bathed in a calcium-free extracellular solution and at the point indicated is exposed to 10^{-5}M ACh. An initial increase in both outward and inward currents is not sustained in the absence of extracellular calcium. Reintroduction of extracellular calcium in the continued presence of the agonist (ACh) again gives a sustained increase in outward current amplitude. (E) Removal of extracellular calcium has no effect on whole-cell currents in the unstimulated lacrimal acinar cell. Stimulation with 10^{-5}M ACh is again associated with a transient increase in current amplitudes, and ACh removal after 4 min gives only a small further decrease in whole-cell currents. In the absence of ACh, the reintroduction of extracellular calcium produces no change in current amplitudes, but the cell responds upon readmission of ACh. (Adapted from Morris et al., 1987).

the Ca^{2+}-dependent Cl^- current (Petersen & Wakui, 1990). One of the main pieces of evidence for this theory (Goldbeter et al., 1990) is shown in Fig. 8. This experiment is the most elaborate patch-clamp whole-cell recording configuration yet devised. As seen in Fig. 8 there are two intrapipette tubes: one for Ca^{2+} and one for EGTA application. In addition to whole-cell Cl^- current recording there is also microfluorimetric single-cell $[Ca^{2+}]_i$ measurement. A quasi-rectangular intracellular Ca^{2+} infusion causes a transient increase in $[Ca^{2+}]_i$ and Ca^{2+}-dependent Cl^- current as expected, but in addition there are shortlasting repetitive Ca^{2+}-activated Cl^- current spikes in the ascending and descending phases. This is Ca^{2+}-induced Ca^{2+} release and the fact that these pulses stop at the peak of $[Ca^{2+}]_i$ probably is an indication of Ca^{2+} inhibition of Ca^{2+}-induced Ca^{2+} release. The picture that emerges from these studies (Osipchuk et al., 1990; Wakui & Petersen, 1990; Petersen & Wakui, 1990) is that a constant intracellular Ca^{2+} infusion can mimick the action of intracellularly applied IP_3 or extracellularly applied ACh. The hypothesis that Ca^{2+} release is a two-stage process involving sequential opening of Ca^{2+} channels in two separate intracellular Ca^{2+} pools has recently been further tested by the intracellular perfusion method. In such studies it has been shown that intracellular application of the IP_3-antagonist heparin abolishes both ACh- and IP_3-evoked Ca^{2+} oscillations but not Ca^{2+}-induced Ca^{2+} release (Petersen & Wakui, 1990). This indicates that the Ca^{2+} channels in the two separate intracellular Ca^{2+} pools have different properties. Additional studies with the powerful internal cell perfusion method will probably be important for further tests of intracellular Ca^{2+} signalling mechanisms.

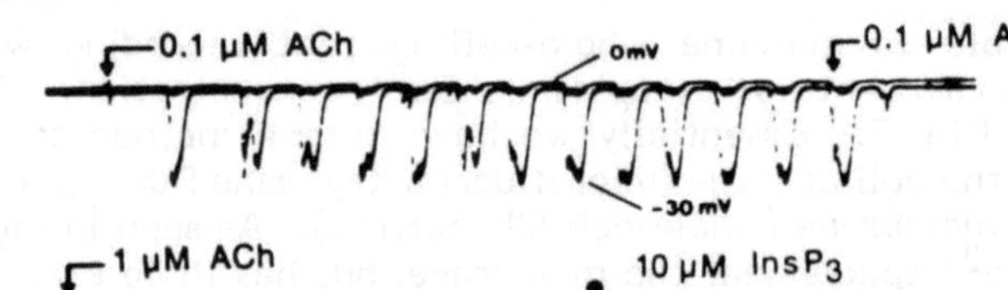

Fig. 6. Effects of acetylcholine (ACh), IP_3 ($InsP_3$) and inositol (1,4,5) trisphosphorothioate (Ins PS_3) on Ca^{2+}-dependent Cl^- current (at -30 mV membrane potential) in single internally perfused pancreatic acinar cells. ACh (0.1 μM) is seen to evoke repetitive pulses of Ca^{2+}-dependent Cl^- current and this effect is blocked by atropine (Atr) and when the cell is filled with a high concentration of the Ca^{2+}-chelator EGTA. Internal perfusion with Ins P_3 also evokes regular pulses of Ca^{2+}-dependent Cl^- current and this effect is also blocked by a high internal EGTA concentration. Inositol (1,3,4) trisphosphate (Ins (1,3,4) P_3) which is known not to cause intracellular Ca^{2+} release has no effect, but Ins PS_3 causes a dose-dependent response consisting of repetitive pulses of Ca^{2+}-activated Cl^- current. (Adapted from Wakui et al., 1989).

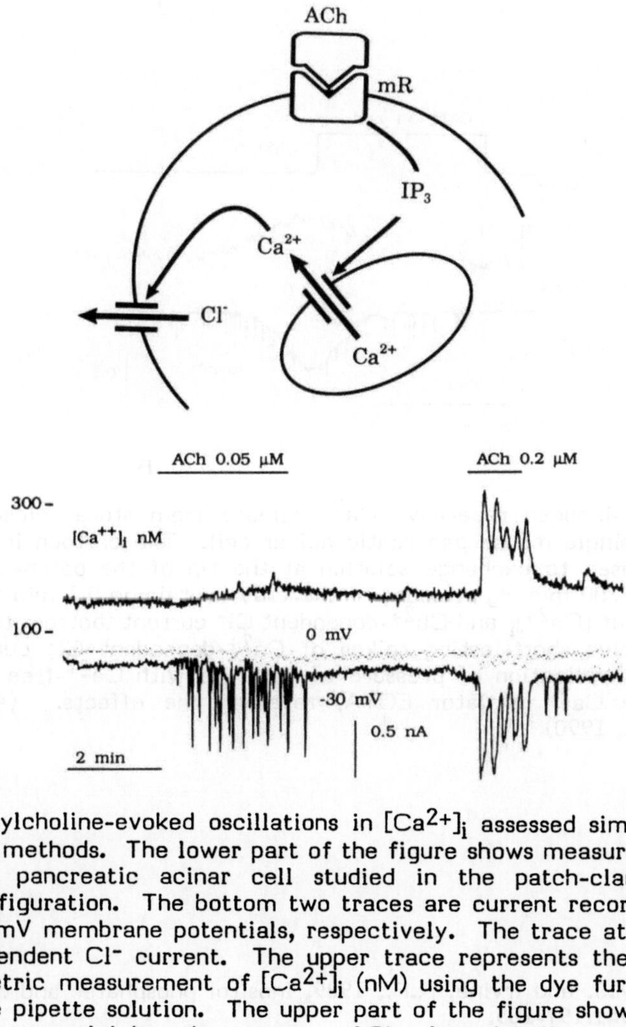

Fig. 7. Acetylcholine-evoked oscillations in $[Ca^{2+}]_i$ assessed simultaneously by two different methods. The lower part of the figure shows measurements from a single mouse pancreatic acinar cell studied in the patch-clamp whole-cell recording configuration. The bottom two traces are current recordings obtained at 0 and -30 mV membrane potentials, respectively. The trace at -30 mV shows the Ca^{2+}-dependent Cl^- current. The upper trace represents the result of the microfluorimetric measurement of $[Ca^{2+}]_i$ (nM) using the dye fura-2 which was present in the pipette solution. The upper part of the figure shows the simplest possible scheme explaining the events. ACh: Acetylcholine; mR; muscarinic receptor; IP_3 inositol (1,4,5) trisphosphate. (Adapted from Osipchuk et al., 1990).

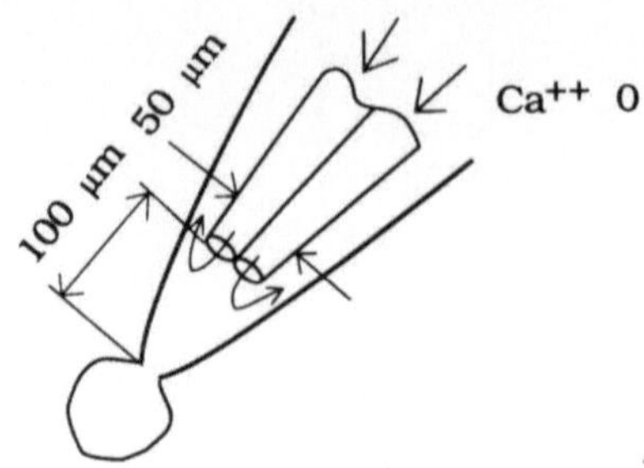

A

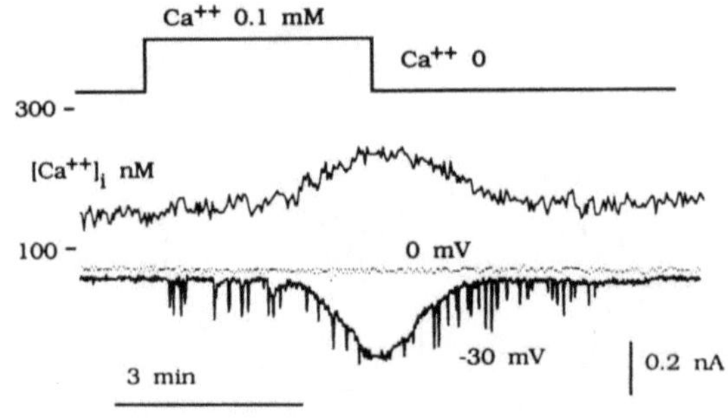

B

Fig. 8. Ca^{2+}-induced repetitive Ca^{2+} release from stores close to the cell membrane in single mouse pancreatic acinar cell. The cartoon in A shows the arrangement used to exchange solution at the tip of the patch-clamp pipette. Application of 200 mm Hg pressure to the tube containing 0.1 mM Ca^{2+} resulted in a slow rise of $[Ca^{2+}]_i$ and Ca^{2+}-dependent Cl^- current (bottom trace in B) and also in repetitive short-lasting spikes of Ca^{2+}-dependent Cl^- current (bottom trace in B). Application of pressure to the tube with Ca^{2+}-free solution (also containing the Ca^{2+} chelator EGTA) reversed the effects. (Adapted from Osipchuk et al., 1990).

REFERENCES

1. Berridge, M.J. and Irvine, R.F., 1989, Inositol phosphates and cell signalling, Nature, 341:197-205.
2. Evans, M.G. and Marty, A., 1986, Calcium-dependent chloride currents in isolated cells from rat lacrimal glands, J. Physiol, 378:437-460.
3. Findlay, I. and Petersen, O.H., 1985, Acetylcholine stimulates a Ca^{2+}-dependent Cl^- conductance in mouse lacrimal acinar cells, Pflugers Arch., 403:328-330.
4. Goldbeter, A., Dupont, G. and Berridge, M.J., 1990, Minimal model for signal-induced Ca^{2+} oscillations and for their frequency encoding through protein phosphorylation. Proc. Natl. Acad. Sci. U.S.A. 87:1461-1654.

5. Hamill, O.P., Marty, A., Neher, E., Sakmann, B. and Sigworth, F.J., 1981, Improved patch-clamp techniques for high resolution current recordings from cells and cell-free membrane patches, Pflugers Arch., 391:85-100.
6. Iwatsuki, N. and Petersen, O.H., 1977, Acetylcholine-like effects of intracellular calcium application in pancreatic acinar cells, Nature, 268:147-149.
7. Iwatsuki, N. and Petersen O.H., 1985, Action of tetraethylammonium on calcium-activated potassium channels in pig pancreatic acinar cells studied by patch-clamp single-channel and whole-cell current recording, J. Membr. Biol., 86:139-144.
8. Jauch, P., Petersen, O.H. and Lauger, P., 1986, Electrogenic properties of the sodium-alanine cotransporter in pancreatic acinar cells: I. Tight-seal whole-cell recordings, J. Membr. Biol., 94:99-115.
9. Laugier, R. and Petersen, O.H., 1980, Effects of intracelllar EGTA injection on stimulant-evoked membrane potential and resistance changes in pancreatic acinar cells, Pflugers Arch., 386:147-152.
10. Lundberg, A., 1958, Electrophysiology of salivary glands, Physiol. Rev., 38:21-40.
11. McCandless, M., Nishiyama, A., Petersen, O.H. and Philpott, H.G., 1981, Mouse pancreatic acinar cells: Voltage-clamp study of acetylcholine-evoked membrane current, J. Physiol., 318:57-71.
12. Marty, A., Tan, Y.P. and Trautmann, A., 1984, Three types of calcium-dependent channel in rat lacrimal glands. J. Physiol., 357:293-325.
13. Maruyama, Y., Nishiyama, A., Izumi, T., Hoshimiya, N. and Petersen, O.H., 1986, Ensemble noise and current relaxation analysis of K^+ current in single isolated salivary acinar cells from rat, Pflugers Arch., 406:69-72.
14. Maruyama, Y. and Petersen, O.H., 1982a, Single-channel currents in isolated patches of plasma membrane from basal surface of pancreatic acini, Nature, 299:159-161.
15. Maruyama, Y. and Petersen, O.H., 1982b, Cholecystokinin activation of single-channel currents is mediated by internal messenger in pancreatic acinar cells, Nature, 300-61-63.
16. Maruyama, Y. and Petersen, O.H., 1983, Voltage clamp study of stimulant-evoked currents in mouse pancreatic acinar cells, Pflugers Arch., 399:54-62.
17. Maruyama, Y. and Petersen, O.H., 1984a, Control of K^+ conductance by cholecystokinin and Ca^{2+} in single pancreatic acinar cells studied by the patch-clamp technique, J. Membr. Biol., 79:293-300.
18. Maruyama, Y. and Petersen, O.H., 1984b, Single calcium-dependent cation channels in mouse pancreatic acinar cells, J. Membr. Biol., 81:83-87.
19. Maruyama, Y., Petersen, O.H., Flanagan, P. and Pearson, G.T., 1983, Quantification of Ca^{2+}-activated K^+ channels under hormonal control in pig pancreatic acinar cells, Nature, 305:228-232.
20. Morris, A.P., Gallacher, D.V., Irvine, R.F. and Petersen, O.H., 1987, Synergism of inositol trisphosphate and tetrakisphosphate in activating Ca^{2+}-dependent K^+ channels, Nature, 330:653-655
21. Neher, E. and Sakmann, B., 1976, Single-channel currents recorded from membrane of denervated frog muscle fibres, Nature, 260:799-802.
22. Osipchuk, Y.V., Wakui, M., Yule, D.I., Gallacher, D.V. and Petersen, O.H., 1990, Cytoplasmic Ca^{2+} oscillations evoked by receptor stimulation, G-protein activation, internal application of inositol trisphosphate or Ca^{2+}: simultaneous microfluorimetry and Ca^{2+}-dependent Cl^- current recording in single pancreatic acinar cells, EMBO J., 9:697-704
23. Petersen, O.H., 1980, The Electrophysiology of Gland Cells, Academic Press: New York.
24. Petersen, O.H. and Findlay, I., 1987, Electrophysiology of the pancreas, Physiol. Rev., 67:1054-1116.
25. Petersen, O.H., and Maruyama, Y., 1984, Calcium-activated potassium channels and their role in secretion, Nature, 307:693-696.

26. Petersen, O.H. and Philpott, H.G., 1980, Pancreatic acinar cells: the anion selectivity of the acetylcholine-opened chloride pathway, J. Physiol., 306:481-492.
27. Petersen, O.H. and Wakui, M., 1990, Oscillating intracellular Ca^{2+} signals evoked by activation of receptors linked to inositol lipid hydrolysis: mechanism of generation, J. Membr. Biol., in the press.
28. Philpott, H.G. and Petersen, O.H., 1979, Separate activation sites for cholecystokinin and bombesin on pancreatic acini. An electrophysiological study employing a competitive antagonist for the action of CCK, Pflugers Arch., 382:263-267.
29. Suzuki, K., Petersen, C.C.H. and Petersen, O.H., 1985, Hormonal activation of single K^+ channels via internal messenger in isolated pancreatic acinar cells, FEBS Letts., 192:307-312.
30. Wakui, M. and Petersen, O.H., 1990, Cytoplasmic Ca^{2+} oscillations evoked by acetylcholine or intracellular infusion of inositol trisphosphate or Ca^{2+} can be inhibited by internal Ca^{2+}, FEBS Letts., 263:206-208.
31. Wakui, M., Potter, B.V.L. and Petersen, O.H., 1989, Pulsatile intracellular calcium release does not depend on fluctuations in inositol trisphosphate concentration, Nature, 339:317-320.

23

APPLICATION OF RAPID DUAL TRACER DILUTION TECHNIQUES FOR THE STUDY OF ENDOTHELIAL CELL AMINO ACID TRANSPORT IN PERFUSED MICROCARRIER CULTURES

G.E. Mann, C-J. Sheriff, V.J. Toothill* and J.D. Pearson*

Biomedical Sciences Division, King's College London, Campden Hill Road, London W8 7AH, UK and *Section of Vascular Biology, MRC Clinical Research Centre, Harrow, UK

INTRODUCTION

Endothelial cells play an important role in the regulation of vascular tone, blood coagulation, platelet aggregation and leukocyte traffic by means of cell surface reactions and secreted mediators such as, prostacyclin (PGI_2) and endothelium-derived relaxing factor (EDRF) (Moncada et al., 1976; Furchgott & Zawadzki, 1980; Pearson & Gordon, 1984; Gordon et al., 1986; Gordon & Pearson, 1987; Van de Voorde et al., 1987; Luscher et al., 1990; see reviews Furchgott, 1990 and Moncada & Higgs, 1990). EDRF induces vascular relaxation and inhibits platelet aggregation and adhesion by activating soluble guanylate cyclase in vascular smooth muscle (see Griffiths et al., 1985). Recent studies with perfused aortic endothelial cell microcarrier cultures have now identified EDRF as nitric oxide (Palmer et al., 1987), and suggested that a novel NADPH-dependent enzyme may be involved in the generation of nitric oxide from the terminal nitrogen atom(s) of L-arginine (Palmer & Moncada, 1989).

With exception of the pulmonary and brain circulation, little is known about the cellular processes regulating nutrient uptake and metabolism in endothelial cells. Our current knowledge is based upon relatively recent studies *in vitro* with isolated microvessels or cultured microvascular cells, derived primarily from brain (see review Betz & Goldstein, 1986; Cancilla & Debault, 1983; Vinters et al., 1985; Cardelli-Cangiano et al., 1987) and to a lesser extent from cultured arterial or venous endothelial cells (see review Pearson & Gordon, 1984; Corkey et al., 1981). The pulmonary endothelium has been identified as one of the primary sites for removal and inactivation of circulating vasoactive substances and adenosine (see reviews Gillis, 1986 and Shepro & Dunham, 1986; Strum & Junod, 1972; Syrota et al., 1982). Similarly, the brain and retinal microvascular endothelium appears to mediate the uptake of hexoses, amino acids, monocarboxylates and purines (see reviews Betz & Goldstein, 1986 and Pardridge, 1983; Maruki et al., 1984; Allen & Gerritsen, 1986; Cancilla & Debault, 1983; Cardelli-Cangiano et al., 1987).

Cell Membrane Transport, Edited by D. L. Yudilevich *et al.*
Plenum Press, New York, 1991

As micro- and macro- vascular endothelium exhibit significant differences in growth control, secretory capacities and hormonal sensitivity (Zetter, 1980; Gerritsen & Cheli, 1983; Gerritsen et al., 1988; King et al., 1983), it was of interest to establish whether this phenotypic variation was reflected by differential rates of nutrient transport. In the present study we have examined uptake and efflux of amino acids, D-glucose, dopamine, serotonin and CGP-12177 (a ß receptor ligand) in perfused microcarrier cultures of vascular endothelial cells derived from human umbilical veins and bovine or porcine

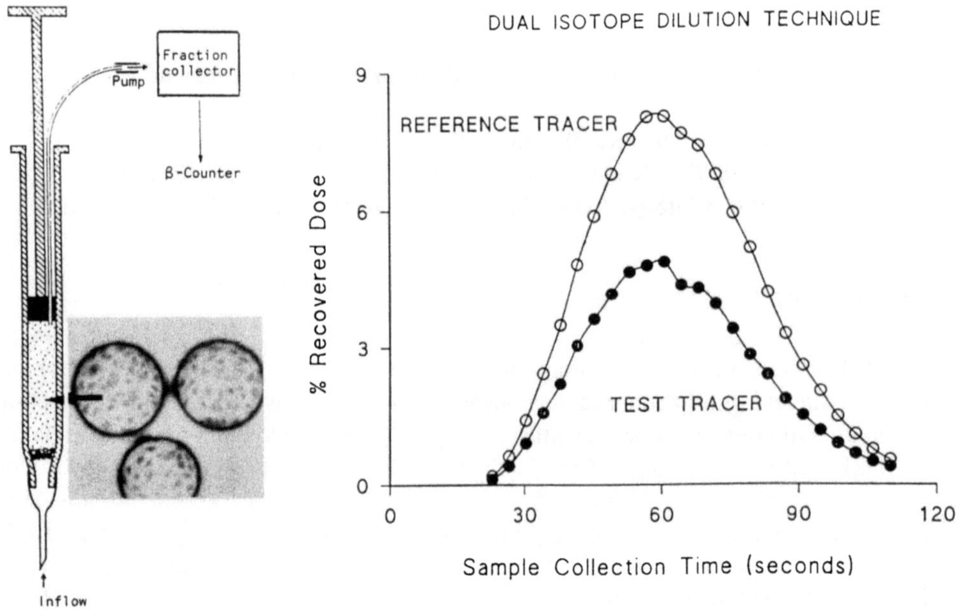

Fig. 1. Perfusion of cultured vascular endothelial cells on microcarrier beads. Perfusate was drawn through the column from below at a constant rate (0.3 - 1.2 ml/min) to avoid packing of the beads. The inset shows confluent umbilical vein endothelial cells on Biosilon microcarrier beads. Taken from Mann et al. (1989b).

aortae. Unidirectional and steady-state transport rates were measured using dual isotope tracer dilution techniques (Yudilevich & Mann, 1982; Mann et al., 1989a). Accounts of the present work have been published (Mann et al., 1988, 1989b, 1990; Sheriff et al., 1990).

METHODS

Isolation and culture of vascular endothelial cells

Endothelial cells were isolated from human umbilical cord veins and bovine/porcine aortae using collagenase (0.5 mg/ml Type II, Sigma, U.K.) (Jaffe et al., 1973; Needham et al., 1987; Mann et al., 1989b; Mann et al., 1990). Umbilical vein endothelial cells were grown in tissue culture Medium 199 (M199, Flow Laboratories, U.K.) and aortic endothelial cells in Dulbecco's Modified Eagles Medium (DMEM, Flow Laboratories, U.K.), supplemented with 10% fetal calf serum (FCS, Flow Laboratories), 10% new born calf serum (NBCS, Flow Laboratories) and 4 mM L-glutamine. Cultures of aortic endothelial cells also contained gentomycin (25 μg/ml) in addition to penicillin and streptomycin. Cells were incubated at 37°C in a 5% CO_2 atmosphere until confluent monolayers were obtained within 5 to 7 days. During this time cells were fed every 2-3 days by removing half the medium and replacing it with fresh medium.

Endothelial cells were grown to confluence on microcarrier beads (see Gordon et al., 1986; Needham et al., 1987). Human umbilical vein endothelial cell monolayers from two T25 flasks were resuspended by brief treatment with 0.1% trypsin and 0.02% EDTA and added to a siliconized 125 ml culture flask (Techne, Duxford, U.K.) containing 40 ml Medium 199, 10% FCS, 10% NBCS and 2 ml gelatin-coated (1%) plastic microcarrier beads (approx. 200 μm diameter, Biosilon, Nunc, Roskilde, Denmark). A similar protocol was used for aortic endothelial cell monolayers, but cells from only one T75 flask were added to a stirrer flask containing 40 ml of supplemented DMEM and 2 ml of non-gelatin coated Biosilon beads. Endothelial cells were then incubated at 37°C in 5% CO_2 atmosphere. A Techne MCS-104S magnetic stirrer base was used to stir venous endothelial cells intermittently at 40 rpm (2 min every 20 min) and aortic endothelial cells continuously at 40 rpm. Beads became confluent within 5-7 days with approximately 30-50 cells/bead.

Perfusion of cultured endothelial cells on microcarrier beads

Microcarrier beads with confluent endothelial cells were carefully packed into the barrel of a 1 ml disposable polypropyline syringe at volumes of 0.25 ml or 0.5 ml for respectively aortic or umbilical vein endothelial cells. As shown in Fig. 1, venous endothelial cell columns were perfused from below (0.3 ml/min) at 37°C with an amino acid-free, HEPES buffered (20 mM) Ringer solution containing 11 mM D-glucose (see Needham et al., 1987; Mann et al., 1989b). In recent experiments with either bovine or aortic endothelial cells the perfusate glucose concentration was reduced to 5 mM.

Measurement of endothelial cell transport rates

Unidirectional substrate transport was measured using a rapid dual isotope dilution technique (Yudilevich & Mann, 1982) employed previously for amino acid transport studies in the perfused pancreas (Mann & Peran, 1986; Mann et al., 1986; Norman & Mann, 1987; Norman et al., 1989a), cultured Vero cells (Peran & McGee, 1986) and more recently cultured human umbilical vein (Mann et al., 1989b) and aortic (Mann et al., 1990) endothelial cells . After a 45-60 min equilibration period with an amino acid-free perfusate, endothelial cell columns were challenged (150 μl in 30 s) with a tritiated (7 μCi) amino acid, D-glucose, dopamine, serotonin or CGP-12177 and D-[^{14}C]mannitol (1.4 μCi). The column effluent was collected sequentially for 3 min and representative tracer dilution profiles are shown in Fig. 2A.

Fractional [^{3}H] uptake was calculated for each of the successive effluent samples shown in Fig. 2A, i.e. uptake = 1 - L-[^{3}H]leucine/D-[^{14}C]mannitol (see Fig. 2B). Unidirectional influx (v) was estimated from the maximal fractional tracer uptake (U_{max}), the perfusion rate (F, ml/min) and the perfusate concentration of unlabelled substrate (C_a, mM): $v = \{- F \times \ln (1 - U_{max})\} \times C_a$ [see Yudilevich & Mann, 1982; Mann et al., 1989b]. After equilibrating cells for 2 min with a given amino acid concentration (0.025 - 1 mM), influx was then measured in the continued presence of substrate. Michaelis-Menten kinetic constants were quantified using a single entry site analysis, which revealed the lowest weighted standard deviation of residuals.

Trans-stimulation of amino acid efflux was studied using a classical dual isotope cell loading and wash-out technique [Boyd & Parsons, 1979], recently applied to the rat exocrine pancreas [Mann et al., 1989a] and perfused microcarrier cultures of venous endothelial cells [Mann et al., 1988]. In these experiments net uptake of a given labelled amino acid was assessed relative to D-[^{14}C]mannitol during a 10-15 min cell loading period. Tracer wash-out was subsequently monitored during perfusion with an isotope-free solution. During the wash-out phase endothelial cells were challenged with unlabelled amino acids and cellular efflux of tritiated L-leucine or L-arginine was monitored.

RESULTS

Application of a rapid dual isotope dilution technique

Fig. 2A shows characteristic effluent isotope dilution profiles obtained with confluent venous endothelial cell monolayers on microcarrier bead columns. Following an isotope challenge into the column inflow (see Fig. 1), the effluent recovery of L-[^{3}H]leucine was lower than that of D-[^{14}C]mannitol, indicating endothelial cell uptake of leucine. Rechallenging the same endothelial cell column in the presence of 1 mM phenylalanine markedly reduced the rapid (15 s) uptake of L-[^{3}H]leucine, suggesting **cis**-inhibition of transport (Fig. 2B).

Validation of D-mannitol as an extracellular tracer

In venous endothelial cell cultures uptakes of L-[^{3}H]glucose, [^{3}H]sucrose and D-[^{14}C]mannitol were compared with the aim of identifying a valid extracellular marker. Negligible losses of L-[^{3}H]glucose (5.6 ± 0.8 %, mean ± S.E., n = 3) and [^{3}H]sucrose (5.2 ± 1.5 %, n = 3) were measured relative to D-[^{14}C]mannitol. When gelatin-coated beads alone were challenged with D-[^{14}C]mannitol and L-[^{3}H]leucine, effluent recoveries of both tracers approximated 96%. On the basis of these findings D-mannitol was selected as an appropriate reference tracer.

Endothelial cell nutrient uptake

Human umbilical vein endothelial cells rapidly transported several neutral and cationic amino acids, although uptake for proline, 2-MeAIB (System A analogue), D-glucose, serotonin and dopamine were negligible (Table 1). Serotonin and dopamine uptakes were measured from perfusates containing 0.3 mM ascorbate. Lack of uptake for 2-MeAIB and proline, suggests that system A may not be expressed in venous endothelial cells. Further studies are in progress to determine whether System A activity in umbilical vein endothelial cells is lost during collagenase treatment.

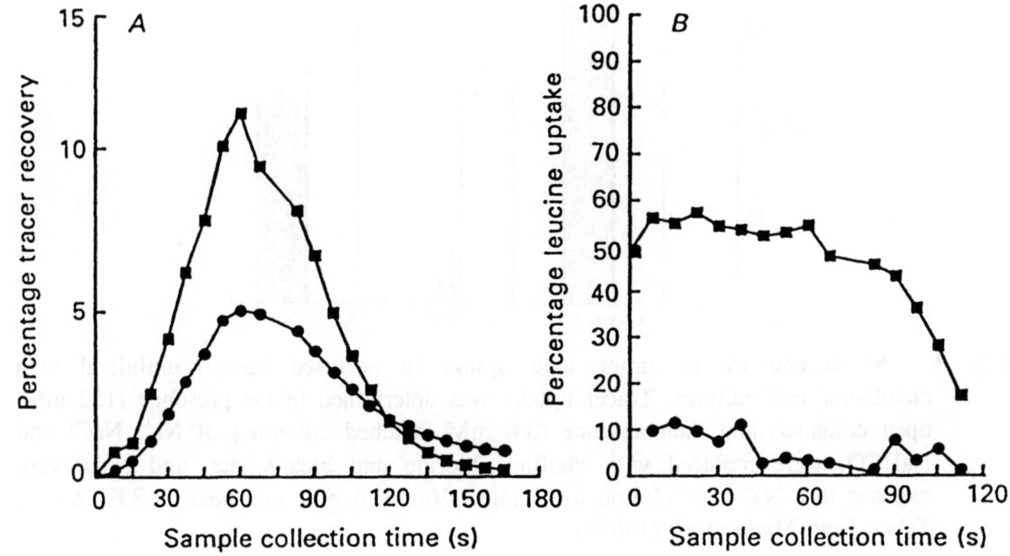

Fig. 2. Dual tracer dilution experiment in a perfused microcarrier culture of umbilical vein endothelial cells. A, paired effluent tracer recoveries for D-[^{14}C]mannitol (■—■) and L-[^{3}H]leucine (●—●) after a rapid isotope challenge (150 µl in 20-30 s). B, time course of L-[^{3}H]leucine uptake in the absence (■—■) and presence of 1 mM extracellular L-phenylalanine (●—●, dilution data not shown) in the perfusate. Taken from Mann et al. (1989b).

Unlike in venous endothelial cells, aortic endothelium exhibits a small but measurable uptake for the non-metabolized System A analogue 2-MeAIB (Table 1). Our initial experiments with bovine and porcine aortic endothelial cells were performed using smaller microcarrier columns. In subsequent studies amino acid transport was found to be similar in perfused endothelial cell cultures derived from umbilical vein and either bovine or porcine aortae, although tracer uptake was dependent upon the volume of packed cells (Mann et al., 1990). Uptake of D-glucose was low in both endothelial cell types, suggesting that venous and aortic endothelium derived from large vessels may have similar glucose transport rates.

Na^+-dependence of amino acid uptake

In venous endothelial cells transport of L-leucine and L-arginine was largely independent of sodium (Mann et al., 1989b), consistent with a major fraction of uptake being mediated by systems L and y+ (see Christensen & Kilberg, 1987). By contrast,

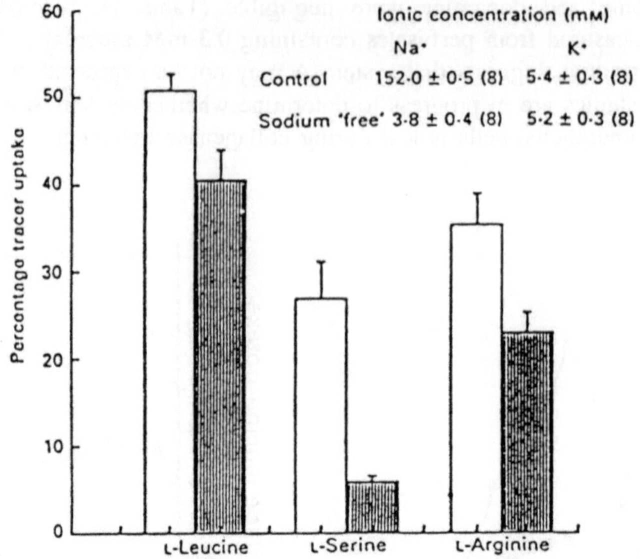

Fig. 3. Na^+-dependence of amino acid uptake in perfused human umbilical vein endothelial cell cultures. Tracer uptake was determined in the presence (152 mM, open columns) and then absence (3.8 mM, hatched columns) of Na^+. NaCl and $NaHCO_3$ were replaced with choline chloride and bicarbonate, and cells were exposed to a Na^+-free solution for 3 min. Values denote the mean ± S.E., n = 3. Taken from Mann et al. (1989b).

uptake of L-serine was inhibited by 77 ± 5 % during perfusion with low sodium. Detection of Na^+-independent and Na^+-dependent pathways for L-arginine uptake merits further study, since L-arginine may be the primary physiological precursor for nitric oxide released from endothelium in response to vasoactive stimuli (Palmer et al., 1987, 1988).

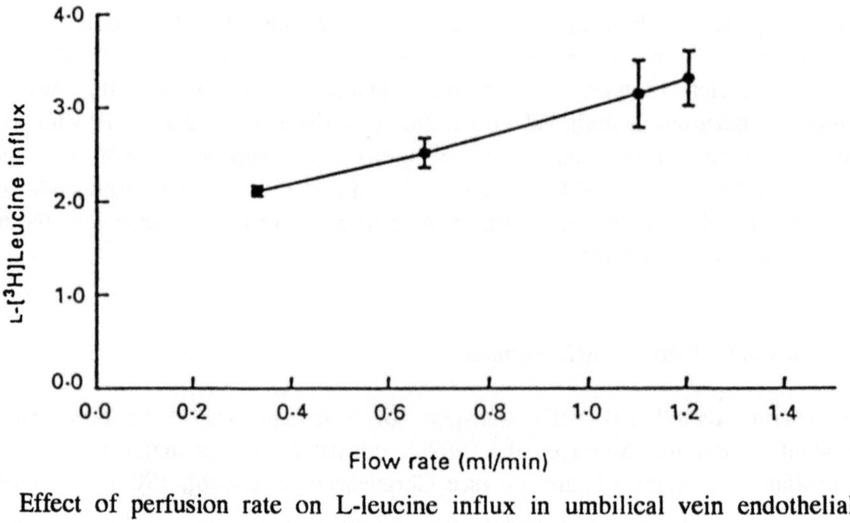

Fig. 4. Effect of perfusion rate on L-leucine influx in umbilical vein endothelial cell cultures. Values denote mean ± S.E., n = 4. Taken from Mann et al. (1989b).

Effects of perfusion flow on amino acid uptake

In these experiments umbilical vein endothelial cells were perfused with 0.025 mM L-leucine and unidirectional transport was assessed at perfusion rates ranging from 0.3 - 1.2 ml/min. The data in Fig. 4 show that L-leucine influx increased by approx. 50% as the perfusion rate through the column was increased 4-fold. It was therefore important to standardize flow rates in these perfused endothelial cell experiments, since changes in the unstirred layer may affect kinetics of transport.

Kinetics of amino acid transport in endothelial cells

In kinetic studies unlabelled L-leucine was added to the control perfusate and isotope challenges at different concentrations (0.025 - 1 mM), and transport was assessed during

Table 1. Vascular endothelial cell uptake of extracellular metabolites and amines. Unidirectional and net uptake values were measured at tracer concentrations in perfused microcarrier cultures of human umbilical vein (HUVEC) and bovine aortic (BAEC) endothelial cells. Differential uptake values may reflect the variable column size used for venous (0.5 ml) and aortic (0.25 ml) endothelial cells. Values denote the mean ±S.E., n = 3 - 13 experiments.

Substrate	Unidirectional uptake (%)		Net uptake (%)	
	HUVEC	BAEC	HUVEC	BAEC
L-[^{3}H]Arginine	42 ± 3	17 ± 3	29 ± 2	13 ± 0.4
L-[^{3}H]Lysine	-	15 ± 1	-	11 ± 2
L-[^{3}H]Ornithine	26 ± 3	15 ± 1	6 ± 0.3	11 ± 2
L-[^{3}H]Phenylalanine	73 ± 2	23 ± 5	52 ± 2	13 ± 1
L-[^{3}H]Leucine	53 ± 2	34 ± 8	37 ± 1	23 ± 6
2-[1^{4}C]MeAIB	-5 ± 3	9 ± 2	-4 ± 3	-3 ± 2
L-[^{3}H]Serine	40 ± 4	27 ± 1	36 ± 3	12 ± 2
L-[^{3}H]Proline	10 ± 1	-	10 ± 1	-
D-[^{3}H]Glucose	8 ± 2	12 ± 1	4 ± 2	6 ± 1
5-[^{3}H]Serotonin	6 ± 2	-	3 ± 1	-
[7,8-^{3}H]Dopamine	7 ± 2	-	-2 ± 4	-
[^{3}H]CGP 12177 (β-receptor ligand)	10 ± 2	-	6 ± 1	-

perfusion with a given L-leucine concentration. At physiological plasma concentrations L-leucine transport was saturable (Fig. 5): K_t = 0.24 ± 0.08, V_{max} = 27.8 ± 4.6 nmol/min per column (approx. 3 x 10^6 cells). In aortic endothelial cells transport of L-arginine (25 - 400 μM) was also saturable (K_t = 206 μM, V_{max} = 8.26 nmol/min per 3 x 10^6 cells), although influx tended to increase at higher substrate levels. Further kinetic studies are necessary to confirm whether uptake of L-arginine is mediated by an additional diffusional pathway described recently in static bovine aortic endothelial cell monolayers (Sneddon et al., 1990).

Selectivity of endothelial cell amino acid transporters

The inhibitory effects of various 'competitor' amino acids were examined initially in perfused venous endothelial cell cultures by adding a given unlabelled amino acid (1 mM) to successive 150 μl isotope challenges containing only tracer doses of L-

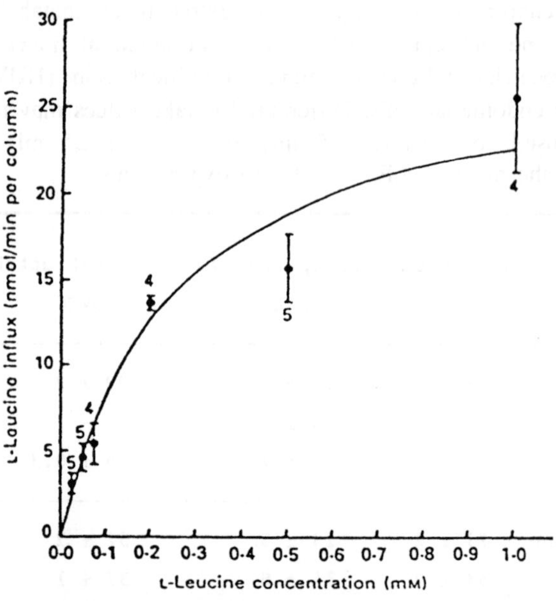

Fig. 5. Kinetics of L-leucine transport in perfused umbilical vein endothelial cell cultures. Influx data were best described by a single weighted rectangular hyperbola. Values denote mean ± S.E. of influx measurements in 4-5 cell cultures. Taken from Mann et al. (1989b).

[^{3}H]leucine and D-[^{14}C]mannitol. L-leucine uptake was significantly inhibited by other large neutral amino acids, including L-phenylalanine, L-methionine, BCH (a System L analogue) and L-DOPA (Fig. 6). L-serine also inhibited L-[^{3}H]leucine uptake, whereas uptake was unaffected by the model System A analogue 2-MeAIB and marginally reduced by L-cysteine, hydroxy-L-proline, glycine, L-proline, L-aspartate, ß-alanine and the cationic substrates L-lysine and L-arginine. These results indicate that L-leucine uptake by venous endothelial cells is mediated by a classical L-type system similar to that identified in brain microvascular endothelium (Cancilla & Debault, 1983; Smith et al., 1987; Takasoto et al., 1985; Wade & Katzman, 1975).

As L-arginine may be one of the precursors for EDRF (nitric oxide) biosynthesis in endothelial cells (Palmer et al., 1987, 1988), we examined the specificity of L-arginine uptake in perfused aortic endothelial cell cultures (Mann et al., 1990). As shown in Fig. 7, uptake of L-[^{3}H]arginine was markedly inhibited by other cationic amino acids, including L-ornithine, L-citrulline and L-lysine, and the branched-chain amino acid L-leucine. Uptake was negligibly affected by the non-metabolized System A analogue 2-MeAIB. N^G-monomethyl-L-arginine (L-NMMA),

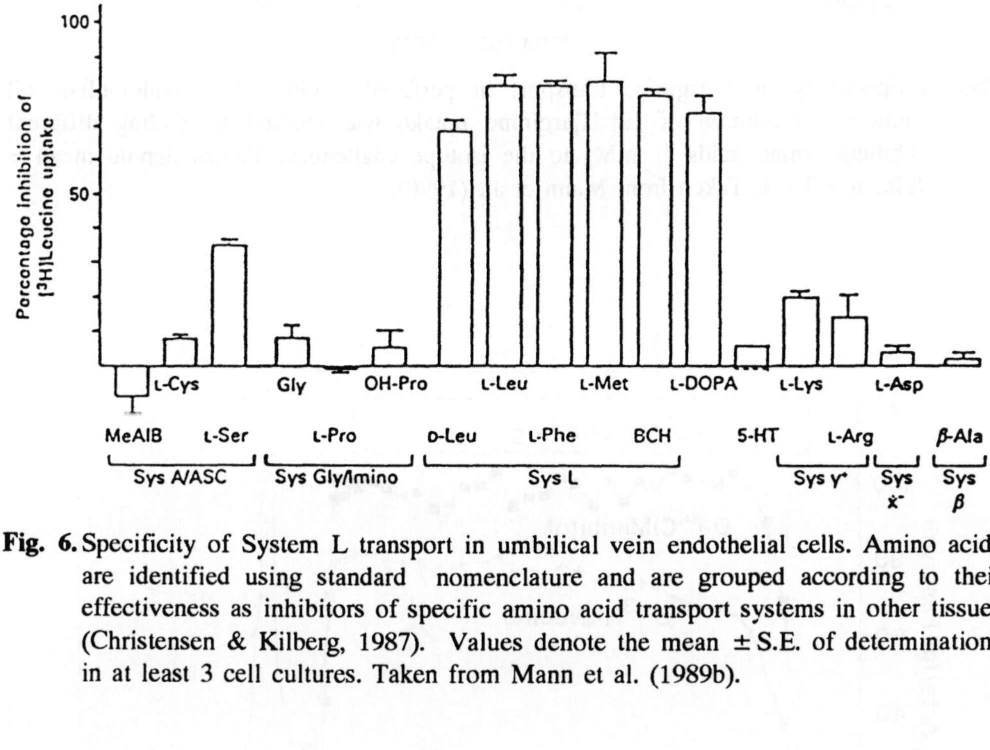

Fig. 6. Specificity of System L transport in umbilical vein endothelial cells. Amino acids are identified using standard nomenclature and are grouped according to their effectiveness as inhibitors of specific amino acid transport systems in other tissues (Christensen & Kilberg, 1987). Values denote the mean ± S.E. of determinations in at least 3 cell cultures. Taken from Mann et al. (1989b).

a specific inhibitor of nitric oxide production by endothelial cells (Palmer et al., 1988), reduced L-[^{3}H]arginine uptake, although a fraction of L-arginine uptake appeared to be insensitive to inhibition by 1 mM L-NMMA.

Selectivity of amino acid efflux mechanisms

In these experiments we investigated whether cellular efflux of labelled amino acids was modulated by substrates acting on either the trans- or cis-face of the endothelial cell membrane. A steady-state loading and wash-out technique recently applied to perfused endothelial cells (Mann et al., 1988) and rat pancreas (Mann et al., 1989a), was used to

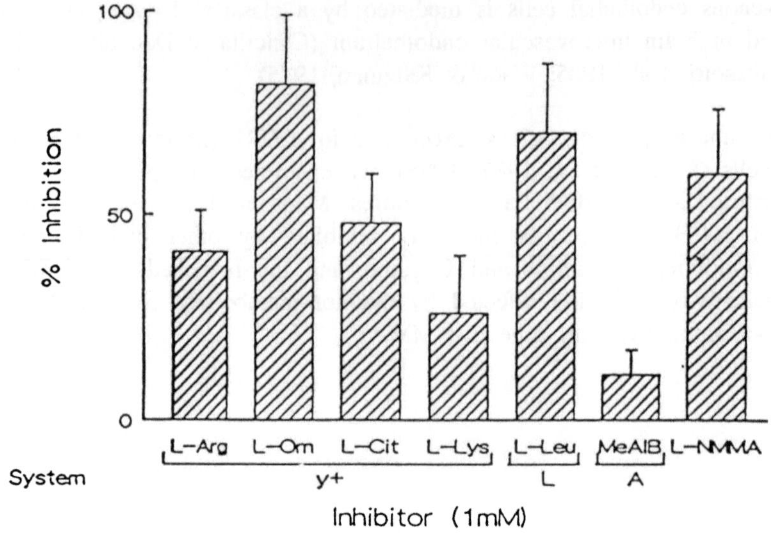

Fig. 7. Specificity of L-arginine transport in perfused bovine aortic endothelial cell cultures. Inhibition of L-[^{3}H]arginine uptake was studied by adding different inhibitor amino acids (1 mM) to the isotope challenges. Values denote mean ± S.E., n = 3 - 4. Taken from Mann et al. (1990).

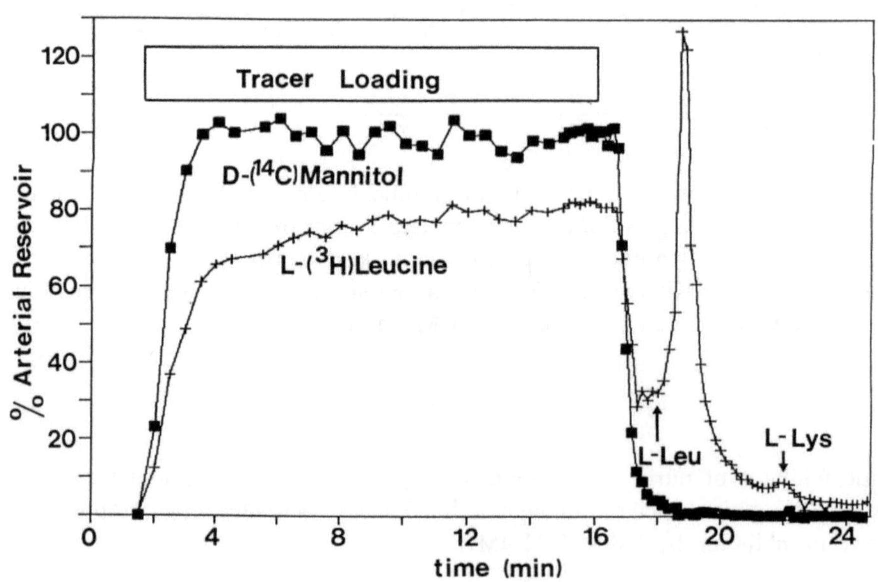

Fig. 8. Trans-stimulation of L-leucine efflux from perfused umbilical vein endothelial cell cultures. Amino acid evoked tracer efflux was studied in cells preloaded for approximately 16 min with L-[^{3}H]leucine and D-[^{14}C]mannitol. Wash-out tracer profiles were then monitored during perfusion with an amino acid-free and isotope-free solution. Cells were challenged successively with 100 mM extracellular L-leucine and L-lysine. Taken from Mann et al. (1988).

monitor unidirectional and net transport, as well as, tracer efflux. Human umbilical vein endothelial cell columns were preloaded (10-15 min) with labelled D-mannitol and L-leucine to assess steady-state washin and wash-out of L-leucine (Fig. 8). Net uptake of L-[^{3}H]leucine (28 ± 4%, n = 5) was significantly lower than the unidirectional uptake measured during a single transit (Table 1), indicating a time-dependent efflux of [^{3}H]-labelled substrate. Tracer wash-out was monitored continuously after switching the perfusate to an isotope-free solution. When cells were challenged during the wash-out phase with unlabelled L-leucine (50 mM) or L-lysine (50 mM), efflux of L-[^{3}H]leucine (or a [^{3}H] metabolite) was **trans**-stimulated (Fig. 8). The wash-out profile for D-[^{14}C]mannitol remained unchanged during challenges with extracellular amino acids or D-mannitol. Although L-lysine appeared to be less effective in this experiment, it was a potent stimulator of tracer leucine efflux in other studies (data not shown). Assuming a large fraction of accumulated L-[^{3}H]leucine exchanged with L-leucine, subsequent exchange with other extracellular substrates may appear less significant.

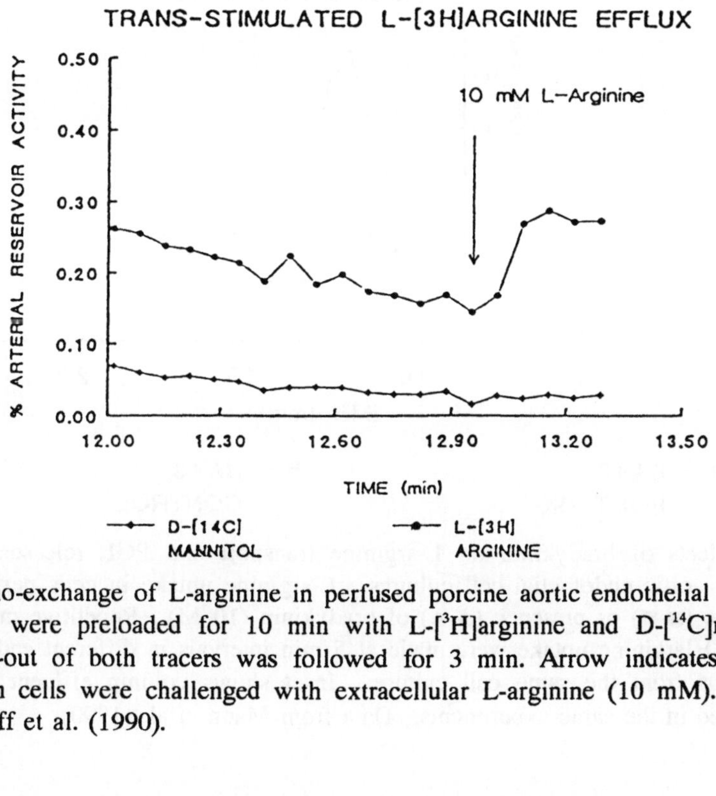

Fig. 9. Homo-exchange of L-arginine in perfused porcine aortic endothelial cell cultures. Cells were preloaded for 10 min with L-[^{3}H]arginine and D-[^{14}C]mannitol and wash-out of both tracers was followed for 3 min. Arrow indicates the time at which cells were challenged with extracellular L-arginine (10 mM). Taken from Sheriff et al. (1990).

Similar experiments were then performed in perfused porcine aortic endothelial cell cultures. Fig. 9 shows a representative wash-out phase for L-[^{3}H]arginine and D-[^{14}C]mannitol following a 10 min cell loading period with both tracers (data not shown). When cells were challenged with 10 mM L-arginine, efflux of L-[^{3}H]arginine (or a [^{3}H] metabolite) was transiently accelerated. Such homo-exchange is a characteristic feature of the cationic transport System y$^+$ described in hepatocytes (see review, Christensen & Kilberg, 1987; see Yudilevich & Boyd, 1987).

Effects of bradykinin on PGI_2 release and L-arginine uptake

Receptor-mediated stimulation of aortic endothelial cells with bradykinin elicits a transient release of two potent vasodilators, EDRF and PGI_2 (Moncada et al., 1976; Furchgott & Zawadzki, 1980; see reviews Busse et al., 1985). Although bradykinin and ATP are known to cause breakdown of phosphatidylinositol and increase intracellular free calcium in these cells, extracellular calcium appears to play an essential role in EDRF but not PGI_2 release (Luckhoff et al., 1988).

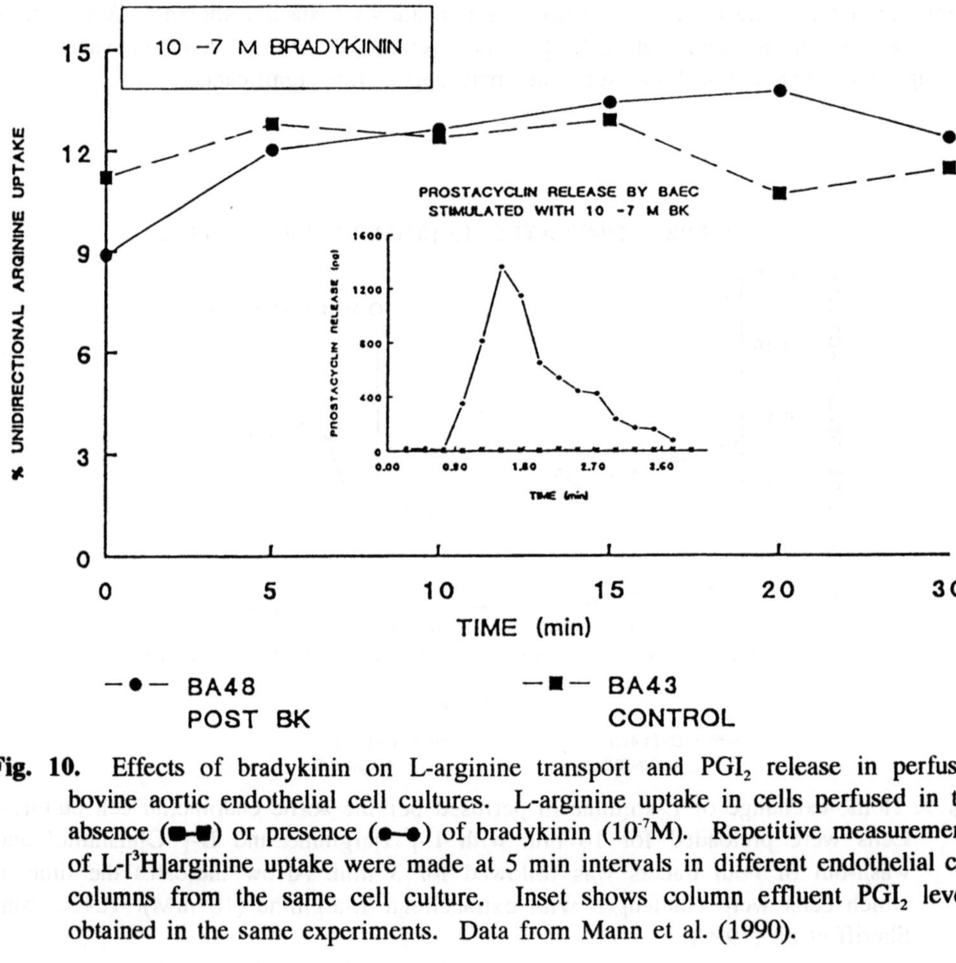

Fig. 10. Effects of bradykinin on L-arginine transport and PGI_2 release in perfused bovine aortic endothelial cell cultures. L-arginine uptake in cells perfused in the absence (■-■) or presence (●-●) of bradykinin (10^{-7}M). Repetitive measurements of L-[^{3}H]arginine uptake were made at 5 min intervals in different endothelial cell columns from the same cell culture. Inset shows column effluent PGI_2 levels obtained in the same experiments. Data from Mann et al. (1990).

Recent studies in bovine aortic endothelial cells revealed that L-arginine uptake was unaffected by 10^{-7}M bradykinin (Fig. 10). When repetitive measurements of L-[^{3}H]arginine uptake were made over a 30 min perfusion interval, uptake appeared to remain relatively constant. The biological activity of bradykinin was assessed by monitoring PGI_2 release from challenged endothelial cells, although release of EDRF was not measured in these experiments. The inset in Fig. 10 shows levels of PGI_2 measured in the effluent of endothelial cell columns perfused in the absence or presence of bradykinin. Measurements of L-[^{3}H]arginine uptake and PGI_2 release were obtained simultaneously in the same experiment. Release of PGI_2 was significantly stimulated during a bradykinin challenge, suggesting that bradykinin was biologically active. In order to establish whether L-arginine is rate-limiting for EDRF (nitric oxide) production during a bradykinin challenge, it may become necessary to examine L-arginine transport in cells cultured in L-arginine-free medium.

DISCUSSION

The dual tracer dilution techniques (Yudilevich & Mann, 1982; Mann et al., 1988, 1989b) applied here to to vascular endothelial cells provide powerful methods for characterizing transport properties of cultured cells (Peran & McGee, 1986). We have demonstrated that umbilical vein and aortic endothelial cells, similar to the microvessel endothelium of the brain, retina and heart (Pardridge, 1983; Allen & Gerritsen, 1986; Gerritsen et al., 1988), rapidly transport several essential nutrients. Although recent studies with static cultured endothelial cell monolayers have confirmed that large vessel endothelium transports extracellular amino acids (Steiger et al., 1987; Deneke et al., 1987; Ewadh et al., 1988; Sneddon et al., 1990), the time course of uptake appears to be much slower than that measured in perfused microcarrier endothelial cell cultures. Table 2 compares kinetic constants measured for L-leucine and L-arginine in perfused microcarrier cultures with values reported for amino acids, hexoses and monocarboxylates in the intact perfused brain, isolated brain microvessels and confluent endothelial cell monolayers.

The rapid time course of substrate uptake in perfused endothelial cell microcarrier cultures parallels that observed previously in intact perfused tissues (Betz et al., 1975; Syrota et al., 1982; Yudilevich & Mann, 1982; Mann & Peran, 1986; Mann et al., 1986; Bassingthwaighte & Sparks, 1986; Norman & Mann, 1987; Norman et al., 1989). Although rapid tracer dilution techniques applied *in vivo* or in isolated perfused organs may not be able to resolve endothelial cell transport from parenchymal cell uptake, such studies should reconsider the possibility that uptake processes in vascular endothelium may account for some of the findings orginally attributed to parenchymal cells.

Our most important finding has been the identification of an endothelial amino acid transporter which mediates the influx of L-arginine against a concentration gradient (Baydoun et al., 1990). We have now established that venous and aortic endothelial cells rapidly take up L-arginine by a saturable system, sensitive to inhibition by L-lysine, L-ornithine and neutral amino acids but insensitive to L-citrulline or D-arginine. In addition, transport was inhibited by N^G-monomethyl-L-arginine, a specific inhibitor of the cytosolic enzyme mediating nitric oxide synthesis from the terminal guanidino nitrogen atom(s) of L-arginine (Palmer et al., 1989; Palmer & Moncada, 1989). Moreover, uptake of L-arginine in endothelial cells (Sheriff et al., 1989) and other cells is sensitive to accelerative countertransport by other intracellular amino acids.

Although the role of an arginine transporter in regulating nitric oxide synthesis is not yet established, a recent study have suggested that nitric oxide synthase might be associated with the endothelial cell plasma membrane (Boje & Fung, 1990). This finding

Table 2. Transport of amino acids, hexoses and monocarboxylates in vascular endothelium. Kinetic constants measured in perfused microcarrier cultures of human umbilical vein and bovine aortic endothelial cells are compared with values reported for the intact perfused brain and isolated brain microvessels or confluent endothelial cell monolayers.

Transport System	Substrate	K_t (mM)	V_{max} (nmol/min/g)	Reference
Neutral amino acids	L-valine	0.21	49	Smith et al. (1987)
		0.51	35	Pardridge (1983)
	L-isoleucine	0.056	60	Takasoto et al. (1985)
		0.28	42	Pardridge (1983)
	L-leucine	*0.24*	*28a*	***Mann et al. (1989)***
		1.58	323	*Betz et al.(1975)
		0.03	59	Smith et al. (1987)
		0.14	0.58†	Cardelli- et al. (1987)
	L-phenylalanine	0.11	28	Pardridge (1983)
		0.011	41	Smith et al.(1987)
	L-tyrosine	0.15	31	Pardridge (1983)
		0.17	0.74†	Cardelli- et al. (1987)
	BCH	0.16	30	Wade & Katzman (1975)
	L-DOPA	0.38	100	Wade & Katzman (1975)
	2-MeAIB	0.28	0.15†	Cardelli- et al. (1987)
	L-cystine	0.044	-	Ewadh et al. (1988)
Basic amino acids	*L-arginine*	*0.206*	*8.3[a]*	***Mann et al. (1990)***
	L-arginine	0.0005	0.08[a]	Sneddon et al. (1990)
	L-arginine	0.092	7.8	Pardridge (1983)
	L-ornithine	0.22	10.5	Pardridge (1983)
	L-lysine	0.11	7.6	Pardridge (1983)
Hexoses	D-glucose	8	1750	Betz et al. (1973)
		11	1420	Pardridge (1983)
	2-deoxyglucose	7.3	1490	Pardridge (1983)
		0.09	0.76†	Betz & Goldstein (1986)
Monocarboxylic acids	L-lactate	1.8	91	Pardridge (1983)
	Pyruvate	0.57	88	Pardridge (1983)

[a] V_{max} expressed in nmol/min per 10^6 cells
† Isolated brain capillaries: V_{max} expressed in nmol/min/mg of protein
* Study in perfused dog brain

together with recent *in vivo* observations (Asika et al., 1989; 1990) and our transport suggest that arginine uptake may become important during sustained nitric oxide synthesis. Despite these convincing observations, recent studies have questioned (i) whether L-arginine alone or other polypeptides rich in arginine are the precursor(s) for nitric oxide synthesis (Schmidt et al., 1988; Thomas et al., 1989; Ignarro et al., 1989) and (ii) whether nitric oxide is the principal EDRF released from bovine aortic endothelial cells (Myers et al., 1989).

Previous studies in macrophages had demonstrated that formation of oxides of nitrogen (NO_2, NO_3) from the guanidino nitrogens atom(s) of L-arginine was involved in macrophage cytotoxic activity (Hibbs et al., 1987; Iyengar et al., 1987). As in endothelial cells, N^G-monomethyl-L-arginine effectively inhibits formation of NO-containing compounds in macrophages. However, endothelial cells and macrophages exhibit unexplained differences in their enzymatic substrate specificity. It is intriguing that nitric oxide has now also been identified as a potential neuromodulator in the central nervous system (Garthwaite et al., 1988; Knowles et al., 1989).

An impairment in endothelial cell mediated control of vascular smooth muscle relaxation may have important clinical implications for the pathogenesis of coronary artery disease, hypertension and atherosclerosis (see reviews Pearson & Gordon, 1984; Busse et al., 1985; Gordon & Pearson, 1987; Luscher et al., 1990). This study provides the basis for exploring whether changes in amino acid availability modulate the endothelial cell arginine-nitric oxide pathway.

ACKNOWLEDGEMENTS

The present study was generously supported by research grants from the British Heart Foundation, King's College Academic Development Fund, Medical Research Council, SERC/CASE Cooperative Award with Dr Salvador Moncada, FRS from the Wellcome Research Laboratories, Beckenham, U.K. Miss C-J. Sheriff was the recipient of a Science and Engineering Research Council PhD studentship.

REFERENCES

Allen, L.A. & Gerritsen, M.E. 1986. Regulation of glucose transport in cultured bovine retinal microvessel endothelium by insulin. Exp. Eye. Res. **43**:679-686.

Aisaka, K., Gross, S.S., Griffith, O.W. & Levi, R. 1990. Role of L-arginine in blood pressure regulation. Arch. Intern. Pharmacodyn. Therap. **305**: 226P.

Bassingthwaighte, J.B. & Sparks, H.V. 1986. Indicator dilution estimation of capillary endothelial transport. Ann. Rev. Physiol. **48**:321-334.

Baydoun, A.R., Emery, P.W., Pearson, J.D. & Mann, G.E. (1990). Modulation of intracellular amino acid concentrations in cultured bovine aortic endothelial cells deprived of exogenous L-arginine. J. Physiol. **(abstract in press).**

Betz, A.L., Gilboe, D.D. & Drewes, L.R. 1975. Kinetics of unidirectional leucine transport into brain: effects of isoleucine, valine and anoxia. Am. J. Physiol. **228**, 895-900.

Betz, A.L., Gibloe, D.D., Yudilevich, D.L. & Drewes, L.R. 1973. Kinetics of unidirectional glucose transport into the isolated dog brain. Am. J. Physiol. **225**:586-592.

Betz, A.L. & Goldstein, G.W. 1986. Specialized properties and solute transport in brain capillaries. Ann. Rev. Physiol. **48**:241-250.

Boje, K.M. & Fung, H-L. 1990. Endothelial cell nitric oxide generating enzyme(s) in the bovine aorta: Subcellular location and metabolic characterization. J. Pharmacol. Exp. Therap. **253**:20-26.

Boyd, C.A.R. & Parsons, D.S. 1979. Movement of monosaccharides between blood and tissue of vascularly perfused small intestine. J. Physiol. **287**:371-391.

Busse, R., Trogisch, G. & Bassenge, E. 1985. The role of endothelium in the control of vascular tone. Basic Res. Cardiol. **80**:475-490.

Cancilla, M.D. & Debault, L.E. 1983. Neutral amino acid transport properties of cerebral endothelial cells in vitro. J. Neuropathol. Exp. Neurol. **42**:191-199.

Cardelli-Cangiano, P., Fiori, A., Cangiano, C., Barberini, F., Allegra, P., Peresemmpio, V. & Strom, R. 1987. Isolated brain microvessels as in vitro equivalents of the blood-brain barrier: Selective removal by collagenase of the A-system of neutral amino acid transport. J. Neurochem. **49**:1667-1675.

Christensen, H.N. & Kilberg, M. 1987. Amino acid transport across the plasma membrane: role of regulation in interorgan flows. In: **Amino Acid Transport in Animal Cells**, ed. YUDILEVICH, D.L. & BOYD, C.A.R., pp. 1-46. Manchester University Press.

Corkey, R.F., Corkey, B.E. & Gimbone, M.A. 1981. Hexose transport in normal and SV-40 transformed human endothelial cells in culture. J. Cell Physiol. **106**:425-434.

Deneke, S.M., Steiger, V, Fanburg, B.L. 1987. Effect of hyperoxia on glutathione levels and glutamic acid uptake in endothelial cells. J. Appl. Physiol. **63**:1966-1971.

Ewadh, M.J.A., Tuball, N. & Rose, F.A. 1988. Transport of L-cystine in human umbilical vein endothelial cells in culture. Bioscience Reports **8**:449-453.

Furchgott, R.F. 1990. Studies on the endothelium-dependent vasodilatation and the endothelium-derived relaxing factor. Acta Physiol. Scand. **139**:257-270.

Furchgott, R.F. & Zawadski J.V. 1980. The obligatory role of endothelial cells in the relaxation of arterial smooth muscle by acetylcholine. Nature **288**:373-376.

Garthwaite, J., Charles, G.L. & Chess-Williams, R. 1988. Endothelium-derived relaxing factor release on activation of NMDA receptors suggests role as intracellular messenger in the brain. Nature **336**:385-388.

Gerritsen, M.E., Burke, T.M. & Allen, L.A. 1988. Glucose starvation is required for insulin stimulation of glucose uptake and metabolism in cultured microvascular endothelial cells. Microvasc. Res. **35**:153-166.

Gerritsen, M.E. & Cheli, C.D. 1983. Arachidonic acid and prostaglandin endoperoxide metabolism in isolated rabbit coronary microvessels and isolated and cultivated coronary microvessel endothelial cells. J. Clin. Invest. **72**:1658-1671.

Gillis, C.N. 1986. Pharmacological aspects of metabolic processes in the pulmonary circulation. Ann. Rev. Pharmacol. Toxicol. **26**:183-200.

Gold, M.E., Bush, P.A. & Ignarro, L.J. 1989. Depletion of arterial L-arginine causes reversible tolerance to endothelium-derived relaxation. Biochem. Biophys. Res. Commun. **164**:714-721.

Gordon, J.L. & Pearson, J.D. 1987. Biology of the vascular endothelium. In: **Haemostasis and Thrombosis**, eds. Bloom, A.L. & Thomas, D.P., pp. 303-311, Churchill Livingstone.

Gordon, E.L., Pearson, J.D. & Slakey, L.L. 1986. The hydrolysis of extracellular adenine nucleotides by cultured endothelial cells from pig aorta. Feed-forward inhibition of adenosine production at the cell surface. J. Biol. Chem. **261**:15496-15504.

Griffiths, T.M., Edwards, D.H., Lewis, M.J. & Henderson, A.H. 1985. Evidence that cyclic guanosine monophosphate (cGMP) mediates endothelium-dependent relaxation. Eur. J. Pharmacol. **112**:195-202.

Hibbs, J.B., Taintor, R.R. & Vavrin, Z. 1987. Macrophage cytotoxicity: role of L-arginine deiminase and imino nitrogen oxidation to nitrate. Science **235**:473-476.

Ignarro, L.J., Gold, M.E., Buga, G.M., Byrns, R.E., Wood, K.S., Chaudhuri, G. & Frank, G. 1989. Basic polyamino acids rich in arginine, lysine or ornithine cause both enhancement of and refractoriness to formation of endothelium-derived nitric oxide in pulmonary artery and vein. Circ. Res. **64**:315-329.

Iyengar, R., Stuehr, D.J. & Marletta, M.A. 1987. Macrophage synthesis of nitrite, nitrate and N-nitrosamines: precursors and role of respiratory burst. Proc. Natl. Acad. Sci. USA **84**:6369-6373.

Jaffe, E.A., Nachman, R.L., Becker, C.G. & Minick, C.R. 1973. Culture of human endothelial cells derived from umbilical veins. Identification by morphologic and immunologic criteria. J. Clin. Invest. **52**:2745-2756.

King, G.L., Buzney, S.M., Kahn, C.R., Hetu, N., Buchwald, S., Macdonald, S.G. & Rand, L.I. 1983. Differential responsiveness to insulin of endothelial and support cells from micro- and macrovessels. J. Clin. Invest. **71**:974-979.

Knowles, R.G., Palacios, M., Palmer, R.M.J. & Moncada, S. 1989. Formation of nitric oxide from L-arginine in the central nervous system: A transduction mechanism for stimulation of the soluble guanylate cyclase. Proc. Natl. Acad. Sci. USA **86**:5159-5162.

Luckhoff, A., Pohl, U., Mulsch, A. & Busse, R. 1988. Differential role of extra- and intracellular calcium in the release of EDRF and prostacyclin from cultured endothelial cells. Br. J. Pharmacol. **95**:189-196.

Luscher, T.F., Richard, V., Tschudi, M., Yang, Z. & Boulanger, C. 1990. Endothelial control of vascular tone in large and small coronary arteries. J. Am. Coll. Cardiol. **15**:519-527.

Mann, G.E., Munoz, M. & Peran, S. 1986. Fasting and refeeding modulate neutral amino acid transport activity in the basolateral membrane of the rat exocrine pancreatic epithelium: fasting-induced insulin insensitivity. Biochim. Biophys. Acta **862**:119-126.

Mann, G.E., Norman, P.S.R. & Smith, I.C.H. 1989a. Amino acid efflux in the isolated perfused rat exocrine pancreas: Trans-stimulation by extracellular amino acids. J. Physiol. **416**:485-502.

Mann, G.E., Pearson, J.D., Sheriff, C-J. & Toothill, V.J. 1988. Characterization of amino acid and glucose transport in cultured endothelial cells. Pflugers Arch. **411**:R42.

Mann, G.E., Pearson, J.D., Sheriff, C-J. & Toothill, V.J. 1989b. Expression of amino acid transport systems in cultured human umbilical vein endothelial cells. J. Physiol. **410**:325-339.

Mann, G.E. & Peran, S. 1986. Basolateral amino acid transport systems in the perfused exocrine pancreas: sodium-dependency and kinetic interactions between influx and efflux mechanisms. Biochim. Biophys. Acta **858**:263-274.

Mann, G.E., Sheriff, C-J. & Pearson, J.D. 1990. N^G-monomethyl-L-arginine and cationic amino acids inhibit L-arginine transport in perfused microcarrier cultures of aortic endothelial cells. In: **Nitric Oxide from L-Arginine: A Bioregulatory System**, ed. Moncada, S. & Higgs, E.A., pp. 331-339. Amsterdam: Excerpta Medica.

Maruka, C., Spatz, M., Ueki, Y., Nagatsu, I. & Bembray, J. 1984. Cerebrovascular endothelial cell culture: Metabolism and synthesis of 5-hydroxytryptamine. J. Neurochem. **43**:316-319.

Moncada, S. & Higgs, E.A. 1990. Nitric Oxide from L-Arginine: A Bioregulatory System. pp. 1-512 Excerpta Medica, Amsterdam.

Moncada, S., Gryglewski, R.J., Bunting, S. & Vane, J.R. 1976. An enzyme isolated from arteries transforms prostaglandin endoperoxides to an unstable substance that inhibits platelet aggregation. Nature **263**:663-665.

Myers, P.R., Guerra, R. & Harrison, D.G. 1989. Release of NO and EDRF from cultured bovine aortic endothelial cells. Am. J. Physiol. **256**:H1030-H1037.

Needham, L., Cusack, N.J., Pearson, J.D. & Gordon, J.L. 1987. Characteristics of the P_2 purinoceptor that mediates prostacyclin production by pig aortic endothelial cells. Eur. J. Pharmacol. **134**:199-209.

Norman, P.S.R., Habara, Y. & Mann, G.E. 1989. Paradoxical effects of endogenous and exogenous insulin on amino acid transport activity in the isolated rat pancreas: somatostatin-14 inhibits insulin action. Diabetologia **32**:177-184.

Norman, P.S.R. & Mann, G.E. 1987. Ionic dependence of amino acid transport in the exocrine pancreatic epithelium: calcium dependence of insulin action. J. Membrane Biol. **96**:153-163.

Palmer, R.M.J. & Moncada, S. 1989. A novel citrulline-forming enzyme implicated in the formation of nitric oxide by vascular endothelial cells. Biochem. Biophys. Res. Comm. **158**:348-353.

Palmer, R.M.J., Ferrige, A.G. & Moncada, S. 1987. Nitric oxide release accounts for the biological activity of endothelium-derived relaxing factor. Nature **327**:524-526.

Palmer, R.M.J., Rees, D.R., Ashton, D.S. & Moncada, S. 1988. L-arginine is the physiological precursor for the formation of nitric oxide in endothelium-dependent relaxation. Biochem. Biophys. Res. Comm. **153**:1251-1256.

Pardridge, W.M. 1983. Brain metabolism: a perspective from the blood-brain barrier. Physiol. Rev. **63**:a1481-1535.

Pearson, J.D. & Gordon, J.L. 1984. Metabolism of serotonin and adenosine. In: **Biology of Endothelial Cells**, ed. Jaffe, E.A., pp. 330-342. Martinus Nijholt.

Peran, S. & McGee, M.P. 1986. Unidirectional flux of phenylalanine into Vero cells. Measurement using paired tracers in perfused cultures. Biochim. Biophys. Acta **856**:231-236.

Schmidt, H.H.H.W., Klein, M.M., Niroomand, F. & Bohme, E. 1988. Is arginine a physiological precursor of endothelium-derived nitric oxide? Eur. J. Pharmacol. **148**:293-295.

Shepro, D. & Dunham, B. 1986. Endothelial cell metabolism of biogenic amines. Ann. Rev. Physiol. **48**, 335-345.

Sheriff, C-J., Pearson, J.D. & Mann, G.E. 1990. Characteristics of amino acid transporters in perfused endothelial cell microcarrier cultures: Role of L-arginine in nitric oxide synthesis. J. Physiol. **423**:6P.

Smith, Q.R., Momma, S., Aoyagi, M. & Rapoport, S.I. 1987. Kinetics of neutral amino acid transport across the blood-brain barrier.
J. Neurochem. **49**:1651-1658.

Sneddon, J.M., Bearpark, T.M., Galton, S.A. & Vane, J.R. 1990. Transport and metabolism of L-arginine by bovine aortic endothelial cells. In: **Nitric Oxide from L-Arginine: A Bioregulatory System**, eds. Moncada, S. & E.A. Higgs, pp. 457-461. Amsterdam: Excerpta Medica.

Steiger, V., Deneke, S.M. & Fanburg, B.L. 1987. Characterization of glutamic acid uptake by bovine pulmonary arterial endothelial cells. J. Appl. Physiol. **63**:1961-1965.

Strum, J.M. & Junod, A.F. 1972. Radioautographic demonstration of 5-hydroxytryptamine-[^{3}H] uptake by pulmonary endothelial cells. J. Cell Biol. **54**:456-467.

Syrota, A., Girault, M., Pocidalo, J-J. & Yudilevich, D.L. 1982. Endothelial uptake of amino acids, sugars, lipids and prostaglandins in rat lung. Am. J. Physiol. **243**:C20-C26.

Takasoto, Y., Momma, S. & Smith, Q.R. 1985. Kinetic analysis of cerebrovascular isoleucine transport from saline and plasma. J. Neurochem. **45**:1013-1020.

Thomas, G., Hecker, M. & Ramwell, P.W. 1989. Vascular activity of polycations and basic amino acids: L-arginine does not specifically elicit endothelium-dependent relaxation. Biochem. Biophys. Res. Comm. **158**:177-180.

Van de Voorde, J., Vanderstichele, H. & Leusen, I. 1987. Release of endothelium-derived relaxing factor from human umbilical vessels. Circ. Res. **60**:517-522.

Vinters, H.V., Beck, D.W., Bready, J.V., Maxwell, K., Berliner, J.A., Hart, M.N. & Cancilla, P.A. 1985. Uptake of glucose analogues into cultured cerebral microvessel endothelium. J. Neuropathol. Exp. Neurol. **5**:445-458.

Wade, L.A. & Katzman, R. 1975. Synthetic amino acids and the nature of L-DOPA transport at the blood-brain barrier. J. Neurochem. **25**:837-842.

Yudilevich, D.L. & Boyd, C.A.R. 1987. **Amino Acid Transport in Animal Cells**, Manchester: Manchester Univ. Press.

Yudilevich, D.L. & Mann, G.E. 1982. Unidirectional uptake of substrates at the blood side of secretory epithelia: stomach, salivary gland and pancreas. Fed. Proc. **41**:3045-3053.

Zetter, B.R. 1980. Migration of capillary endothelial cells is stimulated by tumour-derived factors. Nature **285**:41-43.

CONTRIBUTORS

Barros, Felipe. Departamento de Fisiología y Biofísica, Facultad de Medicina, Universidad de Chile, Casilla 70055, Santiago 7, Chile.

Bezanilla, Francisco. Department of Physiology, University of California, Los Angeles, CA 90024, USA.

Biber, Jurg. Institute of Physiology, University of Zurich, Winterthurerstr. 190, CH-8057 Zurich, Switzerland.

Boyd, C.A. Richard. Department of Human Anatomy, University of Oxford, South Parks Road, Oxford OX1 3QX, UK.

Cabantchik, Ioav. Department of Biological Chemistry, The Hebrew University of Jerusalem, Jerusalem, 91904, Israel.

Cotton, Calvin. Department of Physiology and Biophysics, University of Texas Medical Branch, Galveston, Texas, USA.

Davis, Elaine. Department of Biochemistry, University of Cambridge, Tennis Court Road, Cambridge CB2 1QW, UK.

Devés, Rosa. Departamento de Fisiología y Biofísica, Facultad de Medicina, Universidad de Chile, Casilla 70055, Santiago, Chile.

Eidelman, Ofer. Department of Biological Chemistry, The Hebrew University of Jerusalem, Jerusalem, 91904, Israel.

Eisner, David. Department of Veterinary Preclinical Sciences, University of Liverpool, Liverpool L69 3BX, U.K.

García-Díaz, Fernando. Department of Physiology, Boston University, School of Medicine, Boston, Massachusetts, USA.

Giráldez, Fernando. Departamento de Bioquímica y Biología Molecular y Fisiología, Universidad de Valladolid, Valladolid, Spain.

Henderson, Peter. Department of Biochemistry, University of Cambridge, Tennis Court Road, Cambridge CB2 1QW, UK.

Ilundaín, Ana. Departamento de Fisiología y Biología Animal, Universidad de Sevilla, Sevilla, Spain.

Jarvis, Simon. Biological Laboratory, University of Kent, Canterbury, Kent CT2 7NJ, UK.

Kanno, Tomio. Department of Physiology, Faculty of Veterinary Medicine, Hokkaido University, Sapporo 060, Japan.

King, Glenn. Department of Biochemistry, University of Sidney, Sidney NSW 2006, Australia.

Lopez-Barneo, José. Departamento de Fisiología y Biofisíca, Facultad de Medicina, Universidad de Sevilla, Sevilla, Spain.

Lowenstein, Werner. Department of Physiology and Biophysics, University of Miami, School of Medicine, P.O. Box 016430, Miami, Florida 33101, USA.

Maiden, Martin. Department of Biochemistry, University of Cambridge, Tennis Court Road, Cambridge CB2 1QW, UK.

Mann, Giovanni. Physiology, Biomedical Sciences Division, King's College, University of London, Campden Hill Road, London W8 7AH, UK.

McKeown, Brian. Department of Biochemistry, University of Cambridge, Tennis Court Road, Cambridge CB2 1QW, UK.

Muñoz, Manuel. Department of Biochemistry and Molecular Biology, Faculty of Medicine, University of Málaga, Málaga, Spain.

Murer, Heini. Institute of Physiology, University of Zurich, Winterthurerstr. 190, CH-8057 Zurich, Switzerland.

Naftalin, Richard. Physiology, Biomedical Sciences Division, King's College, University of London, London WC2R 2LS, UK.

Pearson, Jeremy. Section of Vascular Biology, MRC Clinical Research Centre, Harrow, UK.

Perán, Salvador. Department of Biochemistry and Molecular Biology, Faculty of Medicine, University of Málaga, Málaga, Spain.

Petersen, Ole. The Physiological Laboratory, University of Liverpool, Liverpool L69 3BX, UK.

Reuss, Luis. Department of Physiology and Biophysics, University of Texas Medical Branch, Galveston, Texas, USA.

Saíz, María. Department of Biochemistry and Molecular Biology, Faculty of Medicine, University of Málaga, Málaga, Spain.

Sepúlveda, Francisco. AFRC Institute of Animal Physiology and Genetics Research, Babraham Hall, Cambridge CB2 4AT, UK.

Sherrif, Cherry-Jane. Physiology, Biomedical Sciences Division, King's College, University of London, Campden Hill Road, London W8 7AH, UK.

Smith, I. Christopher H. Physiology, Biomedical Sciences Division, King's College, University of London, London W8 7AH, UK.

Sobrevía, Luis. Departamento de Fisiología y Biofísica, Facultad de Medicina, Universidad de Chile, Casilla 70055, Santiago 7, Chile.

Stein, Wilfred. Department of Biochemistry, The Hebrew University of Jerusalem, Jerusalem, 91904, Israel.

Toothill, Valerie. Section of Vascular Biology, MRC Clinical Research Centre, Harrow, UK.

Valdeolmillos, Miguel. Departamento de Fisiología, Facultad de Medicina, Universidad de Alicante, Spain.

Wray, Susan. Physiology Department, The University of Liverpool, P.O. Box 147, Liverpool L69 3BX, UK.

Yudilevich, David L. Physiology, Biomedical Sciences Division, King's College, University of London, London W8 7AH, UK. and Departamento de Fisiología y Biofisíca, Facultad de Medicina, Universidad de Chile, Casilla 70055, Santiago 7, Chile.

INDEX

Numbers in italics indicate data in tables.

MIX
Papier aus verantwortungsvollen Quellen
Paper from responsible sources
FSC® C105338

If you have any concerns about our products,
you can contact us on
ProductSafety@springernature.com

In case Publisher is established outside the EU,
the EU authorized representative is:
Springer Nature Customer Service Center GmbH
Europaplatz 3, 69115 Heidelberg, Germany

Printed by Libri Plureos GmbH
in Hamburg, Germany